Chemical Signals in Vertebrates 11

Jane L. Hurst, Robert J. Beynon, S. Craig Roberts
and Tristram D. Wyatt

Editors

Chemical Signals
in Vertebrates 11

 Springer

Jane L. Hurst
Department of Veterinary Preclinical Science
University of Liverpool, Leahurst
Neston CH64 7TE, UK
jane.hurst@liverpool.ac.uk

Robert J. Beynon
Department of Veterinary Preclinical Science
University of Liverpool,
Crown Street, Liverpool, L69 7ZJ, UK
r.beynon@liverpool.ac.uk

S. Craig Roberts
School of Biological Sciences
University of Liverpool,
Crown Street, Liverpool, L69 7ZB, UK
craig.roberts@liverpool.ac.uk

Tristram D. Wyatt
Office of Distance and Online Learning
University of Oxford
Oxford OX2 7DD, UK
tristram.wyatt@continuing-
education.oxford.ac.uk

ISBN: 978-0-387-73944-1   e-ISBN: 978-0-387-49835-5

Library of Congress Control Number: 2007934764

Printed on acid-free paper.

9 8 7 6 5 4 3 2 1

springer.com

Preface

This volume reports the proceedings of the eleventh triennial meeting of the Chemical Signals in Vertebrates International Symposium and thus, is the 30th anniversary of the informal grouping of scientists who convene to discuss their common interests in the ways in which vertebrates use chemical signals. Previous meetings were held in Saratoga Springs, New York; Syracuse, New York; Sarasota, Florida; Laramie, Wyoming; Oxford, England; Philadelphia, Pennsylvania; Tubingen, Germany; Ithaca, New York; Kraków, Poland and Corvallis, Oregon.

The eleventh meeting was hosted by the Faculty of Veterinary Science at the University of Liverpool, and was held in Chester, England. CSiV 11 was the latest in a well regarded series of meetings, and was attended by about 80 scientists, with nearly 120 further co-authors, all with a common interest in vertebrate chemical signalling, and its role in vertebrate behaviour. The species range was, as ever, remarkable – from lion to salamander, from mouse to elephant, from salmon to human, a biodiversity matched by the range of the substances used for communication. As might be expected from such diversity, we enjoyed a broad ranging programme that included sessions on olfactory assessment, pheromone delivery, sexual selection (human and animal), urinary proteins, anti-predator responses, scent organs and their function, individual recognition, species recognition, sexual development and sexual communication (the full programme can still be viewed on the CSiV website). The meeting was launched by a thoughtful and far-reaching presentation from Milos Novotny, on "Genetic and environmental control of volatile mammalian chemosignals: structural and quantitative aspects" in which he introduced a future in which sophisticated 'omics technologies would feature more strongly. We closed with an equally thoughtful perspective from another of the leading scientists in this field, Dietland Müller-Schwarze.

As has been a feature of previous meetings, the group was enthusiastic, very willing to make huge species jumps and also to listen and support the younger investigators who often make their first international presentation at this meeting. This collegiate spirit extended to the meeting dinner at the Blue Planet Aquarium, where several members of the group unwittingly donated their watches, courtesy of the magician 'Magic Matt', to offset the meeting costs! CSiV has no funds of its own (a small float is carried from one meeting to the next) and we are totally reliant on sponsorship to supplement the registration fees. We are particularly pleased therefore to acknowledge the support of the European Chemoreception Organisation (http://ecro.cesg.cnrs.fr/) and also, the Dean of the Faculty of Veterinary Science at the University of Liverpool (http://www.liv.ac.uk/vets/).

The group is increasingly proud of its acronym (CSiV), in part at least because of the television series featuring analytical forensic chemistry that almost shares the same name! In the period up to the meeting a low key web presence was created (http://www.csiv.org) that will act as a home for the group, but there are still no membership fees, and no restrictions on attendance at the meetings. This website will be the source of information about the twelfth meeting, at Dalhousie University in Halifax, Nova Scotia, Canada from July 28th-31st, 2009, organised by Heather Schellinck and colleagues.

One name in particular was noticeably absent from the attendees at CSiV 11, that of the wonderfully passionate and insightful Bets Rasmussmen, who was sadly too ill to attend the meeting and who subsequently died on September 17th, 2006. We will remember her fine abilities as a scientist who truly straddled the biochemical/behavioural divide, and who did much to allow us to see one another 'across the gap'. We remember with fondness her enthusiasm, her sense of fun, and her willingness to engage in discussions about the whole field. A tribute chapter from Tom Goodwin and Bruce Shulte captures her unique and fascinating contribution to the field and we are pleased to dedicate this volume to her memory.

Finally, it is a pleasure to acknowledge the very considerable 'behind the scenes" help from all those who helped make the conference a success. We would like to thank Anne Tuson and Marg Hedges, as well as all the members of the Mammalian Behaviour and Evolution Group and the Proteomics and Functional Genomics Group at the University of Liverpool. We would also like to express our thanks to the University of Chester Conference Team for their support and efforts at the venue.

Jane L. Hurst
Robert J. Beynon
S. Craig Roberts
Tristram D. Wyatt

Contents

Part V   Maternal - Offspring Communication

Part VI   Communication between Species, Predators and Prey

Part VII   Applications

Contributors

Paul F. Alewood
Institute for Molecular Bioscience
University of Queensland
St. Lucia QLD 4072
Australia
p.alewood@imb.uq.edu.au

Stuart D. Armstrong
Proteomics and Functional
    Genomics Group
Department of Veterinary
Preclinical Science
University of Liverpool
Crown Street
Liverpool L69 7ZJ
UK

Stevan J. Arnold
Department of Zoology
Oregon State University
Corvallis, OR 97331
USA
arnoldst@onid.orst.edu

Lourdes Arteaga
Centro Tlaxcala de Biología
    de la Conducta
Universidad Autónoma de Tlaxcala
Carretera Tlaxcala-Puebla Km 1.5
CP 90070 Tlaxcala
Mexico
lourdesac@cci.uatx.mx

Yves Aubut
Department of Life Science
& Chemical Análisis
Agilent Technologies Inc.
USA

Carmen Agustín-Pavón
Departament de Biologia Funcional
    i Antropologia Fisica
Universitat de València
C/ Dr Moliner 50
ES-46100 Burjassot, València
Spain
M.carmen.agustin@uv.es

Kathryn Bagley
Department of Biology
Georgia Southern University
Statesboro, GA 30460-8042
USA

Julie Bakker
Centre for Cellular and
    Molecular Neurobiology
University of Liège
B-36 L'Avenue de l'H?pital
Liege B-4000
Belgium
jbakker@ulg.ac.be

Alexis Barrett
Institute for Molecular Bioscience
University of Queensland

St. Lucia QLD 4072
Australia
a.barrett@imb.uq.edu.au

Michael J. Baum
Department of Biology
Boston University
5 Cummington Street
Boston, MA 02215
USA
baum@bu.edu

Stuart D. Becker
Mammalian Behaviour
 and Evolution Group
Department of Veterinary
 Preclinical Science
University of Liverpool
Leahurst, Neston CH64 7TE
UK
Stuart.becker@liv.ac.uk

Laura Been
Department of Psychology
Georgia State University
24 Peachtree Center Ave NE
Atlanta, GA 30302
USA
Lbeen1@student.gsu.edu

Robert J. Beynon
Proteomics and Functional
 Genomics Group
Department of Veterinary
 Preclinical Science
University of Liverpool
Crown Street
Liverpool L69 7ZJ
UK
r.beynon@liv.ac.uk

R. Bhar
Department of Instrumentation Science
Jadaypur University
Kolkata 700072
India
Rbusic32@yahoo.co.in

Camille Blake
Department of Biological Science
Florida State University
Tallahassee, FL 32306-4340
USA
blake@neuro.fsu.edu

R. L. Brahmachary
Flat 10, 21B Motijheel
Dumdum
Kolkata 700078
India

Peter Brennan
Department of Physiology
University of Bristol
University Walk
Bristol BS8 1TD
UK
p.brennan@bristol.ac.uk

Patrick A. Brown
Department of Chemistry
Hendrix College
1600 Washington Avenue
Conway, AR 72032
USA

Nicole Burgener
Leibniz Institute for Zoo
 and Wildlife Research
Postfach 601103
D-10252 Berlin
Germany
burgener@izw-berlin.de

Robert J. Capon
Institute for Molecular Bioscience
University of Queensland
St. Lucia QLD 4072
Australia
r.capon@imb.uq.edu.au

Barbara Caspers
Leibniz Institute for Zoo
    and Wildlife Research
Postfach 601103
D-10252 Berlin
Germany
caspers@izw-berlin.de

Ashish Chakroborty
327/1 Rammohan Park
Behala
Kolkata 700008
India
Asish_ck@yahoo.co.in

Mark Challis
Belfast Zoological Gardens
Antrim Road
Newtownabbey BT36 7PN
UK
ChallisM@belfastcity.gov.uk

Sarah A. Cheetham
Mammalian Behaviour
    and Evolution Group
Department of Veterinary
    Preclinical Science
University of Liverpool
Leahurst
Neston CH64 7TE
UK
sacheet@liv.ac.uk

C. Joi Chen
Department of Chemistry
Hendrix College
1600 Washington Avenue
Conway, AR 72032
USA

Fay Clark
Institute of Zoology
Zoological Society of London
Regents Park
London NW1 4RY
UK
prolixiat@aol.com

Dwyer Coleman
School of Psychology
Queen's University of Belfast
Belfast BT7 1NN
UK

Gérard Coureaud
Equipe d'Ethologie et de
    Psychobiologie Sensorielle
Centre Européen des Sciences
    du Goût
UMR 5170 CNRS
Université de Bourgogne / Inra
15 rue Hugues Picardet
21000 Dijon
France
coureaud@cesg.cnrs.fr

Martin Dehnhard
Leibniz Institute for Zoo
    and Wildlife Research
Postfach 601103
D-10252 Berlin
Germany
dehnhard@izw-berlin.de

Hans Distel
Institut für Medizinische
    Psychologie
Universität München
Goethestrasse 31
D-80336 Munich
Germany
Hans.distel@med.uni-muenchen.de

Sébastien Doucet
Equipe d'Ethologie et de
    Psychobiologie Sensorielle
Centre Européen des Sciences
    du Goût
UMR 5170 CNRS
Université de Bourgogne / Inra
15 rue Hugues Picardet
21000 Dijon
France
doucet@cesg.cnrs.fr

Christine M. Drea
Dept of Biological Anthropology
  and Anatomy
Duke University
Box 90383
08 Biological Science Building
Durham, NC 27708-0383
USA
cdrea@duke.edu

Marion East
Leibniz Institute for Zoo
  and Wildlife Research
Postfach 601103
D-10252 Berlin
Germany
east@izw-berlin.de

Jeff Eggert
Department of Chemistry
Hendrix College
1600 Washington Avenue
Conway, AR 72032
USA

Mindy S. Eggert
Department of Chemistry
Hendrix College
1600 Washington Avenue
Conway, AR 72032
USA

Maria G. Evola
Department of Chemistry
Hendrix College
1600 Washington Avenue
Conway, AR 72032
USA

Pamela W. Feldhoff
Department of Biochemistry
  and Molecular Biology
University of Louisville
319 Abraham Flexner Way
Louisville, KY 40202
USA
Pamela.feldhoff@louisville.edu

Richard C. Feldhoff
Department of Biochemistry
  and Molecular Biology
University of Louisville
319 Abraham Flexner Way
Louisville, KY 40202
USA
Rick.feldhoff@louisville.edu

Michael H. Ferkin
Department of Biology
University of Memphis
Ellington Hall
Memphis, TN 38152
USA
mhferkin@memphis.edu

Wittko Francke
Institute of Organic Chemistry
University of Hamburg
Martin-Luther-King-Platz 6
20146 Hamburg
Germany
francke@chemie.uni-hamburg.de

Stephan Franke
Institute of Organic Chemistry
University of Hamburg
Martin-Luther-King-Platz 6
20146 Hamburg
Germany
Franke@chemi.uni-hamburg.de

Hans von Gizycki
Academic Computing Center
SUNY Downstate Medical Center
450 Clarkson Avenue
Brooklyn, NY 11203
USA
Hans.vongizycki@downstate.edu

Frank Goeritz
Leibniz Institute for Zoo
  and Wildlife Research
Postfach 601103
D-10252 Berlin
Germany
goeritz@izw-berlin.de

Thomas E. Goodwin
Department of Chemistry
Hendrix College
1600 Washington Avenue
Conway, AR 72032
USA
Goodwin@hendrix.edu

Gordon C. Grigg
School of Integrative Biology
University of Queensland
St. Lucia QLD 4072
Australia
g.grigg@uq.edu.au

Matthew Groover
Department of Biology
Georgia Southern University
Statesboro, GA 30460-8042
USA

Mimi Halpern
Department of Anatomy
    and Cell Biology
SUNY Downstate Medical Center
450 Clarkson Avenue
Brooklyn, NY 11203
USA

Jan Havlicek
Department of Anthropology
Charles University
Husnikova 2075
158 00 Prague 13
Czech Republic
Jan.havlicek@fhs.cuni.cz

R. Andrew Hayes
Institute for Molecular Bioscience
University of Queensland
St. Lucia QLD 4072
Australia
r.hayes@imb.uq.edu.au

Anna L. Heckla
Department of Zoology
203 Natural Science Building
Michigan State University
East Lansing, MI 48824-1115
USA
hecklaan@msu.edu

Sigrid R. Heise-Pavlov
PavEcol Wildlife Management
    Consultancy
211 Turpentine Road
Diwan, via Mossman, 4873
Australia
ryparosa@bigpond.com.au

Anne Henrot
Service de Réanimation pédiatrique
Hôpital Clocheville
CHU Tours
France

Peter Hepper
School of Psychology
Queen's University of Belfast
Belfast BT7 1NN
UK
p.hepper@qub.ac.uk

Thomas B. Hildebrandt
Leibniz Institute for Zoo
    and Wildlife Research
Postfach 601103
D-10252 Berlin
Germany
hildebrand@iz-berlin.de

Heribert Hofer
Leibniz Institute for Zoo
    and Wildlife Research
Postfach 601103
D-10252 Berlin
Germany

Kay E. Holekamp
Department of Zoology
203 Natural Science Building
Michigan State University
East Lansing, MI 48824-1115
USA
holekamp@msu.edu

Lynne Houck
Department of Zoology
Oregon State University
Corvallis, OR 97331
USA
houckl@science.oregonstate.edu

Sam J. House
Department of Chemistry
Hendrix College
1600 Washington Avenue
Conway, AR 72032
USA

Robyn Hudson
Instituto de Investigaciones Biomédicas
Universidad Nacional Autónoma
    de México
70228, Ciudad Universitaria
04510 Distrito Federal
Mexico
rhudson@biomedicas.unam.mx

Richard E. Humphries
Mammalian Behaviour
    and Evolution Group
Department of Veterinary
    Preclinical Science
University of Liverpool
Leahurst
Neston CH64 7TE
UK
r.humphries@liv.ac.uk

Jane L. Hurst
Mammalian Behaviour
    and Evolution Group

Department of Veterinary
    Preclinical Science
University of Liverpool
Leahurst
Neston CH64 7TE
UK
Jane.hurst@liv.ac.uk

Stephen R. Jackson
Department of Chemistry
Hendrix College
1600 Washington Avenue
Conway, AR 72032
USA

Matthieu Keller
Centre for Cellular and
    Molecular Neurobiology
University of Liège
B-36 L'Avenue de l'H?pital
Liege B-4000
Belgium
keller@tours.inra.fr

Andrew J. King
Institute of Zoology
Zoological Society of London
Regents Park
London NW1 4RY
UK
Andrew.King@ioz.ac.uk

John Kubie
Department of Anatomy
    and Cell Biology
SUNY Downstate Medical Center
450 Clarkson Avenue
Brooklyn, NY 11203
USA
John.kubie@downstate.edu

Antonieta Labra
CEES, University of Oslo
Kristine Bonnevies Hus
Blinderveien 31

Oslo
Pb 1066 – Blinder 0316
Norway
Antonieta.labra@bio.uio.no

Enrique Lanuza
Departament de Biologia Funcional
    i Antropologia Fisica
Universitat de València
C/ Dr Moliner 50
ES-46100 Burjassot, València
Spain
Enrique.lanuza@uv.es

Michael P. LeMaster
Department of Biology
Western Oregon University
345 North Monmouth Avenue
Monmouth, OR 97361
USA
lemastm@wou.edu

Pavlina Lenochova
Department of Anthropology
    and Human Genetics
Charles University
Husnikova 2075
158 00 Prague 13
Czech Republic
p.lenoska@seznam.cz

Stuart T. Leonard
Department of Pharmacology
Health Science Center
Louisiana State University
1901 Perdido Street
New Orleans, LA 70112
USA
sleona@lsuhsc.edu

Anthony C. Little
Department of Psychology
University of Stirling
Stirling FK9 4LA
UK
Anthony.little@stir.ac.uk

James G. Logan
Biological Chemistry Division
Rothamsted Research, Harpenden
Hertfordshire AL5 2JQ
UK
James.logan@bbsrc.ac.uk

Helen Loizi
Department of Biology
Georgia Southern University
Statesboro, GA 30460-8042
USA

Jennifer Louie
Monell Chemical Senses Center
3500 Market Street
Philadelphia, PA 19104-3308
USA
jlouie@monell.org

Alan MacNicoll
Central Science Laboratory
DEFRA, Sand Hutton
York YO41 1LZ
UK
a.macnicoll@cso.gov.uk

Pamela Maras
Georgia State University
Center for Behavioral Neuroscience
P.O. Box 3966
Atlanta, GA 30302-3966
USA
Pmaras1@student.gsu.edu

Fernando Martínez-García
Departament de Biologia Funcional
i Antropologia Fisica
Universitat de València
C/ Dr Moliner 50
ES-46100 Burjassot, València
Spain
Fernando.mtnez-garcia@uv.es

Margarita Martínez-Gómez
Instituto de Investigaciones Biomédicas
Universidad Nacional Autónoma
    de México
70228, Ciudad Universitaria
04510 Distrito Federal
Mexico
marmag@garza.uatx.mx

Jose Martínez-Hernández
Departament de Biologia Funcional
    i Antropologia Fisica
Universitat de València
C/ Dr Moliner 50
ES-46100 Burjassot, València
Spain

Joana Martínez-Ricós
Departament de Biologia Funcional
    i Antropologia Fisica
Universitat de València
C/ Dr Moliner 50
ES-46100 Burjassot, València
Spain
Joana.martinez@uv.es

Robert T. Mason
Department of Zoology
Oregon State University
Covallis, OR 97331
USA
masonr@science.oregonstate.edu

Patrick McArdle
School of Psychology
Queen's University of Belfast
Belfast BT7 1NN
UK

Louise Mead
National Center for Science Education
420 40th Street Suite 2
Oakland, CA 94609-2509
USA
mead@ncseweb.org

Michael Meredith
Department of Biological Science
Florida State University
Tallahassee, FL 32306-4340
USA
mmered@neuro.fsu.edu

Christen Merte
Department of Biology
Georgia Southern University
Statesboro, GA 30460-8042
USA

Jordana M. Meyer
Department of Biology
Georgia Southern University
Statesboro, GA 30460-8042
USA

Masao Miyazaki
Sphingolipid Expression Laboratory
RIKEN Frontier Research System
Hirosawa 2-1
Wako, Saitama 351-0198
Japan
mmiyazaki@riken.jp

Jose Moncho-Bogani
Departament CC Médicas
Universitat Castilla la Mancha
Alabcete 02006
Spain
Josevaleriano.moncho@uclm.es

Rodrigo Morales
Institute for Molecular Bioscience
University of Queensland
St. Lucia QLD 4072
Australia

R. Grant Morshedi
Department of Chemistry
Hendrix College
1600 Washington Avenue
Conway, AR 72032
USA

Erek Napora
Department of Biology
Georgia Southern University
Statesboro, GA 30460-8042
USA

Amparo Novejarque
Departament de Biologia Funcional
i Antropologia Fisica
Universitat de València
C/ Dr Moliner 50
ES-46100 Burjassot, València
Spain
Amparo.Novejarque@uv.es

Milos V. Novotny
Institute for Pheromone Research
Department of Chemistry
Indiana University
Bloomington, IN 47405
USA

Catherine A. Palmer
Department of Biology
Portland State University
P.O. Box 751
Portland, OR 97207-0751
USA
palmerc@pdx.edu

Aras Petrulis
Center for Behavioral Neuroscience
Department of Psychology
Georgia State University
P.O. Box 3966
Atlanta, GA 30302-3966
USA
apetrulis@gsu.edu

John A. Pickett
Biological Chemistry Division
Rothamsted Research
Harpenden
Hertfordshire AL5 2JQ
UK
John.pickett@bbsrc.ac.uk

Mousumi Poddar-Sarkar
Department of Botany
Surendranath College
24/2 M. G. Road
Kolkata 700009
India
Mpsarkar1@rediffmail.com

Richard Porter
UMR 6175
INRA-PRC
37380 Nouzilly
France
Porter.rh@gmail.com

Stephen R. Price
Department of Psychology
    and Institute of Neuroscience
Dalhousie University
Halifax, Nova Scotia B3H 4J1
Canada
pricesr@dal.ca

Chantal Raimbault
Université François Rabelais
Unité INSERM 619
Tours
France
Ch.chantal@numericable.fr

Bets Rasmussen
Department of Environmental
    Biomolecular Systems
Oregon Health and Science
    University
20000 NW Walker Rd
Beaverton, OR 97006
USA

Matthias Rietdorf
Institute of Organic Chemistry
University of Hamburg
Martin-Luther-King-Platz 6
20146 Hamburg
Germany
Matthias.rietdorf@varianinc.com

S. Craig Roberts
School of Biological Sciences
University of Liverpool
Crown Street
Liverpool L69 7ZB
UK
Craig.roberts@liverpool.ac.uk

Duncan H. L. Robertson
Proteomics and Functional
    Genomics Group
Department of Veterinary
    Preclinical Science
University of Liverpool
Crown Street
Liverpool L69 7ZJ
UK
dhlr@liv.ac.uk

Carolina Rojas
Instituto de Investigaciones
    Biomédicas
Universidad Nacional Autónoma
    de México
70228, Ciudad Universitaria
04510 Distrito Federal
Mexico
rocastan@yahoo.com.mx

Elie Saliba
Université François Rabelais
Unité INSERM 619
Tours
France
saliba@med.univ-tours.fr

Chad Samuelsen
Department of Biological Science
Florida State University
Tallahassee, FL 32306-4340
USA
samuelsen@neuro.fsu.edu

Tamsin Saxton
School of Biological Sciences
University of Liverpool
Crown Street
Liverpool L69 7ZB
UK
Tamsin.saxton@liverpool.ac.uk

Benoist Schaal
Equipe d'Ethologie et de
    Psychobiologie Sensorielle
Centre Européen des Sciences
    du Goût
UMR 5170 CNRS
Université de Bourgogne / Inra
15 rue Hugues Picardet
21000 Dijon
France
schaal@cesg.cnrs.fr

Heather M. Schellinck
Department of Psychology
    and Institute of Neuroscience
Dalhousie University
Halifax, Nova Scotia B3H 4J1
Canada
heathers@dal.ca

Bruce Schulte
Department of Biology
Georgia Southern University
Statesboro, GA 30460-8042
USA
bschulte@georgiasouthern.edu

Takisha G. Schulterbrandt
Department of Anatomy
    and Cell Biology
SUNY Downstate Medical Center
450 Clarkson Avenue
Brooklyn, NY 11203
USA
schultert@mail.nih.gov

Elizabeth Scordato
Committee on Evolutionary Biology
University of Chicago
1025 E. 57th Street
Culver Hall 402
Chicago, IL 60637
USA
Escordato@uchicago.edu

Mark J. T. Sergeant
Division of Psychology
Nottingham Trent University
Burton Street
Nottingham NG1 4BU
UK
Mark.sergeant@ntu.ac.uk

Richard Shine
School of Biological Sciences
University of Sydney
New South Wales 2006
Australia
rics@bio.usyd.edu.au

Deborah Simpson
Proteomics and Functional
    Genomics Group
Department of Veterinary
    Preclinical Science
University of Liverpool
Crown Street
Liverpool L69 7ZJ
UK
dsimpson@liv.ac.uk

Helena A. Soini
Institute for Pheromone Research
Department of Chemistry
Indiana University
Bloomington, IN 47405
USA
hsoini@indiana.edu

Robert Soussignan
Laboratoire Vulnerabilite,
    Adaptation, et
    Psychopathologie
CNRS, Hopital de la Salpetriere
Paris
France
rsoussignan@tele2.fr

Lauren Stanley
Department of Biology
Georgia Southern University
Statesboro, GA 30460-8042
USA

Amber Stefani
Department of Biology
Western Oregon University
345 North Monmouth Avenue
Monmouth, OR 97361
USA
lemastm@wou.edu

Paula Stockley
Mammalian Behaviour
    and Evolution Group
Department of Veterinary
    Preclinical Science
University of Liverpool
Leahurst
Neston CH64 7TE
UK

Akemi Suzuki
Sphingolipid Expression Laboratory
RIKEN Frontier Research System
Hirosawa 2-1
Wako, Saitama 351-0198
Japan
aksuzuki@riken.jp

Hideharu Taira
Department of Agro-Bioscience
Iwate University
3-18-8 Ueda, Morioka
Iwate 020-8550
Japan
tiara@iwate-u.ac.jp

Kevin Theis
Department of Zoology
203 Natural Science Building
Michigan State University
East Lansing, MI 48824-1115
USA
theiskev@msu.edu

Michael D. Thom
Mammalian Behaviour
    and Evolution Group
Department of Veterinary
    Preclinical Science
University of Liverpool
Leahurst
Neston CH64 7TE
UK
mthom@liv.ac.uk

Michael J. Turton
Department of Veterinary
    Preclinical Science
University of Liverpool
Crown Street
Liverpool L69 7ZJ
UK

Joseph R. Verge
Department of Zoology
203 Natural Science Building
Michigan State University
East Lansing, MI 48824-1115
USA
vergejoe@msu.edu

Barbara R. Vogler
Leibniz Institute for Zoo
    and Wildlife Research
Postfach 601103
D-10252 Berlin
Germany
vogler@izw-berlin.de

Christian C. Voigt
Leibniz Institute for Zoo
    and Wildlife Research
Postfach 601103
D-10252 Berlin
Germany
voigt@izw-berlin.de

Dhaval K. Vyas
Department of Biology
Georgia Southern University
Statesboro, GA 30460-8042
USA

Richard A. Watts
Department of Zoology
Oregon State University
Corvallis, OR 97331
USA

Margaret E. Weddell
Department of Chemistry
Hendrix College
1600 Washington Avenue
Conway, AR 72032
USA

Deborah Wells
School of Psychology
Queen's University of Belfast
Belfast BT7 1NN
UK
d.wells@qub.ac.uk

Gunnar Weibchen
Laboratoire Vulnerabilite,
    Adaptation, et
Psychopathologie
CNRS, Hopital de la Salpetriere
Paris
France
Gunnar.weibchen@varianinc.com

Jenne Westberry
Department of Physiology
University of Kentucky
800 Rose Street
Lexington, KY 40536-0298
USA
Jwestberry@uky.edu

Kimberly Wollett
Department of Biology
Georgia Southern University
Statesboro, GA 30460-8042
USA

Michael J. Wong
Department of Psychology
    and Institute of Neuroscience
Dalhousie University
Halifax, Nova Scotia B3H 4J1
Canada
Mc729345@dal.ca

Tristram Wyatt
Office of Distance and
Online Learning
University of Oxford
Oxford OX2 7DD

UK
Tristram.wyatt@continuing-
Education.oxford.ac.uk

Charles J. Wysocki
Monell Chemical Senses Center
3500 Market Street
Philadelphia, PA 19104-3308
USA
wysocki@monell.org

Tetsuro Yamashita
Department of Agro-Bioscience
Iwate University
3-18-8 Ueda, Morioka
Iwate 020-8550
Japan
yamashit@iwate-u.ac.jp

Ido Zuri
Department of Anatomy
    and Cell Biology
SUNY Downstate Medical Center
450 Clarkson Avenue
Brooklyn, NY 11203
USA
izuri@mail.rockefeller.edu

This volume is dedicated to the memory of Dr. Bets Rasmussen (1938-2006). Photograph supplied by Thomas E. Goodwin (photographer unknown).

A Tribute to L.E.L. "Bets" Rasmussen (1938–2006)

Thomas E. Goodwin and Bruce A. Schulte

Abstract A pioneer in the study of chemical signals in vertebrates, Dr. L.E.L. "Bets" Rasmussen, died on September 17, 2006. While Bets had a wide variety of research interests and accomplishments, her passion was the study of chemical communication among elephants. She was the driving force behind the discovery of the estrous pheromone in Asian elephants, the musth signal of frontalin and the relationship of its enantiomeric ratios to signal meaning, the chemistry of adolescent musth, and numerous other advances in our understanding of elephant biochemistry, anatomy, chemical ecology, and applications of this knowledge for elephant conservation. It is appropriate that we pay homage to her with this overview of some of her seminal accomplishments. She will be greatly missed and long remembered.

1 Introduction

On September 17, 2006, the world lost a great scientist, the world's elephants lost a tireless benefactor, and we lost a superb collaborator, wise mentor, and dear friend. Dr. L.E.L. "Bets" Rasmussen was a wonderful and remarkable human being, with an infectious joie de vivre, boundless energy, a brilliant and probing intellect, an insatiable curiosity about nature, and an impish wit. She was clearly the world's leading authority on chemical communication among elephants, and a leader in the study of elephant biochemistry.

Bets received her Bachelor's degree in biology from Stanford University and her PhD in neurochemistry from Washington University (St. Louis). She was a postdoctoral fellow at the U.S. National Institutes of Health and after stints at the Dow-Corning Corporation and Washington State University, she moved to the Oregon Graduate Institute (OGI) in 1977. At the time of her death, Bets was a Research Professor in the Department of Environmental and Biomolecular Systems

Thomas E. Goodwin
Hendrix College, Department of Chemistry
goodwin@hendrix.edu

at OGI School of Science & Engineering (OHSU), Department of Environmental and Biomolecular Systems.

2 Elephant Research Accomplishments

2.1 The Beginnings

Over the years, Bets had a variety of animal research interests ranging from coelacanths (Rasmussen 1979) to elasmobranchs (Rasmussen and Schmidt 1992; Rasmussen and Crow 1993); elephants, however, were her passion. Bets' love of elephants was engendered during her collaboration with Dr. Irven Buss (Department of Zoology, Washington State University), with whom she published her first elephant research paper (Buss, Rasmussen and Smuts 1976). An excerpt from this paper is illustrative of the sometimes unorthodox yet "tasteful" lengths to which Bets and her colleagues would go for scientific discovery: "Numerous published statements on physical characteristics of temporal gland secretion such as 'the oily pungent musth secretion' and 'a dark strong-smelling oily substance', caused us to seek objective information on odor, taste, and viscosity of the secretion. Five persons independently conducted the following simple tests. Fresh samples of clean secretion were taken at least five steps upwind from the donor elephant and then smelled, tasted, and finally watched as the secretion was allowed to flow over a slanting piece of paper. Based on these tests the secretion was appraised as: (1) having a barely detectable smell to a 'musky' odor..., (2) having no taste to very little taste—possibly producing a very slight burning or acid-like sensation, and (3) not being oily but being decidedly watery or serous and drying on paper without leaving an oily mark." In an expansion of this work in later years, Bets and her collaborators found the first chemical evidence for musth in male African elephants, revealing characteristic chemical compounds in the temporal gland secretions (TGS), as well as high serum levels of testosterone (Rasmussen, Hall-Martin and Hess 1996).

2.2 Impact on Young Scientists

Bets was enamored with the signals involved in musth, and in their evolution (Rasmussen 1998; Rasmussen and Schulte 1998; Rasmussen 1999). She collaborated with many young researchers in her pursuit of a better understanding of musth in Asian and African elephants (e.g., Perrin and Rasmussen 1994; Rasmussen and Perrin 1999; Schulte and Rasmussen 1999a; Rasmussen and Wittemyer 2002). As a Research Professor at OGI, Bets had the intellectual freedom to pursue a variety of research topics and collaborate with numerous individuals; however, she was away on research ventures so often that overseeing students was not easy. One of the authors of this tribute (B.A.S.) was fortunate to have a recommendation from his major professor, Dietland Müller-Schwarze, which swayed Bets into expanding her laboratory group through a post-doctoral position in 1993. Schulte and Rasmussen initiated, and for the next 12 years continued, studies on the interaction between sender and receiver hormonal state, as well as chemical signal production and response (e.g., Schulte and Rasmussen 1999a,b; Rasmussen and Schulte 1999; Schulte, Bagley, Correll, Gray, Heineman, Loizi, Malament, Scott, Slade, Stanley, Goodwin and Rasmussen 2005). Bets also served on thesis committees or acted as a helpful adviser on projects for many students (for example, see Perrin, Rasmussen, Gunawardena and Rasmussen 1996; Slade, Schulte and Rasmussen 2003; Scott and Rasmussen 2005). Wherever she went, Bets left a lasting impression on young researchers, including both graduate and undergraduate students—knowledge, enthusiasm, creativity and credibility were resounding features of her tutelage.

2.3 The Asian Elephant Estrous Pheromone

Perhaps Bets is best known for her 15-year quest to identify the estrous pheromone in Asian elephants. In this search she used classical bioassay-guided fractionation of an elephantine volume of female Asian elephant urine ("liquid gold", she often called it). Most of this early work was carried out in close collaboration with the elephant managers and the elephants at the Metro Washington Park Zoo (now the Oregon Zoo). Countless hours of observation led to the discovery of the elephant's unique version of a flehmen response: the trunk tip, after exploration of an environmental stimulus, is placed on the openings of the vomeronasal ducts in the roof of the mouth (Rasmussen, Schmidt, Henneous, Groves and Daves 1982). This behavior was to play a pivotal role in the bioassays necessary to test pheromone candidate compounds and also led her into research on the anatomy of the structures related to chemosensory behaviors (e.g., the trunk tip finger: Rasmussen and Munger 1996; and VNO development: Rasmussen and Hultgren 1990; Johnson and Rasmussen 2002).

Eventually hard-won success revealed the estrous pheromone to be Z-7-dodecen-1-yl acetate (Rasmussen, Lee, Roelofs, Zhang and Daves 1996; Rasmussen, Lee, Zhang, Roelofs and Daves 1997; Rasmussen 2001). This was a remarkable breakthrough, for it was astounding that a single non-steroidal pheromone in a mammal would have such profound importance to reproduction. Further, many moths use the same compound as part of their pheromone blend, a fascinating case of convergent evolution. Years later, Bets discovered another surprising insect/elephant connection: frontalin, a well-known beetle aggregation pheromone, was found to be a multi-purpose chemical signal in the TGS of male Asian elephants in musth (Rasmussen and Greenwood 2003). Even more exciting was the discovery by Bets and her colleagues that the nature of this chemical signal depends upon the precise enantiomeric ratio of frontalin that is present (Greenwood, Comesky, Hunt and Rasmussen 2005). This phenomenon, long recognized in insects, had never before been demonstrated in mammals.

2.4 A Study of Olfaction

Bets believed that elephants were ideal subjects for an in-depth study of olfaction per se, an area in which she had long harbored a deep interest. She was determined to uncover how, on a molecular level, Z-7-dodecen-1-yl acetate was transported from female Asian elephant urine to conspecific male trunk mucus, then to the vomeronasal organ duct mucus, and finally to the vomeronasal organ itself. She and her collaborators made significant progress in this arena as they discovered and characterized proteins that are part of the pheromonal transport system (Lazar, Greenwood, Rasmussen and Prestwich 2002; Lazar, Rasmussen, Greenwood, Bang and Prestwich 2004). With support from the U.S. National Science Foundation, Bets and Dave Greenwood (HortResearch; University of Auckland) were actively continuing this study until her untimely demise.

2.5 The Queen of Elephant Secretions

The search for putative elephant pheromones was not limited to a single biological matrix. Indeed, in view of the breadth of Bets' exploration of elephant chemical communication, we might anoint her "the queen of elephant secretions, excretions and exhalations". In these studies, she was a master at identifying and attracting collaborators, and at catalyzing fruitful partnerships. This led, *inter alia*, to investigations of liquid expelled from African elephant ears (Riddle, Riddle, Rasmussen and Goodwin 2000), interdigital sweat glands on Asian elephant feet (Lamps, Smoller, Rasmussen, Slade, Fritsch and Goodwin 2001; Lamps, Smoller, Goodwin and Rasmussen 2004), and even analysis of elephant breath (Rasmussen and Riddle 2004a).

One of the more interesting discoveries began when Heidi Riddle (Riddle's Elephant Sanctuary; www.elephantsanctuary.org) noticed that early temporal gland secretions from a young male Asian elephant smelled sweet like honey, rather than acrid and unpleasant as is characteristic of TGS from mature males. A fruitful liaison with Bets and the late Dr. V. Krishnamurthy (a highly respected elephant veterinarian in India) revealed the differing characteristic chemical contents of adolescent Asian male musth TGS versus that of mature males (Rasmussen, Riddle and Krishnamurthy 2002). Interestingly, Edgerton (1931) states that ancient Hindu poetry is "full of allusions of bees coming and gathering sweetness from the temples of musth elephants".

2.6 Prevention of Elephant Crop-Raiding

A strong interest in global elephant welfare and long-term survival, and a desire to ameliorate elephant/human conflicts fueled and sustained Bets' boundless and far-flung research efforts. One practical outgrowth of this interest can be illustrated by her ongoing collaboration with Scott Riddle (Riddle's Elephant Sanctuary) on the use of naturally occurring chemical repellents to alleviate crop raiding by elephants in the wild. An understanding that one repellent modality alone would likely not suffice led Bets and Scott to couple these repellents with a unique mechanical platform, easily and inexpensively constructed in the field using indigenous labor and materials. Testing of this promising multi-sensory approach was done initially with captive elephants, then with wild elephants in southern India, and soon will be extended to field testing in Sri Lanka and elsewhere (Rasmussen and Riddle 2004b; a full account entitled "A Multisensory, Learning Approach to Reducing Asian Elephant Crop Raiding", is in preparation).

2.7 African Elephant Chemical Signaling

In recent years, Bets and the authors of this tribute forged an alliance (a biochemist, an animal behaviorist, and an organic chemist) to study the chemical ecology of African elephants in a project funded by the U.S. National Science Foundation. One of the co-investigators is her past post-doctoral student (B.A.S.). The last member of the trio (T.E.G.) was made aware of Bets through a chance encounter with a mutual acquaintance. A two-hour Sunday afternoon phone conversation soon followed, and the friendship and collaboration were thus begun (for example, see: Goodwin, Brown, Counts, Dowdy, Fraley, Hughes, Liu, Mashburn, Rankin, Roberson, Wooley, Rasmussen, Riddle, Riddle and Schulz 2002; Goodwin, Rasmussen, Schulte, Brown, Davis, Dill, Dowdy, Hicks, Morshedi, Mwanza and Loizi 2005).

Our recent NSF-funded collaboration with Bets has yielded, for example, the first statistically significant evidence that male African elephants can distinguish conspecific female urine collected at the time of ovulation from urine obtained at the mid-luteal time of the estrous cycle (Bagley, Goodwin, Rasmussen and Schulte 2006). Additionally, we have published the first report of insect pheromones in the urine of female African elephants (Goodwin, Eggert, House, Weddell, Schulte and Rasmussen 2006). These findings bode well for the eventual discovery of the first African elephant pheromones.

3 A Few Words from the Matriarch

Bets was widely sought after as a speaker, due not only to the excellence of her research, but also to the joyous enthusiasm and clarity with which she presented her results. In late 2005, she gave the keynote address, "Mirrored in the Footprints of Elephants: Communication by Chemical Signals", at the annual meeting of the Elephant Managers Association in her hometown of Portland, Oregon (USA). It is fitting here that we include a message from Bets by printing the abstract from her talk at that meeting.

"In the early morning softly squishing through the moisty, misty Asian forest, some elephant foot puddles of water reflect faces of calves anxiously following their mothers, other water-pockets mirror visages of bigger aunties, while still other shimmering spots show the features of big musth males following female pheromone trails. In actuality, chemical communication investigations on elephants began in Oregon–the inspiration of Irven Buss, based in turn on his observations of African elephants in the wild. For 27 years my research has followed his footsteps and the footprints of elephants. A brief historical perspective will lead into the essentiality in this sensory discipline of basing initial studies on elephants on their behaviors, but also on individual animals with known histories and currently measured physiological parameters. Only then can chemistry be coupled to behavior to decipher the wily elephant. I consider it a life-long responsibility to at least partially figure out how elephants conduct inter-conspecific relationships. Glimpses into the rewards to both elephants and humans from such deciphering will be given both with studies of the role of musth among wild Asian elephants and with ways to alter crop-raiding behaviors. The ability to effectively conduct such programs directly stems from knowledge learned among captive elephants. For just as our selves are mirrored in our eyes, elephants' countenances are mirrored in their footprints, guided by chemical signals and pheromones."

4 Conclusion

Appearances perhaps to the contrary, Bets did not spend every waking minute on research in the laboratory or field. She was an inveterate traveler, and an intrepid adventurer. Bets loved to hike, scuba dive, swim (two miles each day at 5 a.m.!), and water ski. She was also renowned for her healthy diet of fruits and vegetables. At a memorial service for Bets at the Oregon Graduate Institute, one of us (T.E.G.) toasted Bets' memory with a big bite from one of her favorite treats, a large orange bell pepper. Then in a gesture that we know would give her pleasure, the remainder of the pepper was fed to Packy, the large male Asian elephant at the Oregon Zoo and an early pachyderm collaborator in Bets' behavioral bioassay searches for the estrous pheromone.

In the elite company of pioneers pursuing chemical signals in vertebrates, Professor L.E.L. "Bets" Rasmussen has clearly earned a place of honor. Bets will be greatly missed. She loved life and lived it to the fullest. We humbly dedicate this chapter to her memory, with gratitude for her guidance, friendship, and immense legacy.

Acknowledgments Bets and all of us elephant researchers owe a large debt of gratitude to many members of the community of humans who study, care for, and work for the survival of the majestic, magnificent mammals: the elephants. She would want us to thank all those she has worked with over the years. While many organizations and institutions have played major roles in her research activities and successes, we are sure she would want especially to thank Riddle's Elephant and Wildlife Sanctuary, the Ringling Center for Elephant Conservation, and the Oregon Zoo.

References

Bagley, K.R., Goodwin, T.E., Rasmussen, L.E.L. and Schulte, B.A. (2006) Male African elephants (*Loxodonta africana*) can distinguish oestrous status via urinary signals. Anim. Behav. 71, 1439–1445.

Buss, I.O., Rasmussen, L.E. and Smuts, G.L. (1976) The role of stress and individual recognition in the function of the African elephant's temporal gland. Mammalia 40, 437–451.

Edgerton, F. (translated from the original Sanskrit) (1931) *The Elephant-Lore Of The Hindus: The elephant-sport (Matanga-Lila) of Nilakantha.* Yale University Press, New Haven, p. 34 (reprinted, 1985, Motilal Banarsidass, Delhi).

Goodwin, T.E., Brown, F.D., Counts, R.W., Dowdy, N.C., Fraley, P.L., Hughes, R.A., Liu, D.Z., Mashburn, C.D.; Rankin, J.D., Roberson, R.S., Wooley, K.D., Rasmussen, E.L., Riddle, S.W., Riddle, H.S. and Schulz, S. (2002) African elephant sesquiterpenes. II. Identification and synthesis of new derivatives of 2,3-dihydrofarnesol. J. Nat. Prod. 65, 1319–1322.

Goodwin, T.E., Eggert, M.S., House, S.J., Weddell, M.E., Schulte, B.A. and Rasmussen, L.E.L. (2006) Insect pheromones and precursors in female African elephant urine. J. Chem. Ecol. 32, 1849–1853.

Goodwin, T.E., Rasmussen, L.E.L., Schulte, B.A., Brown, P.A., Davis, B.L., Dill, W.M., Dowdy, N.C., Hicks, A.R., Morshedi, R.G., Mwanza, D. and Loizi, H. (2005) Chemical analysis of African elephant urine: A search for putative pheromones. In: R.T. Mason, M.P. LeMaster and D. Müller-Schwarze (Eds.), *Chemical Signals in Vertebrates 10.* Springer Press, New York, pp. 128–139.

Greenwood, D.R., Comesky, D., Hunt, M.B. and Rasmussen, L.E.L. (2005) Chirality in elephant pheromones. Nature 438, 1097–1098.

Johnson, E.W. and Rasmussen, L.E.L. (2002) Morphological characteristics of the vomeronasal organ of the newborn Asian elephant, *Elephas maximus.* Anat. Rec. 267, 252–59.

Lamps, L.W., Smoller, B.R., Goodwin, T.E. and Rasmussen, L.E.L. (2004) Hormone receptor expression in interdigital glands of the Asian elephant *(Elephas maximus).* Zoo Biol. 23, 463–469.

Lamps, L.W., Smoller, B.R., Rasmussen, L.E.L., Slade, B.E., Fritsch, G. and Goodwin, T.E. (2001) Characterization of interdigital glands in the Asian elephant (*Elephas maximus*). Res. Vet. Sci. 71, 1–4.

Lazar, J., Greenwood, D.R., Rasmussen, L.E.L. and Prestwich, G.D. (2002) Molecular and functional characterization of an odorant binding protein of the Asian elephant, *Elephas maximas*: Implications for the role of lipocalins in mammalian olfaction. Biochemistry 41, 11786–11794.

Lazar, J., Rasmussen, L.E.L., Greenwood, D.R., Bang, I-S. and Prestwich, G.D. (2004) Elephant albumin: A multipurpose shuttle. Chem. & Biol. 11, 1093–1100.

Perrin, T.E. and Rasmussen, L.E.L. (1994) Chemosensory responses to female Asian elephants (*Elephas maximus*) to cyclohexanone. J. Chem. Ecol. 20, 2577–2586.

Perrin, T.E., Rasmussen, L.E.L., Gunawardena, R. and Rasmussen, R.A. (1996) A method for the collection, long-term storage, and bioassay of labile volatile chemosignals. J. Chem. Ecol. 22, 207–221.

Rasmussen, L.E. (1979) Some biochemical parameters in the coelacanth, *Latimeria chalumnae*, ventricular and notochordal fluids. In: J. McCosker and M. Lagios (Eds.), *The Biology and Physiology of the Living Coelacanth.* Calif. Acad. Sci. 134, 94–111.

Rasmussen, L.E.L. (1998) Chemical communication: an integral part of functional Asian elephant (*Elephas maximus*) society. Ecoscience 5, 410–426.

Rasmussen, L.E.L. (1999) Evolution of chemical signals in the Asian elephant, *Elephas maximus*: behavioural and ecological influences. J. Biosci. (Bangalore) 24, 241–251.

Rasmussen, L.E.L. (2001) Source and cyclic release pattern of Z-7-dodecenyl acetate, the pre-ovulatory pheromone of the female Asian elephant. Chem. Senses 26, 611–623.

Rasmussen, L.E.L. and Crow, G.L. (1993) Serum corticosterone concentrations in immature captive whitetip reef sharks, *Triaenodon obesus.* J. Exp. Biol. 267, 283–287.

Rasmussen, L.E.L. and Greenwood, D.R. (2003) Frontalin: A chemical message of musth in Asian elephants (*Elephas maximus*). Chem. Senses 28, 433–446.

Rasmussen, L.E.L. and Hultgren, B. (1990) Gross and microscopic anatomy of the vomeronasal organ in the Asian elephant, *Elephas maximus*. In: D. McDonald, D. Müller-Schwarze, and S. E. Natynczuk (Eds), *Chemical Signals in Vertebrates 5*. Oxford University Press, Oxford, pp. 154–161.

Rasmussen, L.E.L. and Munger, B.L. (1996) The sensorineural specializations of the trunk tip (finger) of the Asian elephant, *Elephas maximus*. Anat. Rec. 246, 127–134.

Rasmussen, L.E.L. and Perrin, T.E. (1999) Physiological correlates of musth: lipid metabolites and chemical composition of exudates. Physiol. Behav. 67, 539–549.

Rasmussen, L.E.L. and Riddle, H.S. (2004a) Elephant breath: Clues about health, disease, metabolism and social signals. J. Eleph. Manag. Assoc. 15, 24–33.

Rasmussen, L.E.L. and Riddle, S.W. (2004b) Development and initial testing of pheromone-enhanced mechanical devices for deterring crop raiding elephants: A positive conservation step. J. Eleph. Manag. Assoc. 15, 30–37.

Rasmussen, L.E.L. and Schmidt, M.J. (1992) Are sharks chemically aware of crocodiles? In: R.L. Doty and D. Müller-Schwarze (Eds.), *Chemical Signals in Vertebrates*. Plenum Press, New York, pp. 335–342.

Rasmussen, L.E.L. and Schulte, B.A. (1998) Chemical signals in the reproduction of Asian (*Elephas maximus*) and African (*Loxodonta africana*) elephants. Anim. Reprod. Sci. 53, 19–34.

Rasmussen, L.E.L. and Schulte, B.A. (1999) Ecological and biochemical constraints on pheromonal signaling systems in Asian elephants and their evolutionary implications. In: R. Johnston, D. Müller-Schwarze and P. Sorenson (Eds.), *Advances in Chemical Communication in Vertebrates*, Vol. 8 Kluwer Academic/Plenum Press, New York, pp. 46–62.

Rasmussen, L.E.L. and Wittemyer, G. (2002) Chemosignaling of musth by individual wild African elephants, *(Loxodonta africana)*: implications for conservation and management. Proc. Royal Soc. London 269, 853–860.

Rasmussen, L.E.L., Hall-Martin, A.J. and Hess, D.L. (1996) Chemical profiles of male African elephants, *Loxodonta africana*: Physiological and ecological implications. J. Mammal. 77, 422–439.

Rasmussen, L.E.L., Lee, T.D., Roelofs, W.L., Zhang, A. and Daves, G.D. Jr. (1996) Insect pheromone in elephants. Nature 379, 684.

Rasmussen, L.E.L., Lee, T.D., Zhang, A., Roelofs, W.L. and Daves, G.D. Jr. (1997) Purification, identification, concentration and bioactivity of Z-7-dodecen-1-yl acetate: sex pheromone of the female Asian elephant, *Elephas maximus*. Chem. Senses 22, 417–437.

Rasmussen, L.E.L., Riddle, H.S. and Krishnamurthy, V. (2002) Mellifluous matures to malodorous in musth. Science 415, 975–976.

Rasmussen, L.E., Schmidt, M.J., Henneous, R., Groves, D. and Daves, Jr., G.D. (1982) Asian bull elephants: Flehmen-like responses to extractable components in female estrous urine. Science 217, 159–162.

Riddle, H.S., Riddle, S.W., Rasmussen, L.E.L. and Goodwin, T.E. (2000) The first disclosure and preliminary investigation of a liquid released from the ears of African elephants. Zoo Biol. 19, 475–480.

Schulte, B.A. and Rasmussen, L.E.L. (1999a) Musth, sexual selection, testosterone and metabolites. In: R.E Johnston, D. Müller-Schwarze and P. Sorensen (Eds.), *Advances in Chemical Communication in Vertebrates*, Plenum Press, New York, pp. 383–397.

Schulte, B.A. and Rasmussen, L.E.L. (1999b) Signal-receiver interplay in the communication of male condition by Asian elephants. Anim. Behav. 57, 1265–1274.

Schulte, B.A., Bagley, K., Correll, M., Gray, A., Heineman, S.M., Loizi, H., Malament, M., Scott, N.L., Slade, B.E., Stanley, L., Goodwin, T.E. and Rasmussen, L.E.L. (2005) Assessing chemical communication in elephants. In: R.T. Mason, M.P. LeMaster and D. Müller-Schwarze (Eds.), *Chemical Signals in Vertebrates* 10. Springer Press, New York, pp. 140–151.

Scott, N.L. and Rasmussen, L.E.L. (2005) Chemical communication of musth in captive Asian elephants, *Elephas maximus*. In: R.T. Mason, M.P. LeMaster and D. Müller-Schwarze (Eds.), *Chemical Signals in Vertebrates 10.* Springer Press, New York, pp. 118–127.

Slade, B.E., Schulte, B.A. and Rasmussen, L.E.L. (2003) Oestrous state dynamics in chemical communication by captive female Asian elephants. Anim. Behav. 65, 813–819.

Part I
New Directions in Semiochemistry

Chapter 1
Volatile Mammalian Chemosignals: Structural and Quantitative Aspects

Milos V. Novotny and Helena A. Soini

Abstract Current investigations of mammalian chemical signals are providing biologically useful information. Precision of profiling data has become substantially improved due to better sampling methodologies, selective detection techniques and vastly improved mass-spectrometric instrumentation. This, in turn, facilitates the use of powerful chemometric methodologies in evaluation of large data sets and a future integration of volatile profiling data into the systems biology knowledge.

1.1 Introduction

In molecular terms, we live in the complex world involving chemosignaling in many events, both internal and external. In morphologically and functionally sophisticated mammalian systems, a balanced chemical communication among their biological constituent cells is a standard of normalcy, as are the intracellular events determined through transmembrane signaling. In the "outside world", a living organism is constantly being subjected to chemical signals, both spurious and naturally programmed, which are both consciously and subconsciously perceived. The biological selectivity and sensitivity of perception are inherent and important to the processes of chemical communication both within a species and among different species. Consequently, the inherited traits for chemical communication signals (pheromones and allomones) and their receptors, and the learned responses to other chemosignals in the environment are widely and effectively utilized by different animals. While it would seem that the large number of species and their regulatory needs for hormonal and reproductive function could result in a staggering number of chemical messengers to fulfill these tasks, there are some limitations imposed by the nature of biosynthetic processes. Consequently, in the mammalian pheromones identified

Milos V. Novotny
Institute for Pheromone Research, Department of Chemistry
novotny@indiana.edu

J.L. Hurst et al., *Chemical Signals in Vertebrates 11.*
© Springer 2008

thus far, we see a great deal of structural redundancy, or at least small structural modifications, with the more scientifically explored world of insects. Naturally, this observation has been confined to the volatile molecules used in a distance communication, while the larger biomolecules, if employed in chemical communication, can present a wider range of structural/biochemical diversity to satisfy the requirements for biological selectivity.

We still know globally relatively little above the chemical nature of the individual chemosignals and the biochemical events involved in their perception. Since identification of the first definitive mammalian pheromones in the house mouse (Novotny, Harvey, Jemiolo and Alberts 1985; Jemiolo, Harvey and Novotny 1986), substantial improvements were made in bioanalytical methodologies (combined capillary gas chromatography/mass spectrometry and liquid chromatography/mass spectrometry) to yield additional chemosignal structures. Moreover, with the more recent capabilities of quantifying precisely the ratios of different volatile components of a biological stimulus, there now appear new exciting possibilities for comparative studies of the genetic correlates of chemical communication, as exemplified by the recent studies of MHC-congenic mice, (Novotny, Soini, Koyama, Bruce Wiesler, and Penn 2006). The strategy of looking at the ratios or patterns of organic volatiles as a distinct signal interpretable by a mammal's sensory pathways has been amply precedented in modern biology: for example, in a situation of comparable complexity, cell-to-cell communication in sophisticated mammalian systems is likely based on the entire reservoir of different glycan (oligosaccharide) structures which guard a cellular function in disease or health (Lowe and Marth 2003). To interpret properly such multicomponent (pattern) signals, chemical ecologists should turn increasingly to modern chemometric procedures (Brereton 2003) that can deal effectively with complex analytical data. At a more readily understandable level, optical activity and other subtle forms of substance isomerism can play a role in conveying different messages. This has recently been shown by Rasmussen and co-workers (Greenwood, Comeskey, Hunt and Rasmussen 2005) with the utilization of frontalin in different ratios of its enantiomers. In this view, many other cases where biological responses were elicited from the racemic mixtures of putative pheromone need to be re-examined.

During the last decade, the availability of new analytical methodologies and instrumentation has revolutionized modern biology and biochemistry. Starting with the Human Genome Project and the field of genomics, new separation techniques and biomolecular mass spectrometry have paved the way to conquering the proteomics of different species, including their sophisticated posttranslational modifications and non-covalent biomolecular complexation. This quest for additional molecular information continues unabated, toward the fields of glycomics, lipidomics and metabolomics. The unprecedented wealth of molecular information can now be tied, through the advanced computational techniques (bioinformatics) into the holistic approach of systems biology (Kitano 2002) to provide an unprecedented understanding of an organism in its different conditions and environment. Such an approach and capabilities provide a unique opportunity for solving typical problems in chemical communication.

1.2 Advances in Analytical Methodologies

A search for mammalian pheromones in different laboratories during the last two decades followed the methodologies used with the studies of juvenile hormones, ecdysones and pheromones in the insect world (McCaffery and Wilson 1990). Solvent extractions and trapping of volatiles are still commonly used in this type of research, although the methodologies based on dynamic headspace extraction (Novotny, Lee and Bartle 1974) or solid-phase microextraction (SPME) using coated silica fibers (Zhang and Pawliszyn 1993) are more typically seen in today's laboratories pursuing pheromone identification through GC-MS. Solvent extraction is seemingly beneficial for handling moderately volatile compounds at relatively high concentrations (Burger, Tien, LeRoux and Mo 1996). Preconcentration of volatiles on the surface of adsorptive materials, such as organic porous polymers, with a subsequent desorption for a GC-MS analysis was for a long time a standard procedure in our laboratory, used in structural elucidation of the first mouse pheromones (Schwende, Wiesler, Jorgenson, Carmack and Novotny 1986). While this method referred to as the "purge-and-trap" or "dynamic headspace' technique, may be used with satisfaction in semiquantitative studies to observe large differences in metabolite excretion, it may not be the best approach when small quantitative differences in compounds may account for recognition of a biological message.

Current developments in analytical sampling seem to favor *absorption* rather than *adsorption* preconcentration principles. Using newer techniques in this direction, the organic polymers (absorption media) are attached to a mechanical device, such as a silica fiber (Pawliszyn 1997) or a glass-coated stir bar (Baltussen, Sandra, David and Cramers 1999) to allow sample sorption from a medium (e.g., urine or a glandular secretion). The stir bar approach has further advantages over the SPME approach in a better dynamic concentration range of the trapped organics, better preconcentration of trace volatiles, and analytical reproducibility (Bicchi, Iori, Rubiolo and Sandra 2002). In contrasting the purge-and-trap sampling with a stir bar preconcentration (Fig. 1.1) some additional advantages of the latter approach become at once evident: a system setup allows high analytical throughput and more reproducible sample transfer.

Quantitative aspects of stir bar sorptive extraction in the measurements pertaining to various applications in chemical ecology have recently been demonstrated. This novel methodology has been found highly reproducible and linear in aqueous sampling of volatile and semivolatile organic compounds from mammalian urine and tissue extracts. In comparison with the adsorbent-based preconcentration procedures, the stir bar extraction technique has shown vastly improved repeatability (Soini, Bruce, Wiesler, David, Sandra and Novotny 2005). A rolling stir bar sampling procedure has recently been found effective in the in-situ surface sampling of biological objects such as human skin (Soini, Bruce, Klouckova, Brereton, Penn and Novotny 2006) or bird feathers. As demonstrated with nearly 200 human volunteers, whose skin volatile profiles were recorded as "odor signatures" of individuality (examples shown in Fig. 1.2) (Penn, Oberzaucher, Grammer, Fischer, Soini, Wiesler, Novotny, Dixon, Xu and Brereton 2006), a high degree of analytical reproducibility

Dynamic Head-Space Sampling

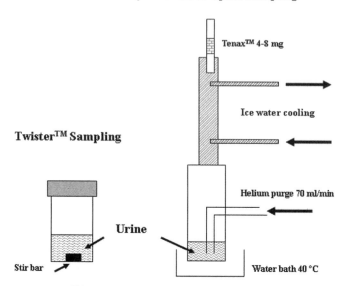

Fig. 1.1 Aqueous Twister™ and purge- and-trap Tenax™ sampling devices for trapping volatiles from urine

can be ascertained through the use of internal standards embedded onto the stir bar surface before sampling. This sampling variant provides opportunities for sample acquisition in a location geographically distant from the analytical laboratory.

Since the first uses of GC-MS in pheromone research about four decades ago, the instrumentation has made significant gains in performance and sensitivity. Highly efficient capillary columns are almost exclusively used, while sophisticated and automated inlet systems ensure the analytical reliability required for the meaningful comparative profile studies. The structural elucidation (solute identification) often remains the most difficult task requiring a chemical knowledge. The newer and more affordable high-resolution mass spectrometers can be most useful in aiding the structural identification. This can be exemplified by the case of implicating a possible structure of 2-*sec*-butyl-4,5-dihydrothiazole (a male mouse pheromone) from its MS fragmentation pattern and nominal molecular weight ($m/z = 143$) with a conventional quadrupole instrument, as compared to its unequivocal identification through the exact molecular mass ($m/z = 143.0819$) on a high-resolution time-of-flight mass spectrometer (Fig. 1.3). Naturally, additional spectrometric techniques based on IR and NMR can provide further, structurally important data whenever applicable.

Element-specific detection combined with capillary GC has become a key technique in the chemical communication studies of our laboratory. An effective detector of this type is based on the microwave plasma emission (Wylie and Quimby 1989), with a tunable selectivity for several elements and a prominent sensitivity for sulfur-containing compounds, which is significantly greater than

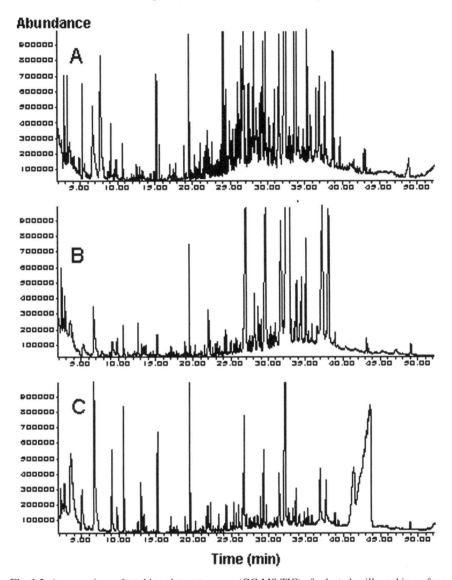

Fig. 1.2 A comparison of total ion chromatograms (GC-MS TIC) of selected axillary skin surface compounds for A: a female , B: a male (1), C: a male (2) human subjects

that of ordinary GC-MS. As many such compounds, including certain known chemosignals, are particularly odoriferous, sulfur-sensitive detection remains an important tool. Representative applications of the element-specific detection include mouse urinary profiles (Novotny et al. 2006), male-female comparisons of the ferrets (Zhang, Soini, Bruce, Wiesler, Woodley, Baum and Novotny 2005), junco songbird preen oils (Soini, Bruce, Klouckova, Brereton, Penn and Novotny 2006)

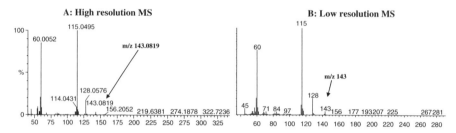

Fig. 1.3 Illustration of the mass spectra for the GC-MS TIC compound eluting at retention time of 23.4 min (2-sec-butyl-4,5-dihydrothiazole). A: high resolution MS spectrum, B: low resolution MS spectrum

and sulfur-containing human skin volatiles (Soini, Bruce et al. 2006). Using the sulfur detection mode, we have also demonstrated large quantitative and qualitative differences in male and female cat urinary volatile constituents containing sulfur (Soini, Wiesler and Novotny 2006).

The recent advances in sampling reproducibility and the availability of high-throughput GC-MS instrumentation have made it now feasible to conduct biological observations and experiments at large scale, at which hundreds of specimens can be quantitatively profiled under different sets of circumstances. This has recently been demonstrated in the studies of individual and gender "fingerprints" in human body odor involving a repeated collection of axillary sweat samples from 197 adults and their subsequent GC/MS analysis (Penn et al. 2006). To assist such large-sale investigations, it has been essential to implement new methodologies for peak detection and data set matching and computer-aided pattern recognition (Dixon, Brereton, Soini, Novotny and Penn 2007; Dixon, Brereton, Soini, Novotny and Penn 2007; Penn et al. 2007).

1.3 Chemosignaling Diversity in Mammalian Species: Different Structures or Their Proportions?

Genetic and biochemical diversities in different species have their reflections in their use of chemical communication: Are they gregarious or solitary animals? Noctur-nal scant chemical evidence thus far, we know that the compounds both "chem-ically sophisticated" and simple can serve as chemical messengers in mammals when perceived in a proper behavioral context. Examples of some sophistication are the stereospecifically determined chemosignals in the house mouse (Novotny, Xie, Harvey, Wiesler, Jemiolo and Carmack 1995) and the Asian elephant (Greenwood et al. 2005), while the behaviorally fascinating mammary pheromone of the rabbit happens to be a relatively simple organic compound (Schaal, Coureaud, Langlois, Giniès, Sémon and Perrier 2003). The biosynthetic pathways leading to what we know today are the behaviorally distinct chemosignals in both insects and mammals, which are seemingly preserved across different species. Thus, the terpenic structures such as α- and β- farnesene are pheromones in the house mouse (Novotny, Harvey and Jemiolo 1990), ovulatory Asian elephant (Goodwin, Eggert, House, Weddell, Schulte and Rasmussen 2006) and several insect species (Bowers, Nault, Webb and Dutky 1972), while (Z)-7-dodecen-1-yl acetate is the pre-ovulatory pheromone of female Asian elephant (Rasmussen, Lee, Roelofs, Zhang and Daves 1996) and several moth species alike. The mouse pheromone, dehydro-*exo*-brevicomin (Novotny, Schwende, Wiesler, Jorgenson and Carmack 1984) differs only in the presence of its double bond from the pheromone of the bark beetle and Asian elephant (brevicomin).

While 2,5-dimethylpyrazine has previously been implicated as the puberty-delay primer pheromone in the house mouse (Jemiolo and Novotny 1994), we find many other pyrazine derivatives in additional species: in deermice *Peromyscus californicus* (Jemiolo, Gubernick, Yoder and Novotny 1994), *Peromyscus maniculatus* (Ma, Wiesler and Novotny 1999) and hamsters (*Phodopus campbelli*) (Soini, Wiesler, Apfelbach, König, Vasilieva and Novotny 2005). Interestingly, substituted pyrazines are absent in *Phodopus roborovski*, which feature uniquely alkylquinaxolines instead.

While the earlier studies in chemical ecology of mammals were preoccupied with relatively gross measurements (e.g., "presence" or "absence" of a chemical/pheromone in an olfactory stimulus), the new quantitative capabilities make it now imperative to evaluate more accurately the volatile compound ratios or patterns under different biological circumstances.

1.4 Genetic Comparisons

Mice have often been referred to as the reproductively most successful mammals on Earth, while their sophisticated chemical communication systems are viewed as highly important in this success (Bronson 1979). Yet within the *Mus* genus group, there is a wide geographical distribution of behavioral and reproductive attributes that justifies comparative studies of different mouse types (Patris and Baudoin 2000). While the phylogeny is well established (Kikkawa, Miura, Takahama, Wakana, Yamazaki, Moriwaki, Shiroishi and Yonekava 2001), there have been no consistent chemical investigations which would characterize their genetic differences in terms of chemical signaling. For example, two groups which differ substantially in their nesting behavior and breeding are *Mus musculus domesticus* (common house mouse, which has a polygamous mating system) and *Mus spicilegus* (mound-building mouse, which is monogamous, Patris and Baudoin 1998). In a recent collaborative effort (Soini, Wiesler, Bruce, Koyama, Ferón, Baudoin and Novotny, 2006), we have observed some fundamental differences between the two mouse groups (Fig. 1.4): the key male pheromone of the house mouse, a thiazoline derivative, is totally absent in the urine of *M. spicilegus* (as is its characteristic smell), while the mound-building male mouse appears to feature different volatiles of its own, which are also under testosterone control.

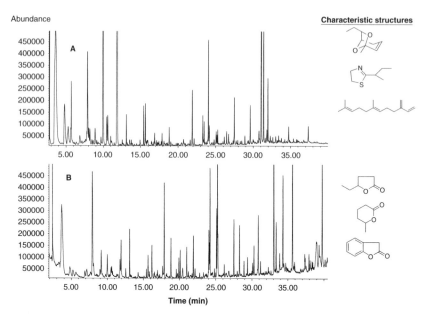

Fig. 1.4 Comparative male mouse urinary volatile profiles for (A) Mus domesticus and (B) Mus spicilegus by GC-MS with characteristic chemical structures

The house mouse has been an extremely important species due to its wide utilization in biomedical experimentation. It has also been the most documented case for studying the pheromonal effects in both behavioral terms (Hurst and Beynon 2004) and chemistry (Novotny 2003). Among the most recent investigations of the genetic influences on the mating behavior, there has been a renewed interest in a role of the major histocompatibility complex (MHC) (Penn and Potts 1998; Beauchamp and Yamazaki 2003). Our recent data (Novotny et al. 2006) for a large group of mouse chemosignaling compounds provide evidence that even minute genetic variations in MHC can be reflected in the quantitative differences of selected volatile profile constituents of the mouse urine. The results also indicate that concentrations of these compounds are not solely determined by this gene complex, but can be linked to other gene regions (for different background strains). While *Mus domesticus* is an attractive model for additional gene-behavior studies, some recent studies suggest that the MHC-dependent mating preferences may exist in fish, lizards, birds, and even humans (reviewed in Penn 2002).

1.5 Conclusions

Whereas the role of olfaction in chemical ecology is gaining increased attention, new bioanalytical methodologies and instrumentation provide unprecedented opportunities besides the structural elucidation of pheromones and other chemosignals, their release into the environment and perception at the level of olfactory neurons. Precise comparative chemical measurements can also be helpful in phylogeny studies and developmental biology.

References

Baltussen, E., Sandra, P., David, F. and Cramers, C. (1999) Stir bar extraction (SBSE), a novel extraction technique for aqueous samples: Theory and principles. J. Microcolumn Sep. 11, 737–747.

Beauchamp, G.K. and Yamazaki, K. (2003) Chemical signaling in mice. Biochem. Soc. Trans. 31, 141–151.

Bicchi, C., Iori, C., Rubiolo, P. and Sandra, P. (2002) Headspace sorptive extraction (HSSE), stir bar sorptive extraction (SBSE) and solid phase microextraction (SPME) applied to the analysis of roasted Arabica coffee and coffee brew. J.Agric. Food Chem. 50, 449–459.

Bowers, W.S., Nault, L.R., Webb, R.E. and Dutky, S.R. (1972) Aphid alarm pheromone: isolation, idenfication, synthesis. Nature 177, 1121–1122.

Brereton, R.G. (2003) *Chemometrics: Data Analysis for the Laboratory and Chemical Plant*, Wiley, Chichester, UK.

Bronson, F.H. (1979) Reproductive ecology of the house mouse. Quart. Rev.Biol. 54, 265–299.

Burger, B.V., Tien, F.-C., LeRoux , M. and Mo, W.-P. (1996) Mammalian exocrine secretions: X. Constituents of preorbital secretion of grysbok, *Raphicerus melanotis*. J.Chem.Ecol. 22, 739–764.

Dixon, S.J., Brereton, R.G, Soini, H.A., Novotny, M.V. and Penn, D.J. (2007) An automated method for peak detection and matching in large gas chromatography-mass spectrometry data sets**.** Journal of Chemometrics In press.

Dixon, S.J., Xu, Y., Brereton, R.G., Soini, H.A., Novotny, M.V., Oberzaucher, E., Grammer, K. and Penn, D.J. (2007) Pattern recognition of gas chromatography mass spectrometry of human volatiles in sweat to distinguish the sex of subjects and determine potential markers. Chemometrics Intell. Lab. Systems. In press.

Goodwin, T.E., Eggert, M.S., House, S.J., Weddell, M.E., Schulte, B.A. and Rasmussen L.E.L. (2006) Insect pheromones and precursors in female African elephant urine. J.Chem. Ecol. 32, 1849–1853.

Greenwood, D.R., Comeskey, D., Hunt, M.B. and Rasmussen L.E.L. (2005) Chirality in elephant pheromones. Nature 438, 1097–1098.

Hurst, J.L. and Beynon, R.J. (2004) Scent wars: the chemobiology of competitive signalling in mice. BioEssays 26, 1281–1298.

Jemiolo, B., Harvey, S. and Novotny M. (1986) Promotion of Whitten effect in female mice by the synthetic analogs of male urinary constituents. Proc. Natl. Acad. Sci. USA 83, 4576–4579.

Jemiolo, B., Gubernick, D.J., Yoder, M., C. and Novotny, M. (1994) Chemical characterization of urinary volatile compounds of Peromyscus californicus, a monogamous biparental rodent. J. Chem. Ecol. 20, 2489–2500.

Jemiolo, B. and Novotny, M. (1994) Inhibition of sexual maturation in juvenile male and female mice by a chemosignal of female origin. Physiol. Behav. 55, 1119–1122.

Kikkawa, Y., Miura, S., Takahama, S., Wakana, S., Yamazaki, Y., Moriwaki, K., Shiroishi, T. and Yonekava, H. (2001) Microsatellite database for MSM/Ms and JF1/Ms, molossinus derived inbred strains. Mammalian Genome 12, 750–752.

Kitano, K. (2002) Systems biology: a brief overview. Science 295, 1662–1664.

Lowe, J.B. and Marth, J. D. (2003) A genetic approach to mammalian glycan function. Ann. Rev. Biochem. 72, 643–691.

Ma, W., Wiesler, D. and Novotny, M.V. (1999) Urinary volatile profiles of the deermouse (Peromyscus maniculatus) pertaining to gender and age. J. Chem. Ecol. 25, 417–431.

McCaffery, A.R. and Wilson, I.D. (Eds.) (1990) *Chromatography and Isolation of Insect Hormones and Pheromones*, Plenum, NY

Novotny, M., Lee, M.L. and Bartle, K.D. (1974) Analytical aspects of the chromatographic headspace concentration methods using a porous polymer. Chromatographia 7, 333–338.

Novotny, M., Schwende, F.J., Wiesler, D., Jorgenson, J.W. and Carmack, M. (1984) Identification of a testosterone-dependent unique volatile constituent of male mouse urine: 7-*exo*-ethyl-5-methyl-6,8-dioxabicyclo[3.2.1]-3-octene. Experientia 40, 217–219.

Novotny, M., Harvey, S., Jemiolo, B. and Alberts, J. (1985) Synthetic pheomones that promote inter-male aggression in mice. Proc. Natl. Acad. Sci. USA 82, 2059–2061.

Novotny, M., Harvey, S. and Jemiolo, B. (1990) Chemistry of male dominance in the house mouse, *Mus domesticus*. Experientia 46, 109–113.

Novotny, M.V., Xie, T.-M., Harvey, S., Wiesler, D., Jemiolo, B. and Carmack, M. (1995) Stereoselectivity in mammalian chemical communication: male mouse pheromones. Experientia 51, 738–743.

Novotny, M.V. (2003) Pheromones, binding proteins and receptor responses in rodents. Biochem.Soc. Trans. 31, 117–122.

Novotny, M.V., Soini, H.A., Koyama, S., Wiesler, D., Bruce, K.E. and Penn , D.J. (2007) Chemical identification of MHC-influenced volatile compounds in mouse urine. I: Quantitative proportions of major chemosignals. J. Chem. Ecol. 33, 417–434.

Patris, B. and Baudoin, C. (1998) Female sexual preferences differ in *Mus spicilegus* and *Mus musculus domesticus*. Anim.Behav. 56, 1465–1470.

Patris, B. and Baudoin, C. (2000) A comparative study of parental care between two rodent species: implications for the mating system of the mound-building mouse *Mus spicilegus*. Behav.Processes 51, 35–43.

Pawliszyn, J. (1997) *Solid-Phase Microextraction – Theory and Practice*, Wiley, New York.

Penn, D.J. and Potts, W. K. (1998) How do major histocompatibility complex genes influence odor and mating preferences ? Adv.Immunol. 69, 411–436.

Penn, D.J. (2002) The scent of genetic compatibility: sexual selection and the major histocompatibility complex. Ethology 108, 1–21.

Penn, D.J., Oberzaucher, E., Grammer, K., Fischer, G., Soini, H.A., Wiesler, D., Novotny, M.V., Dixon, S.J., Xu, Y. and Brereton, R.G. (2007) Individual and gender fingerprints in body odour. J.R. Soc. Interface 4, 331–340. .

Rasmussen, L.E.L., Lee, T.D., Roelofs, W.L., Zhang, A. and Daves, G.D. Jr. (1996) Insect pheromone in elephants. Nature 379, 684.

Schaal, B., Coureaud, G., Langlois, D., Giniès, C., Sémon, E. and Perrier, G. (2003) Chemical and behavioural characterization of the rabbit mammary pheromone. Nature 424, 68–72.

Schwende, F.J., Wiesler, D., Jorgenson, J.W., Carmack, M. and Novotny, M. (1986) Urinary volatile constituents of the house mouse (*Mus musculus*) and their endocrine dependency. J. Chem. Ecol. 12, 277–296.

Soini, H.A., Bruce, K.E., Wiesler, D., David, F., Sandra, P. and Novotny, M.V. (2005) Stir bar sorptive extraction: a new quantitative and comprehensive sampling technique for determination of chemical signal profiles from biological media J. Chem. Ecol., 31, 377–392.

Soini, H.A., Wiesler, D., Apfelbach, R., König, P., Vasilieva, N.Y. and Novotny, M.V. (2005) Comparative investigation of the volatile urinary profiles of different *Phodopus* hamster species. J. Chem. Ecol. 31, 1125–1143.

Soini, H.A., Bruce, K.E., Klouckova, I., Brereton, R.G., Penn, D.J. and Novotny, M.V. (2006) In-situ surface sampling of biological objects and preconcentration of their volatiles for chromatographic analysis. Anal. Chem. 78, 7161–2168.

Soini, H.A., Schrock, S.E., Bruce, K.E., Wiesler, D., Ketterson, E.D. and Novotny, M.V. (2007) Seasonal variation in volatile compound profiles of preen gland secretions of the dark-eyed junco (*Junco hyemalis*). J. Chem. Ecol. 33, 183–198.

Soini, H.A., Wiesler, D. and Novotny, M.V. (2006) Unpublished data.

Soini, H.A., Wiesler, D., Bruce, K.E., Koyama, S., Ferón, C. Baudoin, B. and Novotny, M.V. (2006) Unpublished data.

Wylie, P.L. and Quimby, B.D. (1989) Applications of gas chromatography with an atomic emission detector. J. High Resol. Chromatogr. 12, 813–818.

Zhang, J.X., Soini, H.A., Bruce, K.E., Wiesler, D., Woodley, S.K.,.Baum, M.J. and Novotny, M.V. (2005) Putative chemosignals of the ferret (*Mustela furo*) associated with individual and gender recognition Chem. Senses 30, 727–737.

Zhang, Z. and Pawliszyn, J. (1993) Headspace solid-phase microextraction. Anal.Chem. 65, 1843–1852.

Chapter 2
Use of Automated Solid Phase Dynamic Extraction (SPDE)/GC-MS and Novel Macros in the Search for African Elephant Pheromones[‡]

Thomas E. Goodwin, Patrick A. Brown, Mindy S. Eggert, Maria G. Evola, Sam J. House, R. Grant Morshedi, Margaret E. Weddell, C. Joi Chen, Stephen R. Jackson, Yves Aubut, Jeff Eggert, Bruce A. Schulte, and L. E. L. Rasmussen[†]

Abstract A relatively small number of mammalian pheromones has been identified, in contrast to a plethora of known insect pheromones, but two remarkable Asian elephant/insect pheromonal linkages have been elucidated, namely, (Z)-7-dodecen-1-yl acetate and frontalin. In addition, behavioral bioassays have demonstrated the presence of a chemical signal in the urine of female African elephants around the time of ovulation. Our search for possible ovulatory pheromones in the headspace over female African elephant urine has revealed for the first time the presence of a number of known insect pheromones. This search has been facilitated by the use of a powerful new analytical technique, automated solid phase dynamic extraction (SPDE)/GC-MS, as well as by novel macros for enhanced and rapid comparison of multiple mass spectral data files from Agilent ChemStation®. This chapter will focus on our methodologies and results, as well as on a comparison of SPDE and the more established techniques of solid phase microextraction (SPME) and stir bar sorptive extraction (SBSE).

2.1 Introduction

Compared to the large number of chemical signals identified in insects, only a small number of proven mammalian pheromones are known, two of which have been identified in elephants (Albone 1984; Brown and Macdonald 1985; Wyatt 2003; Burger 2005). Female Asian elephants (*Elephas maximus*) release a

Thomas E. Goodwin
Hendrix College, Department of Chemistry,
goodwin@hendrix.edu

[†] Deceased

[‡] This chapter is dedicated to the memory of Dr. Bets Rasmussen

preovulatory pheromone in their urine as a mating signal to conspecific males (Rasmussen, Lee, Roelofs, Zhang and Daves 1996; Rasmussen, Lee, Zhang, Roelofs and Daves 1997; Rasmussen 2001). This compound, (Z)-7-dodecen-1-yl acetate (**1**), is also a sex pheromone in over 100 species of Lepidoptera (Kelly 1996). Additionally, a Coleopteran aggregation pheromone, frontalin (**2**), was shown to be a multifaceted chemical signal in the temporal gland secretions (TGS) of male Asian elephants in musth (a state characterized by high testosterone levels) (Rasmussen and Greenwood 2003). Additionally, the precise ratio of frontalin enantiomers in TGS varies with the age of the male elephant and affects the nature of the chemical signal (Greenwood, Comesky, Hunt and Rasmussen 2005).

Though male African elephants (*Loxodonta africana*) can distinguish conspecific female urine from different times in the estrous cycle (Bagley, Goodwin, Rasmussen and Schulte 2006), specific chemical signals have not yet been verified. Of interest, however, is the recent discovery of the insect pheromones frontalin (**2**), *exo*-brevicomin (**3**), *endo*-brevicomin (**4**), (E)-β-farnesene (**5**), and (E, E)-α-farnesene (**6**) (Fig. 2.1) in the headspace over female African elephant urine (Goodwin, Eggert, House, Weddell, Schulte and Rasmussen 2006).

The search for African elephant pheromones requires research at the biology/chemistry interface, and has benefited from the complementary skills of an interdisciplinary research team: an animal behaviorist (B. A. Schulte), a biochemist (L. E. L. Rasmussen), and an organic chemist (T. E. Goodwin), and their students (Schulte, Bagley, Groover, Loizi, Merte, Meyer, Napora, Stanley, Vyas, Wollett, Goodwin, and Rasmussen, this volume). In order to carry out this search, one obviously needs a source of female African elephant urine that could be collected at well-defined points in the estrous cycle. Fortunately, we established productive liaisons with various facilities (see Acknowledgments) that were carefully tracking estrous cycles via blood hormone analyses, and thus were able to furnish urine that was collected at the mid-luteal phase of the cycle, as well as at the anovulatory luteinizing hormone surge (LH1) and at the ovulatory LH surge (LH2). Procedures must then be developed for extracting the volatile organic compounds from the urine, for separating

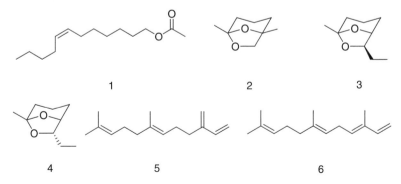

Fig. 2.1 Compounds identified in female African elephant urine headspace that are known insect pheromones

them one from another (gas chromatography; GC), for determining their identity (mass spectrometry; MS), and then for selecting and assessing putative pheromones by extensive behavioral bioassays. This chapter will focus on our use of a relatively new, highly efficient, and environmentally benign (solvent-free) technique (solid phase dynamic extraction) for extraction of organic compounds from the urinary matrix, as well as the implementation of novel macros for analysis and organization of mass spectral data generated by Agilent's ChemStation® software.

2.2 Automated Solid Phase Dynamic Extraction (SPDE)

2.2.1 Background: SPME, SBSE, and SPDE

Traditionally, extraction of dissolved organic chemicals from dilute aqueous media was carried out by using an organic solvent and a separatory funnel or an apparatus for continuous extraction (for a recent use of solvent extraction, see Zhang, Ni, Ren, Sun, Zhang and Wang 2003). For extraction of gaseous chemicals from headspace (the vapor phase in equilibrium with a liquid mixture), passage of the vapor over a porous polymer (usually Tenax®), was followed by thermal desorption, cryocon-centration (cryo-focusing), and finally the use of gas chromatography-mass spec-trometry (GC-MS) for analysis (see, for example, Zlatkis, Bertsch, Lichtenstein, Tishbee, Shunbo, Liebich, Coscia and Fleischer 1973).

In recent years, several solvent-free, environmentally benign ("green"), headspace sampling techniques have been developed. These offer several advan-tages over older methodologies, including the following: pre-GC concentration of a relatively large quantity of analytes onto a very small area, ease of use, ease of automation, and faster extraction times. We shall give only a brief overview here, however these innovations ("SPME","SBSE" and "SPDE") are compared and contrasted in more detail elsewhere (see, for example, Baltussen, Cramers and Sandra 2002; Bicchi, Cordero and Rubiolo 2004).

The first of the new "acronym techniques" to be developed was solid phase microextraction (SPME; Pawliszyn 1997). SPME (Supelco, Inc.) relies on a small glass fiber coated with an adsorbant polymer that can be exposed to the headspace over an aqueous sample, or immersed directly in the aqueous solution. Due in part to its ease of use, SPME has been employed for an immense variety of applications (for example, see Theodoridis, Koster and de Jong 2000), as is easily demonstrated by an online literature search. A variety of fiber coatings is available, and selection is based upon the properties of the analytes of interest. SPME is easily automated using the versatile, rugged and immensely useful Combi PAL GC autosampler (CTC Ana-lytics). Unlike the older Tenax® dynamic methodology mentioned earlier, SPME is a static technique (the vapor does not move across the adsorbant). In addition, the fiber is somewhat fragile, and has a relatively low polymer loading.

Stir bar sorptive extraction (SBSE) is a more recent development, and offers some advantages over SPME (Baltussen, Sandra, David and Cramers 1999;

Baltussen et al. 2002). SBSE, as its name implies, has the adsorbant polymer coated onto a small stir bar (marketed by Gerstel, Inc. as Twister®). SBSE originally was developed for immersion use, but more recently has been adapted for headspace extraction, wherein the stir bar becomes a static (non-dynamic) extractor suspended above the liquid matrix of interest. SBSE is more expensive to use than SPME, requiring specialized add-ons for the GC-MS, but with a much larger polymer coating it can extract a higher concentration of analytes. Some impressive results using SBSE to analyze biological media have been reported recently (Tienpont, David, Desmet and Sandra 2002; Soini, Bruce, Wiesler, David, Sandra and Novotny 2005; Zhang, Soini, Bruce, Wiesler, Woodley, Baum and Novotny 2005).

The newest of these extraction innovations is solid phase dynamic extraction (SPDE). SPDE features concentration of headspace analytes by repetitive dynamic flow back and forth over a polymer coating on the inside wall of a stainless steel syringe needle (Lipinski 2001). SPDE utilizes a 2.5 mL gas-tight syringe, with a 55 mm, 22 gauge stainless steel needle (Chromsys, Inc.; Chromtech, Inc.), and has more adsorbant polymer coating than SPME but less than SBSE. A 74 mm SPDE needle is also available, as is an extraction cooler for very volatile analytes. Currently, both SPME and SPDE have a larger variety of polymer coatings available than SBSE. Like SPME and SBSE, SPDE is easily automated by use of the Combi PAL robot. Unlike SPME and SBSE, SPDE is a dynamic technique for headspace analysis, and appears to offer some advantages for extraction of volatile organic compounds. The SPDE needle is much more robust than the SPME fiber, has more extraction capacity, and for most applications can be used for hundreds of extractions before replacement. SPDE/GC-MS is useful in a variety of applications (see, for example, Musshoff, Lachenmeier, Kroener and Madea 2002; Bicchi, Cordero, Liberto, Rubiolo and Sgorbini 2004).

2.2.2 Use of SPDE/GC-MS for Analysis of African Elephant Urine Headspace

In the Introduction, we presented a brief overview of our research interests and research strategies. A more detailed account has been published, which also includes a description of our early work using SPME/GC-MS, the specifics on our GC-MS instrumentation, and a typical set of GC-MS operating conditions (Goodwin, Rasmussen, Schulte, Brown, Davis, Dill, Dowdy, Hicks, Morshedi, Mwanza and Loizi 2005).

For a typical SPDE/GC-MS analysis of the headspace over female African elephant urine, a 10 mL aliquot of urine and a small stir bar were sealed in a 20 mL screw-top vial (Viton® septum). Multiple samples were programmed to run automatically using the Combi PAL robot and associated SPDE hardware and software (single magnet mixer/heater, needle bake-out station, etc.; Chromsys). Normally, we used a SPDE AC-PDMS syringe (activated charcoal (Carboxen®)-polydimethylsiloxane), incubated the stirred sample at 37 °C for 15 minutes, and

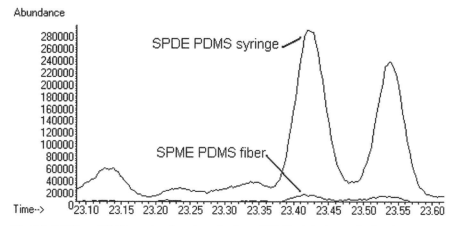

Fig. 2.2 SPDE vs. SPME: *exo*-brevicomin peaks from GC-MS of elephant urine headspace

extracted for 13 minutes (200 up-and-down 1 mL strokes of the syringe). Desorption was at 250 °C in the GC inlet. For our applications, we have not found cryo-focusing to be necessary. We have not carried out a systematic or extensive comparison of SPME and SPDE (and have not used SBSE at all), however a typical result in our laboratory for SPME versus SPDE extraction of the headspace over elephant urine headspace (identical 10 mL aliquots) is shown in Fig. 2.2. Clearly, SPDE is extracting a larger amount of *exo*-brevicomin.

Use of automated headspace SPDE/GC-MS not only enabled the identification in female African elephant urine of a number of known insect pheromones (compounds **2-6**, Fig. 2.1), but also revealed the presence of the beetle biochemical precursors to frontalin (**2**), *exo*-brevicomin (**3**) and *endo*-brevicomin (**4**), thus suggesting a common biosynthetic pathway (Goodwin et al. 2006). Extensive behavioral bioassays must be performed to determine whether any of these compounds is functioning as a pheromone among African elephants.

2.3 New "FindPeak" Macros for Organization and Comparison of Mass Spectral Data Output from Agilent ChemStation® Software

2.3.1 Background

In our search for a female African elephant urinary pheromone that functions as a chemical signal of ovulation to conspecific males, we wanted to compare the volatile organic chemicals present in the urine at the time of ovulation (the LH2 surge) with those present at the mid-luteal phase of the estrous cycle. The use of automated SPDE/GC-MS facilitated the generation of a large number of mass spectral data files, typically representing hundreds of components in the total ion chromatogram

(TIC) for each urine sample. For comparison of TICs, Agilent ChemStation® software allows one easily to stack (overlay) several chromatograms in the same computer window (Fig. 2.2). This feature helps to identify compounds with common retention times among several samples, but does not contain any mass spectral data for verification of a common identity for these peaks and is most useful for comparison of only a few TICs or a few compounds of interest. In these cases, mass spectral comparisons can be made quickly by visual inspection of the relevant spectra.

Of course one may employ automated library searches ("library percent reports") to check for compound identities, but algorithms for library matching are not infallible, and mass spectral libraries are not exhaustive, thus some compounds of interest will likely not be identified. Additional dilemmas are presented by mere reliance on retention times and library percent reports to ascertain the presence of common or unique peaks from among multiple mass spectral data files. As illustrated in Table 2.1, the TICs from the GC-MS of urine from four elephants evidence a peak at essentially the same retention time, but the library search results are inconclusive as to their common identity or lack thereof. As will be seen below, our novel macros can assist in making such decisions for a large number of peaks.

As a final test for comparing and contrasting mass spectra, an experienced and cautious investigator will always make a visual comparison. If there are a large number of data files, however, this process is tedious and time-consuming. Therefore, the new macros described below may be used to streamline the decisions about which putative peak matches require visual inspection of the actual mass spectra.

2.3.2 Development of New Macros

For reasons discussed above, we needed a complementary, ancillary tool for comparison of the mass spectra of components from multiple urine samples. We desired that the procedure have several characteristics: (1) requires little if any manual data entry by the operator; (2) utilizes data automatically generated by ChemStation® and organized into Microsoft Excel® spreadsheets; (3) displays both retention times and mass spectral data in the same window; (4) minimizes subjective operator judgments; and (5) is simple and rapid to use. What emerged after several iterative improvements are the "FindPeak" macros discussed below. These are largely due to the expertise of Y. Aubut, with valuable input from J. Eggert.

In brief, the macros carry out the following steps: (1) Agilent Chemstation®

Table 2.1 Example of problematic match decision from a library search report

File name	Retention Time	Library ID	Match Quality
Kiba305	23.8546	5-methyl-2-heptanone	78
Tava33	23.8545	5-methyl-2-heptanone	83
Tombil22	23.8579	2-methylpentane	9
Kiba304	23.877	7-octen-2-one	64

Macro—"FindPeak.mac" - exports GC-MS data to newly created .CSV files; (2) Microsoft Excel® Macro—"FindPeak.xls" - imports and sorts GC-MS data from the .CSV files into an Excel® spreadsheet; and (3) Processing Imported Data— "FindPeak.xls"—imported data are processed by matching retention times of TIC components. Compounds from multiple GC-MS data files are grouped by common retention times. For each compound in each group, the ten most abundant mass spectral ions are displayed in descending order of abundance, along with the base peak and apparent molecular ion. A more detailed description follows below. A different, elegant, and complex approach to comparison of the mass spectra of urinary volatiles has been reported (Willse, Belcher, Preti, Wahl, Thresher, Yang, Yamazaki and Beauchamp 2005).

2.3.3 Step 1: Agilent ChemStation® Macro (FindPeak.mac)

Macro "FindPeak.mac" automatically exports the mass spectral data, including ion masses and abundances for each peak in the gas chromatogram, to a .CSV (comma separated value) file. Each .CSV file contains MS data for one peak from the gas chromatogram.

2.3.4 Step 2: Microsoft Excel® Macro (FindPeak.xls)

The Excel® macro "findpeak.xls" imports the data from the .CSV files into an Excel® spreadsheet. This imports all of the .CSV files containing all the mass spectra of one Total Ion Chromatogram. The user may import as many data files as desired for comparisons.

2.3.5 Step 3: Final Data Processing (FindPeak.xls)

Imported data can be sorted so that GC peaks are grouped according to retention time. The user defines the allowable retention time difference, if any, for two peaks to be considered as a match. (We normally use $+/- 0.05$ min.) MS data are compared with respect to the number of similar ions, the base peak, and the apparent molecular ion. The ten most abundant ions in each mass spectrum are listed in descending order of abundance from left to right. The user defines how strictly the order of the ten ions must agree among spectra under comparison in order to be deemed a match. An example of the final output is shown in Table 2.2.

The macro output excerpt shown in Table 2.2 represents the same four TIC peaks from the same four elephant urine samples as in Table 2.1. Now, however, it is more clear that the first, second and fourth entries represent identical compounds, but the third entry may not be the same. The low abundance for the compound in the third entry, however, may lead to a false conclusion and thus warrant a visual inspection

Table 2.2 Typical final output from the FindPeak macros

DataFile	Area	R.T	Base Peak	Molecular Ion	Masses									Similar Ions	Same Base Peak	Same Molecular Ion
Kiba305.D	81247	23.855	43.1	128	58.1	71.1	41.1	70.1	55.1	57.1	59.1	39.1	95.1	10	Yes	No
Tava33.D	89501	23.855	43.1	128.1	58.1	71.1	70.1	41.1	55.1	57.1	59	39.1	95.1	6	Yes	No
Tombi122.D	3144	23.858	43.1	71.1	58	41.1	52.8	71.1	59	70	39.1	65.1	57.3	6	Yes	No
Kiba304.D	183329	23.877	43.1	128.2	58.1	55.1	41.1	71.1	39.1	70.1	67.1	68.1	108.1	0	No	No

of the mass spectrum. One can imagine that as the number of data files entered into the macros is increased and peak matching decisions escalate in complexity, the information gain increases accordingly to facilitate this task.

In summary, the benefits of the macros include the following. (1) Compound matches among mass spectral data files that are not confidently made using library search reports alone can be made using the macros. (2) Matches can be verified by visual inspection of the most abundant MS fragment ions. Chromatogram overlays are a quick way to compare the retention times of GC peaks, but provide no MS data. (3) The macros provide a quick alternative to manual inspection of all the MS data for each GC peak. (4) The macros are time-efficient, less tedious, and more accurate than the manual manipulation of spreadsheet data, which is prone to operator error. Two caveats should be mentioned for prospective macro users. (1) Since the macros rely upon retention times for grouping of peaks from multiple mass spectral data files, one must determine via standard runs that retention times for known compounds have not changed for files that are to be compared. (2) During analysis of MS data files that are to be compared, setting consistent and reasonable integration thresholds is necessary to avoid overlooking compounds that are present in low abundance. Otherwise, they will appear to be absent. A copy of the FindPeak macros and instructions for their use are available from the corresponding author (TEG).

2.4 Conclusion

Automated headspace SPDE/GC-MS is a powerful analytical technique that which will undoubtedly play an increasingly major role in the search for chemical signals in vertebrates. The novel macros described herein provide an important alternative for facile manipulation and comparison of multiple mass spectral data files. These new tools will complement existing procedures for the identification of organic compounds in biological media. We are currently applying them to an investigation of headspace volatiles in the glandular secretions of ringtailed lemurs (*Lemur catta*) and the urine of maned wolves (*Chrysocyon brachyurus*) in collaboration with C. Drea (Duke University; Drea and Scordato, this volume) and N. Songsasen/B. Baker (Smithsonian National Zoological Park/Little Rock Zoo), respectively.

The search for African elephant pheromones is continuing in our laboratories as we seek to identify the following: (1) compounds that are present exclusively or largely around the time of ovulation; (2) groups of compounds with relative concentration patterns that are characteristic of various times in the estrous cycle; (3) compounds that are known to be pheromones in other species; and (4) pheromonal activity that is dependent on enantiomeric ratios of putative chemical signals. This search will continue to rely on chemical analyses in the laboratory, coupled with extensive behavioral bioassays with both captive and wild elephants. We hope that by learning more about these magnificent and endangered mammals, we can contribute to ensuring their existence.

Acknowledgments We are grateful to Scott and Heidi Riddle (Riddle's Elephant and Wildlife Sanctuary) for valuable advice and assistance. Elephant urine samples and/or bioassay opportunities were provided by the following organizations: Addo Elephant National Park (AENP; South Africa), Baltimore Zoo, Bowmanville Zoo, Cameron Park Zoo, Indianapolis Zoo, Knoxville Zoo, Lion Country Safari, Louisville Zoo, Memphis Zoo, Miami Metro Zoo, Nashville Zoo (R. and C. Pankow), North Carolina Zoo, Riddle's Elephant and Wildlife Sanctuary, Sedgwick County Zoo, Seneca Park Zoo, Six Flags Marine World, Toledo Zoo, West Palm Beach Zoo, and Wildlife Safari Park. We thank Dr. J. Brown and her staff at the Elephant Endocrine Laboratory at the Conservation and Research Center, Smithsonian National Zoological Park, for providing estrous data that aided in timing urine collections and bioassays. Ingo Christ and Kiran Chokshi (Chromsys, Inc.) provided valuable SPDE assistance. Larry O'Kane (Agilent, Inc.) was very helpful with mass spectrometer problems. We thank J. P. Lafontaine of Pherotech for providing pheromone samples. T.E.G., B.A.S. and L.E.L.R. are grateful to the National Science Foundation (NSF-DBI-02-17068, -17062, and -16862, respectively) for financial support. We appreciate additional funding from Hendrix College, Georgia Southern University, and Biospherics Research Corporation.

References

Albone, E.S. (1984) *Mammalian Semiochemistry*. Wiley-Interscience, New York.

Bagley, K.R., Goodwin, T.E., Rasmussen, L.E.L. and Schulte, B.A. (2006) Male African elephants (*Loxodonta africana*) can distinguish oestrous status via urinary signals. Anim. Behav. 71, 1439–1445.

Baltussen, E., Sandra, P., David, F. and Cramers, C. (1999) Stir bar sorptive extraction (SBSE), a novel extraction technique for aqueous samples: theory and principles. J. Microcolumn Sep. 11, 737–747.

Baltussen, E., Cramers, C.A. and Sandra, P.J.F. (2002) Sorptive sample preparation-A review. Anal. Bioanal. Chem. 373, 3–22.

Bicchi, C., Cordero, C. and Rubiolo, P. (2004) A survey on high-concentration-capability headspace sampling techniques in the analysis of flavors and fragrances. J. Chromatogr. Sci. 42, 402–409.

Bicchi, C., Cordero, C., Liberto, E., Rubiolo, P. and Sgorbini, B. (2004) Automated headspace solid-phase dynamic extraction to analyse the volatile fraction of food matrices. J. Chromatogr. A 1024, 217–226.

Brown, R.E. and Macdonald, D.W. (Eds.) (1985) *Social Odours in Mammals, Vol. 1 & 2*. Clarendon Press, Oxford.

Burger, B.V. (2005) Mammalian Semiochemicals. In: S. Schulz (Ed.), *The Chemistry of Pheromones and Other Semiochemicals II* (Topics in Current Chemistry 240). Springer-Verlag, Heidelberg, pp. 231–278.

Drea, C.M. and Scordato, E.S. (2007) Olfactory communication in the ringtailed lemur (*Lemur catta*): form and function of multimodal signals. In: J. Hurst, R. Beynon, C. Roberts and T. Wyatt (Eds.), *Chemical Signals in Vertebrates 11*. Springer Press, New York, pp. 80–90.

Goodwin, T.E., Eggert, M.S., House, S.J., Weddell, M.E., Schulte, B.S. and Rasmussen, L.E.L. (2006) Insect pheromones and precursors in female African elephant urine. J. Chem. Ecol. 32, 1849–1853.

Goodwin, T. E., Rasmussen, L.E.L., Schulte, B. A., Brown, P. A., Davis, B. L., Dill, W. M., Dowdy, N. C., Hicks, A. R., Morshedi, R. G., Mwanza, D. and Loizi, H. (2005) Chemical analysis of preovulatory female African elephant urine: a search for putative pheromones. In: R.T. Mason, M.P. LeMaster, and D. Müller-Schwarze (Eds.), *Chemical Signals in Vertebrates 10*. Springer, New York, pp. 128–139.

Greenwood, D. R. Comesky, D., Hunt, M. B. and Rasmussen, L.E.L. (2005) Chirality in elephant pheromones. Nature 438, 1097–1098.

Kelly, D.R. (1996) When is a butterfly like an elephant? Chem. & Biol. 3, 595–602.

Lipinski, J. (2001) Automated solid phase dynamic extraction-Extraction of organics using a wall coated syringe needle. Fresenius' J. Anal. Chem. 369, 57–62.

Musshoff, F., Lachenmeier, D.W., Kroener, L. and Madea, B. (2002) Automated headspace solid-phase dynamic extraction for the determination of amphetamines and synthetic designer drugs in hair samples. J. Chromatogr. A 958, 231–238.

Pawliszyn, J. (1997) Solid Phase Microextraction: Theory and Practice. Wiley-VCH, New York.

Rasmussen, L.E.L. (2001) Source and cyclic release pattern of (Z)-7-dodecenyl acetate, the pre-ovulatory pheromone of the female Asian elephant, Chem. Senses 26, 611–623.

Rasmussen, L.E.L. and Greenwood, D.R. (2003) Frontalin: a chemical message of musth in Asian elephants (*Elephas maximus*). Chem. Senses 28, 433–446.

Rasmussen, L.E.L., Lee, T.D., Roelofs, W.L., Zhang, A. and Daves, Jr., G.D. (1996) Insect pheromones in elephants. Nature 379, 684.

Rasmussen, L.E.L., Lee, T.D., Zhang, A., Roelofs, W.L. and Daves, Jr., G.D. (1997) Purification, identification, concentration and bioactivity of (Z)-7-dodecen-1-yl acetate: sex pheromone of the female Asian elephant, *Elephas maximus*. Chem. Senses 22, 417–437.

Schulte, B.A., Bagley, K.R., Groover, M., Loizi, H., Merte, C., Meyer, J.M., Napora, E., Stanley, L., Vyas, D.K., Wollett, K., Goodwin, T.E. and Rasmussen, L.E.L. (2007) Comparisons of state and likelihood of performing chemosensory event behaviors in two populations of African elephants (*Loxodonta africana*). In: J. Hurst, R. Beynon, C. Roberts and T. Wyatt (Eds.), *Chemical Signals in Vertebrates 11*. Springer Press, New York, pp. 70–79.

Soini, H.A., Bruce, K.E., Wiesler, D., David, F., Sandra, P. and Novotny, M.V. (2005) Stir bar sorptive extraction: a new quantitative and comprehensive sampling technique for determination of chemical signal profiles from biological media. J. Chem. Ecol. 31, 377–392.

Theodoridis, G., Koster, E.H.M. and de Jong, G.J. (2000) Solid-phase microextraction for the analysis of biological samples. J. Chromatogr. B 745, 49–82.

Tienpont, B., David, F., Desmet, K. and Sandra, P. (2002) Stir bar sorptive extraction-Thermal desorption-capillary GC-MS applied to biological fluids. Anal. Bioanal. Chem. 373, 46–55.

Willse, A., Belcher, A.M., Preti, G., Wahl, J.H., Thresher, M., Yang, P., Yamazaki, K. and Beauchamp, G.K. (2005) Identification of major histocompatibility complex-regulated body odorants by statistical analysis of a comparative gas chromatography/mass spectrometry experiment. Anal. Chem. 77, 2348–2361.

Wyatt, T.D. (2003) *Pheromones and Animal Behaviour*. Cambridge University Press, Cambridge.

Zhang, J-X., Ni, J., Ren, X-J., Sun, L., Zhang, Z-B. and Wang, Z-W. (2003) Possible coding for recognition of sexes, individuals and species in anal gland volatiles of *Mustela eversmanni* and *M. sibirica*. Chem. Senses 28, 381–388.

Zhang, J.X., Soini, H.A., Bruce, K.E., Wiesler, D., Woodley, S.K., Baum, M.J. and Novotny, M.V. (2005) Putative chemosignals of the ferret (*Mustela furo*) associated with individual and gender recognition. Chem. Senses 30, 727–737.

Zlatkis, A., Bertsch, W., Lichtenstein, H.A., Tishbee, A., Shunbo, F., Liebich, H.M., Coscia, A.M. and Fleischer, N. (1973) Profile of volatile metabolites in urine by gas chromatography-mass spectrometry. Anal. Chem. 45, 763–767.

Chapter 3
Urinary Lipocalins in Rodenta:
is there a Generic Model?

Robert J. Beynon, Jane L. Hurst, Michael J. Turton, Duncan H. L. Robertson, Stuart D. Armstrong, Sarah A. Cheetham, Deborah Simpson, Alan MacNicoll, and Richard E. Humphries

Abstract It is increasingly clear that mediation of chemical signals is not the exclusive domain of low molecular volatile or water soluble metabolites. Pheromone binding proteins play an important role in mediating the activity of low molecular weight compounds, while proteins and peptides can also act as information molecules in their own right. Understanding of the role played by proteins in scents has been derived largely from the study of Major Urinary Proteins (MUPs) in the mouse (*Mus musculus domesticus*) and the rat (*Rattus norvegicus*). As part of an ongoing programme to explore the diversity and complexity of urinary proteins in rodents, we have applied a proteomics-based approach to the analysis of urinary proteins from a wider range of rodents. These data suggest that many species express proteins in their urine that are structurally similar to the MUPs, although there is considerable diversity in concentration, in sexual dimorphism and in polymorphic complexity. This is likely to reflect a high degree of species-specificity in communication and the information that these proteins provide in scent signals.

3.1 Introduction

Early views of pheromone chemistry were shaped in part by precedents derived from the insect world. Thus, semiochemicals were considered to be low molecular weight, volatile molecules that were transmitted through the atmosphere from sender to receiver. The (somewhat alliterative) "simple, single signal" model has served well, but in higher animals it is necessary to invoke additional complexity. First, it becomes more critical that the receiver of the signal is able to identify the individual that transmitted the signal together with its status. The ability to recognise individual conspecifics and/or kin and associate this with information about that individual's status and behaviour is likely to be critical to most social interactions

Robert J. Beynon
University of Liverpool, Proteomics & Functional Genomics Group
r.beynon@liv.ac.uk

within vertebrate species, including competitor assessment, mate assessment, and the development of relationships within social groups. Scents thus need to provide a range of information that is clearly discriminable, and require different qualities to transmit variable information about an animal's current status (e.g. social status, reproductive status, health status) and invariant information about the animal's identity (species, sex, relatedness, individual). Second, as scents are often deposited in the environment in the form of scent marks or odour plumes to provide information over a period of time, some information in scent needs to be sustained, whilst other components will reflect temporal changes as the scent ages.

It might be expected, *a priori*, that information about the variable qualities of an animal is encoded via metabolites that provide for plasticity of expression, while invariant information about identity is more directly encoded in the genome. The most obvious candidates for directly encoded components that signal owner identity are proteins derived directly from the scent owner's genes, or peptides generated indirectly by proteolytic degradation of proteins encoded in the genome (note that any such peptides would need to be distinguishable from degradation products derived from food sources or from infectious agents that would not provide invariant identity information). Indeed, the emergent literature is providing increasing evidence for the presence of proteins in scent marks, and the ability of the vomeronasal system to respond to proteins or short peptides.

The family of proteins most commonly associated with the processes of chemical communication are the lipocalins, a large and diverse family of small extracellular β-barrel proteins with a hydrophobic calyx suitable for the transportation of small hydrophobic molecules (Akerstrom, Flower and Salier 2000). Although there is a low pairwise conservation of the specific amino acid sequence of lipocalins (often < 20%), the structure of these proteins is a highly conserved eight-stranded antiparallel β-barrel with an internal hydrophobic calyx. The structure of most lipocalins is stabilised by a disulphide bond linking the main β-barrel to the carboxyl terminus of the protein. Lipocalins exhibit a wide specificity of natural ligand binding as the dimensions of the hydrophobic calyx are highly variable and the parts of the protein sequence responsible for ligand binding can tolerate a wide variety of amino acid side chains (Skerra 2000). Individual lipocalins are classified according to a number of highly conserved short sequences or typical structurally conserved regions (SCRs; Flower 1996).

3.2 A protein-based experimental approach

The lipocalin family is characterised by a high rate of evolution and substantial sequence divergence. As such, a genome based approach to lipocalin identification in other species is less satisfactory as sequences derived from genomic data from the rat or the mouse are unlikely to generate useful probes, for example, for PCR amplification of genomic or cDNA. Moreover, the known genomes are populated with many lipocalin genes, not all of which are involved in chemical communication, or are expressed in scent secretions. As such, our approach has been protein based, targeting the emerging methodologies of proteomics to the proteinaceous components of scent marks. The advantages of such an approach are that the proteins are observed directly in the scent secretion, it is possible to quantify and examine the complexity of the scent proteins and, by mass spectrometry, to assess the heterogeneity, sequence conservation and primary sequence data for each protein. Once primary sequence data are obtained, even for short runs of amino acids, the sequences can be used to search databases using alignment tools such as BLAST or to specify the sequence of PCR primers. This approach has been exemplified by our work on urinary lipocalins from the house mouse (Darwish Marie, Veggerby, Robertson, Gaskell, Hubbard, Martinsen, Hurst and Beynon 2001; Beynon, Veggerby, Payne, Robertson, Gaskell, Humphries and Hurst 2002; Beynon and Hurst 2004; Armstrong, Robertson, Cheetham, Hurst and Beynon 2005; Robertson, Hurst, Searle, Gunduz and Beynon 2007), which has provided a paradigm for the analysis of similar proteins from other species.

3.3 Urinary lipocalins in *Mus musculus domesticus*

House mice (*Mus musculus domesticus*), thought to have become commensal some 10,000 years ago, live in territorial social groups in which the ranges of many individuals overlap (Hurst 1987a; Barnard, Hurst and Aldhous 1991). In this species the urinary protein concentration can reach ~30 mg/ml in males, and over 99% of the protein content is attributable to members of the lipocalin family: the major urinary proteins or MUPs (Beynon and Hurst 2003; Beynon and Hurst 2004; Hurst and Beynon 2004). MUPs are the product of a multigene family of approximately 30 genes and pseudogenes located on chromosome 4 (Bennett, Lalley, Barth and Hastie 1982; Bishop, Clark, Clissold, Hainey and Francke 1982). Urinary MUPs are synthesised in the liver and secreted into serum where they are rapidly excreted by the kidneys. The synthesis of MUPs in the liver is sex-dependent, resulting in a urinary protein concentration three to four times higher in post-pubescent male mice than female mice (Beynon & Hurst 2004), and an even more pronounced sexual dimorphism in expression in the closely related sub-species *Mus musculus musculus* (Stopkova, Stopka, Janotova and Jedelsky 2007). The sexual dimorphism extends beyond the total amount of protein—there are several proteins that are expressed in a male-specific pattern (Armstrong et al. 2005). The expression of MUP mRNA has also been detected in a number of secretory tissues including the submaxillary, lachrymal, mammary, parotid, sublingual and nasal glands (Shahan, Denaro, Gilmartin, Shi and Derman 1987; Utsumi, Ohno, Kawasaki, Tamura, Kubo and Tohyama 1999). In urine, multiple MUPs are expressed simultaneously, leading to complex protein profiles. These profiles are highly polymorphic in wild-caught mice, such that the overall MUP pattern expressed by each unrelated individual is unique although the polymorphism is not evident in inbred strains that are geneticially homogenous (Robertson, Cox, Gaskell, Evershed and Beynon 1996; Robertson, Hurst, Bolgar, Gaskell and Beynon 1997; Beynon et al. 2002).

MUPs bind pheromones within the hydrophobic calyx of their structure (Bocskei, Groom, Flower, Wright, Phillips, Cavaggioni, Findlay and North 1992; Zidek, Stone, Lato, Pagel, Miao, Ellington and Novotny 1999; Timm, Baker, Mueller, Zidek and Novotny 2001), and delay their release from scents into the air (Robertson, Beynon and Evershed 1993). A number of pheromones in mouse urine show sex or status-specific expression. These have a number of reproductive priming and behavioural effects including acceleration of female puberty onset (Novotny, Jemiolo, Wiesler, Ma, Harvey, Xu, Xie and Carmack 1999) or puberty delay (Novotny, Jemiolo, Harvey and et 1986), extension of oestrus (Jemiolo, Harvey and Novotny 1986), inter-male aggression and male-female attraction (Jemiolo, Alberts, Sochinski-Wiggins, Harvey and Novotny 1985; Novotny, Harvey, Jemiolo and Alberts 1985).

Two pheromonally active ligands in mouse urine, 2,3-dehydro-exo-brevicomin (brevicomin) and 2-sec-butyl-4,5-dihydrothiazole (thiazole) (Bacchini, Gaetani and Cavaggioni 1992; Novotny, Ma, Wiesler and Zidek 1999) are associated with urinary MUPs following purification. In addition to the role of binding the pheromonally active ligands *in vivo*, which may be important for transporting pheromones to receptors in the vomeronasal organ, MUPs extend the duration of scent signals by delaying the release of thiazole and brevicomin from urine marks after deposition (Robertson et al. 1993; Hurst, Robertson, Tolladay and Beynon 1998).

Individual mice express a combinatorial pattern of MUPs (typically at least 7–12 isoforms) reflecting multiple allelic variants and multiple expressed loci (Robertson et al. 1997). Among wild mice, individuals each express a different pattern even when captured from the same population (Payne, Malone, Humphries, Bradbrook, Veggerby, Beynon and Hurst 2001; Beynon et al. 2002), with the exception of very closely related animals that have inherited the same haplotypes from their parents (a 25% chance among outbred sibs, similar to MHC type sharing). The extreme heterogeneity in the sequence of MUPs is mostly confined to strands B, C and D and the intervening turns of the ß-barrel structure (Beynon et al. 2002).

Recent work has indicated a number of potential chemical communication roles for MUPs, as opposed to their ligands, in deposited urine (Beynon and Hurst 2004). There is persuasive evidence that the MUP themselves are a source of olfactory signals; stimulating increased competitive scent marking (Humphries, Robertson, Beynon and Hurst 1999), puberty acceleration (Mucignat Caretta, Caretta and Cavaggioni 1995) and pregnancy block (Peele, Salazar, Mimmack, Keverne and Brennan 2003). More critically, it is clear that the pattern of MUPs expressed in the urine encodes an individual ownership signal that allows individuals to distinguish their own scent marks from those of other males (Hurst, Payne, Nevison, Marie, Humphries, Robertson, Cavaggioni and Beynon 2001), and allows females to recognise individual males (Cheetham, Thom, Jury, Ollier, Beynon and Hurst 2007). Although airborne volatiles emanating from scent marks induce mice to investigate the scent more closely, they only countermark when they can contact the scent (Nevison, Armstrong, Beynon, Humphries and Hurst 2003) and then only when the scent contains MUPs that are different from their own (Hurst, Beynon, Humphries, Malone, Nevison, Payne, Robertson and Veggerby 2001). This suggests that owner recognition involves detection of involatile MUPs through the vomeronasal system.

Evolution of a MUP expression profile as a signal of individuality and kinship is appealing, given the high sequence heterogeneity, stable expression patterns and non-volatile nature of proteins. The high concentration of MUPs in urine and the resistance of the β-barrel structure to denaturation or degradation are consistent with a dual role of delivery and slow release of volatile signals, and stable encoding of identity of the owner. It is increasingly important to explore the nature, complexity and use of urinary lipocalins in other species, to assess the extent to which the subtleties of the process in the house mouse may be generalised. In the remainder of this chapter, we report an overview of our recent work on other rodent species

3.4 Urinary lipocalins in *Mus macedonicus*

Three other *Mus* species (*M. macedonicus, M. spretus and M. spicilegus*) are closely related to and occur sympatrically with *M. m. domesticus* in Europe and the Middle East. These species live independently of humans, utilizing more scattered food resources and thus live at much lower densities. We therefore sought to characterise MUPs from *M. macedonicus* for comparison with the well characterised MUPs from *M. m. domesticus*. Urine from *M. macedonics* individuals demonstrated a MUP-sized band on gel electrophoresis. However, when the samples were analysed by mass spectrometry, the urine from each male *M. macedonicus* contained a single major protein species of mass 18742Da and all individuals were the same, in marked contrast to *M. m. domesticus*. A combination of peptide mass fingerprinting and tandem mass spectrometry/*de novo* sequencing revealed that this protein was a kernel lipocalin, containing all three SCRs (Flower 1996), and differed by only seven amino acid changes to the most similar protein that has been characterized from *M. m. domesticus*. All of the amino acid changes were located at the surface of the molecule and molecular modeling of the predicted protein of the *M. macedonicus* sequence demonstrated that the amino acid substitutions had little effect on the tertiary structure—this protein was indubitably a MUP (Robertson et al. 2007). At present, we lack data on *M. macedonicus* females.

3.5 Urinary lipocalins in *Mus spretus*

In common with *M. macedonicus*, urine from male *M. spretus* also demonstrated a MUP-sized band following gel electrophoresis. The proteins within this band were analysed by high resolution anion exchange chromatography and electrospray ionisation mass spectrometry (ESI-MS). The former technique produced an elution profile consisting of just three peaks, in contrast to both the more complex patterns observed previously from *M. m. domesticus* and the single major peak in *M. macedonicus*. Furthermore, similar analyses from five individual males resulted in near identical profiles, in terms of relative peak area and chromatographic retention time. The molecular mass of the proteins within the anion exchange peaks was subsequently determined by ESI-MS. Each peak was found to contain a single protein, the masses of which were 18666Da, 18687Da and 18758Da, with the 18758Da protein being the most abundant. A peptide mass fingerprinting experiment performed on the 18758Da protein confirmed that it shared considerable sequence identity to MUPs from *M. m. domesticus* but also contained some differences in the amino acid chain. *De-novo* sequencing of two Lys-C peptides from the 18758Da *M. spretus* MUP characterised two such changes: these were $A_{103}T$ and $E_{49}D$ (*M. m. domesticus/M.spretus*). Both substitutions involve amino acid residues on the surface of the molecule and in the light of the *M. macedonicus* investigation, are thought to have little effect on the structure. The MUP status in *M. spretus* females is not yet known but under investigation.

3.6 Urinary lipocalins in the Norway rat, *Rattus norvegicus*

Rattus norvegicus developed within the rodent family Muridae about 5–6 million years ago (Verneau, Catzeflis and Furano 1998) and now is a commensal presence virtually worldwide. As in the *Mus* species, *Rattus norvegicus* excrete a great deal of protein in their urine (20 mg/day for mature males), most of which is lipocalin formerly known as α_{2U} globulin (Chatterjee, Hopkins, Dutchak and Roy 1979) but which is now more properly referred to as rat major urinary protein (rat MUP). Rat MUPs are tissue and sex specific proteins under complex multihormonal and developmental control (Kulkarni, Gubits and Feigelson 1985). They migrate to a similar position as mouse MUPs on SDS PAGE gels and are structurally very similar (Bocskei et al. 1992). Rat MUPs bind small hydrophobic ligands (Lehman-McKeeman, Caudill, Rodriguez and Eddy 1998), although no endogenous ligand has been identified as yet, and male urine has been implicated in puberty acceleration in female rats (Vandenbergh 1976) and the timing of lactational estrous in dams (Schank and McClintock 1997). The rat MUPs belong to a multigene family with more than 20 closely related isoforms (McFadyen, Addison and Locke 1999; McFadyen and Locke 2000). As with the house mouse, rat MUPs are expressed in salivary, lachrymal and mammary glands, but the highest concentration and complexity is found in preputial glands which do not secrete MUPs in mice. Further, only male rats express MUPs in liver, corresponding to the male-specific expression of urinary MUPs in this species (MacInnes, Nozik and Kurtz 1986).

Most previous work on rat MUPs has been conducted with inbred or relatively inbred laboratory rat strains that are likely to exhibit considerably reduced phenotypic variation relative to the wild population, as we see in mice. As an initial exploration of MUP expression, we analysed urine from nine wild-caught male rats captured from several different populations in northern UK by isoelectric focusing electrophoresis (IEF). The protein banding pattern was very similar between individuals, consisting of two major and several minor bands. Peptide mass fingerprinting (PMF) of the two main bands revealed them to be strong matches to rat MUPs. Electrospray ionisation mass spectrometry (ESI-MS) demonstrated that the urine of each individual contained two principal proteins of 18714Da and 18730Da. The ESI-MS and PMF data allowed unambiguous identification of the two main proteins as the rat MUPs AAA40642 (18714Da) and P02761 (18730Da), both synthesised in the liver. One of the minor bands was identified as the rat MUP Q63213 (18340 Da) which is also expressed in preputial and salivary glands (Bayard, Holmquist and Vesterberg 1996; Saito, Nishikawa, Imagawa, Nishihara and Matsuo 2000). The other minor bands are novel, previously unknown rat MUPs and are currently being characterised. The overall pattern of rat urinary MUPs by IEF and ESI-MS is remarkably consistent between individuals, contrasting that of the wild caught *M. m. domesticus* urinary MUP profiles. Additional wild individuals are being investigated to see if the rat urinary MUP pattern remains invariant.

3.7 Urinary lipocalins in *Phodopus roborovskii*

The Roborovski hamster is closely related to the other dwarf hamster species-the dwarf winter white hamster (*Phodopus sungorus*) and the Djungarian hamster (*Phodopus campbelli*). All three dwarf hamster species live in extreme environments: *P. roborovskii* inhabits desert and semi-desert regions with little vegetation in Russia, China, Manchuria and Mongolia, whereas both *P. sungorus* and *P. campbelli* are native to the forest-steppe zone of central Asia. Dwarf hamsters are nocturnal and live in a system of subterranean tunnels and nests formed by burrowing. The extreme physical conditions in their natural habitats has caused dwarf hamsters to adapt physiologically to conserve heat and water, while the harsh conditions also limit the opportunities for breeding, resulting in a highly compressed reproductive cycle that enables rapid maturation of their offspring. Dwarf hamsters have adapted to the limited water availability in their natural habitat by developing a highly effective renal mechanism to concentrate urine and limit the volume of water lost. The desert environment of *P. roborovskii* is the most extreme habitat of the dwarf hamsters, consequently they are able to highly concentrate their urine to a volume significantly less than that of *P. sungorus* and *P. campbelli* (Natochin Iu, Meshcherskii, Goncharevskaia, Makarenko, Shakhmatova, Ugriumov, Feoktistova and Alonso 1994). Male dwarf hamsters respond to urine and other scents emitted by females during different reproductive states, suggesting a combined set of odours could provide precise information about female reproductive state (Lai and Johnston 1994). Males can discriminate between male and female odour, and investigate scent marks from males and females in a sex dependent manner (Reasner and Johnston 1987). The frequency of urine marking is greater in males, particularly when within a female's home area, while females mark at a constant rate irrespective of location in the habitat.

Urinary protein output was assessed by measuring total protein and creatinine concentration for six male and six female captive-bred *P. roborovskii* hamsters.

The protein:creatinine ratio was very similar for males (12.0 ± 0.8) and females (13.5 ± 0.8). The similarity between the sexes was maintained when urinary proteins were resolved by 1D SDS PAGE. For all individuals, two low molecular weight proteins were apparent, one migrating at approximately 21 kDa and a second, smaller protein migrating at approximately 6 kDa. The intensity and the relative abundance of the 21 kDa and 6kDa bands were remarkably consistent across individuals. Proteins were subjected to in-gel digestion with trypsin, followed by MALDI-ToF mass spectrometry of the resultant peptides. The mass spectrum of the tryptic peptides from the male and female 21 kDa protein were virtually identical, demonstrating that the 21 kDa protein in male and female urine is likely to consist of the same protein(s). Similarly, the 6kDa protein yields the same mass spectrum in both sexes. However, the lack of similar peptides derived from the 6 kDa protein and the 21 kDa protein mean that the smaller protein is not a degradation product of the larger. Peptide mass fingerprint analysis of the 21 kDa protein did not identify any statistically significant matching protein sequences from non-redundant protein sequence databases. However, comprehensive mass spectrometric analysis and *de novo* peptide sequencing have allowed us to define virtually all of the protein sequence of the 21 kDa protein. From this, it is clear that the protein is a lipocalin (of similar length, possessing all of the SCRs), and that is has sequence and structural features that mean that it is most similar to the vaginal protein aphrodisin from the Syrian or Golden hamster *Mesocricetus auratus*, a degree of sequence similarity that permits the construction of a molecular model using aphrodisin as a template (M. J. Turton, J. L. Hurst and R. J. Beynon *unpublished data*). The 21 kDa protein was present in cage washes, in urine samples obtained by bladder massage and by direct recovery from bladder urine—it is most unlikely that this is due to vaginal fluid contamination, especially since the same protein is present in equal amounts in males!

3.8 Urinary lipocalins in the bank vole, *Clethrionomys glareolus*

The bank vole, *Clethrionomys glareolus* is the smallest of the vole species in Britain. The habitat of *C. glareolus* is woodland and thick undergrowth, where they travel along a system of worn routes either forced through the undergrowth or in shallow tunnels to avoid attack from predators (e.g. owls, stoats and weasels). *C. glareolus* is a polygamous rodent species, the mating season is early spring—summer and over winter they form a mixed sex group of 2–4 females with some of the last litter young and 1–2 males. During the mating season, this group breaks up and mature females inhabit non-overlapping solitary home ranges close to the over wintering site while males form hierarchical groups with larger home ranges that overlap (Bujalska 1973). The size of female home ranges is determined by their litter size and availability of food (Koskela, Mappes and Ylonen 1997; Kapusta and Marchlewska-Koj 1998). The increased aggression and territoriality of mature females during pregnancy and lactation decreases the size of home ranges and increases the distance between neighbouring females, preventing home range boundaries overlapping. Male *C. glareolus* form stable dominance hierarchies in the mating season through brief fighting episodes and each inhabits a separate burrow. Some subordinates relocate to vacant areas and immature males live on the breeding territories of females. Bank voles are nocturnal animals using scent from urine, faeces and several skin glands for intraspecific communication.

Wild male *C. glareolus* scent mark their territories by depositing small urine droplets or fine traces using the long brushlike hairs on the prepuce (Johnson 1975). These scent marks appear similar to those of house mice, and indicate a specific and controlled function for marking with urine. Paired male bank voles repeatedly urine mark and over-mark in a new environment and, after the establishment of a hierarchical order, urine marking by the submissive vole is diminished while the dominant vole urine marks the subordinate's burrow and nest area, consistent with a role for the urine marks as territorial markers within a stable hierarchy (Rozenfeld 1987). Females show heightened activity and interest in marking urine and preputial secretions from dominant males. Protein in bank vole urine was identified at 13–14 kDa in sexually mature males. The expression of this protein is thought to be androgen-dependent as the protein was absent or weakly expressed in urine from females, immature males and castrated males (Kruczek and Marchlewska-Koj 1985).

We characterized the urinary protein of *C. glareolus*. There was clear evidence for a strong sexual dimorphism in adults (protein:creatinine ratio in males: 45.4 ± 3.2; females: 3.7 ± 0.9) such that males secreted approximately 10 times as much urinary protein as females. The majority of the protein in male-derived samples migrated at approximately 16 kDa, but in females a similarly sized protein was apparent when samples were concentrated. However, the peptide mass fingerprint for the two sexes yielded unique sets of peptides with very little overlap, from which we can infer that the urinary proteins of either sex are different gene products. Thus far, we have characterized the male 16 kDa protein. On intact mass analysis, two proteins, of average mass 16930Da and 16625Da were present in urine from both laboratory bred and wild caught *C. glareolus*. A detailed *de novo* sequence analysis of overlapping peptides obtained by digestion of the predominant 16930Da protein with different endopeptidases allowed assembly of over 95% of the protein sequence, and clear identification of this protein as a kernel lipocalin, in which all three SCRs were present. The primary sequence showed greatest similarity to aphrodisin, rather than MUP type sequences, and a model could be readily built using the three dimensional structure of aphrodisin as a template (M. J. Turton, J. L. Hurst and R. J. Beynon *unpublished data*).

3.9 Conclusions

Within the limitations of sample size and extent of characterization, several general statements can be made in relation to the species described here, all of which express substantial concentrations of urinary lipocalins. First, these urinary proteins appear to be widespread across species that are not very closely related and occupy different niches but use urine for scent communication. The urinary lipocalins seem to exhibit less plasticity of sequence and structure than the broader lipocalin family, which implies that they fulfil a specific role in chemical communication. Considering those proteins that have been characterized in detail (from *Mus* species, brown rat, bank vole, Roborovski hamster), the emergent picture is of a protein between 150 and 170 amino acids that can readily be modelled onto the structures of either mouse MUP or aphrodisin. This does not of course guarantee that the proteins fold in a typical lipocalin beta barrel, but the modelling data are of sufficient quality to suggest that this is a valid presumption. Second, sexual dimorphism in expression of urinary lipocalins varies considerably between species, from a lack of any observed dimorphism in *P. roborovskii,* through greater investment in MUPs among male house mice with only some MUP isoforms being male-specific, to entirely male-specific expression of urinary MUPs in brown rats. This suggests that the role of urinary lipocalins in sexual communication is strongly species specific and MUP genes may be subject to strong sexual selection and rapid evolution.

A third compelling feature to emerge from these studies is the surprising lack of polymorphic heterogeneity in species other than *M. m. domesticus*. The pattern more commonly seen is of the expression of one or a few lipocalin variants with a similar pattern of expression between individuals of the same sex. Even though relatively small numbers of individuals have been examined, a comparable set of samples obtained from *M. m. domesticus* reveals multiple urinary MUPs expressed per individual with substantial inter-individual variation in the MUP profile. The simpler lipocalin pattern in other species examined so far means that there is inadequate polymorphism in these proteins to provide an individual ownership signal in urine. This may reflect differences in the population ecology of aboriginal species such as *M. macedonicus* where individual recognition may be less important than in commensal house mice, and might imply rapid expansion of the genome and of the role of MUPs in commensal house mice to meet a species-specific requirement for individual and kin recognition. In these mice in particular, multiple males and females live within close territorial social groups with extensive spatial overlap between neighbours such that borders need to be vigorously defended (Hurst 1987b; Barnard et al. 1991). The need to maintain territorial dominance scent marks and advertise a stable signal of ownership may then be driven by such high density populations and have led to selection for polymorphic MUP expression. However, it is also clear that MUPs and MUP-like proteins are expressed in other glands involved in scent communication, with similarities and differences between species. As yet, there has been little exploration of individual heterogeneity in these proteins and their functions in sent communication. Further exploration of urinary lipocalins will do much to expand our understanding of the role of these proteins in behavioural ecology. However, that exploration should focus as much on the proteins themselves as their putative ligands.

Acknowledgments The research described in this chapter was supported by grants awarded to RJB & JLH by the Biotechnology and Biological Sciences Research Council.

References

Akerstrom, B., Flower, D. R. and Salier, J. P. (2000) Lipocalins: unity in diversity. Biochim. Biophys. Acta 1482, 1–8.

Armstrong, S. D., Robertson, D. H. L., Cheetham, S. A., Hurst, J. L. and Beynon, R. J. (2005) Structural and functional differences in isoforms of mouse major urinary proteins: a male-specific protein that preferentially binds a male pheromone. Biochem. J. 391, 343–350.

Bacchini, A., Gaetani, E. and Cavaggioni, A. (1992) Pheromone binding proteins of the mouse, *Mus musculus*. Experientia 48, 419–21.

Barnard, C. J., Hurst, J. L. and Aldhous, P. (1991) Of mice and kin - the functional significance of kin bias in social behavior. Biol. Rev. 66, 379–430.

Bayard, C., Holmquist, L. and Vesterberg, O. (1996) Purification and identification of allergenic alpha (2u)-globulin species of rat urine. Biochim. Biophys. Acta 1290, 129–34.

Bennett, K. L., Lalley, P. A., Barth, R. K. and Hastie, N. D. (1982) Mapping the structural genes coding for the major urinary proteins in the mouse: combined use of recombinant inbred strains and somatic cell hybrids. Proc. Natl. Acad. Sci. U S A 79, 1220–4.

Beynon, R. J. and Hurst, J. L. (2003) Multiple roles of major urinary proteins in the house mouse, *Mus domesticus*. Biochem. Soc. Trans. 31, 142–6.

Beynon, R. J. and Hurst, J. L. (2004) Urinary proteins and the modulation of chemical scents in mice and rats. Peptides 25, 1553–1563.

Beynon, R. J., Veggerby, C., Payne, C. E., Robertson, D. H., Gaskell, S. J., Humphries, R. E. and Hurst, J. L. (2002) Polymorphism in major urinary proteins: molecular heterogeneity in a wild mouse population. J. Chem. Ecol. 28, 1429–46.

Bishop, J. O., Clark, A. J., Clissold, P. M., Hainey, S. and Francke, U. (1982) Two main groups of mouse major urinary protein genes, both largely located on chromosome 4. EMBO J. 1, 615–20.

Bocskei, Z., Groom, C. R., Flower, D. R., Wright, C. E., Phillips, S. E. V., Cavaggioni, A., Findlay, J. B. C. and North, A. C. T. (1992) Pheromone binding to two rodent urinary proteins revealed by X-ray crystallography. Nature 360, 186–188.

Bujalska, G. (1973) The role of spacing behavior among females in the regulation of reproduction in the bank vole. J. Reprod. Fertil. Suppl. 19, 465–74.

Chatterjee, B., Hopkins, J., Dutchak, D. and Roy, A. K. (1979) Superinduction of alpha 2u globulin by actinomycin D: evidence for drug-mediated increase in alpha 2u mRNA. Proc. Natl. Acad. Sci. U S A 76, 1833–7.

Cheetham, S. A., Thom, M. D., Jury, F., Ollier, W. E. R., Beynon, R. J. and Hurst, J. L. (2007) MUPs, not MHC provide a specific signal for individual recognition in wild house mice. (unpublished manuscript).

Darwish Marie, A., Veggerby, C., Robertson, D. H., Gaskell, S. J., Hubbard, S. J., Martinsen, L., Hurst, J. L. and Beynon, R. J. (2001) Effect of polymorphisms on ligand binding by mouse major urinary proteins. Protein Sci. 10, 411–7.

Flower, D. R. (1996) The lipocalin protein family: structure and function. Biochem. J. 318, 1–14.

Humphries, R. E., Robertson, D. H. L., Beynon, R. J. and Hurst, J. L. (1999) Unravelling the chemical basis of competitive scent marking in house mice. Anim. Behav. 58, 1177–1190.

Hurst, J. L. (1987a) Behavioral variation in wild house mice *Mus domesticus* Rutty - a quantitative assessment of female social organization. Anim. Behav. 35, 1846–1857.

Hurst, J. L. (1987b) The functions of urine marking in a free-living population of house mice, *Mus domesticus* Rutty. Anim. Behav. 35, 1433–1442.

Hurst, J. L. and Beynon, R. J. (2004) Scent wars: the chemobiology of competitive signalling in mice. Bioessays 26, 1288–98.

Hurst, J. L., Beynon, R. J., Humphries, R. E., Malone, N., Nevison, C. M., Payne, C. E., Robertson, D. H. L. and Veggerby, C. (2001). Information in scent signals of competitive social status: the interface between behaviour and chemistry. In: A. Marchelewska-Koj, D. Muller-Schwarze and J. Lepri (Eds.), *Chemical Signals in Vertebrates 9*, Plenum Press, New York, pp. 43–50.

Hurst, J. L., Payne, C. E., Nevison, C. M., Marie, A. D., Humphries, R. E., Robertson, D. H., Cavaggioni, A. and Beynon, R. J. (2001) Individual recognition in mice mediated by major urinary proteins. Nature 414, 631–4.

Hurst, J. L., Robertson, D. H. L., Tolladay, U. and Beynon, R. J. (1998) Proteins in urine scent marks of male house mice extend the longevity of olfactory signals. Anim. Behav. 55, 1289–97.

Jemiolo, B., Harvey, S. and Novotny, M. (1986) Promotion of the Whitten effect in female mice by synthetic analogs of male urinary constituents. Proc. Natl. Acad. Sci. U S A 83, 4576–9.

Jemiolo, D., Alberts, J., Sochinski-Wiggins, S., Harvey, S. and Novotny, M. (1985) Behavioural and endocrine responses of female mice to synthetic analogs of volatile compounds in male urine. Anim. Behav. 33, 1114–1118.

Johnson, R. P. (1975) Scent marking with urine in two races of the bank vole (*Clethrionomys glareolus*). Behaviour 55, 81–93.

Kapusta, J. and Marchlewska-Koj, A. (1998) Interfemale aggression in adult bank voles (*Clethrionomys glareolus*). Aggressive Behavior 24, 53–61.

Koskela, E., Mappes, T. and Ylonen, H. (1997) Territorial behaviour and reproductive success of bank vole *Clethrionomys glareolus* females. J. Anim. Ecol. 66, 341–349.

Kruczek, M. and Marchlewska-Koj, A. (1985) Androgen-dependent proteins in the urine of bank voles (*Clethrionomys glareolus*). J. Reprod. Fertil. 75, 189–92.

Kulkarni, A. B., Gubits, R. M. and Feigelson, P. (1985) Developmental and hormonal regulation of alpha 2u-globulin gene transcription. Proc. Natl. Acad. Sci. U S A 82, 2579–82.

Lai, S. C. and Johnston, R. E. (1994) Individual Odors in Djungarian Hamsters (*Phodopus campbelli*). Ethology 96, 117–126.

Lehman-McKeeman, L. D., Caudill, D., Rodriguez, P. A. and Eddy, C. (1998) 2-sec-butyl-4,5-dihydrothiazole is a ligand for mouse urinary protein and rat alpha 2u-globulin: physiological and toxicological relevance. Toxicol. Appl. Pharmacol. 149, 32–40.

MacInnes, J. I., Nozik, E. S. and Kurtz, D. T. (1986) Tissue-specific expression of the rat alpha 2u globulin gene family. Mol. Cell. Biol. 6, 3563–7.

McFadyen, D. A., Addison, W. and Locke, J. (1999) Genomic organization of the rat alpha 2u-globulin gene cluster. Mamm. Genome 10, 463–70.

McFadyen, D. A. and Locke, J. (2000) High-resolution FISH mapping of the rat alpha2u-globulin multigene family. Mamm. Genome 11, 292–9.

Mucignat Caretta, C., Caretta, A. and Cavaggioni, A. (1995) Pheromonally accelerated puberty is enhanced by previous experience of the same stimulus. Physiol. Behav. 57, 901–3.

Natochin Iu, V., Meshcherskii, I. G., Goncharevskaia, O. A., Makarenko, I. G., Shakhmatova, E. I., Ugriumov, M. V., Feoktistova, N. and Alonso, G. (1994) [A comparative study of the osmoregulating system in the hamsters *Phodopus roborovskii* and *Phodopus sungorus*]. Zh. Evol. Biokhim. Fiziol. 30, 344–57.

Nevison, C. M., Armstrong, S., Beynon, R. J., Humphries, R. E. and Hurst, J. L. (2003) The ownership signature in mouse scent marks is involatile. Proc. R. Soc. Lond. B 270, 1957–1963.

Novotny, M., Harvey, S., Jemiolo, B. and Alberts, J. (1985) Synthetic pheromones that promote inter-male aggression in mice. Proc. Natl. Acad. Sci. U S A 82, 2059–61.

Novotny, M., Jemiolo, B., Harvey, S., Wiesler, D. and Marchlewska-Koj, A. (1986) Adrenal-mediated endogenous metabolites inhibit puberty in female mice. Science 231, 722–5.

Novotny, M. V., Jemiolo, B., Wiesler, D., Ma, W., Harvey, S., Xu, F., Xie, T. M. and Carmack, M. (1999) A unique urinary constituent, 6-hydroxy-6-methyl-3-heptanone, is a pheromone that accelerates puberty in female mice. Chem. Biol. 6, 377–83.

Novotny, M. V., Ma, W., Wiesler, D. and Zidek, L. (1999) Positive identification of the puberty-accelerating pheromone of the house mouse: the volatile ligands associating with the major urinary protein. Proc. R. Soc. Lond. B 266, 2017–22.

Payne, C. E., Malone, N., Humphries, R. E., Bradbrook, C., Veggerby, C., Beynon, R. J. and Hurst, J. L. (2001). Heterogeneity of major urinary proteins in house mice: population and sex differences. In: A. Marchelewska-Koj, D. Muller-Schwarze and J. Lepri (Eds.), *Chemical Signals in Vertebrates 9*, Plenum Press, New York, pp. 233–240.

Peele, P., Salazar, I., Mimmack, M., Keverne, E. B. and Brennan, P. A. (2003) Low molecular weight constituents of male mouse urine mediate the pregnancy block effect and convey information about the identity of the mating male. Eur. J. Neurosci. 18, 622–8.

Reasner, D. S. and Johnston, R. E. (1987) Scent marking by male dwarf hamsters (*Phodopus sungorus campbelli*) in response to conspecific odors. Behav. Neural Biol. 48, 43–48.

Robertson, D. H., Cox, K. A., Gaskell, S. J., Evershed, R. P. and Beynon, R. J. (1996) Molecular heterogeneity in the major urinary proteins of the house mouse *Mus musculus*. Biochem. J. 316, 265–72.

Robertson, D. H., Hurst, J. L., Bolgar, M. S., Gaskell, S. J. and Beynon, R. J. (1997) Molecular heterogeneity of urinary proteins in wild house mouse populations. Rapid Commun. Mass Spectrom. 11, 786–90.

Robertson, D. H., Hurst, J. L., Searle, J. B., Gunduz, I. and Beynon, R. J. (2007) Characterization and comparison of major urinary proteins from the house mouse, *Mus musculus domesticus*, and the aboriginal mouse, *Mus macedonicus*. J. Chem. Ecol. (*in press*).

Robertson, D. H. L., Beynon, R. J. and Evershed, R. P. (1993) Extraction, characterization and binding analysis of two pheromonally active ligands associated with major urinary protein of house mouse (*Mus musculus*). J. Chem. Ecol. 19, 1405–1416.

Rozenfeld, F. M. (1987) Urine marking by male bank voles (*Clethrionomys glareolus* Schreber, 1780 - Microtidae) in relation with their social status. Mammalia 51, 476–477.

Saito, K., Nishikawa, J., Imagawa, M., Nishihara, T. and Matsuo, M. (2000) Molecular evidence of complex tissue- and sex-specific mRNA expression of the rat alpha(2u)-globulin multigene family. Biochem. Biophys. Res. Commun. 272, 337–44.

Schank, J. C. and McClintock, M. K. (1997) Ovulatory pheromone shortens ovarian cycles of female rats living in olfactory isolation. Physiol. Behav. 62, 899–904.

Shahan, K., Denaro, M., Gilmartin, M., Shi, Y. and Derman, E. (1987) Expression of six mouse major urinary protein genes in the mammary, parotid,sublingual, submaxillary, and lachrymal glands and in the liver. Mol. Cell. Biol. 7, 1947–1954.

Skerra, A. (2000) Lipocalins as a scaffold. Biochim. Biophys. Acta 1482, 337–50.

Stopkova, R., Stopka, P., Janotova, K. and Jedelsky, P. L. (2007) Species-specific expression of major urinary proteins in the house mice (*Mus musculus musculus* and *Mus musculus domesticus*). J Chem Ecol 33, 861–9.

Timm, D. E., Baker, L. J., Mueller, H., Zidek, L. and Novotny, M. V. (2001) Structural basis of pheromone binding to mouse major urinary protein (MUP-I). Protein Sci. 10, 997–1004.

Utsumi, M., Ohno, K., Kawasaki, Y., Tamura, M., Kubo, T. and Tohyama, M. (1999) Expression of major urinary protein genes in the nasal glands associated with general olfaction. J Neurobiol. 39, 227–36.

Vandenbergh, J. G. (1976) Acceleration of sexual maturation in female rats by male stimulation. J. Reprod. Fertil. 46, 451–3.

Verneau, O., Catzeflis, F. and Furano, A. V. (1998) Determining and dating recent rodent speciation events by using L1 (LINE-1) retrotransposons. Proc. Natl. Acad. Sci. U S A 95, 11284–9.

Zidek, L., Stone, M. J., Lato, S. M., Pagel, M. D., Miao, Z., Ellington, A. D. and Novotny, M. V. (1999) NMR mapping of the recombinant mouse major urinary protein I binding site occupied by the pheromone 2-sec-butyl-4,5-dihydrothiazole. Biochemistry 38, 9850–61.

Chapter 4
The Biological Function of Cauxin, a Major Urinary Protein of the Domestic Cat (*Felis catus*)

Masao Miyazaki, Tetsuro Yamashita, Hideharu Taira and Akemi Suzuki

Abstract A major protein component of domestic cat urine is the carboxylesterase family member termed cauxin. Cauxin is secreted into the urine from the proximal straight tubular cells of the kidney, and the level of cauxin excretion is species-, sex-, and age-dependent. Cauxin is excreted in large amounts in the closely related members of the Felidae lineage, the cat (*Felis catus*), bobcat (*Lynx rufus*), and lynx (*Lynx lynx*). Male and female immature cats begin excreting cauxin about 2.5 months after birth, and excretion levels increase with age. In mature cats, cauxin excretion is significantly higher in intact males than in castrated males or female cats. The physiological function of cauxin is to provide species-, sex-, and age-dependent regulation of 2-amino-7-hydroxy-5,5-dimethyl-4-thiaheptanoic acid (felinine) production. Cauxin hydrolyzes the peptide bond of the felinine precursor, 3-methylbutanol-cysteinylglycine, to produce felinine and glycine. The sulfur-containing volatile compounds, 3-mercapto-3-methyl-1-butanol, 3-mercapto-3-methylbutyl formate, 3-methyl-3-methylthio-1-butanol, and 3-methyl-3-(2-methyl-disulfanyl)-1-butanol, are identified as species-specific odorants and candidates of felinine derivatives from the headspace gas of cat urine. These cat-specific volatile compounds may represent pheromones used as territorial markers for conspecific recognition or reproductive purposes by mature cats. The elucidation of cauxin-dependent felinine production provides new evidence for the existence of species-specific odorants and pheromones produced by species-specific biosynthetic mechanisms in mammalian species.

4.1 Introduction

Proteinuria, the excretion of large amounts of proteins into the urine, is often considered to be an indication of pathological event of the kidney because mammals only excrete small amounts of proteins into the urine under physiological conditions

Masao Miyazaki
RIKEN Frontier Research System, Sphingolipid Expression Laboratory
mmiyazaki@riken.jp

to prevent proteins loss from the body (D'Amico and Bazzi 2003). However, it is known that some mammalian species, including mice (*Mus musculus*) and rats (*Rattus norvegicus*), exhibit proteinuria under physiological conditions and this type of proteinuria is involved in chemical communication. Mouse urine contains a high concentration (10–15 mg/ml) of the 19-kDa lipocalin family proteins termed major urinary proteins (MUPs) that remain stable in urine marks over many weeks without degradation (Cavaggioni and Mucignat-Caretta 2000; Beynon and Hurst 2004). Many polymorphic variants of MUPs are expressed in the liver and secreted into the blood. The secreted MUPs pass through glomerular barriers of kidneys and are excreted into the urine. In male mice, MUPs bind male-specific volatile pheromones such as 2-*sec*-butyl-4,5-dihydrothiazole and 3,4-dehydro-*exo*-brevicomin (Bacchini, Gaetani and Cavaggioni 1992; Robertson, Beynon and Evershed 1993) and release them from deposited urine over several hours (Hurst, Robertson, Tolladay and Beynon 1998). In addition, polymorphic variants of MUPs are known to play a direct role in individual recognition during male mice (Hurst, Payne, Nevison, Marie, Humphries, Robertson, Cavaggioni and Beynon 2001; Nevison, Armstrong, Beynon, Humphries and Hurst 2003).

In this chapter, we review a new type of physiological proteinuria found in the domestic cat (*Felis catus*) and illustrate the biological relevance of this process. Chronic renal diseases represent the leading cause of illness and death among aging cat populations. Therefore, we carried out a precise analysis of cat urine to diagnose early-stage renal diseases. In this process, we determined that male cats exhibit proteinuria under physiological conditions. Since previously identified MUPs were not present in cat urine, we investigated the protein contents of cat urine and found that the major protein component of cat urine was the carboxylesterase family member termed cauxin involved in the production of felinine, a putative precursor of cat pheromones.

4.2 Cauxin Discovery and Biochemical Characterization

Urinary proteins were analyzed by SDS–polyacrylamide gel electrophoresis (PAGE), and a 70-kDa protein was identified as the major component of cat urine (Fig. 4.1A). Comparative analysis of urinary proteins in several other mammals such as humans, mice, dogs, and cattle did not detect a 70-kDa protein. Therefore, the 70-kDa protein was purified from cat urine and characterized by biochemical methods (Miyazaki, Kamiie, Soeta, Taira and Yamashita 2003). Analysis of tissue distribution indicated that the 70-kDa protein is expressed in the kidney in a tissue-specific manner and secreted from the proximal straight tubular cells of the kidney into the urine (Fig. 4.1B). A full-length cDNA for a 70-kDa protein was cloned from a cat kidney cDNA library. The cDNA clone encoded a polypeptide of 545 amino acid residues. The deduced amino acid sequence shared 47% identity with cat carboxylesterase (CES, EC 3.1.1.1), and contained both the CES family protein motif (EDCLY) and a conserved active site motif (GESAG) associated with

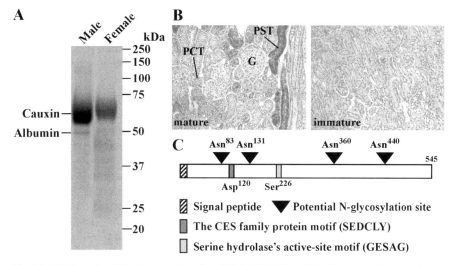

Fig. 4.1 (A) Cat urine (10 µl) was analyzed by SDS-PAGE followed by Coomassie blue staining. A single heavily stained band of cauxin was detected at 70 kDa in the urine of mature intact male and female cats. (B) Immunohistochemical images of cauxin in paraffin-embedded renal sections of 2-year-old mature and 1-month-old immature male cats. Cauxin-positive proximal straight tubular cells (PST) are observed between the inner cortex and outer medulla regions in the mature cat, but not in the immature cat. PCT, proximal converted tubules; G, the renal glomerulus. (C) A schematic image of the primary structure of cauxin

serine hydrolase family members (Fig. 4.1C). *In vitro* enzyme assays confirmed that the 70-kDa protein has esterase activity and can hydrolyze the artificial substrates *p*-nitrophenylacetate and 1-naphthylacetate. Therefore, we concluded that the 70-kDa protein is a novel member of the mammalian CES family and named the protein cauxin (carboxylesterase-like urinary excreted protein).

Cauxin is markedly different from previously reported mammalian CESs in term of urinary excretion. Other mammalian CESs comprise multigene families, and CES isozymes are highly and ubiquitously expressed in tissues such as the brain, liver, kidney, lung, and small intestine (Satoh and Hosokawa 1998). Our work on cauxin was the first description of a carboxylesterase excreted in urine.

4.3 Sex and Age Dependence of Cauxin Excretion

Although cauxin is excreted in the urine of both male and female cats, the average urinary concentration of cauxin was found to be significantly higher in intact males (0.87 ± 0.19 g/L) than in females (0.23 ± 0.12 g/L) (Miyazaki, Yamashita, Hosokawa, Taira and Suzuki 2006a). RT–PCR indicated that the expression of cauxin mRNA in the kidney was about fivefold higher in intact males than females (Miyazaki, Yamashita, Suzuki, Soeta, Taira and Suzuki 2006b). In addition, we found that cauxin expression decreased in the kidney proximal straight tubular cells immediately after castration of intact males, and that the average urinary concentration of cauxin in castrated males (0.13 ± 0.10 g/L) was significantly lower than in intact males. These results suggest that sex hormones such as testosterone are important regulatory factors of cauxin production in the proximal straight tubular cells.

Additionally, we found that cauxin excretion levels are dependent on the age of the cat (Miyazaki et al. 2006a). Cauxin was not detected in the urine of cats of less than 3 months of age by SDS–PAGE followed by Coomassie blue staining. Consistent with this finding, the expression of cauxin in the proximal straight tubular cells was not detected with an anti-cauxin antibody using immunohistochemistry (Fig. 4.1C). Temporal analysis of cauxin excretion by Western blotting with the anti-cauxin antibody indicated that immature male and female cat begin excreting cauxin about 2.5 months after birth, and the excretion level increases with age.

4.4 Species-Specific Excretion of Cauxin

The cauxin gene is conserved in several mammals, including humans, mice, rats, dogs, and cattle, although only the cat excretes cauxin. Cauxin mRNA expresses in the liver and kidney of humans and mice (M. Miyazaki, T. Yamashita, H. Taira and A. Suzuki, unpublished data). Ecroyd et al. reported that cauxin is present in ram (*Ovis aries*) reproductive fluids (Ecroyd, Belghazi, Dacheux, Miyazaki, Yamashita and Gatti 2006). Ram cauxin was secreted from the caudal epididymis and associated with epididymal soluble prion protein (Ecroyd, Belghazi, Dacheux and Gatti 2005). Based on these findings, we investigated whether additional Felidae species excrete cauxin into the urine. Urinary proteins were analyzed by SDS–PAGE followed by Coomassie blue staining and Western blotting with an antibody raised against cauxin purified from cat urine. Our analysis indicated that cauxin is a major component in bobcat (*Lynx rufus*) and Siberian lynx (*Lynx lynx*) urine, as well as in domestic cat urine (Miyazaki et al. 2006a). We could not detect cauxin in the urine of the puma (*Puma concolor*), leopard (*Panthera pardus*), tiger (*P. tigris*), jaguar (*P. onca*), snow leopard (*P. uncia*), or lion (*P. leo*). However, McLean and colleagues detected cauxin in the urine of the Asiatic lion (*P. leo persica*), Amur tiger (*P. tigris altaica*), Persian leopard (*P. pardus saxicolor*), clouded leopard (*Neofelis nebulosa*), and jaguar by MALDI–ToF mass spectrometry (L. McLean, J. Lewis, J. Hurst and R. Beynon, personal communication). Interestingly, in these Felidae animals, cauxin is excreted in part as a disulfide linked multimer. At present, Felidae animals are classified into eight lineages including *Panthera*, *bay cat*, *caracal*, *ocelot*, *Lynx*, *puma*, *leopard cat*, and *domestic cat* (Johnson, Eizirik, Pecon-Slattery, Murphy, Antunes, Teeling and O'Brien 2006). The bobcat and Siberian lynx are same members of the *Lynx* lineage and near to the cat in the phylogenetic tree of Felidae animals. These results suggest that over time, each Felidae animal developed unique machinery for the kidney-specific expression and the urinary excretion of cauxin, and the closely related members, the domestic cat, bobcat, and lynx, posses the most highly developed machinery. Cauxin would play species-specific physiological roles in each mammalian species.

4.5 Physiological Function of Cauxin

The CES family of proteins is characterized by the ability to hydrolyze a wide variety of aromatic and aliphatic substrates containing ester, thioester, and amide bonds (Heymann 1980, 1982). Cauxin is a member of the CES family, and is secreted from the proximal straight tubular cells into the urine in a species-, sex-, and age-dependent manner. Therefore, we postulated that cauxin was involved in an enzymatic reaction in cat urine and the products made by the reaction should vary with species, sex, and age. Based on this hypothesis, we searched for physiological substrates and products of cauxin in cat urine and identified 2-amino-7-hydroxy-5,5-dimethyl-4-thiaheptanoic acid, also known as felinine.

Felinine is excreted in a sex- and age-dependent manner in the urine of the cat, ocelot (*Felis pardalis*), and bobcat (Hendriks, Tarttelin and Moughan 1995c). Felinine is suggested to be a putative pheromone precursor in the cat (MacDonald, Rogers and Morris 1984; Hendriks, Moughan, Tarttelin and Woolhouse 1995b) because chemically synthesized felinine is odorless but develops a characteristic catty odor during storage (Hendriks, Woolhouse, Tarttelin and Moughan 1995a). Previously, it had been suggested that felinine is synthesized in the kidney because felinine is not present in blood or other cat tissues. However, Rutherfurd et al. found that 3-methylbutanol-glutathione (3-MBG) identified from cat blood contains the chemical structure of felinine (Rutherfurd, Rutherfurd, Moughan and Hendriks 2002). They suggested that the synthesis of the felinine precursor is formed via a glutathione S-conjugation reaction between glutathione and isopentenylpyrophosphate, an intermediate of cholesterol biosynthesis. Hendriks et al. found that cat urine contains 3-MBG, 3-methylbutanol-cysteinylglycine (3-MBCG), and *N*-acetyl felinine, in addition to felinine, and suggested that the breakdown of 3-MBG occurs in a similar manner to other glutathione S-conjugates (Hendriks, Harding and Rutherfurd-Markwick 2004). It is well known that glutathione S-conjugates are hydrolyzed to cysteinylglycine S-conjugates by γ-glutamyl transferase (EC 2.3.2.2) in the kidney, and the cysteinylglycine S-conjugates are hydrolyzed by renal dipeptidase (EC 3.4.13.11) (Lohr, Willsky and Acara 1998; Wang and Ballatori 1998). In this degradation pathway, the cysteine S-conjugates like felinine are reabsorbed into the renal tubular cells, *N*-acetylated by *N*-acetyltransferase (EC 2.3.1.5), and ultimately excreted as *N*-acetyl cysteine S-conjugates in the urine. However, the excretion level of *N*-acetyl felinine is much lower than that of felinine in cat urine (Hendriks et al. 2004; Rutherfurd-Markwick, McGrath, Weidgraaf and Hendriks 2006). Therefore, we hypothesized that the metabolic pathway from 3-MBG to felinine is cat-specific and cauxin is involved in this pathway. Thus, we tested whether cauxin hydrolyzed the felinine precursors, 3-MBG and 3-MBCG *in vitro*. Enzymatic assays determined that cauxin hydrolyzes the peptide bond of 3-MBCG to felinine and glycine (Fig. 4.2) (Miyazaki et al. 2006b). This finding suggested that cauxin regulates the production of felinine in a species-, sex-, and age-dependent manner *in vivo* and this suggestion was consistent with *in vivo* data correlating cauxin and felinine levels in the urine with the sex and age of the cat.

Figure 4.2 shows the proposed metabolic pathways for the conversion of 3-MBCG to felinine in the cat kidney. The 3-MBG is filtered through the glomerulus of kidneys and converted to 3-MBCG by γ-glutamyl transferase localized at the brush border of proximal tubular cells. A small portion of the 3-MBCG is hydrolyzed to felinine and glycine by renal dipeptidase localized in the proximal tubular cells. Felinine thus formed is absorbed by the proximal tubular cells, where it is converted to *N*-acetyl felinine by *N*-acetyltransferase. Most of the remaining

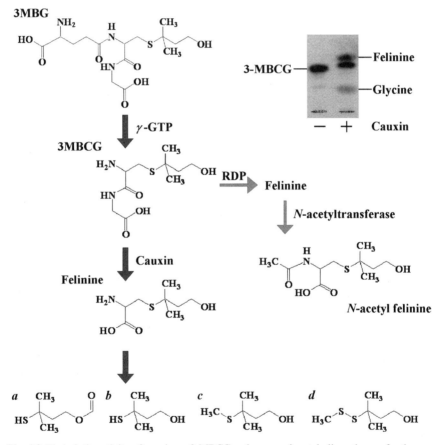

Fig. 4.2 Hydrolytic activity of cauxin on 3-MBCG and proposed metabolic pathways for the conversion of 3-MBG to felinine in the cat kidney. The 3-MBCG (20 mM) was incubated with or without cauxin (1.5 mg/ml) at 38°C for 6 h, and the reaction mixtures were analyzed by thin layer chromatography with ninhydrin staining. γ-GTP, γ-glutamyl transferase; RDP, renal dipeptidase; *a*, 3-mercapto-3-methylbutyl formate; *b*, 3-mercapto-3-methyl-1-butanol; *c*, 3-methyl-3-methylthio-1-butanol; and *d*, 3-methyl-3-(2-methyldisulfanyl)-1-butanol

3-MBCG is hydrolyzed to felinine and glycine by cauxin in the renal tubules and the bladder and/or *ex vivo*. The felinine produced in this manner is excreted into the urine.

4.6 Biological Relevance of Cauxin-Dependent Felinine Production

To elucidate the biological relevance of cauxin-dependent felinine production, it was necessary to determine the bioactivity of felinine and/or felinine derivatives. Felinine was postulated to be biologically important as a territorial marker for intraspecies communication (MacDonald et al. 1984; Hendriks et al. 1995b), or, although the direct evidence remains to be demonstrated, a putative precursor pheromone involved in attracting females (Tarttelin, Hendriks and Moughan 1998). Hendriks et al. (1995b) hypothesized that felinine degradated to 3-mercapto-3-methyl-1-butanol contributed to cat-specific urinary odor. Joulain and Laurent (1989) suggested that 3-mercapto-3-methyl-1-butanol, 3-methyl-3-methylthio-1-butanol, and 3-methyl-3-(2-methyldisul-fanyl)-1-butanol were present

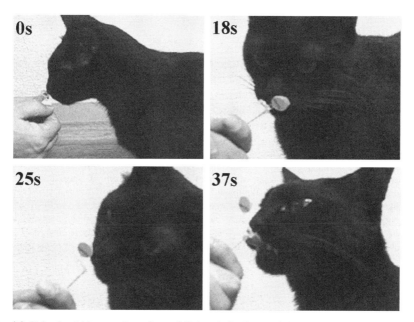

Fig. 4.3 Behavioral bioassay by using a felinine derivative. Felinine purified from cat urine by HPLC was dissolved in water at a concentration of 10 mg/ml, and 200 μl of the solution was stored in a 1.5-ml eppendorf tube at room temperature for 5 days. GC–MS analysis detected 3-mercapto-3-methyl-1-butanol in the headspace gas of the tube. The cat (6-year-old castrated male) was able to sniff the opening of the tube, but not contact the felinine solution. The cat sniffed 3-mercapto-3-methyl-1-butanol with considerable interest (18s and 25s) and then licked his lips five times (37s)

in cat urine. These compounds were identified in bobcat urine (Mattina, Pignatello and Swiharat 1991), but Rutherfurd et al. (2004) could not detect these compounds in cat urine. To identify the volatile compound responsible for cat-specific urinary odor, we analyzed the components of volatile compounds in the headspace gas of cat urine using a gas chromatograph mass spectrometry (GC–MS). Our analysis detected four sulfur-containing volatile compounds, 3-mercapto-3-methyl-1-butanol, 3-mercapto-3-methylbutyl formate, 3-methyl-3-methylthio-1-butanol, and 3-methyl-3-(2-methyl-disulfanyl)-1-butanol, as candidates of felinine derivatives (Miyazaki et al. 2006b). The levels of these compounds were found to be sex- and age-dependent. Based on these results, we are now designing studies to determine the bioactivity of felinine derivatives with cats. Since sulfur-containing volatile compounds would give species-, sex-, and age-specific odor to cat urine, it is possible that these compounds are putative pheromones used for conspecific recognition or reproductive purposes in the cat. Our preliminary behavioral bioassays indicated that cats sniffed 3-mercapto-3-methyl-1-butanol with considerable interest and sometimes licked their lips (Fig. 4.3), but we could not identify a functional behavioral response. Further studies using behavioral bioassays are needed to clarify the biological relevance of cauxin-dependent felinine excretion.

4.7 Conclusion

Although proteinuria is often considered to be a pathological event, we demonstrated that this is not the case for the domestic cat. Male cat urine contains a large amount of the mammalian carboxylesterase family member termed cauxin. Cauxin is excreted in a species-, sex-, and age-dependent manner and regulates the production of felinine, a putative pheromone precursor. This finding provides an example of a previously unknown type of proteinuria involved in chemical communication.

Acknowledgments Masao Miyazaki was supported by a fellowship of Special Postdoctoral Researchers Program from RIKEN. We thank Dr. Katsuyoshi Kamiie of Aomori University for molecular cloning of a cauxin cDNA, Dr. Satoshi Soeta of Nippon Veterinary and Life Science University for tissue-distribution analysis, Dr. Yusuke Suzuki of RIKEN for identification of cauxin substance, Mr Yoshihiro Saito of Shimadzu Corporation for GC–MS analysis, and Ms Tamako Miyazaki of Iwate University for behavioral bioassay. This work was supported in part by a FY2006 DRI Research Grant from RIKEN to M.M.

References

Bacchini, A., Gaetani, E. and Cavaggioni, A. (1992) Pheromone binding proteins of the mouse, *Mus musculus*. Experientia 48, 419–421.

Beynon, R.J. and Hurst, J.L. (2004) Urinary proteins and the modulation of chemical scents in mice and rats. Peptides 25, 1553–1563.

Cavaggioni, A. and Mucignat-Caretta, C. (2000) Major urinary proteins, α_{2U}-globulins and aphrodisin. Biochim. Biophys. Acta 1482, 218–228.

D'Amico, G. and Bazzi, C. (2003) Pathophysiology of proteinuria. Kidney Int. 63, 809–825.

Ecroyd, H., Belghazi, M., Dacheux, J.L. and Gatti, J.L. (2005) The epididymal soluble prion protein forms a high-molecular-mass complex in association with hydrophobic proteins. Biochem. J. 392, 211–219.

Ecroyd, H., Belghazi, M., Dacheux, J.L., Miyazaki, M., Yamashita, T. and Gatti, J.L. (2006) An epididymal form of cauxin, a carboxylesterase-like enzyme, is present and active in mammalian male reproductive fluids. Biol. Reprod. 74, 439–447.

Hendriks, W.H., Woolhouse, A.D., Tarttelin, M.F. and Moughan, P.J. (1995a) Synthesis of felinine, 2-amino-7-hydroxy-5,5-dimethyl-4-thiahep-tanoic acid. Bioorg. Chem. 23, 89–100.

Hendriks, W.H., Harding, D.R. and Rutherfurd-Markwick, K.J. (2004) Isolation and characterisation of renal metabolites of γ-glutamylfelinylglycine in the urine of the domestic cat (*Felis catus*). Comp. Biochem. Physiol. B Biochem. Mol. Biol. 139, 245–251.

Hendriks, W.H., Moughan, P.J., Tarttelin, M.F. and Woolhouse, A.D. (1995b) Felinine: a urinary amino acid of Felidae. Comp. Biochem. Physiol. B Biochem. Mol. Biol. 112, 581–588.

Hendriks, W.H., Tarttelin, M.F. and Moughan, P.J. (1995c) Twenty-four hour felinine excretion patterns in entire and castrated cats. Physiol. Behav. 58, 467–469.

Heymann, E. (1980) Carboxylesterases and amidases. In: W.B. Jakoby (Ed.), *Enzymatic Basis Detoxification, Vol. II*, Academic Press, New York, pp. 291–323.

Heymann, E. (1982) Hydrolysis of carboxylic esters and amides. In: W.B. Jakoby, J.R. Bend and J. Caldwell (Eds.), *Metabolic Basis Detoxification*, Academic Press, New York, pp. 229–245.

Hurst, J.L., Payne, C.E., Nevison, C.M., Marie, A.D., Humphries, R.E., Robertson, D.H., Cavaggioni, A. and Beynon, R.J. (2001) Individual recognition in mice mediated by major urinary proteins. Nature 414, 631–634.

Hurst, J.L., Robertson, D.H.L., Tolladay, U. and Beynon, R.J. (1998) Proteins in urine scent marks of male house mice extend the longevity of olfactory signals. Anim. Behav. 55, 1289–1297.

Johnson, W.E., Eizirik, E., Pecon-Slattery, J., Murphy, W.J., Antunes, A., Teeling, E. and O'Brien, S.J. (2006) The late Miocene radiation of modern Felidae: a genetic assessment. Science 311, 73–77.

Joulain, D. and Laurent, R. (1989). The catty odour in blackcurrant extracts versus the blackcurrant odour in the cat's urine? In: (S.C. Bhattacharyya, N. Sen and K.L. Sethi (Eds.), *11th International Congress of Essential Oils, Fragrances, and Flavours*, Oxford IBH Publishing. New Delhi, India, pp. 89.

Lohr, J.W., Willsky, G.R. and Acara, M.A. (1998) Renal drug metabolism. Pharmacol. Rev. 50, 107–141.

MacDonald, M.L., Rogers, Q.R. and Morris, J.G. (1984) Nutrition of the domestic cat, a mammalian carnivore. Annu. Rev. Nutr. 4, 521–562.

Mattina, M.J.I., Pignatello, J.J. and Swiharat, R.K. (1991) Identification of volatile components of bobcat (*Lynx rufus*) urine. J. Chem. Ecol. 17, 451–462.

Miyazaki, M., Kamiie, K., Soeta, S., Taira, H. and Yamashita, T. (2003) Molecular cloning and characterization of a novel carboxylesterase-like protein that is physiologically present at high concentrations in the urine of domestic cats (*Felis catus*). Biochem. J. 370, 101–110.

Miyazaki, M., Yamashita, T., Hosokawa, M., Taira, H. and Suzuki, A. (2006a) Species-, sex-, and age-dependent urinary excretion of cauxin, a mammalian carboxylesterase family. Comp. Biochem. Physiol. B Biochem. Mol. Biol. 145, 270–277.

Miyazaki, M., Yamashita, T., Suzuki, Y., Soeta, S., Taira, H. and Suzuki, A. (2006b) A major urinary protein of the domestic cat regulates the production of felinine, a putative pheromone precursor. Chem. Biol. 13, 1071–1079.

Nevison, C.M., Armstrong, S., Beynon, R.J., Humphries, R.E. and Hurst, J.L. (2003) The ownership signature in mouse scent marks is involatile. Proc. Biol. Sci. 270, 1957–1963.

Robertson, D.H.L., Beynon, R.J. and Evershed, R.P. (1993) Extraction, characterization and binding analysis of two pheromonally active ligands associated with major urinary protein of house mouse (*Mus musculus*). J. Chem. Ecol. 19, 1405–1416.

Rutherfurd, K.J., Rutherfurd, S.M., Moughan, P.J. and Hendriks, W.H. (2002) Isolation and characterization of a felinine-containing peptide from the blood of the domestic cat (*Felis catus*).

J. Biol. Chem. 277, 114–119.

Rutherfurd, S.M., Zhang, F., Harding, D.R., Woolhouse, A.D. and Hendriks, W.H. (2004) Use of capillary (zone) electrophoresis for determining felinine and it's application to investigate the stability of felinine. Amino Acids 27, 49–55.

Rutherfurd-Markwick, K.J., McGrath, M.C., Weidgraaf, K., and Hendriks, W.H. (2006) γ-glutamylfelinylglycine metabolite excretion in the urine of the domestic cat (*Felis catus*). J. Nutr. 136, 2075S–2077S.

Satoh, T. and Hosokawa, M. (1998) The mammalian carboxylesterases: from molecules to functions. Annu. Rev. Pharmacol. Toxicol. 38, 257–288.

Tarttelin, M.F., Hendriks, W.H. and Moughan, P.J. (1998) Relationship between plasma testosterone and urinary felinine in the growing kitten. Physiol. Behav. 65, 83–87.

Wang, W. and Ballatori, N. (1998) Endogenous glutathione conjugates: occurrence and biological functions. Pharmacol. Rev. 50, 335–356.

Chapter 5
Putative Pheromones of Lion Mane and Its Ultrastructure

M. Poddar-Sarkar, A. Chakroborty, R. Bhar and R. L. Brahmachary

Abstract The significance of mane in the lion as a putative source of pheromone is discussed. Since head-rubbing on the tree trunks is as frequent as spraying marking fluid (MF) both MF and mane are likely to leave two osmic signals simultaneously at two neighbouring sites. The chemical analysis of mane reveals the presence of C9–C24 straight-chain branched and unsaturated fatty acids which may play the role of pheromones. The presence of erucic acid in hair is a new observation. Scanning electron micrographs of cross sections of mane hair show "pockets" which could store pheromones.

5.1 Introduction

Schaller (Schaller 1972) pointed out that lions rub their head against tree trunks or other prominent objects before spraying marking fluid (MF), generally upwards and backwards but also horizontally backwards and sometimes downwards. This variation in the direction of ejection has also been noted in the Asiatic lion (Brahmachary, Singh and Rajput 1999). This spray, also known as scent-marking as opposed to ordinary urination, has been considered as a source of semiochemicals for a long time and the chemistry and ethology of MF have been studied in the big cats, mostly in the tiger (Brahmachary, Poddar-Sarkar, Mousumi and Dutta 1990; Poddar-Sarkar, Brahmachary and Dutta 1994). In contrast, head rubbing of lions has never been seriously explored as a behaviour pattern that might mediate chemical communication. The very fact that head rubbing is as frequent as that of spraying MF warrants a closer look at head (mane) rubbing.

5.2 Material and Methods

Black mane lions were observed to head rub and spray MF in a lion breeding centre, Transvaal (Northern Province), South Africa. A black maned lion, a lioness at the onset of oestrous (only when she sprays MF) and subadult cubs were all observed

M. Poddar-Sarkar
Surendranath College, Dept. of Botany
mpsarkar1@rediffmail.com

to perform this action. The lion head-rubbed a tree trunk before spraying MF, the lioness frequently head-rubbed the lion and cubs head rubbed the lioness, just as Schaller (1972) described. When the lion approached the fence, it was allowed to drink water from a bottle at which time a lock of mane hair (black) was cut off with a pair of scissors and sealed in a plastic bag. White lion mane hair was collected from fences on which a white lion had repeatedly rubbed his head in another breeding centre in Transvaal. Black mane and white mane were treated separately for study of ultrastructure as well as for chemical analysis.

5.2.1 Electron Microscopy

A minute portion of a single hair of both types were mounted on brass stubs and covered with a thin layer of gold by using sputter coating and viewed under a Scanning Electron Microscope (JEOL, JSM-5200, using 15–20 KV accelerating potential ~20 mm working distance and vacuum ~10 -5 torr.).

5.2.2 Chemical analysis

Lipid were extracted from the black and white mane following the method of Bligh and Dyer (1959). One fraction of chloroform extract was spotted on Thin Layer Chromatographic (TLC) plate and run in solvent mixture Hexane:Diethyl ether:Acetic acid (80:20:1.5) for characterizing the lipid nature. The second fraction was derivatized to methyl ester for characterizing the fatty acid profile. The fatty acid methyl esters (FAME) were purified by TLC and extracted with hexane. The hexane extracts were subjected to Gas Chromatography Mass Spectrometry (GCMS, Varian CP 3800, MS 2200) with DB5 capillary column under a temperature programming 70-200 °C with 10 °C rise/min keeping MS ion trap 200 °C, manifold at 40 °C and transfer line at 260 °C. Fatty acid methyl esters (FAME) were identified from NIST mass spectral data bank and comparative retention time (Rt).

5.3 Results

5.3.1 Scanning Electron Microscopy

Figure 5.1 shows that the dorsal surface of mane, both black and white, is covered with the usual scaly structures found in a large number of species of animal hair. No significant difference was observed between white and black mane. The cross section of mane (Fig. 5.2) reveals a network of tubular spaces inside the hair. Small grooves or pockets within the concave scales are distinguishable at a magnification of 5000X.

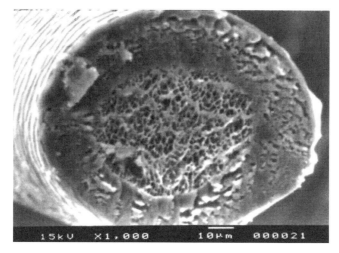

Fig. 5.1 Cross section through mane (1000X) of white lion showing dorsal surface and net work

5.3.2 Chromatogram

TLC spots with marker reveal the presence of free fatty acids (FFA), diglyceride (DG), monoglyceride (MG) but negligible amount of TG. GCMS of fatty acid—methyl esters (FAME) from lion mane presented evidence for fatty acids ranging from C9-C24 (Figs. 5.3– 5.6). Low volatility molecules like nonanedioic acid (Fig. 5.3), tridecanoic acid (Fig. 5.4), 12-methyl tridecanoic acid were also present in lion hair lipids. In addition fatty acids such as myristic, pentadecanoic, palmitic, heptanoic, stearic and octadecenoic acids (Fig. 5.5) have also been detected. Erucic

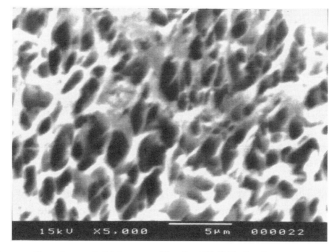

Fig. 5.2 Cross section of mane in higher magnification (5000X) showing grooves

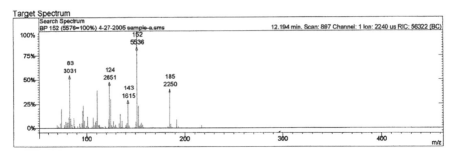

Fig. 5.3 Mass fragments of nonanedioic acid

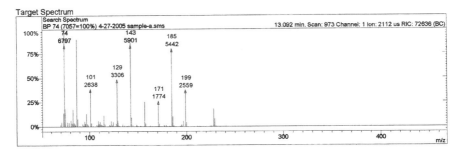

Fig. 5.4 Mass fragments of tridecanoic acid

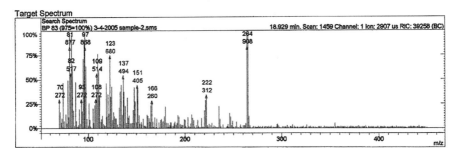

Fig. 5.5 Mass fragments of octadecenoic acid

acid (MW ~338), a comparatively high molecular compound, has also been identified (Fig. 5.6). These were also confirmed in a second sample from the same animal.

5.4 Discussion

The function of lion mane is not clear. West and Packer (2002) are of the opinion that the classic concept of mane as a defensive barrier is not quite tenable. On the other hand the ethological aspect of head rubbing in the lion is just as important as that of spraying MF, indeed, the two follow in close succession. This ethological coupling ensures that both are equally frequent and so head rubbing should have

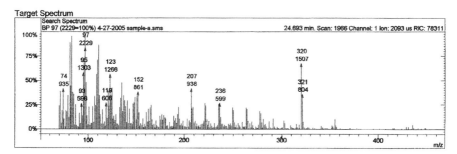

Fig. 5.6 Mass fragments of erucic acid

evolved for some purpose, just as scent-marking is believed to be functional. If head rubbing leaves osmic messages, it must be through some volatile molecules. Thus, just as in MF, the volatiles released by mane are candidates for putative pheromones. Theoretically, the products of sebum, ecrine glands etc. are stored in and released through the hair (Albone 1983), thus leaving osmic signals which might be decoded by conspecifics. SEM images show that lion hair (mane) has the normal scaly structure on the surface and has not evolved into osmetrichia, i.e. specialized hair which can efficiently store pheromones within them. The classic examples are the honey combed surface of the crest hair of the African crested rat (Stoddart 1979) and of the black-tailed deer (Muller-Schwarze, Volkman and Zemanek 1977). More recently a number of other examples have been found, e.g., in a marsupial (Toftegaard and Bradley 1999) and in bats (Scully, Fenton and Saleuddin 2000). Although lion hair is not specialized, the tubular cavities inside might facilitate storing of lipid molecules (Fig. 5.1). The grooves or pockets which have concavities could be the appropriate structure for retaining the osmic signalling material containing informational cues (Fig. 5.2). Shaggy hair is smelly even in human beings and in two unrelated ancient cultures, Biblical and Vedic, head smelling among close kin seems to have been practised. It is therefore probable that the lion, the only maned cat, is leaving two chemical signals simultaneously at two neighboring sites through mane rubbing and MF ejected during scent-marking. Thus, hair volatiles demand our attention. One of the classic investigations on hair chemistry is that on sheep wool hair wax (lanolin), sterol, fatty acids, aliphatic alcohols etc. were identified (Demanze 1996) whereas lipids such as triglycerides were not found, possibly because they had already been oxidized. In human hair, in addition to waxes and free fatty acids monoglycerides, diglycerides and triglycerides were also identified (Aitzetmuller and Koch 1978). A later study revealed that hexa and octadecenoic acids were the chief fatty acid components in human hair (Florkin and Mason 1962).

We thus note that some fatty acids which we have studied in the MF of tiger, leopard and cheetah occur in hair, too (Poddar-Sarkar and Brahmachary 1997, **?**). Moreover, the lipids of wool have been shown to be largely uninfluenced by dietary changes thus suggesting that these are stable chemical fingerprints that could function as identification tags at the species and individual level. Fats and oils of animals generally have an even number of carbon atoms (Deuel 1951) but the occurrence of

odd carbon number fatty acids like tridecanoic acid, pentadecanoic acid and hep-
tadecanoic acid may be of special interest.

Our GCMS analysis of lion hair shows a number of fatty acids ranging from
C9–C24 of which C9–C18 with low volatility might have a pheromonal significance. The higher fatty acids of relatively low volatility will last longer and an animal may be able to distinguish fresh and old marks (osmic signals) on a molecular
basis, - perceiving the gradual loss of low volatility molecules. Indeed, Lin, Zhang,
Shao-Zhong, Block and Katz (2005) pointed out that a garlic-smell compound, hith-
erto unrecorded in male mouse urine, is more volatile than the other pheromones
and might indicate freshness. That transient animals trespassing into new territories
distinguish fresh from old marks and behave accordingly is a known fact. A spectac-
ular example is that of cheetah (Caro 1994) suffering from deleterious physiological
effects partly induced by osmic signals of the resident animal. As this is based on
the reception of pheromone molecules, the old marks with their concomitant loss
of molecules will necessarily be less active. Non-resident cheetahs wander widely
through others' territories and suffer from elevated cortisol levels concomitant with
other stress effects. As these animals are not physiologically sub-normal, the poor
state of health is not likely to be due to long wanderings alone, but also due to
perception of fear of being located by the resident. This fear may be based on the
evidence that the resident is close by and this is largely due to the perception of the
pheromonal signals.

In the TLC lipid profiles we find little or no triglyceride which might have
been oxidized by the time we processed it. Alternatively, the triglyceride content
might have been initially lower than that of monoglyceride or diglyceride. Slowly
oxidiation might control the rate of appearance of free fatty acids, the putative
pheromones, thus prolonging the life of pheromone emission. This is, of course,
equally valid for MF markings. The difference in the composition of black and white
mane is likely to be due to individual differences, rather than racial or strain-specific.
The white lion is expected to be deficient in testosterone production. Erucic acid,
being comparatively large (molecular weight 338), may not be a pheromone. It has
been known to occur in the edible mustard and similar oil, and recently has also been
detected in the oil of some marine animals (Technical report, 2003). The presence
of erucic acid in mane lipid may be a new observation.

Acknowledgments We thank Messers Ross, Lamson, Warren and West of Transvaal, South Africa
without whose help collection of lion mane would not have been possible. We acknowledge the
kind help of the Principal, Surendranath College for access to different facilities.

References

Aitzetmuller, K. and Koch, J. (1978) Liquid Chromatographic analysis of sebum lipids, and other
    lipids of medical interest. J. Chromatog. 145, 195–202.
Albone, E.S. (1983) *Mammalian Semiochemistry*. John Wiley & Sons. Ltd., UK.
Bligh, E.G. and Dyer, W.J. (1959) A rapid method of total lipid extraction and purification.
    Can. J. Biochem. Physiol. 37, 911–917.

Brahmachary, R.L., Poddar-Sarkar, Mousumi and Dutta, J. (1990) The aroma of rice and tiger. Nature 344, 26.

Brahmachary, R.L. Singh, M. and Rajput, Manisha (1999) Scent Marking in the Asiatic Lion. Current Science 76(4), 25 Feb., 480–481.

Caro, T.M. (1994) Cheetah of the Serengeti plains, Chicago Univ. Press, USA. Demanze, C. (1996) *Natural Waxes in Oils and Fats Manual*, Vol. I (original Manual des corps gras, 1992) (Ed. Karleskind, A.), Intercept Ltd., USA.

Deuel, H.J. (1951) *Lipids*, vol. I (Interscience), USA.

Florkin, M. and Mason, H. (1962) Comparative Biochemistry Vol. II, Academic Press, USA.

Lin, D.Y., Zhang, Shao-Zhong, Block, E and Katz, L.C. (2005) Encoding social signals in the mouse main olfactory bulb. Nature 433, 470–477.

Muller-Schwarze, M.D., Volkman, N.J. and Zemanek, K.F. (1977) Osmetrichia: specialised scent hair in black tailed deer. J. Ultrastructural Research. 59, 223–230.

Poddar-Sarkar M., Brahmachary, R.L. and Dutta, J. (1994) Scent marking in the Tiger. In: R. Apfelbach, D. Muller-Schwarze, K. Reutter and E. Weiler (Eds.) *Chemical Signals in Vertebrates VII*. Pergamon Press, UK. pp 339–344.

Poddar-Sarkar M. and Brahmachary, R.L. (1997) Putative semiochemicals in the African Cheetah(Acinonyx jubatus). J.Lipid Mediator and Cell Signalling 15, 285–287.

Poddar-Sarkar M. and Brahmachary, R.L. (2004) Putative chemical signals of Leopard. Animal Biology Netherland J. Zool 54, 255–259.

Schaller, G. (1972) *The Serengeti Lion*, University of Chicago Press, USA.

Scully, W.M.R. Fenton, M.B and Saleuddin, A.S.M. (2000) A histological examination of the holding sacs and glandular scent organs of some bat species (*Emballonuridae, Hipposideridae, Phyllostomidae, Vespertilionidae and Molossidae*). Can. J. Zool. 78(4), 613–623.

Stoddart, D.M. (1979) A specialised scent-releasing hair in the crested rat, *Lophiomys imhausi*. J. Zool. 189, 551–553.

Toftegaard,C.L. and Bradley,A.J. (1999) Structure of specialised osmetrichia in the brown antechinus *Antechinus stuartii* (Marsupialia:Dasuridae). J. Zool. 248, 27–30.

Technical report series no. 21 (June 2003) Erucic acid in food-A toxicological review and risk assessment; Food Standards, Australia, New Zealand.

West, P. M. and Packer, C. (2002) Sexual selection, temperature and the Lion's mane. Science 297, 1339–1343.

Part II
Olfactory Response and Function

Chapter 6
Using Ethologically Relevant Tasks to Study Olfactory Discrimination in Rodents

Heather M. Schellinck, Stephen R. Price and Michael J. Wong

Abstract For most mammals, the ability to detect odours and discriminate between them is necessary for survival. Information regarding the availability of food, the presence of predators and the sex, age and dominance status of conspecifics is odour mediated. Probably because of this extraordinary reliance upon odour cues, mice and rats have developed the ability to learn and remember information associated with olfactory cues as effectively as primates recall visually related cues. As a result, these rodents have become the model of choice to study the neural and cognitive processes involved in olfactory discrimination. In this paper, we describe some of the more ethologically based tasks used in assessing olfactory discrimination and the advantages and disadvantages of the different methodologies employed.

6.1 Introduction

The perception and understanding of olfactory cues dominates the entire life of virtually all mammals other than primates. In rodents, first bouts of suckling have been shown to depend upon olfaction (Pederson and Blass 1982) and odours associated with food or other stimuli very early in life may have long term effects on later preferences (Hepper 1990; Smotherman 1982; Sullivan and Leon 1987). Adult animals can identify the age, sex, territory, and dominance status of a conspecific from odours produced by scent glands or deposited in urine and faeces (Brown and Mcdonald 1985). Changes in reproductive status, including acceleration of puberty, changes in estrous cycle and termination of pregnancy, primarily demonstrated in mice,are also odour-mediated (reviewed in Brown and Mcdonald, 1985). Similarly, the recognition and avoidance of predators is largely a result of the perception of odours (Apfelbach, Blanchard, Blanchard, Hayes and McGregor 2005).

The relevance of odours to mammals is also reflected in their anatomy. The olfactory bulbs are large relative to the rest of the forebrain and projections from

Heather M. Schellinck

Dalhousie University, Department of Psychology and Institute of Neuroscience

heathers@dal.ca

the olfactory bulb extend to most of the cortex below the rhinal fissure. Projections from the olfactory bulb terminate posteriorly in the entorhinal cortex, considered to be the primary entry point to the hippocampus. The piriform cortex and entorhinal cortex also send projections to the amygdala and hypothalamus (Slotnick and Schellinck 2002). It is as a result of these connections that odours have the capacity to become associated with memory for both contextual and emotional events. One outcome of our understanding of the behavioural and anatomical processes of olfaction has been the ability to exploit this knowledge to further understand the odour discrimination process in mammals. In this paper, we describe some of the ethologically relevant tasks for the study of olfactory discrimination in rodents and the factors that may influence the results of investigations that use these paradigms.

6.2 Olfactory Discrimination Tasks

6.2.1 Habituation-dishabituation

The habituation-dishabituation test is based upon the observation that successive presentations of the same stimulus odour will result in a decrease in investigatory behaviour, i.e., habituation. Then, when a different odour stimulus is presented to the same subject, the habituated response will reoccur or will be "dishabituated" (Sundberg, Doving, Novikov and Ursin 1982). Some investigators prefer to call this procedure a habituation-discrimination task because the second stimulus is different from the first odour and thus the animal's response is not considered to be a dishabituation. Others only refer to habituation-discrimination if the second odour is presented simultaneously with the first odour, following the habituation trials (Halpin 1974). In our view, the "dishabituation" refers to the response rather than the stimulus, and we will continue to refer to the test in the context of habituation-dishabituation.

   The test usually consists of brief, consecutive trials: several two minute presentations of one odour, followed by either one or several presentations of the second odour (e.g. Schellinck, Rooney and Brown 1995). The odours may be presented on a filter paper, cotton swab or glass plate (Schellinck et al. 1995; Lee, Emsley, Brown and Hagg 2003; Mayeaux and Johnston 2004) The orienting response and/or the time spent investigating the odour is recorded for each stimulus presentation. This task may be completed in the animal's home cage or in a novel test arena. If the experiment is to be undertaken in a novel environment, we recommend beginning the test with a no-odour adaptation period during which the subject is put in the test chamber and presented with the odour vehicle (Schellinck et al. 1995; Lee et al. 2003). This reduces the possibility that the subject's exploratory or anxiety-related behaviour in the new environment may interfere with its interest in the odour stimuli. In most instances, the subject is allowed to sniff the odour but it is not able to interact with it in any other way. Tests in which the subject can interact with the odour are described in section 6.2.3.

When choosing odours, particularly if you are interested in identication of individuals, it is important to make sure that the stimuli are equally familiar as otherwise the subjects may respond on the basis of novelty rather than cues of individuality (Martin and Beauchamp 1982). Also, the order in which the odours are presented should be counterbalanced so that order effects do not influence the results. Results from a recent study by Brown and colleagues illustrate the typical pattern of sniffing in this task (Fig. 6.1; Lee et al. 2003).

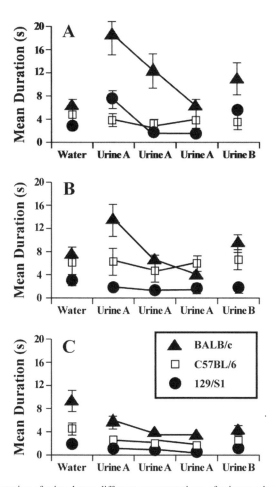

Fig. 6.1 Different strains of mice detect different concentrations of urinary odours of conspecifics with BALB/c>129/S1>C57BL/6. Presented are the mean durations (± SEM) that mice investigated water and three sequential 2-min presentations of the same concentration of urine from one of the other strains at (A) 10^{-2}, (B) 10^{-3} and (C) 10^{-4} concentrations. Immediately after the presentation of Urine A, mice were exposed to a novel urine B (the other foreign strain) at the same concentration.(Reprinted from Neuroscience, 118, Lee, Emsley, Brown, and Hagg, T. Marked differences in olfactory sensitivity and apparent speed of forebrain neuroblast migration in three inbred strains of mice. 263-270, Copyright (2006), with permission from Elsevier)

The principal benefits of the habituation-dishabituation test arise from the speed of testing and the minimal need for equipment. Each animal may complete a nine trial session in less then 30 minutes. An additional advantage comes with the sequential presentation of the stimuli as this methodology avoids the problems associated with the mixing of vapours from different odour stimuli if the odours are presented simultaneously. The test is particularly well suited for field work as the stimuli may be presented on a substrate that the subject normally investigates, thus eliciting a more natural response than one might encounter in a lab environment.

The main difficulties with the habituation-dishabituation test are twofold. First, 'investigative behaviour' is not consistently defined and this in turn makes it hard to compare the results from different studies. Although most investigators assess the time spent sniffing the odours as a measure of habituation and dishabituation, they rarely define sniffing behaviour. For example, they may not indicate how close to a filter paper the animal must be before it is considered to be sniffing the stimulus. In addition, some studies include the time spent orienting toward the odour as a measure of habituation-dishabituation (e.g. Sundberg et al. 1982). These behaviours are not always easy to quantify (Fig. 6.2).

The second problem comes not from discrepancies in measurement but in interpretation of the results. When dishabituation occurs the outcome is clear—the odours are discriminable. If, however, the animal has made a significant response to the first odour but does not respond to the second stimulus, is unclear how to interpret such an outcome. This could reflect a failure to discriminate the two odours or a general lack of motivation resulting from disinterest in the test odour. It is also possible that the test odour is discriminable but not sufficiently different to produce a clear dishabituation effect.

If a null result ensues, subsequent replication of the experiment is one way to feel more confident that the results are reliable. Alternatively, the animals may be trained in a conditioned odour preference task as described below. This would enable the investigator to further understand why the subjects did not respond in the habituation-dishabituation task. Videotaping the results should also become standard practice. As well as creating a permanent record that enables behavioural

Fig. 6.2 The sequence of the orienting response of a rat to an odour presented from above. Before odour presentation the animal is at rest (A). When the odour is presented, the rat lifts his head and forepart of the body (B, C), moves about (D), and finally rears (E) (reprinted with permission from Sundberg et al. (1982)

scores to be recorded accurately, it provides a posthoc method to more easily score a variety of investigatory behaviours that may be indicative of dishabituation. The habituation-dishabituation test is primarily used to assess an animal's response to the odours of conspecifics and with recognition of the limitations noted above is a reliable method of doing so.

6.2.2 Conditioned Odour Preference

Tasks that measure digging behaviour have been developed to assess memory for olfactory related information (e.g., Berger-Sweeney, Libbey, Arters, Junagadhwalla and Hohmann 1998; Mihalick, Langlois, Krienke and Dube 2000; Schellinck, Forestell and LoLordo 2001; Pena, Pitts and Galizio 2006). Generally the animal is trained to dig for a food reward in a cup containing bedding or sand that has a particular scent associated with it. Then, it is tested for its preference for digging in this odour in the absence of reward but in the presence of a second odour.

6.2.2.1 Tests Developed for Mice

The test developed by Schellinck and colleagues is representative of the odour preference tasks developed for mice. The mice are transferred individually to a testing room and trained in a clean animal housing cage. In the training phase, two different odours are each presented in a quasi-random order for three trials a day for several days. The mouse is required to learn to associate one of the odours (CS+) with a sugar reward buried in wood shavings contained in a small odour pot and the second odour (CS−) with absence of reward. In the test phase, carried out in a three chambered box, both the CS+ and CS-odours are presented simultaneously in the absence of sugar and the amount of time spent digging in each odour pot is measured. Animals that have learned that the CS+ signals the availability of sugar, dig consistently and almost exclusively in the wood chips that contain the previously rewarded odour (Fig. 6.3). The test has been used in a number of contexts and has been shown to be useful to assess long-term memory; mice can remember these discriminations for up to six months (Wong and Brown 2006).

Odour associated digging tasks have proven popular for several reasons. Because the procedure takes advantage of the mouse's natural tendency to dig for food, minimal training is required. In addition, measuring such species typical behaviour may also reduce distress in the experimental setting and actually provide environmental enrichment (Deacon 2006). We have also found that the mice need not be food restricted during training; animals that had their food removed for only 12 hours prior to preference testing still demonstrated significantly more digging in the CS+ odour (Forestell, Schellinck, Drumont and Lolordo 2001). As digging is more easily scored and more objectively measured than sniffing, results within and between laboratories have yielded similar results (Wong and Brown 2006). This paradigm is also more appropriate to assess anosmia than the habituation-dishabituation test. For example, when mice missing a neural cell adhesion molecule

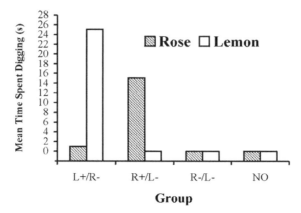

Fig. 6.3 Mean time spent digging in rose and lemon odours for the experimental groups, i.e., those for whom rose was paired with sugar (group R + /L−) or lemon was paired with sugar (group L + /R−) in training, and control groups, i.e., those for whom neither rose nor lemon was paired with sugar (group R − /L−), or who received no odours or sugar (group NO) in training. Reprinted from Chemical Senses, 26, Schellinck, Forestell and LoLordo (2001) A simple and reliable test of olfactory learning and memory in mice. 663-672. Copyright (2006), with permission from Oxford University Press

(NCAM knockout mice) were tested in a habituation-dishabituation test, their lack of response was considered evidence that they were unable to discriminate between two odours (Gheusi, Cremer, McLean, Chazal, Vincent and Lledo 2000). In contrast, Schellinck, Arnold and Rafuse (2004) tested similar NCAM knockout mice in the olfactory preference task and found that their ability to discriminate between odours was unimpaired. The difference in outcome in these two experiments is most likely a result of differences in motivation. The habituation-dishabituation task does not depend upon external motivation, and is thus a more ethologically valid approach making it more a test of an animal's interest in different odours than a test of its ability to discriminate between them based upon receiving a reward. Nonetheless, it will not always provide valid results when olfactory learning ability or acuity is the question being addressed.

As the digging task relies upon external motivation, investigators should be careful not to misinterpret the results. For example, if different strains of mice spend different amounts of time digging in the CS+ during the test phase of the experiment, this should not be interpreted to mean that one strain has learned more than another strain (Schellinck et al. 2001). It is possible that the different strains are more sensitive to food restriction, however mild, or have different levels of activity. Both factors could lead to increased digging. With this caveat, the task is very useful for assessing acuity and long-term memory.

6.2.2.2 Tests developed for rats

The conditioned odour preference tasks for rats were developed to assess more complex cognitive processes. In general, rats appear more suitable than mice for such tasks, not because their cognitive abilities are known to be greater, but because their activity levels are lower and, thus, they are easier to train. In addition, these tasks frequently require mild food restriction for longer periods; because of their high metabolic rate, mice are more difficult to maintain at a reduced weight for long periods of time. Rats have been shown to perform accurately in a matching-to-sample digging task (Pena et al. 2006). After digging in a cup of scented sand for sugar reward, subjects were presented with two comparison cups, a baited cup that contained the same odour and a cup that had a different scent. The rats not only learned to accurately match the appropriate comparator with the sample during training, they also could transfer this learning to novel exemplars. The results of both the matching to sample test and transitive inference demonstrate that rats have an excellent capacity to learn the relationship among a series of odour stimuli. Moreover, the high level of performance on olfactory tasks is in marked contrast to the rat's ability to learn similarly designed tasks in the visual and auditory modalities (Slotnick 2001). These findings demonstrate the value of using an ethologically relevant modality when using any animal as a model for complex learning.

6.2.3 Interactions with Biological Substrates

Changes in behaviour and or neurophysiological state may be produced when an individual has the opportunity to investigate and interact with social odours of conspecifics in a more natural context. Under this category we have included tasks that measure differential behavioural responses to scent marks as well as paradigms that induce physiologically mediated reproductive changes. Although some habituation-dishabituation tasks may involve the discrimination of social odours such as those contained in urine, the odours are frequently presented through a screen and the animal does not usually have the opportunity to have close contact with the odours.

In one of the few circumstances in which odour discrimination has been studied extensively in rodents other than mice or rats, Johnston and his colleagues have assessed the response of hamsters to species-specific stimuli such as flank odours and vaginal secretions (e.g., Heth, Todrank and Johnston 1999; Lai, Vasilieva and Johnston 1996). Depending upon the design of the study, the subject's sniffing responses to accessible scent marks on glass plates or their actual scent marking behaviours in vacant cages have been measured. As both vaginal and flank marking are distinct, stereotyped events and are readily identifiable, the results of these experiments have yielded very reliable results (Petrulis and Johnston 1997). Using these methods, Johnston has contributed greatly to our understanding of both the behavioural and neural bases of the discriminability of sex and individual odours.

A major effort in contemporary olfactory research has focused upon determining the genetic basis of odours of individuality in mice with both major urinary proteins (MUPs) and the major histocompatibility complex (MHC) believed to provide a contribution to discriminable individual differences (Singer, Beauchamp and Yamazaki 1997; Hurst and Beynon 2004). Many of these studies utilized instrumental conditioning tasks in which the animals were rewarded for learning to discriminate between odours. While these experiments demonstrate what an animal is able to do, they do not provide us with information as to whether this is analgous to their behaviour in a natural environment. In contrast, Hurst and colleagues have studied scent marking, primarily in the context of competitive territory marking in seminatural enclosures. Their investigations have revealed that urine from mice that differed genetically in the MUPs produced countermarking responses in territory owners and is particularly relevant in the context of odours of individuality (Hurst, Payne, Nevison, Marie, Humphries, Robertson, Cavaggioni and Beynon 2001). Their work demonstrates that a methodology that measures a natural response to a conspecific's urine is a valid way to determine which scents are readily meaningful to an animal.

Assessing changes in a neurobiological process in response to different stimuli also provides a powerful tool with which to study the relevance of biological odours to an individual. Neurochemical and electrophysiological changes have been shown to correspond to differential odour presentation (Brennan, Schellinck, De La Riva, Kendrick and Keverne 1999; Baum and Keverne 2002; Lin, Zhang, Block and Katz 2005; Kadohisa and Wilson 2006). As well, the changes produced in reproductive status by pheromones may be used to assess the discriminability of stimuli and in this way provide greater understanding of the nature and function of a particular chemical signal. These reproductive changes have primarily been demonstrated in mice and include the Bruce effect in which a female of a species aborts her young as a result of exposure to a male other than the original "stud" male (Brown and Mcdonald 1985).

The function of the MUPs has been investigated in the context of the Bruce effect. Urine from "stud" males to which MUPs from males of a different strain had been added after being stripped of their volatile ligands caused female mice to abort their pregnancy; control urines from males that were familiar to the females did not affect the pregnancies (Brennan, Schellinck and Keverne 1999). Further experiments revealed that low molecular weight (LMW) fractions i.e., those containing volatile ligands, were very effective in inducing pregnancy block and that LMWs from familiar males created an intermediate degree of blocking (Peele, Salazar, Mimmack, Keverne and Brennan 2003). Together the results from these experiments would suggest a role for both high and low molecular weight fractions of urine in pregnancy block and individual identity. The functional relationship between the main and accessory olfactory systems may also be better understood by studying reproductive changes in response to an odour . In our laboratory, we are currently assessing the role of these systems in the processing and discrimination of urinary odours by measuring the response of juvenile females to urinary odours from adult males.

6.3 Conclusions

Ethologically appropriate tasks are extremely useful to study olfactory discrimination in rodents to investigate several areas of current scientific interest. They provide us with the methodology to increase our understanding of cognitive processes related to olfaction in rats and mice. This will subsequently enable us to advance our knowledge of similar learning processes in humans. Furthermore, using the methodologies described in this review, we should be able to clarify the role of specific chemosignals in determining odours of individuality for rodents and other macrosmatic mammals. In the latter context in particular, behavioural and physiological responses to odours should be studied in parallel and hypotheses as to their function made from multidimensional response systems.

Acknowledgments We are extremely grateful for the expertise provided by Vivien Hannon, Academic Computing Services, Dalhousie University in preparing the images for publication.

References

Apfelbach, R., Blanchard, C.D., Blanchard, R.J., Hayes, R.A. and McGregor, I.S. (2005) The effects of predator odors in mammalian prey species: a review of field and laboratory studies. Neurosci. Biobehav. Rev. 29, 1123–1144.

Baum, M.J. and Keverne, E.B. (2002) Sex difference in attraction thresholds for volatile odors from male and estrous female mouse urine. Horm. Behav. 41, 213–219.

Berger-Sweeney, J., Libbey, M., Arters, J., Junagadhwalla, M., and Hohmann, C.F. (1998) Neonatal monoaminergic depletion in mice (*Mus musculus*) improves performance of a novel odor discrimination task. Behav. Neurosci. 112, 1318–1326.

Brennan, P.A., Schellinck, H.M., De La Riva, C., Kendrick, K.M. and Keverne, K.B. (1998) Changes in neurotransmitter release in the main olfactory bulb following an olfactory conditioning procedure in mice. Neurosci. 87, 583–590.

Brennan, P.A., Schellinck, H.M. and Keverne, E.B. (1999) Patterns of expression of the immediate-early gene egr-1 in the accessory olfactory bulb of female mice exposed to pheromonal constituents of male urine. Neurosci. 90, 1463–1470.

Brown, R.E. and Mcdonald, D. W. (1985) *Social Odours in Mammals*. Oxford University Press, Oxford.

Deacon, J. (2006) Burrowing in rodents: a sensitive method for detecting behavioral dysfunction. Nat. Prot. 1, 118–121.

Forestell, C.A., Schellinck, H.M., Drumont, S. and Lolordo, V.M. (2001) Effect of food restriction on acquisition and expression of a conditioned odor discrimination in mice. Physiol. Behav. 72, 559–566.

Gheusi, G., Cremer, H., McLean, H., Chazal, G., Vincent, J.D. and Lledo, P.M. (2000) Importance of newly rly generated neurons in the adult olfactory bulb for odor discrimination. Proc Natl Acad Sci U S A. 97, 1823–8.

Halpin, Z.T. (1974) Individual differences in the biological odors of the Mongolian gerbil (*Meriones unguiculatus*). Behav. Biol. 11, 253–259.

Hepper, P.G. (1990) Foetal Olfaction. In: D.W. Macdonald, D. Muller-Schwarze, and S.E. Natynczuk (Eds.), *Advances in Chemical Signals in Vertebrates*. Oxford University Press, Oxford, pp. 282–286.

Heth, G., Todrank, J. and Johnston, R.E. (1999) Similarity in the qualities of individual odors among kin and species in Turkish (*Mesocricetus brandti*) and golden (*Mesocricetus auratus*) hamsters. J. Comp. Psychol. 113, 321–326.

Hurst, J.L., Payne, C.E., Nevison, C.M., Marie, A.D., Humphries, R.E., Robertson, D.H., Cavaggioni, A. and Beynon, R.J. (2001) Individual recognition in mice mediated by major urinary proteins. Nature 414, 631–634.

Hurst, J.L. and Beynon, R.J. (2004) Scent wars: the chemobiology of competitive signaling in mice. Bioessays 26, 1288–1298.

Kadohisa, M. and Wilson, D.A. (2006) Olfactory cortical adaptation facilitates detection of odors against background. J. Neurophysiol. 95, 1888–1196.

Lai, S.C., Vasilieva, N.Y. and Johnston, R.E. (1996) Odors providing sexual information in Djungarian hamsters: evidence for an across-odor code. Horm. Behav. 30, 26–36.

Lee, A.W., Emsley, J.G., Brown, R.E. and Hagg, T. (2003) Marked differences in olfactory sensitivity and apparent speed of forebrain neuroblast migration in three inbred strains of mice. Neurosci. 118, 263–270.

Lin, D.Y., Zhang, S.G., Block, E. and Katz, L.C. (2005) Encoding social signals in the mouse main olfactory bulb. Nature 434, 470–477.

Martin, I.G. and Beauchamp, G. K. (1982). Olfactory recognition of individuals by male Cavies (*Cavia apera*). J. Chem. Ecol., 8,1241–1249.

Mayeaux, D. J. and Johnston, R.E. (2004) Discrimination of social odors and their locations: role of lateral entorhinal area. Physiol. Behav. 82, 653–662.

Mihalick, S.M., Langlois, J.C., Krienke, J.D. and Dube, W.V. (2000) An olfactory discrimination procedure for mice. J. Exp. Anal. Behav. 73, 305–318.

Peele, P., Salazar, I., Mimmack, M., Keverne, E. B. and Brennan, P. B. (2003). Low molecular weight constituents of male mouse urine mediate the pregnancy block effect and convey information about the identity of the mating male. Eur. J. Neurosc. 18, 622–628.

Pena, T., Pitts, R.C. and Galizio, M. (2006) Identity matching-to-sample with olfactory stimuli in rats. J. Exp. Anal. Behav. 85, 203–221.

Pedersen, P.E. and Blass, E.M. (1982) Prenatal and postnatal determinants of the first suckling episode in albino rats. Dev. Psychobiol. 15, 349–355.

Petrulis, A. and Johnston, R.E. (1997) Causes of scent marking in female golden hamsters (Mesocricetus auratus): specific signals or classes of information? J. Comp. Psychol. 111, 25–36.

Schellinck, H.M., Forestell, C.A. and LoLordo, V. M. (2001) A simple and reliable test of olfactory learning and memory in mice. Chemical Senses 26, 663–672.

Schellinck, H.M., Rooney, E. and Brown, R.E. (1995) Odors of individuality of germfree mice are not discriminated by rats in a habituation-dishabituation procedure. Physiol. Behav. 57, 1005–1008.

Schellinck, H. M., Arnold, A., Rafuse, V. (2004) Neural Cell Adhesion Molecule (NCAM) null mice do not show a deficit in odor discrimination learning. Behav. Brain Res.152, 327–334.

Singer, A.G., Beauchamp, G.K. and Yamazaki, K. (1997) Volatile signals of the major histocompatibility complex in male mouse urine. Proc. Nat. Acad. Sci. 94, 2210–2214.

Slotnick, B.M. (2001) Animal cognition and the rat olfactory system. Trends Cogn. Sci. 5, 216–222.

Slotnick, B.M. and Schellinck, H. M. (2002) Methods in olfactory research with rodents. In: Simon, S.A. and Nicolelis M. (Eds). In *Frontiers and Methods in Chemosenses* CRC Press, New York, pp 21–61.

Smotherman, W.P. (1982) Odor aversion learning by the rat fetus. Physiol. Behav. 29, 769–771.

Sullivan, R.M. and Leon, M. (1987) One-trial olfactory learning enhances olfactory bulb responses to an appetitive conditioned odor in 7-day-old rats. Brain Res. 432, 307–311.

Sundberg, H., Doving, K., Novikov, S. and Ursin, H. (1982) A method for studying re¬sponses and habituation to odors in rats. Behav. Neural. Biol. 34, 113–119.

Wong, A. A. and Brown, R. E. (2006) Age-related changes in visual acuity, learning and memory in C57BL/6J and DBA/2J mice. Neurobiol Aging. Sep 27; [Epub ahead of print].

Chapter 7
Comparisons of State and Likelihood of Performing Chemosensory Event Behaviors in Two Populations of African Elephants (*Loxodonta africana*)[‡]

Bruce A. Schulte, Kathryn R. Bagley, Matthew Groover, Helen Loizi, Christen Merte, Jordana M. Meyer, Erek Napora, Lauren Stanley, Dhaval K. Vyas, Kimberly Wollett, Thomas E. Goodwin and L.E.L. Rasmussen[†]

Abstract The demonstration of a species-level chemical signal assumes that the same chemical signal serves a similar purpose across the range of the species. Yet, the response to putative chemical signals varies with social setting, environmental conditions, age, sex, and reproductive status of the individuals. Through observations and biological assays with African elephants (*Loxodonta africana*), we evaluated variation in state behaviors and the likelihood that elephants would perform specific chemosensory event behaviors at our study sites at Addo Elephant National Park South Africa and at Ndarakwai Ranch in Tanzania. We have noted similar time budgets in state behaviors. Post-pubescent (> 9 y) elephants showed similar likelihoods in investigating their environment, including conspecific urine and feces. As we pursue the identity of an estrous pheromone, other chemical signals, and developmental patterns, studies with the two populations will be invaluable in assessing the generality of these findings to savanna African elephants.

7.1 Introduction

Studying the variation in time budgets and chemosensory behavior of a species across its geographic range can improve our understanding of chemical signal dynamics and the ubiquity of function for particular signals. Determining chemical

Bruce A. Schulte

Georgia Southern University, Department of Biology

bschulte@georgiasouthern.edu

[†] Decreased

[‡] This chapter is dedicated to the memory of Dr. Bets Rasmussen

signal activity involves performing biological assays with either parts of a relevant subject animal (e.g., electroantennograms) or whole animals. To facilitate chemical identification, bioassays often are performed in laboratories or at facilities with captive animals. These captive animals represent the first population under study. To verify activity and to elucidate the evolutionary function of the signal, field trials are implemented (Wyatt 2003). The discovery of a preovulatory pheromone, (Z)-7-dodecen-1-yl acetate, in Asian elephants (*Elephas maximus*) followed this protocol (Rasmussen, Lee, Roelofs, Zhang and Daves 1996; Rasmussen, Lee, Zhang, Roelofs and Daves 1997; Rasmussen 2001). While our research with African elephants is not as advanced, bioassays with male African elephants at facilities in North America have shown that males respond more to preovulatory than luteal urine from conspecific females (Bagley, Goodwin, Rasmussen and Schulte 2006). Chemical investigation is ongoing and several promising compounds, known to be pheromones in some invertebrates, have been discovered (Goodwin, Eggert, House, Weddell, Schulte and Rasmussen 2006). In preparation for field bioassays, we have been investigating the activity patterns and chemosensory behaviors of African elephants across age and sex in two populations, one in Northern Tanzania and the other in South Africa.

7.2 Objective

The purpose of the current study was to compare the state and chemosensory behaviors of African elephants at our two African study sites. While similarity in these behaviors does not guarantee that chemical signals will be identical in structure and function, the absence of large differences would suggest that such a situation is probable.

7.3 Study Sites, Elephants and Procedures

7.3.1 Study Sites

One of our field sites involves free-ranging African elephants at Ndarakwai Ranch, Tanzania. Ndarakwai Ranch is a privately owned area of ca. 4300 ha located in northern Tanzania. The habitat is a mix of grassland and mixed acacia woodland. This region typically experiences annually a short and a long wet season separated by dry seasons (Vyas 2006), similar to nearby Amboseli National Park, where rainfall averages 350 mm annually (Poole 1999).

The second field site is at Addo Elephant National Park (AENP), founded in 1931 and located in the Eastern Cape of South Africa near Port Elizabeth. The 14,000 ha fenced study area is composed of sub-tropical succulent thicket and grassland (Low and Rebelo 1996, as cited in Whitehouse, Hall-Martin and Knight 2001; Whitehouse and Hall-Martin 2000). Five main waterholes supply pumped water year round, while numerous natural pans are created during rainy periods. The average rainfall for the region is 445 mm annually (Paley and Kerley 1998).

7.3.2 Elephant Populations

Before 1995, only anecdotal records exist of elephants in Ndarakwai Ranch and the surrounding area. Elephants are thought to travel through the area encompassing Ndarakwai as they move between Amboseli and the West Kilimanjaro area (see Vyas 2006 for details). From 2004–6, over 230 elephants in 26 individual groups were identified, as well as 44 solitary adult males (Vyas 2006; Napora, personal communication). Year of birth was known only for animals born during the observation period. Morphological features (e.g., shoulder height and tusk length approximations) were used to estimate ages (Moss 1996) as follows: calves (0–4 y), juveniles (5–9 y), subadults or pubescents (10–19 y), and adults (> 19 y). These ages reflect the approximate time of major developmental changes for elephants (calf: nursing; juvenile: weaning; pubescent: capability of producing viable gametes and dispersal (males) or reproduction (females); adult: full reproductive status and rise in rank). This classification also was used for the elephants at AENP for comparisons.

In 1954, AENP was fenced to protect elephants and humans (Whitehouse 2001). The selective hunting of tusked elephants before this time resulted in a founder population with skewed genetics such that only a small percentage (2–3%) of the female population of elephants at AENP has tusks currently (Whitehouse 2001; personal records). The population has been cataloged in a photographic database (Whitehouse and Hall-Martin 2000; Merte 2006). In 2003, 60 elephants were moved to a new fenced section of the park (Loizi 2004). Currently, our study population in the main park consists of approximately 360 elephants in six family groups. The year of birth was estimated from photographs for older individuals (Whitehouse 2001) and the month or even day of birth is known for almost all individuals born over the past 10 years (Whitehouse 2001; Bagley 2004; Loizi 2004; Merte 2006).

Elephants from both populations were identified using characteristic ear markings, venation patterns, tusk features and other distinguishing marks such as a broken tail or scar. Photographic identification files were created for all new elephants and updated as needed for previously identified animals. Calves were identified by their association with known adult females (Moss & Poole 1983; Wittemyer, Douglas-Hamilton and Getz 2005; Archie, Moss and Alberts 2006).

7.3.3 Observational Methods

At both locations, all observations were made during the day at waterholes. Observations at waterholes permit an unobstructed view of elephants and their trunk movements. Waterholes serve as points of congregation for elephants, so chemical signals in the form of urine and feces as well as their sources (elephants) were readily available. We used continuous focal observations that ended after 20 min or when the animal went out of sight, whichever occurred first (Altmann 1974). Observations of less than 5 min were not included or were added to observations of the same individual in the same area and within a few weeks of the short observation. The order of individuals to be observed was determined by random selection without replacement for the eight age by sex categories (calf, juvenile, pubescent, adult and the two sexes). Whenever the age and sex of the elephant was not available, we watched an animal from the next age and sex category on the day's list. The particular individual observed for this age and sex was selected based on visibility (Lehner 1996).

We classified behaviors as states with measurable duration, or as chemosensory events, recorded as a frequency (Martin and Bateson 1993). The major states were drink/suckle, dust/mud/wallow, stand, walk or other. Chemosensory events were actions by the trunk tip contacting another elephant, investigating a substrate or performing actions called accessory trunk behaviors. We recorded the part of the body touched and the age and sex of the individual touched. The chemosensory events included sniff, check, place, and flehmen, and 10 accessory trunk behaviors (blow, dig, flick, horizontal sniff, periscope sniff, pinch, rub, suck, wrap and wriggle; see Schulte and Rasmussen 1999; Bagley et al. 2006; Schulte 2006; Vyas 2006).

At the Ndarakwai Ranch study site, we observed elephants from the 6 m high platform adjacent to the waterhole in 2004 and 2005. Elephants generally ignored observers who remained quiet and still while recording data. At AENP from 2003–2006, we observed from a vehicle because the elephants were acclimated to them.

7.3.4 Data Analysis

State behaviors were recorded as duration, while event behaviors were converted from frequency to rate (frequency/hour). These data were analyzed as proportion of time spent in a state or proportion of elephants performing a behavior using a Fisher Exact Test (Sokal and Rohlf 1995). Because of the distance between the observer and the elephant, it was sometimes difficult to distinguish sniff, check and place, and flehmens were not common, so the main chemosensory behaviors were analyzed as a single response variable (SCPF) (Schulte and Rasmussen 1999). The chemosensory behaviors were placed into two categories depending on the substrate type: environmental (general substrate) or elephant origin (conspecific, urine or feces). For consideration of age by sex effects, elephants were classified as either pre-puberty (< 10 y) or post-puberty (≥ 10 y) to increase sample size.

7.4 African Elephant Behavior at Northern Tanzania and AENP

The main state behaviors for elephants of all ages at the waterholes in both populations were walk, stand, drink/suckle and bathing in the mud. At Ndarakwai Ranch in Tanzania, these behaviors comprised 90% of the state activities at the waterhole (Fig. 7.1a). In AENP, the same behaviors comprised 97% of the state activities at the various waterholes (Fig. 7.1b). The elephants at Ndarakwai spent somewhat more time drinking and suckling than standing, which was reversed at AENP, but for the

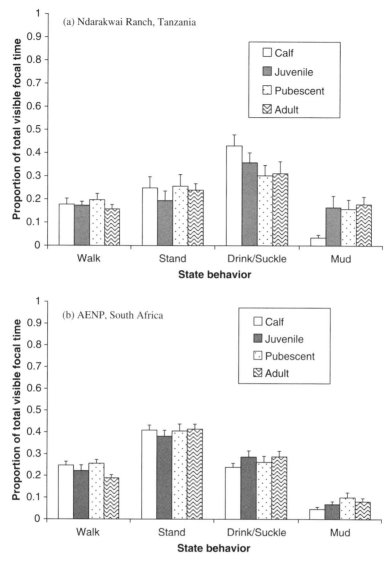

Fig. 7.1 Comparison of the proportion of total time visible that elephants spent in the four major states (a) at Ndarakwai Ranch, Tanzania (90% of all time in all states; sample size was 29, 19, 20 and 32 by age class per behavior from left to right.) and (b) at AENP, South Africa (97% of all time in all states, sample size was 37, 37, 34 and 43 from left to right

two populations these two behaviors comprised 60% and 67%, respectively, of the time at the waterholes. Elephants are not only acting in a similar fashion at waterholes in these two regions, but their activity provides them with similar opportunities to investigate the environment at and around waterholes.

Urine and feces represent a small portion of the total substrate available at a waterhole (Merte 2006), yet they are a source of chemical signals. At both study sites, approximately 50% of the elephants observed investigated the general

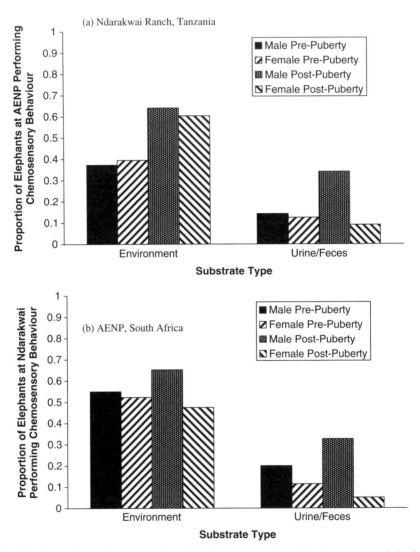

Fig. 7.2 Comparison of the proportion of elephants responding with chemosensory behaviors to the general substrate (environment) and to urine/feces for pre- and post-pubescent males and females. (a) Ndarakwai Ranch, Tanzania; sample size of different elephants from left to right for environment and to urine/feces: 40, 44, 46 and 40. The same animals were observed for response to urine/feces as to the environment. (b) Addo Elephant National Park South Africa; sample size from left to right for environment: 59, 43, 53 and 48. Many of the same animals were observed for response to urine/feces as to the environment. Sample sizes to urine/feces from left to right: 49, 32, 44 and 44

environment with their trunk tip, while 10-30% specifically examined urine or feces. Elephants at the two locations showed very similar trends in their likelihood to investigate. At Ndarakwai, 65% of post-pubescent males performed chemosensory behaviors to the environment, slightly higher than the other age and sex groups (Fig. 7.2a). At AENP, this value was 64% of the post-pubescent males (Fig. 7.2b).

Significantly more pre-pubescent animals investigated the environment with chemosensory behaviors at Ndarakwai (53.5%) than at AENP (38.2%) (Fisher Exact Test, $P = 0.04$), but no such difference existed for post-pubescent elephants (57% and 62%, respectively, $P = 0.46$). The likelihood of performing chemosensory behaviors to urine or feces did not differ between the populations for pre-pubescent ($P = 1.0$) or post-pubescent elephants ($P = 0.85$). The same trends were evident if only the main trunk chemosensory behaviors (SCPF) were considered

7.5 Conclusions and Continuing Research

In the first study by one of our group at AENP (Loizi 2004), we showed that calf activity patterns were similar to those of calves in Amboseli National Park, Kenya (Lee 1986). The results from the current study broaden these findings. In general, the time budgets at waterholes and the likelihood of performing chemosensory behaviors to the environment or to urine and feces did not differ for elephants at the Tanzania and South African study sites. The pre-pubescent elephants at the Ndarakwai Ranch waterhole were more likely to investigate the environment using chemosensory behaviors than this age class of elephants at AENP. Some of the elephants at Ndarakwai are residential in the region, while others are transitory. In either case, the elephants do not visit the waterhole regularly throughout the year. In addition, conflict between humans and elephants occurs in this region, although not on the study site property. Finally, a wide range of other wildlife visit the waterhole, some of which also are transitory (Vyas 2006). In contrast, at AENP the elephants visit the waterholes more regularly. Human-elephant conflict is virtually non-existent and other wildlife is residential. Thus, the younger elephants may have a greater opportunity to experience new odors from conspecifics, potential predators and other organisms at the Ndarakwai waterhole. Interestingly, the likelihood of investigating urine and feces showed very similar patterns across the age and sex classes at the two populations (compare Figs. 7.2a and 7.2b). This pattern was similar whether all chemosensory behaviors or just the main trunk chemosensory behaviors were considered. A comparable pattern was reported for male African elephants at facilities in North American when they were presented with conspecific female urine (Bagley et al. 2006). The main and the accessory trunk chemosensory behaviors showed identical trends in response to luteal and estrous urine. While such results support the contention that African elephants use chemosensory behaviors similarly across their geographic range, until we identify a pheromone in African elephants we will not be able to determine fully the influence of population variation on a specific response.

Our search for an estrous pheromone continues (see Goodwin, Brown, Eggert, Evola, House, Morshedi, Weddell, Chen, Jackson, Aubut, Eggert, Schulte and Rasmussen, this volume). In light of the sexual dimorphism in elephants, we are examining the development of chemosensory behaviors through social play, social interactions and chemosensory exploration (Schulte, Bagley, Correll, Gray, Heineman, Loizi, Malament, Scott, Slade, Stanley, Goodwin and Rasmussen 2005; Meyer 2006; Vyas 2006). Further, at the Tanzanian field site, we are studying elephant-related effects on terrestrial vertebrate biodiversity, the woody habitat and crops (Schulte, Napora, Vyas, Goodwin and Rasmussen 2006). One application of this research may be to use natural chemical signals to reduce human-elephant conflict (Rasmussen and Riddle 2004). Our recent discovery of multiple known insect pheromones in female African elephant urine (Goodwin et al. 2006) has set the stage for future bioassays in North America and in Africa. Identifying such a signal would provide us with a tool to examine further the development of chemosensory behavior and the functional relevance of the signal in two populations of African elephants. The similarities in these two populations of elephants support our hypothesis that they are using chemosensory signals and behaviors for similar purposes.

Acknowledgments We thank our respective colleges and universities for support of our scholarly activities and the University of Chester for hosting the 11th meeting of Chemical Signals in Vertebrates. Funding was provided by the National Science Foundation (B.A.S., L.E.L.R. and T.E.G.: NSF-DBI-02-17062, -16862, -17068, respectively), Biospherics Research Corporation, Georgia Southern University (Academic Excellence Fund, Faculty Research Grant and Graduate Student Research Fund), and Hendrix College. Amy Gray and Maureen Correll provided field assistance in South Africa. We have benefited from research opportunities involving African elephants at Riddle's Elephant & Wildlife Sanctuary and numerous other facilities in North America housing African elephants namely, Baltimore Zoo, Bowmanville Zoo, Cameron Park Zoo, Indianapolis Zoo, Knoxville Zoo, Lion Country Safari, Louisville Zoo, Memphis Zoo, Miami Metro Zoo, Nashville Zoo (R. and C. Pankow), North Carolina Zoo, Riddle's Elephant and Wildlife Sanctuary, Sedgwick County Zoo, Seneca Park Zoo, Six Flags Marine World, Toledo Zoo, West Palm Beach Zoo, and Wildlife Safari Park. We appreciate the assistance of personnel at Addo Elephant National Park, TERU (now ACE) and South Africa National Parks (permit 200(2-6)-12-11 BSCH). We are grateful to Peter & Margot Jones and the staff at Ndarakwai Ranch, and COSTECH and TAWIRI (permit No. 2004-170-N-2004–32) for permission to work in Tanzania. The research met with IACUC approval at Georgia Southern University.

References

Altmann, J. (1974) Observational study of behaviour: sampling methods. Behaviour 49, 227–267.
Archie, E.A., Moss, C.J. and Alberts, S.C. (2006) The ties that bind: genetic relatedness predicts the fission and fusion of social groups in wild African elephants. Proc. Roy. Soc. B 273, 513–522.
Bagley, K.R. (2004) Chemosensory behavior and development of African male elephants (*Loxodonta africana*). M.Sc. thesis, Georgia Southern University.
Bagley, K.R., Goodwin, T.E., Rasmussen, L.E.L. and Schulte, B.A. (2006) Male African elephants (*Loxodonta africana*) can distinguish oestrous status via urinary signals. Anim. Behav. 71, 1439–1445.

Goodwin, T.E., Brown, P., Eggert, M., Evola, M., House, S., Morshedi, G., Weddell, M., Chen, J., Jackson, S.R., Aubut, Y., Eggert, J., Schulte, B.A. and Rasmussen, L.E.L. (2007) Use of automated solid phase dynamic extraction (SPDE)/GC-MS and novel macros in the search for African elephant pheromones. In: J. Hurst, R. Beynon C. Roberts and T. Wyatt (Eds.), *Chemical Signals in Vertebrates 11*. Springer Press, New York, pp. 20–29.

Goodwin, T.E., Eggert, M.S., House, S.J., Weddell, M.E., Schulte, B.S. and Rasmussen, L.E.L. (2006) Insect pheromones and precursors in female African elephant urine. J. Chem. Ecol. 32, 1849–1853.

Lee P.C. (1986) Early social development among African elephant calves. Nat. Geog. Res. 2, 388–401.

Lehner, P.N. (1996) *Handbook of Ethological Methods*. 2nd edition. Cambridge University Press, Cambridge.

Loizi, H. (2004) The development of chemosensory behaviors in African elephants (*Loxodonta africana*) and male responses to female urinary compounds. M.Sc. thesis, Georgia Southern University.

Martin, P. and Bateson, P. (1993) *Measuring Behavoiur: An Introductory Guide*. 2nd edition. Cambridge University Press, Cambridge.

Merte, C. E. (2006) Age effects on social and investigative behaviors in a closed population of African elephants. M.Sc. thesis, Georgia Southern University.

Meyer, J.M. (2006) Sexually dimorphic social development and female intrasexual chemical signaling of African elephants (*Loxodonta africana*). M.Sc. thesis, Georgia Southern University.

Moss, C.J. 1996. Getting to know a population. In: K. Kangwana (Ed.), *Studying Elephants*. African Wildlife Foundation, Nairobi, Kenya, pp. 58–74.

Moss, C.J. and Poole, J.H. (1983) Relationships and social structure of African elephants. In: R. Hinde (Ed.), *Primate Social Relationships: An Integrated Approach*. Blackwell Scientific, Oxford, pp. 315–325.

Paley, R.G.T. and Kerley, G.I.H. (1998) The winter diet of elephants in Eastern Cape Subtropical Thicket, Addo Elephant National Park. Koedoe, 41:37–45.

Poole, J.H. (1999) Signals and assessment in African elephants: evidence from playback experiments. Anim. Behav. 58, 185–193.

Rasmussen, L.E.L. (2001) Source and cyclic release pattern of (Z)-7-dodecenyl acetate, the pre-ovulatory pheromone of the female Asian elephant, Chem. Senses 26, 611–623.

Rasmussen, L.E.L. and Riddle, S.W. (2004) Development and initial testing of pheromone-enhanced mechanical devices for deterring crop raiding elephants: a positive conservation step. J. Eleph. Managers Assoc. 15, 30–37.

Rasmussen, L.E.L., Lee, T.D., Roelofs, W.L., Zhang, A. and Daves, Jr., G.D. (1996) Insect pheromones in elephants. Nature 379, 684.

Rasmussen, L.E.L., Lee, T.D., Zhang, A., Roelofs, W.L. and Daves, Jr., G.D. (1997) Purification, identification, concentration and bioactivity of (Z)-7-dodecen-1-yl acetate: sex pheromone of the female Asian elephant, *Elephas maximus*. Chem. Senses 22, 417–437.

Schulte, B.A. (2006) Elephant behavior. In: M.E. Fowler & S.K. Mikota (Eds.), *The Biology, Medicine and Surgery of Elephants*. Blackwell Publishing, Ames, Iowa, pp. 35–43.

Schulte, B.A. and Rasmussen, L.E.L. (1999) Signal-receiver interplay in the communication of male condition by Asian elephants. Anim. Behav. 57, 1265–1274.

Schulte, B.A., Bagley, K. Correll, M., Gray, A. Heineman, S.M., Loizi, H. Malament, M., Scott, N.L., Slade, B.E., Stanley, L., Goodwin, T.E. and Rasmussen, L.E.L. (2005) Assessing chemical communication in elephants. In: R.T. Mason, M.P. LeMaster, D. Müller-Schwarze (Eds.), *Chemical Signals in Vertebrates 10*. Springer Press, New York, pp. 140–150.

Schulte, B.A., Napora, E., Vyas, D.K., Goodwin, T.E. and Rasmussen, L.E.L. (2006) African elephant chemical communication and humans in Tanzania. J. Eleph. Managers Assoc. 17, 28–36.

Sokal, R.R. and Rohlf, F.J. (1995) *Biometry: The Principles and Practice of Statistics in Biological Research*. W.H. Freeman and Company, New York.

Vyas, D. (2006) Sexually dimorphic developmental patterns of chemosensory behaviors in African elephants (*Loxodonta africana*). M.Sc. thesis, Georgia Southern University.

Whitehouse, A.M. (2001) The Addo elephants: conservation biology of a small, closed population. PhD thesis, University of Port Elizabeth.

Whitehouse, A.M. and Hall-Martin, A.J. (2000) Elephants in Addo Elephant National Park, South Africa: reconstruction of the population's history. Oryx 34, 46–55.

Whitehouse, A.M., Hall-Martin, A.J. and Knight, M.H. (2001) A comparison of methods used to count the elephant population of the Addo Elephant National Park, South Africa. Afr. J. Ecol. 39, 140–145.

Wittemyer, G., Douglas-Hamilton, I. and Getz, W.M. (2005) The socioecology of elephants: analysis of the processes creating multitiered social structures. Anim. Behav. 69, 1357–1371.

Wyatt, T.D. (2003) *Pheromones and Animal Behaviour: Communication by Smell and Taste.* Cambridge University Press, Cambridge.

Chapter 8
Olfactory Communication in the Ringtailed Lemur (*Lemur catta*): Form and Function of Multimodal Signals

Christine M. Drea and Elizabeth S. Scordato

Abstract To better understand the relation between form and function in the complex olfactory communication system of the ringtailed lemur (*Lemur catta*), we integrated observational, experimental, and chemical approaches applied to a population of semi free-ranging animals at the Duke Lemur Center in Durham, North Carolina. Our aim was to examine sex-role reversal in the expression and function of scent marking and unravel the contribution of multimodal components of information transfer, with the unifying framework for all three avenues of our research being that multiplicity of form implies multiplicity of function.

8.1 Introduction

Our research program focuses on the ringtailed lemur (*Lemur catta*)—a predominantly diurnal and terrestrial Malagaszy primate that lives in socially integrated, multimale-multifemale groups. Unusually among mammals, these groups are characterized by unambiguous female dominance over males (Jolly 1966; Kappeler 1990a). Like other strepsirrhines, ringtailed lemurs are macrosmatic, arguably displaying the most complex array of scent-marking behavior within the Order Primate (Schilling 1974). Both sexes have apocrine and sebaceous gland fields in their genital regions, and males boast two additional specialized glands on their wrists and pectoral surfaces, referred to as the antebrachial and brachial organs, respectively (Montagna and Yun 1962). The secretions from these glands (see Fig. 8.1 in Scordato and Drea 2007) are used singly or in combination (e.g. when wrist marking is preceded by shoulder rubbing or when tail fur is impregnated with brachial and antebrachial secretions prior to 'stink fighting': Jolly 1966). The various odorants are deposited using unique displays that produce composite visual and olfactory cues. Wrist marking additionally produces an audible clicking sound as the antebrachial spur scars the substrate being marked. The implication of this broad scent-marking array is that, despite having complex vocal and social

Christine M. Drea
Duke University, Department of Biological Anthropology and Anatomy
cdrea@duke.edu

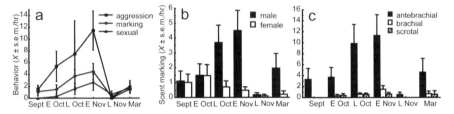

Fig. 8.1 Mean hourly rates of *L. catta* behavior during the pre-breeding (Sep, early and late Oct), breeding (early and late Nov), and post-breeding (Mar) seasons. (a) Rates of scent marking in relation to aggressive and sexual behavior, combined for both sexes. (b) Sex differences in scent marking. (c) Differential usage of male scent glands

repertoires, ringtailed lemurs rely heavily on olfactory communication. Whereas some odor signals (e.g. long-lasting cues deposited at territorial borders or as resource labels) may function equivalently in social and asocial species, others may be more tailored to group living. For instance, ephemeral signals may convey meaning only when used in specific social contexts or in conjunction with visual, behavioral, and auditory cues (Rowe 1999; Candolin 2003). Moreover, in female-dominant species, some functions may be role reversed. We examine these possibilities in three olfactory studies of *L. catta*.

8.2 Observational Study: Within-Group Variation in Scent Marking

8.2.1 Background

Based on the distribution of scent marks and on broad sex, seasonal, and rank-related differences in rates of scent marking, various researchers have proposed that *L. catta* use odor cues to demarcate territories, advertise resource ownership, signal reproductive state, and maintain intrasexual dominance (Jolly 1966; Kappeler 1990b, 1998; Mertl-Milhollen 2006). Here, we elaborate on similar findings, our primary question being whether differentiated gland usage reveals differentiated function. Thus, we additionally examined the relationship between type of scent marking and intra-group aggressive or reproductive behavior, specifically focusing on the sexual, social, and seasonal modulation of the different forms of olfactory signaling.

8.2.2 Methods

Our subjects were the adult members of three mixed-sex social groups that semi free-range in large forested enclosures (3–7 ha). We present data from preliminary analyses of an on-going study, based on ∼100 hours of observation (Sep 2003-Mar 2004). During this period, the groups comprised 25 animals of all ages, but the focal subjects were five males and seven females, aged 6-18 years. We observed each subject year round (weather permitting), during two 20-min focal periods per week. We recorded their behavior in real time, using handheld computers. Our ethogram included the frequency and duration of all forms of social and scent-marking behavior.

In the Northern Hemisphere, the breeding season spans Nov-Feb. Female *L. catta* are polyestrous, cycling up to three times per year at roughly 40-day intervals (Evans and Goy 1968), and show some degree of estrus synchrony (Jolly 1966; Sauther 1991). Thus, the breeding season encompasses three peaks of sexual activity rather than representing one continuous peak; nevertheless, the majority of females conceive in the first cycle (in early-mid Nov: Drea 2007). Our observation period, therefore, encompassed the pre-breeding, peak breeding, and post-breeding seasons.

8.2.3 Results and Discussion

Male and female *L. catta* showed a dramatic increase in aggression ($F_{5,50} = 4.51$, $P < 0.005$), corresponding to seasonal patterns of sexual activity ($F_{5,50} = 2.61$, $P < 0.05$; Fig. 8.1a). Because females have infrequent cycles with narrow windows of receptivity (as little as 24 hrs), male competition for access to females is intense. Similarly, female-initiated aggression escalates concurrently. This period of heightened sexual and aggressive activity is characterized, in both sexes, by increased scent marking ($F_{5,50} = 3.38$, $P = 0.01$; Fig. 8.2a). Consistent with prior suggestions (Kappeler 1998), olfactory signaling is intricately tied to reproductive and competitive behavior.

Gross seasonal patterns in scent marking were nevertheless sexually differentiated, both in frequency, i.e., with males marking at higher rates than females ($F_{1,10} = 7.18$, $P < 0.025$), and in timing ($F_{5,50} = 2.91$, $P < 0.025$), i.e., with peak female behavior (in early Oct) preceding peak male behavior (in early Nov) by about one month (Fig. 8.1b). These different schedules could be functionally significant: Although females may scent mark to assert resource ownership at the onset of a highly competitive period (Mertl-Milhollen 2006), they also may be advertising their impending reproductive state to potential mates, including males transferring between groups. By contrast, resident males may mark during the height of the breeding season to assert their dominance status relative to other males, thereby gaining immediate female preference (through female mate choice) or priority of access to receptive females (through intraspecific competition). Although females often mate multiply (Sauther 1991), dominant males may mate first (Parga 2006). It remains unknown if mating order affects sperm competition or if cryptic mechanisms of female choice may be in operation.

We next asked if males use one type of mark over another to advertise their quality as a competitor or potential mate. Comparing rates of shoulder rubbing, wrist marking, and scrotal marking revealed strong differences ($F_{2,8} = 11.32$,

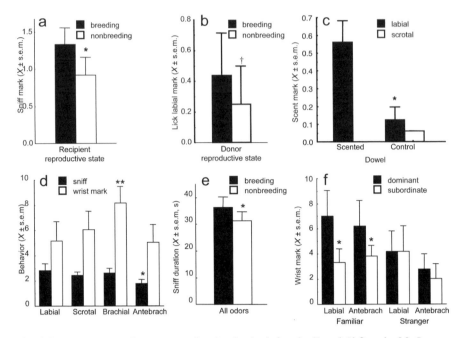

Fig. 8.2 Mean response frequency or duration by (a-c) female, F, and (d-f) male, M, *L. catta* to conspecific glandular secretions. (a) F sniffing all odorants as a function of her reproductive state (breed > non: $F_{1,3} = 28.57$, $P = 0.013^*$). (b) F licking labial odorant as a function of the donors' reproductive state (breed > non: $t_3 = 3.00$, $P = 0.58$, *n.s.*). (c) F frequency and site-specificity of scent marking as a function of odorant type: Fs counter marked the unscented dowel in response to scrotal scent, but over-marked scented dowels in response to labial scent ($t_3 = 3.87$, $P = 0.030^*$). (d) M response as a function of odorant type (antebrachial was sniffed least: $F_{3,9} = 6.75$, $P = 0.011^*$; brachial was wrist marked most: $F_{3,9} = 7.16$, $P = 0.009^{**}$). (e) M sniff duration as a function of donor reproductive state (main effect across dowels: $F_{2,6} = 57.63$, $P = 0.000$; breed > non: $P < 0.05^*$). (f) M wrist marking as a function of donor social status and familiarity: When donors were familiar (expt 1), Ms wrist marked more in response to dominant than subordinate scent ($F_{1,3} = 15.41$, $P = 0.029^*$); when donors were strangers (expt 2), Ms did not distinguish odorants by donor status ($F_{1,4} = 0.09$, $P = 0.775$)

$P = 0.005$), with the males' seasonal pattern being largely attributable to wrist marking ($F_{5,50} = 2.89$, $P < 0.01$; Fig. 8.1c). Scrotal and brachial marking occurred at low levels throughout the period of study. Lastly, although our small sample precluded statistical reliability, dominant males tended to scent mark more frequently than did subordinate males ($F_{1,3} = 6.86$, $P = 0.079$), consistent with prior reports (Jolly 1966; Kappeler 1990b). Thus, it may be that wrist marking, specifically, signals dominance in males – information that would be particularly relevant during the highly competitive breeding season. By contrast, maintaining low levels of genital or brachial marking may be effective in advertising one's continued presence throughout a territory.

8.3 Experimental Studies: Extra-Group Information Transfer

8.3.1 Background

As a complement to observing scent-marking behavior, experimentalists test animals' responses to the controlled presentation of odorants. The habituation-dishabituation paradigm has been used profitably to show that, based on scent alone, lemurs discriminate between the sexes and between unknown individuals (Mertl 1975; Harrington 1976; Dugmore, Bailey and Evans 1984; Palagi and Dapporto 2006). From our prior observations, we suspected that different kinds of scent marks might contain different messages. We also were interested in knowing if the reproductive state of the signal receiver might bear on the efficacy or value of olfactory cues. We used as our behavioral bioassay a discrimination paradigm. In a series of choice trials that modeled extra-group communication (i.e., between familiar non-group members or complete strangers), we manipulated multiple variables (e.g. sex, reproductive state, gland, age, social status, and familiarity) in both the signaler and receiver to better understand the influence of these variables on information transfer.

8.3.2 Methods

In experiment 1, four male and four female *L. catta* (two dominant and subordinate members per sex) served as adult 'recipients' during 'choice' trials in which we presented secretions derived from the various glands (labial, scrotal, brachial, and antebrachial) of 18 adult conspecific 'donors.' Most of these animals were from the same three social groups as before; however, in all cases, we presented the recipients with the odorants of non-group donors. Although housed in separate forested pens, the lemurs were nonetheless familiar with one another. Our design thus modeled the encounters that might occur between animals occupying adjacent territories. By contrast, we ran experiment 2 using five adult males that were unfamiliar with any of the donors. Using cotton swabs (precleaned with methanol and pentane), we collected monthly odorant samples from manually restrained donors. We stored the samples in similarly precleaned vials at $-80\,°C$, and parsed them between these experiments and our subsequent GC-MS study (see section 8.4). Using the GS-MS procedures described later, we verified that storage did not affect sample integrity.

In experiment 1, we tested recipients during their breeding and nonbreeding seasons. In each period, we presented them with two series of four trials, one series each for odorants derived from dominant versus subordinate donors, and one trial each per type of glandular secretion. Within trials, we presented the recipients with a choice between two odorants, both obtained from the same type of gland, from animals of equivalent social status, but from donors in breeding versus nonbreeding seasons. In experiment 2, we tested the recipients only during their nonbreeding season and presented them with two series of four trials, one series each for the donors' breeding versus nonbreeding season, and each of the four trials again representing each of the four glands. Within each trial, the two odorants presented had been obtained from the same type of gland, but from animals that differed in social status. In both experi-ments, we videotaped the 10-min trials, which were later scored by two observers blind to the goals of the study. The behavior of interest included approach frequency, time in proximity, sniff frequency and duration (which we interpreted as investigation of the volatile fraction of odorants), lick frequency (which we interpreted as investigation of the nonvolatile fraction of odorants), and counter- or over-marking (which we interpreted as competitive; for more details, see Scordato et al. (in press).

8.3.3 Results and Discussion

The presentation of different odorants produced varying levels of investigation or responsiveness in female and male recipients. The interest of female recipients was directed primarily to the scent of other females, and varied with their own reproductive state (Fig. 8.2a) and that of the female donors (Fig. 8.2b). Thus, mainly during the breeding season, females sniffed (Fig. 8.2a), licked (Fig. 8.2b), and countermarked (Fig. 8.2c) odorants derived from other reproductive females. From these findings, we conclude that female scent marking functions primarily in reproductive advertisement and intra-sexual competition, and that male scent marks are minimally functional in female assessment of mates. By contrast, male recipients were interested in all secretions, although their responses varied by odorant type (Fig. 8.2d). Males appeared to use odorants to monitor the reproductive state of both sexes (Fig. 8.2e). Thus, as is the case for females, male scent marking appears to function in intrasexual competition, but unlike the case for females, males may derive olfactory information relevant to mate assessment. Lastly, we found that males discriminated between the odorants of dominant versus subordinate donors if the donors were familiar (expt 1), but not if they were strangers (expt 2; Fig. 8.2f). Thus, we suspect that individual identity, but not dominance status, may be encoded within the chemical matrix of scent secretions and that the pairing of social and olfactory signals requires associative learning.

8.4 Chemical Study: Information Content in Glandular Secretions

8.4.1 Background

Behavioral bioassays are inextricably linked with chemical studies to decipher the information content of olfactory signals (Albone 1984). As a complement to the experimental approach described above, several research groups have applied chemical approaches, particularly gas chromatography and mass spectrometry (GC-MS), to the analysis of strepsirrhine glandular secretions. These researchers have reported differences in the semiochemistry of various glandular secretions (Hayes, Morelli and Wright 2004, 2005), as well as seasonal (Hayes et al. 2006) and individual differences (Palagi and Dapporto 2006). Likewise, we have used GC-MS analysis of scent composition to test for chemical differences between all types of glandular secretions in *L. catta*. We asked if the chemical composition of these secretions varied by gland, season (i.e., reproductive state), individual, and/or the dominance status of the signaler. From the responses we had observed in our behavioral bioassays, we predicted that differentiated chemical patterns would emerge for each gland, as well as across seasons and between individuals; however, we did not expect that social status would be encoded within the chemical matrix of scent secretions.

8.4.2 Methods

We obtained secretions ($\sim$ 25 samples per gland type) from 14 adult *L. catta* (seven per sex) during the prebreeding, breeding, and nonbreeding seasons. We adapted the solvent extraction technique (using methyl-*tert*-butyl ether) described by Safi and Kerth (2003) and ran our samples using an HP 5890 series II GC, fitted with a double-focusing JEOL JMS-SX 102A high-resolution MS and an HP-5MS fused silica (5% phenyl)-methylpolysiloxane column (30 m × 0.25 mm × 0.25 μm, Agilent). The injector temperature was set at 280 °C, the ion source at 190 °C, and helium was the carrier gas. We used the following temperature protocol (after 3-min solvent delay): 80–180 °C at a rate of 20 °C/min (held 2 min at 80 °C); 180–320 °C at 7 °C/min (held 5 min at 320 °C). The entire run lasted 32 min; no compounds eluted after 27 min. Our standards included 15μl each of squalene, farnesol, farnesal, and seven alkenes (C12-C18) in 50 ml of solvent. Using both retention time and mass spectra, we used the NIST 2002 Mass Spectral Library to tentatively identify peaks. We used integration software to calcualate the relative area of each component and principal component (PC) analysis to reduce the dimensionality of the data. We then performed linear discriminate analysis (LDA) on PCs with eigenvalues > 1, using gland, season, animal identity or social status as the independent variable. We performed these analyses in JEOL MS analysis software and JMP 6, and report Wilkes' lamda as our test of group differences. We present the preliminary results from these analyses. For methodological details and results on a larger data set, see Scordato et al. (in press).

8.4.3 Results and Discussion

Antebrachial secretions were comprised primarily of low-molecular weight compounds. Because of the high volatility of these secretions, we could not produce reliable chromatograms using our routine extraction procedure. We are currently analyzing these secretions using solid phase dynamic extraction techniques (in col-

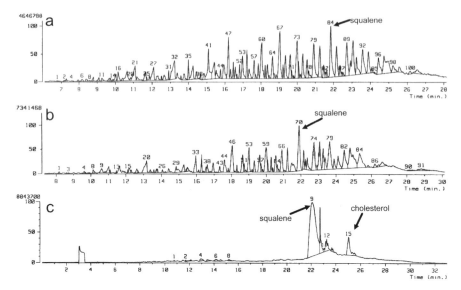

Fig. 8.3 Gas chromatograms of *L. catta* (a) labial, (b) scrotal, and (c) brachial secretions

laboration with Thomas Goodwin, Hendrix College). The remaining three types of odorant for which we obtained reliable results had 102 compounds: 57 for labial, 52 for scrotal, and 39 for brachial secretions. From the mass spectra, it can be seen that labial (Fig. 8.3a) and scrotal (Fig. 8.3b) secretions are composed primarily of organic acid esters, whereas brachial secretions (Fig. 8.3c) are composed predominantly of squalene ($C_{30}H_{50}$, $MW = 410.7$) and cholesterol ($C_{27}H_{46}O$, $MW = 386.7$).

The chemical composition of the various secretions provides further clues as to their differentiated functions: Genital and brachial secretions contain greasy, high-molecular weight compounds, such as squalene, which is a recognized fixative that may increase signal longevity (Alberts 1992). By contrast, antebrachial secretions are low molecular weight compounds, deposited primarily during social displays. Such ephemeral signals may be deployed primarily for within-group communication because they require integration of multiple sensory modalities for full efficacy of signal transmission. If, however, the volatile antebrachial secretions are combined with the squalene-based brachial secretions, as is the case for a percentage of wrist marking and during tail marking, they may produce a more durable signal that could function long after the signal sender is gone.

Despite similarity between the secretions of the two genital glands, differences between the chemical profiles of all three types of secretions emerged in the LDA (Fig. 8.4a). Likewise, reliable seasonal variation appeared in the chemical profiles derived from all three sources (labial: Fig. 8.4b; scrotal: Wilks' lambda $= 0.018$, $P < 0.01$; brachial: Wilks' lambda $= 0.136$, $P < 0.05$). We also found the antic-ipated individual-specific signatures in scent secretions derived from the genital glands (labial: Wilks' lambda $= 0.000$, $P < 0.01$; scrotal: Fig. 8.4c), but not from

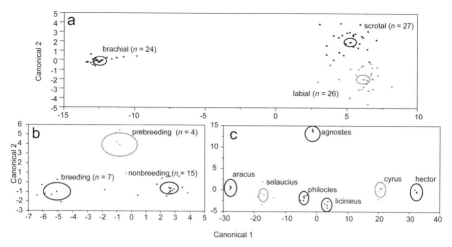

Fig. 8.4 Discriminant analyses of the principal chemical components in *L. catta* scent secretions by (a) gland, (b) season, and (c) individual. (a) Accurate classification of 97.5% of labial, scrotal, and brachial samples ($n = 77$) by gland of origin (Wilks' lambda $= 0.003$; $P < 0.001$). (b) Reliable differentiation of 100% of labial samples ($n = 26$) into prebreeding, breeding, and nonbreeding seasons (Wilks' lambda $= 0.018$, $P < 0.01$). (c) Individual 'scent signatures' in the scrotal secretions from seven males. LDA performed on 17 principal components correctly classified 100% of these samples to the individuals from which they were collected (Wilks' lambda $= 0.000$, $P < 0.002$)

the brachial gland (Wilks' lamba $= 0.013$, $P > 0.10$). The latter finding contrasts with a prior report (Palagi and Dapporto 2006) and may reflect differences between the sampling regimens or analytical procedures used in the two studies (Scordato et al. 2007). Lastly, we found no evidence that social status reliably predicted the chemical profiles of any of the glandular secretions (data not shown). Coupling these findings, with our previous bioassay results, we have shown that lemurs can use olfactory cues to recognize individual conspecifics and modulate their response to a signal based on a combination of factors, including their own physiological state, the physiological state of the signal sender, and their prior experience with that animal.

8.5 Conclusion

Through the information revealed in a series of integrated observational, experimental, and chemical studies of olfactory communication, we suggest that scent marking in the ringtailed lemur produces composite olfactory, visual, and sometimes auditory signals that contain a complex combination of ephemeral and long-lasting cues and serve differentiated functions. Consistent with prior suggestions (Kappeler 1998; Mertl-Milhollen 2006), one primary function of scent marking, in *both* sexes, involves intrasexual competition – to advertise resource ownership, assert status, and maintain intrasexual dominance hierarchies; another involves mediating reproductive behavior. Whereas female interest in, and countermarking of, scent marks is most prominently directed toward labial marks, particularly during the breeding season, male scent generates little female interest or competitive response. Thus, within this female-dominated society, males appear to pose minimal threat to females, whereas females closely monitor one another. Unlike for other female mammals, female mate choice in lemurs may not be heavily influenced by male olfactory cues.

Males, on the other hand, show great interest in the odorants of both sexes, even though they respond differently by odorant type, e.g. predominantly licking labial secretions, but primarily wrist marking in response to male scent. The style of response may reflect the information conveyed in the scent mark: males may lick a female's mark to assess her reproductive state, whereas countermarking a male's mark may reflect intrasexual competition. Based on differential response, we suggest that the various male glands serve specialized functions, some of which may require social experience for effective transmission. We propose that the stable content of scrotal marks lay claim to resources, the act of antebrachial marking displays dominance to onlookers, and brachial compounds function as fixatives of highly volatile chemicals, prolonging the otherwise ephemeral signals in antebrachial marks.

As we suspected, multiplicity of form reflects multiplicity of function; however, the diversity of form also encompasses diverse mechanisms of transmission. For instance, some information, such as dominance status, is conveyed solely by the observable (and sometimes audible) behavioral component of wrist marking and, therefore, may function primarily for instantaneous intra-group communication. Because dominance hierarchies are relational, the communication of rank is most relevant to other group members. Moreover, because the hierarchies of male *L. catta* can be transient, long-lasting signals of dominance could become irrelevant over time. Both of these features are reflected in the rank-differentiated usage and ephemeral chemistry of wrist marks. Nevertheless, if a longer-lasting signal of male competence is required, antebrachial secretions might gain longevity by being mixed with the fixative components of brachial secretions. Such is the case during a subset of wrist marking, but also during tail marking and stink fighting, which occur most commonly during aggressive, inter-group encounters.

Some information (e.g. reproductive state) appears to be contained solely within the chemical matrix of certain marks, particularly within the stable genital secretions. Thus, longer-lasting signals may be broadcast to any animal that comes in contact with labial or scrotal marks. Such a scenario seems particularly applicable to females that scent mark most frequently prior to the onset of estrus cycles, but nonetheless when their sex steroids are on the rise (Drea 2007). Such advertisement may encourage male immigration at a time that would maximize the opportunity of female mate choice, even if the mechanism of, or criteria for, selection remain obscure.

Lastly, some information (e.g. gland of origin and individual identity) is duplicated in the behavior and chemistry of scent marking, suggesting that transmission of certain information targets both group and non-group members. Communicating individual identity would facilitate claiming ownership of resources, and it would benefit the signaler to broadcast such information in a long-lasting cue. It should come as no surprise, especially with respect to a socially complex primate, that effective transmission of information, particularly when conveyed through multiple channels, requires some degree of associative learning. The integration of multi-modal information is evidenced, for instance, by the pairing of unique and presumably permanent 'scent signatures' with more transient observable qualities, such as behavioral displays of dominance status. Further studies promise to shed additional light on the differentiated functions and multimodal mechanisms of lemur scent marking.

Acknowledgments We are indebted to Jim Camden, Sarah Cork, Courtney Fitzpatrick, Julie Rushmore, and Anne Starling for their assistance with data collection. We are grateful to George Dubay for training and advice on GC-MS. The following staff members of the Duke Lemur Center are gratefully acknowledged for all aspects of animal care and handling: David Brewer; Stephanie Combes; Kelly Glenn; Jennifer Hurley, DVM; Julie Ives; Bobbie Schopler, DVM; Cindy Stoll; Julie Taylor; Cathy Williams, DVM.

These studies were funded by NSF grant BCS-0409367 and the Duke University Arts & Science Research Council (to C.M.D.), and by the Molly Glander Award, Howard Hughes Biology Forum, and Duke Undergraduate Research Support (to E.S.S.). All protocols were approved by the Duke University Institutional Animal Care and Use Committee (A245-03-07). This is DLC publication 1018.

References

Alberts, A.C. (1992) Constraints on the design of chemical communication systems in terrestrial vertebrates. Am. Nat. 139 supp, s62–s89.

Albone, E.S. (1984) *Mammalian Semiochemistry*. Wiley and Sons, New York.

Candolin, U. (2003) The use of multiple cues in mate choice. Biol. Rev. 78, 575–595.

Drea, C.M. (2007) Sex differences and seasonal patterns in steroid secretion in *Lemur catta*: Are socially dominant females hormonally 'masculinized'? Horm. Behav. In press.

Dugmore, S.J., Bailey, K. and Evans, C.S. (1984) Discrimination by male ring-tailed lemurs (*Lemur catta*) between the scent marks of male and those of female conspecificis. Int. J. Primatol. 5, 235–245.

Evans, C.S. and Goy, R.W. (1968) Social behaviour and reproductive cycles in captive ringtailed lemurs (*Lemur catta*). J. Zool, Lond. 156, 181–197.

Harrington, J.E. (1976) Discrimination between individuals by scent in *Lemur fulvus*. Anim. Behav. 24, 207–212.

Hayes, R.A., Morelli, T.L. and Wright, P.C. (2004) Anogenital gland secretions of *Lemur catta* and *Propithecus verreauxi coquereli*: A preliminary chemical examination. Am. J. Primatol. 63, 49–62.

Hayes, R.A., Morelli, T.L. and Wright, P.C. (2005) The chemistry of scent marking in two lemurs: *Lemur catta* and *Propithecus verreauxi coquereli*. In: R.T. Mason, M.P. LeMaster and D. Müller-Schwarze, (Eds.), *Chemical Signals in Vertebrates, X*. Springer Press, New York, pp. 159–167.

Hayes, R.A., Morelli, T.L. and Wright, P.C. (2006) Volatile components of lemur scent secretions vary throughout the year. Am. J. Primatol. 68, 1–6.

Jolly, A. (1966) *Lemur Behavior: A Madagascar Field Study*. University of Chicago Press, Chicago.

Kappeler, P.M. (1990a) Female dominance in *Lemur catta*: more than just female feeding priority?

Folia Primatol. 55, 92–95.

Kappeler, P.M. (1990b) Social status and scent-marking behaviour in *Lemur catta*. Anim. Behav. 40, 774–776.

Kappeler, P.M. (1998) To whom it may concern: the transmission and function of chemical signals in *Lemur catta*. Behav. Ecol. Sociobiol. 42, 411–421.

Mertl, A.S. (1975) Discrimination of individuals by scent in a primate. Behav. Biol. 14, 505–509.

Mertl-Milhollen, A.S. (2006) Scent marking as resource defense by female *Lemur catta*. Am. J. Primatol., 68, 605–621.

Montagna, W. and Yun, J.S. (1962) The skin of primates. X. The skin of the ringtailed lemur (*Lemur catta*). Am. J. Phys. Anthropol. 20, 95–117.

Palagi, E. and Dapporto, L. (2006) Beyond odor discrimination: demonstrating individual recognition by scent in *Lemur catta*. Chem. Senses 31, 437–443.

Parga, J.A. (2006) Mulitiple mating and female mate choice in *Lemur catta*: does it pay to be a dominant male? Am. J. Phys. Anthropol. 42 supp, 144–144.

Rowe, C. (1999) Receiver psychology and the evolution of multicomponent signals. Anim. Behav. 58, 921–931.

Safi, K. and Kerth, G. (2003) Secretions of the interaural gland contain information about individuality and colony membership in the Bechstein's bat. Anim. Behav. 65, 363–369.

Sauther, M. (1991) Reproductive behavior of free-ranging *Lemur catta* at Beza Mahafaly special reserve, Madagascar. Am. J. Phys. Anthropol. 84, 463–477.

Schilling, A. (1974) A study of marking behaviour in *Lemur catta*. In: R.D. Martin, G.A. Doyle and A.C. Walker (Eds.), *Prosimian Biology*. University of Pittsburgh Press, Pittsburgh, pp. 347–362.

Scordato, E.S. and Drea, C.M. (2007) Scents and sensibility: information content of olfactory signals in the ringtailed lemur (*Lemur catta*). Anim. Behav. 73, 301–314.

Scordato, E.S., Dubay, G. and Drea, C.M. (2007) Chemical composition of scent marks in the ringtailed lemur, *Lemur catta*: glandular differences, individual signatures, and seasonal variation. Chem. Senses. In press.

Chapter 9
Olfaction in the Gorilla

Peter Hepper, Deborah Wells, Patrick McArdle, Dwyer Coleman, and Mark Challis

Abstract It has been argued that the great apes have poor olfactory capabilities and that olfactory stimuli convey little useful information for them. Thus odours and scents have little functional significance for the great apes in guiding their behaviour. Part of this belief has been based on the observation that brain volume in those areas dealing with olfactory stimuli relative to other brain areas is greatly reduced in these apes. Moreover, naturalistic observations report little obvious olfactory guided behaviour. We have initiated a programme of work (Project SOAP) investigating the olfactory abilities, and the functions of smell, in the great apes. Here we report an initial study examining whether the western lowland gorilla (*Gorilla gorilla gorilla*) is able to detect and discriminate between different odours presented on cloths. The study revealed that gorillas can both detect the presence of an odour and discriminate between different odours. An incidental observation demonstrated one trial olfactory guided taste aversion learning which persisted for over 7 weeks. We conclude that gorillas have a functioning olfactory sense that they use in the investigation of their environment, and that olfaction may not be as irrelevant in great apes as has been suggested.

9.1 Introduction

There has been little study of the olfactory abilities of the great apes. Olfactory stimuli play an important role in guiding behaviour in many species (Stoddart 1980). In the great apes, however, the role of olfaction has been questioned due to the perceived primacy of vision and audition in influencing behaviour (King and Forbes 1974; Dominy, Ross and Smith 2004). This has led to the label of microsmatic being applied to the apes. For most vertebrates, e.g. the dog, there is much evidence of the importance of olfaction in behaviour, and these species are termed macrosmatic (Smith and Bhatnagar 2004).

Peter Hepper
Queen's University of Belfast, School of Psychology
p.hepper@qub.ac.uk

The terms micro- and macro-smatic are not only restricted in their use to the role of olfaction in behaviour, but also refer to the anatomical structures involved in olfaction. For example, the relative size of brain areas dealing with olfactory stimuli when compared to overall brain size is greatly decreased in microsmatic species compared to macrosmatic species (0.01% in man, 0.07% in great apes, cf. prosimians 1.75% and insectivores 8.88%, Stephan, Bauchot and Andy 1970). This has led to the suggestion of poorer olfactory abilities in those animals with reduced relative size of brain structures associated with olfaction. However, whilst there is anatomical evidence of a reduction in relative size of olfactory structures in apes, it is unclear whether this is a good indicator of olfactory function and there has been little experimental study to examine whether this is the case (Smith and Bhatnagar 2004).

Further suggestions of poorer olfactory abilities in great apes have been derived from studies mapping olfactory receptor genes. These studies have revealed a progressive increase in the number of olfactory receptor psuedogenes in apes, including humans, resulting in a decrease in the number of functional olfactory receptor genes. Estimates indicate that humans have approximately 350 functional olfactory receptor genes, old world primates 700 and new world primates 1000 (e.g. Gilad, Wiebe, Przeworski, Lancet and Pääbo 2004). This has also lead to a view that olfactory abilities will be consequently reduced in species with fewer olfactory receptor genes.

Much of the argument regarding the olfactory abilities of great apes has taken place in the absence of experimental evidence. However, studies on humans have challenged this. Whilst humans may have poorer acuity than macrosmatic species, they have good discriminatory abilities and excellent abilities to detect changes in concentration (e.g. Stoddart 1980). More recently, experimental techniques have been derived to examine olfactory abilities in monkeys and have revealed that these supposedly microsmatic primates have good olfactory abilities (e.g. squirrel monkeys: Laska, Seibt and Weber 2000; spider monkeys: Laska, Salazar and Luna 2003; pigtailed macaques: Laska, Wieser and Salazar 2005).

This raises the question of olfactory abilities in the great apes. We have found no experimental studies of olfactory abilities in the great apes other than two abstracts. Oeda, Ueno, Hasegawa and Tomonaga (2002) report a study which found that two infant chimps preferred strawberry and lavender odour and exhibited an aversion to pyridine. An increase in chimpanzee activity was reported when peppermint odour was added into the chimps' enclosure by diffuser (Struthers and Campbell 1996). Naturalistic studies of great ape behaviour have observed apes touching an object with their fingers and then bringing these fingers to the nose (orangutans: Rijksen 1978; chimpanzees: Blackman 1947). Sniffing the genital region may be involved in pre-mating behaviour in gorillas (Dixson 1981). However reports of olfactory behaviour do not feature highly amongst naturalistic observations of the great apes. There are reports of silverback gorillas emitting a pungent odour, especially when stressed (Cousins 1990; Schaller 1963). Examination of the skin of great apes reveals the presence of apocrine glands and axillary organ (e.g. gorilla: Ellis and Montagna 1962; chimpanzee: Montagna and Yun 1963), both involved in scent production (Montagna and Parakkal 1974; Stoddart 1991). Thus, there may be a role for scent in the social behaviour of great apes.

9.2 Project SOAP

On reviewing the literature it appeared to us that the view that olfactory stimuli play little role in the behaviour of the great apes is based on assumptions from theoretical arguments rather than on experimental evidence that has examined olfactory guided behaviour. We thus decided to establish a research programme examining Scent and Olfaction in Apes (project SOAP). This chapter presents some initial research examining olfaction in the gorilla. The aim of this study could be considered somewhat basic, as it simply sought to determine whether gorillas respond to olfactory stimuli and possess the ability to discriminate between different odours. However, given the dearth of studies in great apes, and the apparent lack of any study examining olfaction in the gorilla, it was important to establish whether the gorilla possessed basic olfactory abilities.

9.3 Method

9.3.1 Subjects

Six western lowland gorillas (*Gorilla gorilla gorilla*) housed at the Belfast Zoological Gardens in Co. Antrim, Northern Ireland, were studied. Information regarding the sex, age and origins of the subjects, is provided in Table 9.1. The group had existed in its present form for over two years and all social relationships were well established and stable.

All of the gorillas were housed together in an exhibit consisting of an outdoor arena (60 m long × 40 m wide) containing grass and climbing apparatus, and a large indoor den (20 m long × 12 m wide × 7 m high). The latter consisted of a straw-covered concrete floor and a complex climbing apparatus constructed of logs. The gorillas' individual sleeping quarters were at the rear of the indoor den. The animals had free access between the indoor and outdoor enclosures during the day, although were confined to the indoor area during the night (5.00 pm-8.00 am). The indoor den was cleaned thoroughly every morning. During this time the gorillas were locked out in the outdoor enclosure. Testing took place in the indoor enclosure and commenced following cleaning. The animals were fed once a day, following testing, with a variety of fruit and vegetables. Visitors to the zoo were able to view the gorillas' exhibit

Table 9.1 Description of subjects observed in this study

Name	Sex	Age (in years)	Origins
Jeremiah	M	14	Captive-born
Kukuma	M	10	Captive-born
Gugas	M	4	Wild born
Delilah	F	35	Wild born
Kamili	F	12	Captive born
Bakira	F	3	Captive born

between 10 am - 5 pm every day. A glass barrier (7 m high × 12 m wide) separated the visitors from the gorillas in the indoor enclosure, whilst a concrete wall served as the divide in the outdoor enclosure.

9.3.2 Olfactory Stimuli

The stimuli used as odours in this study were: almond food flavouring (Supercook, Leeds, England); cK1 unisex perfume (PVH Corp., New York, USA); and distilled water. The odours were introduced into the gorillas' environment on sterilised white flannel towels (100% natural cotton, 450 gsm, 70 cm × 140 cm).

9.3.3 Procedure

The towels were impregnated with one odour 60 min before their introduction into the animals' enclosure by spraying the stimulus over the entire surface of the cloths (2 ml almond, 3 ml cK1). Both scented cloths (almond, cK1) smelled equally strong to human subjects. No odour, other than that of the cloth, was detectable from the towel sprayed with distilled water (3 ml). Following odour impregnation, each cloth was sealed in a plastic bag and handled with plastic gloves worn by the experimenters.

On any particular trial gorillas were only presented with cloths bearing the same odour, e.g. almond. Overall, the gorillas were given three presentations of each odour, nine trials in all. Trials were divided into three blocks, such that in each block each odour was presented. Within each block the order of presentation of the three odours was randomised. Eight cloths were used for each trial. Pilot studies had indicated providing more objects than the number of subjects ensured all the gorillas were able to access a cloth and overcame the potential problem of hording. The cloths were used only for a single trial and then discarded. Each trial was separated by a period of seven days.

The gorillas were locked in the outside enclosure to enable cleaning of the indoor den where testing took place. About 5 min before the door was opened to re-admit the gorillas to the den, eight cloths were randomly distributed around the enclosure. Once the door was opened, the gorillas entered their den within 60 s. Each trial started once the door was opened and lasted for 60 min. The gorillas' response to the cloths was noted. Three behaviours were recorded: Hold (the cloth was touched or held in one or both hands); Sniff (the cloth was brought within approximately 10 cm of the animal's nose, or the animal placed its nose within 10 cm of the cloth); and, Lick/Chew (the cloth was placed in the animal's mouth or licked). Each time a gorilla engaged in one of the above behaviours it was noted. The number of occurrences of each behaviour was summed across the group for each trial.

9.3.4 Analysis

The results were analysed by a within-subjects analysis of variance for factors of trial (1,2,3) and odour type (almond, cK1, water). The data were normally distributed (Shapiro-Wilks tests). Planned contrasts compared each of the two odours with the control odour. Only significant results are reported. The results for each behaviour were analysed separately. Four of the trials were videotaped to enable intra-and inter-rater reliability to be assessed for the observations. There was 100% concordance for both intra- and inter-rater codings.

9.3.5 Results

Hold : There was a significant effect of odour type ($F_{2,10} = 17.57, p < 0.01$) on the number of times the cloth was held (see Fig. 1a). Planned contrasts revealed the gorillas held the almond scented cloth significantly more than the cK1 scented cloth ($p = 0.008$) and the unscented (water) cloth ($p = 0.005$). There was no difference between the time holding cK1 scented and control cloth

    Sniff: Odour type had a significant effect ($F_{2,10} = 4.73, p < 0.05$) on the number of times the cloth was sniffed by the gorillas (see Fig. 1b). The gorillas sniffed the almond scented cloth significantly more than the unscented (water) cloth ($p = 0.048$). There was no difference between the time sniffing cK1 scented and control cloth.

    Chew/Lick: There was a significant effect of odour type ($F_{2,10} = 8.97, p < 0.01$) on the number of times the cloth was chewed / licked (see Fig. 1c). There was also a significant effect of trial ($F_{2,10} = 30.46, p < 0.001$). This can be explained by the gorillas' response to the cK1 scented cloth. Following an initial 'chew/lick' on the first presentation (trial 1), no gorillas subsequently 'tasted' the cK1 scented cloth. For the almond and water scented cloth the gorillas continued to lick/chew them on their second and third presentation following an initial sampling on trial 1. There were significant differences between the amount of chew/lick exhibited between the almond and control cloths ($p = 0.043$) and cK1 and control cloth ($p = 0.038$).

9.4 Discussion

The results of these studies indicate that the western lowland gorilla responds to olfactory stimuli. The gorillas in this study held and sniffed the almond scented cloths significantly more than unscented and cK1 scented cloths. Further, the gorillas differentially chewed/licked the almond and cK1 scented cloths, suggesting an ability to discriminate between the two scents. Together, the results indicate that the gorilla has a functioning olfactory sense able to detect, and discriminate, some odours.

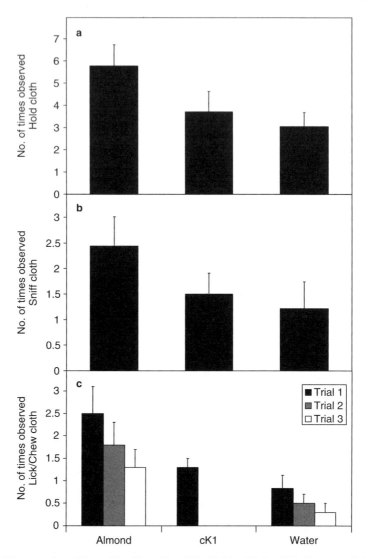

Fig. 9.1 Mean number of times (± s.d.) gorillas (a) held, (b) sniffed and (c) chewed or licked each of the scented cloths across all three trials (a,b) or for each trial (c)

One interesting observation was the decrease in chewing or licking the cloth scented with the perfume. Gorillas licked or chewed this once and after this did not engage in this behaviour. This is suggestive of one trial aversion learning. The gorillas retained a memory of this for at least 14 days, the time between trial one and two, and for a further 35 days, the time between trial 2 and trial 3. Presumably the gorillas had associated the smell of the cloth with the taste and 'knew' not to lick/chew this cloth based on its smell. This indicates a functioning sense of taste in the gorilla and a long term memory for disliked or aversive gustatory stimuli.

Observations of the animals' behaviour indicated that the animals displayed interest in the cloths on their initial discovery, but this interest waned with time and, for the last 40 min of an observation period, little attention was paid to the cloths. This is in contrast to behaviour observed in other studies examining dogs where novel olfactory stimuli maintained the animal's interest for longer (Graham, Wells and Hepper 2005). This supports the notion that olfactory stimuli may not be of primary interest to the apes. However, although one scent (almond) was food related, both were novel and probably meaningless to the gorillas. It would be inappropriate to extrapolate from this study to the role of olfactory stimuli in gorilla behaviour, in particular for potentially more ecologically or socially valid stimuli.

In conclusion, gorillas have a functioning olfactory sense that they use in the investigation of their environment. The gorilla's olfactory system is capable of detecting and discriminating olfactory stimuli and also of being associated with aversive gustatory stimuli. The study also demonstrated a functioning memory for a one- trial version learning task of over seven weeks. Project SOAP is continuing its research (e.g. Wells, Hepper, Coleman and Challis 2006) but already seems to be indicating that olfaction is not as irrelevant in great apes as some have suggested.

References

Blackman, T.M. (1947) Observations on sense of smell in chimpanzees. Am. J. Phys. Anthrop. 5, 283–293.

Cousins, D. (1990) *The Magnificent Gorilla*. Book Guild, Sussex.

Dixson, A.F. (1981) *The Natural History of the Gorilla*. Weidenfeld and Nicolson, London.

Dominy, N.J., Ross, C.F. and Smith, T.D. (2004) Evolution of special senses in primates: past, present and future. Anat. Rec., 281A, 1078–1082.

Ellis, R.A. and Montagna, W. (1962) The skin of primates. VI. The skin of the gorilla (*Gorilla gorilla*). Am J. Phys. Anthrop 20, 79–94, 189–203.

Gilad, Y., Wiebe, V., Przeworski, M., Lancet, D and Pääbo, S. (2004) Loss of olfactory receptor genes coincides with the acquistion of full trichromatic vision in primates. PloS Biology, 2, 120–125.

Graham, L., Wells, D.L. and Hepper, P.G. (2005) The influence of olfactory stimulation on the behaviour of dogs housed in a rescue shelter. App. Anim. Behav. Sci. 91, 143–153.

King, J.E. and Forbes, J.L. (1974) Evolutionary changes in primate sensory capacities. J. Human Evol., 3, 435–443.

Laska, M., Salazar, L.T.H. and Luna, E.R. (2003) Succesful acquistion of an olfactory discrimination paradigm by spider monkeys, *Ateles geoffroyi*. Physiology & Behavior 78, 321–329.

Laska, M., Seibt, A. and Weber, A. (2000) 'Microsomatic' primates revisited: olfactory sensitivity in the squirrel monkey. Chem. Sens. 25, 47–53.

Laska, M., Wieser, A. and Salazar, L.T.H. (2005) Olfactory responsiveness to two odorous steroids in three species of nonhuman primates. Chem. Sens. 30, 505–511.

Montagna, W. and Parakkal, P.F. (1974) *The Structure and Function of Skin*. Academic Press, New York.

Montagna, W. and Yun, J.S. (1963) The skin of primates. XV. The skin of the chimpanzee (*Pan satyrus*). Am. J. Phys. Anthrop. 21, 189–203.

Oeda, R., Ueno, Y., Hasegawa, T. and Tomonaga M. (2002) The responses to odors of infant chimpanzees. Anthrop. Sci. 110, 140.

Rijksen, H.D. (1978) A field study on sumatran Orang Utans (*Pongo pygmaeus Abelii lesson* 1827) H. Veenman and Zonen B.V., Wageningen.

Schaller, G.B. (1963) *The Mountain Gorilla*. University of Chicago Press, Chicago.

Stephan, H., Bauchot, R. and Andy, O.J. (1970) Data on size of the brain and of various brain parts in insectivores and primates. In: C. Noback and W. Montagna (Eds.), *The Primate Brain*, Appleton century Crofts, Newe York, pp. 289–297.

Smith, T.D. and Bhatnagar, K.P. (2004) Microsomatic primates: reconsidering how and when size matters. Anat. Rec. 279B, 24–31.

Stephan, H., Bauchot, R. and Andy, O.J. (1970) Data on size of the brain and of various brain parts in insectivores and primates. In: C. Noback and W. Montagna (Eds.), *The Primate Brain*, Appleton century Crofts, Newe York, pp. 289–297.

Stoddart, D.M. (1980) *The Ecology of Vertebrate Olfaction*. Chapman Hall, London.

Stoddart, D.M. (1991) *The Scented Ape. The biology and culture of human odour.* Cambridge University Press, Cambridge.

Struthers E.J. and Campbell, J. (1996) Scent specific behavioural responses to olfactory enrichment in captive chimpanzees (*Pan troglodytes*). XVI Congress International Primatological Society, Madison, WI. Abstracts p. 762.

Wells, D.L., Hepper, P.G., Coleman, D. and Challis, M. (2006) A note of the effect of olfactory stimulation on the behaviour and welfare of zoo-housed gorillas. App. Anim. Behav. Sci. doi:10.1016/j.applanim.2006.07.010

Chapter 10
Ecological Validity in the Study of Human Pheromones

Tamsin K. Saxton, Anthony C. Little and S. Craig Roberts

Abstract Several constituents of human axillary secretions have been proposed as candidate human pheromones, but their influence on human behaviour remains controversial. Here we briefly review the literature on the behavioural effects of candidate compounds, noting that inconsistencies in findings could be due in part to the variation in experimental context and potential lack of ecological validity. We also report results of a pilot study which attempts to overcome these limitations in an ecologically-valid experimental paradigm: a speed-dating event. We tested the effects of 4,16-androstadien-3-one within a single speed-dating evening with 25 female and 22 male participants. We found a significant effect of androstadienone on female judgments of male attractiveness, which is consistent with the proposal that androstadienone could act as a modulatory pheromone in humans.

10.1 The Case for "Human Pheromones"

Pheromones have been defined as substances released by an individual into the external environment which precipitate a particular reaction in a conspecific (Karlson and Lüscher 1959). Pheromones are used by species in a variety of phyla (see e.g. McClintock, Jacob, Zelano and Hayreh 2001), and there exist many examples of pheromone-mediated behaviour in a wide range of mammals, particularly in relation to mating behaviour and maturation (see e.g. Vandenbergh 1983). In humans however, the question of whether pheromones influence behaviour was recently listed by Science magazine as one of the top 100 outstanding questions (Anon 2005). A recent review of behavioural and anatomical studies relating to the function of pheromones in human interactions concluded that while a small number were "unambiguously supportive", none seemed ultimately conclusive (Hays 2003).

One hurdle for proponents of human pheromones is the lack of clear evidence for a functional human vomeronasal organ (VNO). Located within the nasal cavity,

Tamsin K. Saxton
University of Liverpool, School of Biological Sciences
tamsin.saxton@liverpool.ac.uk

the VNO is used by many terrestrial vertebrates to detect pheromones (Johns, Feder, Komisaruk and Mayer 1978; Keverne 1983). Although one study (Grosser, Monti-Bloch, Jennings-White and Berliner 2000) reported measurable autonomic changes following direct introduction of a proposed human pheromone to the VNO, such that this chemical was prevented from reaching the main olfactory epithelium, there is increasing evidence that the system is vestigial in humans. The VNO is not consistently identified in human adults (Garcia-Velasco and Mondragon 1991;Moran, Jafek and Rowley 1991; Stensaas, Lakver, Monti-Bloch, Grosser and Berliner 1991; Trotier, Eloit, Wassef, Talmain, Bensimon, Doving and Ferrand 2000; Bhatnagar and Smith 2001; Knecht, Kuhnau, Huttenbrink, Martin and Hummel 2001); a discernable physiological substrate from the VNO to the brain has not yet been identified; and those VNO receptor genes which we might expect to enable pheromone detection (on the basis of comparative studies) appear to be non-functional or pseudo-genes in humans (Tirindelli, Mucignat-Caretta and Ryba 1998). However, pheromonal reception may be possible through the main olfactory system (see e.g. Dorries, Adkins Regan and Halpern 1997; Restrepo, Arellano, Oliva, Schaefer and Lin 2004; Liberles and Buck 2006), and so continuing debate surrounding the human VNO need not preclude ongoing research into the behavioural effects of potential human pheromones.

To date, work on possible human pheromones has primarily focused on the androgens androstenone, androstadienone, and androstenol, found in the human axilla (Hays 2003). All of these are found in much higher quantities in males than females (see Gower and Ruparelia 1992 for an overview), and at least in respect of androstenol, production rates rise sharply at puberty (Cleveland and Savard 1964) and fall in old age and post-menopausally (Brooksbank and Haslewood 1961), in a pattern of ontogeny and sexual dimorphism consistent with sexual signalling. Metabolites of androgens, 5α-androstenone, androstadienone and the low-odour androstadienol, are secreted from the apocrine glands (Gower, Holland, Mallet, Rennie and Watkins 1994). Androstadienol and androstadienone have been shown, *in vitro*, to be converted to the musky-smelling androstenol and the more prominent, urinous androstenone, under the influence of skin-inhabiting coryneform bacteria (Mallet, Holland, Rennie, Watkins and Gower 1991; Gower et al. 1994), although alternative biotransformational pathways have also been demonstrated, again *in vitro* (Decréau, Marson, Smith and Behan 2003).

The earliest work on 'human pheromones' focused on androstenone and androstenol, in part because of their documented function in other mammals: androstenone, for example, occurs in boar saliva and triggers lordosis (the 'mating stance') in the female pig (Signoret and du Mesnil du Buisson 1961). Yet the results of these earlier experiments do not lead to a consistent picture of the effects of these two chemicals in humans. For example, while some studies purport to show that androstenone or androstenol exposure reduces female ratings of male sexual attractiveness (Filsinger, Braun and Monte 1985), others show that androstenol increases female ratings of character or sexual attractiveness (Cowley, Johnson and Brooksbank 1977; Kirk-Smith, Booth, Carroll and Davies 1978), or that there is no effect of either chemical on judgments (Black and Biron 1982;

Filsinger, Braun, Monte and Linder 1984). In terms of emotional and behavioural responses, female exposure to androstenone has been associated with a decrease in her self-perceived 'sexiness' (Filsinger et al. 1984), but with an increase in her levels of social interactions with males (Cowley and Brooksbank 1991). On the other hand, neither androstenol nor androstenone led to increased female sexual arousal in response to erotic prose (Benton and Wastell 1986; McCollough, Owen and Pollak 1981). Androstenol exposure has been linked with increased mid-cycle self-reported feelings of submissiveness (Benton 1982) and irritability during menses (Cowley, Harvey, Johnson and Brooksbank 1980), and indeed its effects on the menstrual cycle may play a part in the reported menstrual synchrony of women who live together (Shinohara, Morofushi, Funabashi, Mitsushima and Kimura 2000). In two studies, females chose to approach areas which had been dosed with androstenone (Kirk-Smith and Booth 1980; Pause 2004); in a similar study, androstenol had no such effect (Gustavson, Dawson and Bonett 1987). The effects of these chemicals on males are sometimes, but not always, complementary: in Pause's (2004) study, homosexual men, like women, approached areas dosed with androstenone more often (heterosexual men were not tested). In another study, men (whose sexual orientation was not reported) rated other men as more sexually attractive in the presence of androstenone and androstenol (Filsinger et al. 1985). Although few of these studies on pheromones in humans are directly contradictory, they are sufficiently at odds to indicate that we lack enough understanding of the phenomenon to be able to ask coherent questions of its effects.

One insight into these sometimes inconsistent findings notes that only some of these studies have used a musk odour control: it could be that other musk odours would trigger similar responses to the potential pheromones (Gower, Nixon and Mallet 1988). Nevertheless, an attenuated conclusion from the work would merely be that some musks have behavioural effects, and androstenol and androstenone are examples of such musks. An alternative suggestion to help explain the reported inconsistencies is that these chemicals have different effects depending on the contexts in which they are experienced, and furthermore that only within ecologically valid conditions are we most likely to detect their influences. This explanation, which will be discussed in greater detail below, receives support from more recent work on the potential human pheromone androstadienone.

Androstadienone provides some of the best evidence for a chemical with the potential for some kind of intraspecific communicative function in humans. Females exposed to androstadienone have shown an increase in positive moods and feelings of focus, and a decrease in negative moods (Jacob and McClintock 2000; Lundström and Olsson 2005). Neurological studies indicate that its effects in females extend beyond the olfactory system, activating areas of the brain associated with attention, social cognition, and emotional processing (Jacob, Kinnunen, Metz, Cooper and McClintock 2001a; Gulyas, Keri, O'Sullivan, Decety and Roland 2004). Physiologically, androstadienone has been found to lower women's heart and breathing rates, and raise body temperature (Grosser et al. 2000). If androstadienone functions as a human pheromone, we might expect its effects to be sexually dimorphic, but little work has specifically compared the two sexes, concentrating instead on female response. One study which contrasted male with female response found different physiological response according to gender (Monti-Bloch and Grosser 1991), while another found no effect of gender on the psychological responses attributed to the steroid (Jacob, Garcia, Hayreh and McClintock 2002).

Claims of commercial manufacturers notwithstanding, it is evident that pheromones do not function as behavioural releasers in humans in the same way as they do in other species. Instead of searching for specific reactions to purported human pheromones, it may be that these chemicals are better described as 'modulators' (Jacob and McClintock 2000) which influence psychological states and, thereby, also influence behaviour in a variety of fashions depending on the situation in which they are experienced, or the accompanying cues. The co-occurrence of different cues can affect their interpretation (Rowe 1999). In humans, we know that odour cues provide non-redundant information about potential mates because, while both visual and olfactory cues may be used to gauge physical attractiveness, the information in each is not equivalent (Roberts, Little, Gosling, Jones, Perrett, Carter and Petrie 2005).

10.2 "Human Pheromones": Experimental Considerations

Much work on the impact of potential human pheromones has assumed them to be independent or dominant signals which should give rise to measurable and valid responses when presented in isolation, rather than interacting with the other signals normally available. Yet this assumption may be erroneous, and the importance of context on the effects of purported human pheromones receives empirical backing. For example, Bensafi, Brown, Khan, Levenson and Sobel (2004) found that androstadienone and estratetraenol elicited different affective and autonomic responses depending on the emotional context in which they were presented. Lundström and Olsson (2005) suggested that differences in reactions to androstadienone in two groups of female subjects could be due to the fact that one group was tested by a male researcher whose presence provided the ecological validity and concordant cues for the odour, allowing an effect of the odour to be detected, while the other group was tested by a female researcher and here the steroid had no measurable effect. Likewise, women exposed to androstadienone exhibited different physiological and affective responses from controls, but only in sessions run by a male and not by a female tester (Jacob, Hayreh and McClintock 2001b).

The psychological laboratory, the usual testing ground for isolated effects, may not allow us to fully or adequately investigate the influences of pheromones. It will certainly not allow us to carry out the recommendation made implicit by Jacob et al. (2002, p. 282): "If it does not function under normal, nonexperimental social conditions, then it is not a pheromone". With this in mind, we set out to test the effects of androstadienone within as normal a social context as we could.

We conjectured that a "speed-dating" event might be an ecologically valid and theoretically appropriate context to test the effects of potential pheromones involved in human mating behaviour. Speed-dating is a relatively recent form of organised social introduction. Single males and females subscribe to an independently administered event, which is organised to allow each person to interact with each participant of the opposite sex, in a face-to-face meeting, for a pre-defined, limited time period. At the end of each interaction, males and females note covertly whether they would like to meet again. If both parties select the other, their contact details are exchanged by the organisers. Since the aim of participants is to evaluate and attract potential partners, speed-dating provides a suitable arena for the testing of a chemical which might reasonably be implicated in human partner choice.

We used three experimental groups, two of which were exposed to an odour and one to an odourless condition, in order to be able to parse the effects of an odour alone from those of androstadienone. One group was given a solution containing androstadienone at a concentration of $250 \mu M$, to be applied to the region between the upper lip and the nostrils. This concentration and application method is commonly used in such experiments (Jacob and McClintock 2000; Jacob et al. 2001a; Lundström and Olsson 2005), although what constitutes an appropriate quantity of androstadienone, to reflect that which may be experienced during normal non-intimate social contact, is unclear (Wysocki and Preti 2004; Lundström and Olsson 2005). Analysis of androstadienone concentrations within the apocrine sweat of a small sample of individuals has revealed values as high as $1900 \mu M$ (Gower, Holland, Mallet, Rennie and Watkins 1994), although analysis of the quantities of androstadienone deposited upon axillary hair (excluding the skin) over a 24-hour period is suggestive of much lower quantities, up to just over 4000 pmoles of androstadienone (Nixon, Mallet and Gower 1988). Furthermore, natural production of androstadienone would not usually bring it into such close proximity to the olfactory organs, and so the concentration experienced by the women in our study is probably higher than that which might be experienced during normal social intercourse.

Androstadienone sensitivity can vary widely; since this concentration is close to the average odour threshold of the population (Lundström, Hummel and Olsson 2003), we followed the convention of masking its odour in strongly-scented clove oil (1% clove oil in propylene glycol). It is debatable whether something may be classified as a 'pheromone' if it is consciously detected, and so this procedure aimed to prevent conscious or learnt associations with the odour from influencing its effects. The participants were exposed to the solutions by application to the upper lip, from where they were able to inhale the solutions throughout the evening. This does not exclude the possibility that the solutions were transmitted dermally rather than by odour. However, another study has noted mood effects of androstadienone which were not distinguishable according to whether exposure to the chemical was enabled by upper-lip application or passive inhalation (Jacob et al. 2002), and indeed airborne presentation of a chemical is not a necessary condition of its definition as a pheromone; Queen butterflies and salamanders both transfer pheromones directly from one individual to the next (Wyatt 2003).

10.3 Experimental Study

Twenty two males (aged 18–28) and 25 females (aged 19–24) took part in the speed-dating event. Participants were told that the study was concerned with the effects of different odours, including purported human pheromones, on social interactions and judgments, and provided informed consent. Participants applied either water, clove oil (1% clove oil in propylene glycol) or androstadienone (250 μM concentration in the clove oil solution) with a cotton wool pad to the region of skin between the upper lip and the nostrils. Interactions began at least 15 min after initial odorant exposure, and were completed a maximum of 135 min after first exposure. Previous work indicates that measurable effects of androstadienone exposure begin within 6 min, and last for at least 2 h (Jacob and McClintock 2000). Participants noted whether they would like to meet again; females also rated males for attractiveness, on a scale of 1–7. Five attractiveness ratings were missing from the female score cards, and we excluded a further 14 out of a possible 550 ratings because, upon questioning, one or both of the participants indicated prior familiarity with the other. These 19 omitted or excluded ratings (three being the maximum from any female) were replaced by the average score given to the male in question, calculated from the remaining valid data. This is a conservative approach, reducing between-conditions variance.

The female participants rated each male for physical attractiveness on a 1-7 Likert scale (1 = unattractive, 7 = attractive). Physical attractiveness judgments tend to be remarkably homogeneous between subjects (Langlois, Kalakanis, Rubenstein, Larson, Hallam and Smoot 2000), presenting an appropriate assay of judgment modulation within a small sample. The ability to judge the physical attractiveness of members of the opposite sex is thought to exist in order to enable reproductively successful partner choice (Grammer, Fink, Moller and Thornhill 2003; Rhodes 2006), and so requesting that people assess physical attractiveness in effect provides an insight into their mating decisions. In this regard, other factors are known to have the potential for a systematic effect on female ratings of male attractiveness. These include female hormonal status (Penton-Voak, Perrett, Castles, Kobayashi, Burt, Murray and Minamisawa 1999), self-rated attractiveness (Little, Burt, Penton-Voak and Perrett 2001) and other-rated attractiveness (Penton-Voak, Little, Jones, Burt, Tiddeman and Perrett 2003); relative age of the two participants is also likely to affect judgments. We took details of all of these for control purposes in our analyses.

Analyses were conducted with SPSS version 13.0, and data satisfied the requirements of the statistical tests used; non-normally distributed data were analysed with the Kruskal-Wallis test. We compared how highly the males were rated in terms of their attractiveness by the women in each of the three experimental conditions.

Firstly, we confirmed that there were no significant differences between the three groups of women in terms of other factors which are known to influence attractiveness ratings (number of normally-cycling women in the phase of the menstrual cycle of highest conception risk, 15 to 24 days prior to her next expected menses $H_2 = 0.07$, $p = 0.97$; self-rated attractiveness $H_2 = 0.54$, $p = 0.69$; attractiveness as rated by a panel of 15 males from photographs $F_{2,22} = 0.78$, $p = 0.47$; ages of the women in each group $F_{2,22} = 0.12$, $p = 0.89$).

An ANCOVA was carried out on the average attractiveness rating given by each female to all 22 males, with other-rated attractiveness and age entered as co-variates. We could not include conception risk or self-rated attractiveness data due to non-normal distribution, but these varied little between groups; one woman from each group was at the highest conception risk, and 20 out of 25 females self-rated their attractiveness as four out of seven. The women in the three conditions gave significantly different attractiveness ratings ($F_{2,20} = 5.73$, $p = 0.01$) and androstadienone exposure co-occurred with significantly higher ratings (paired-samples t-tests: androstadienone/clove: $t_{21} = -0.58$, $p < 0.001$, androstadienone/water: $t_{21} = -5.41$, $p < 0.001$, water/clove: $t_{21} = -0.73$, $p = 0.47$). Experimental condition (water, clove or pheromone) was a more influential factor on female ratings than other-rated attractiveness ($F_{1,20} = 6.25$, $p = 0.021$) or age ($F_{1,20} = 3.84$, $p = 0.064$).

10.4 The Future of "Human Pheromone" Research

Our study is suggestive of chemically-induced judgment modulation. Women exposed to androstadienone, compared with both similar-odour and odourless controls, rated males as more attractive. This result is particularly interesting because a recent study (Lundström and Olsson 2005) of the effects of androstadienone on female attractiveness ratings of a set of male images, cropped from the shoulders up and presented on a computer, found no effect when comparing androstadienone in clove oil to clove oil alone, in concentrations identical to those used in our work. If, as discussed above, ecological validity is required to put into motion the "pheromonal" effects, it could be that images on a computer are not sufficient to affect reactions. Alternatively, it could be that the androstadienone was affecting how the women interacted with and assessed the men in relation to both physical traits and behaviour overall, and that this impacted upon their assessments of attractiveness. Such unavoidable generalisation of personal assessments across attributes is known as the 'halo effect'.

One of the most under-investigated aspect of human pheromone research is that of production. It is not enough to show that purported pheromones have effects; it is also necessary to show that these chemicals are secreted in sufficient quantities to influence others in normal circumstances. Although we believe we have demonstrated the efficacy of a speed-dating context as an experimental paradigm for assessing putative pheromonal effects, the question of whether the dosage and mode of presentation is in any way realistic awaits further investigation. In this sense, our experiment is a form of proof of principle, which now requires more detailed enquiry.

Acknowledgments We wish to thank Nicola Koyama, Minna Lyons, Cathal O'Siochru and Sanjeevani Perera for their help with participant recruitment; Tom Heyes for his technical and laboratory expertise; Christopher Hassall and Alexandra Wall for their assistance with data collection; Jan Havlicek, Jane Hurst and Tristram Wyatt for helpful comments and advice on an earlier draft; and the University of Liverpool for funding this work.

References

Anon (2005) So much more to know . . . . Science 309, 78–102.

Bensafi, M., Brown, W. M., Khan, R., Levenson, B. and Sobel, N. (2004) Sniffing human sex-steroid derived compounds modulates mood, memory and autonomic nervous system function in specific behavioral contexts. Behav. Brain Res. 152, 11–22.

Benton, D. (1982) The influence of androstenol - a putative human pheromone – on mood throughout the menstrual cycle. Biol. Psychol. 15, 249–256.

Benton, D. and Wastell, V. (1986) Effects of androstenol on human sexual arousal. Biol. Psychol. 22, 141–147.

Bhatnagar, K. P. and Smith, T. D. (2001) The human vomeronasal organ. III. Postnatal development from infancy to the ninth decade. J. Anat. 199, 289–302.

Black, S. L. and Biron, C. (1982) Androstenol as a human pheromone: No effect on perceived physical attractiveness. Behav. Neural Biol. 34, 326–330.

Brooksbank, B. W. L. and Haslewood, G. A. D. (1961) The estimation of androst-16-en-3α-ol in human urine. Partial synthesis of androstenol and of its β-glucosiduronic acid. Biochem. J. 80, 488–496.

Brooksbank, B.W.L, Wilson, D.A.A. and MacSweeney, D.A. (1972) Fate of androsta-4,16-dien-3-one and the origin of 3α-hydroxy-5α-androst-16-ene in man. J. Endocr. 52, 239–251.

Cleveland, W. W. and Savard, K. (1964) Studies of excretion of androst-16-en-3α-ol. J. Clin. Endocr. Metab. 24, 983–987.

Cowley, J. J. and Brooksbank, B. W. L. (1991) Human exposure to putative pheromones and changes in aspects of social behaviour. J. Steroid Biochem. 39, 647–659.

Cowley, J. J., Johnson, A. T. and Brooksbank, B. W. L. (1977) The effect of two odorous compounds on performance in an assessment-of-people test. Psychoneuroendocrinol. 2,159–172.

Cowley, J. J., Harvey, F., Johnson, A. T. and Brooksbank, B. W. L. (1980) Irritability and depression during the menstrual cycle – possible role for an exogenous pheromone? Irish J. Psychol. 3, 143–156.

Decreau, R. A., Marson, C. M., Smith, K. E., & Behan, J. M. (2003) Production of malodorous steroids from androsta-5,16-dienes and androsta-4,16-dienes by corynebacteria and other human axillary bacteria. J. Steroid Biochem. 87, 327–336.

Dorries, K. M., Adkins Regan, E. and Halpern, B. P. (1997) Sensitivity and behavioural responses to the pheromone androstenone are not mediated by the vomeronasal organ in domestic pigs. Brain Behav. Evolut. 49, 53–62.

Filsinger, E. E., Braun, J. J. and Monte, W. C. (1985) An examination of the effects of putative pheromones on human judgments. Ethol. Sociobiol. 6, 227–236.

Filsinger, E. E., Braun, J. J., Monte, W. C. and Linder, D. E. (1984) Human (Homo sapiens) responses to the pig (sus scrofa) sex pheromone 5 alpha-androst-16-en-3-one. J. Comp.Psychol. 98, 219–222.

Garcia-Velasco, J. and Mondragon, M. (1991) The incidence of the vomeronasal organ in 1000 human subjects and its possible clinical significance. J. Steroid Biochem. 39, 561–563.

Gower, D. B., Holland, K. T., Mallet, A. I., Rennie, P. J., & Watkins, W. J. (1994) Comparison of 16-androstene steroid concentrations in sterile apocrine sweat and axillary secretions: Interconversions of 16-androstenes by the axillary microflora-a mechanism for axillary odour production in man? J. Steroid Biochem. 48, 409–418.

Gower, D.B., Holland, K.T., Mallet, A.I., Rennie, P.J. and Watkins, W.J. (1994) Comparison of 16-androstene steroid concentrations in sterile apocrine sweat and axillary secretions: Interconversions of 16-androstenes by the axillary microflora – a mechanism for axillary odour production in man? J. Steroid Biochem. 48, 409–418.

Gower, D. B., Nixon, A. and Mallet, A. I. (1988) The significance of odorous steroids in axillary odour. In: S. V. Toller & G. H. Dodd (Eds.), *Perfumery: The psychology and biology of fragrance*. Chapman & Hall, London, pp. 47–76.

Gower, D. B., & Ruparelia, B. A. (1993) Olfaction in humans with special reference to odorous 16-androstenes: Their occurrence, perception and possible social, psychological and sexual impact. J. Endocrinol. 137, 167–187.

Grammer, K., Fink, B., Moller, A. P. and Thornhill, R. (2003) Darwinian aesthetics: Sexual selection and the biology of beauty. Biol. Rev. 78, 385–407.

Grosser, B. I., Monti-Bloch, L., Jennings-White, C. and Berliner, D. L. (2000) Behavioral and electrophysiological effects of androstadienone, a human pheromone. Psychoneuroendocrinol. 25, 289–299.

Gulyas, B., Keri, S., O'Sullivan, B. T., Decety, J. and Roland, P. E. (2004) The putative pheromone androstadienone activates cortical fields in the human brain related to social cognition. Neurochem. Int. 44, 595–600.

Gustavson, A. R., Dawson, M. E. and Bonett, D. G. (1987) Androstenol, a putative human pheromone, affects human (*Homo sapiens*) male choice performance. J. Comp. Psychol. 101, 210–212.

Hays, W. S. T. (2003) Human pheromones: Have they been demonstrated? Behav. Ecol. Sociobiol. 54, 89–97.

Jacob, S., Garcia, S., Hayreh, D. and McClintock, M. K. (2002) Psychological effects of musky compounds: Comparison of androstadienone with androstenol and muscone. Horm. Behav. 42, 274–283.

Jacob, S., Hayreh, D. J. S. and McClintock, M. K. (2001b) Context-dependent effects of steroid chemosignals on human physiology and mood. Physiol. Behav. 74, 15–27.

Jacob, S., Kinnunen, L. H., Metz, J., Cooper, M. and McClintock, M. K. (2001a) Sustained human chemosignal unconsciously alters brain function. NeuroReport 12, 2391–2394.

Jacob, S. and McClintock, M. K. (2000) Psychological state and mood effects of steroidal chemosignals in women and men. Horm. Behav. 37, 57–78.

Johns, M. A., Feder, H. H., Komisaruk, B. R. and Mayer, A. D. (1978) Urine-induced ovulation in anovulatory rats may be a vomeronasal effect. Nature 272, 446–448.

Karlson, P. and Luscher, M. (1959) Pheromones: A new term for a class of biologically active substances. Nature 183, 55–56.

Keverne, E. B. (1983) Pheromonal influences on the endocrine regulation of reproduction. Trends Neurosci 6, 381–384.

Kirk-Smith, M., Booth, D. A., Carroll, D. and Davies, P. (1978) Human social attitudes affected by androstenol. Res. Commun. Psych. Psy. 3, 379–384.

Kirk-Smith, M. and Booth, D.A. (1980) Effects of androstenone on choice of location in others' presence. In: H. van der Starre (Ed), *Olfaction and Taste VII*. IRL Press, London, pp. 397–400.

Knecht, M., Kuhnau, D., Huttenbrink, K.-B., Martin, W. and Hummel, T. (2001) Frequency and localization of the putative vomeronasal organ in humans in relation to age and gender. Laryngoscope 111, 448–452.

Langlois, J. H., Kalakanis, L., Rubenstein, A. J., Larson, A., Hallam, M. and Smoot, M. (2000) Maxims or myths of beauty? A meta-analytic and theoretical review. Psychol. Bull. 126, 390–423.

Liberles, S. D. and Buck, L. B. (2006) A second class of chemosensory receptors in the olfactory epithelium. Nature 442, 645–650.

Little, A. C., Burt, D. M., Penton-Voak, I. and Perrett, D. I. (2001) Self-perceived attractiveness influences human female preferences for sexual dimorphism and symmetry in male faces. P. Roy. Soc. Lond. B Bio. 268, 39–44.

Lundstrom, J. N., Hummel, T. and Olsson, M. J. (2003) Individual differences in sensitivity to the odor of 4,16-androstadien-3-one. Chem. Senses 28, 643–650.

Lundström, J. N. and Olsson, M. J. (2005) Subthreshold amounts of social odorant affect mood, but not behavior, in heterosexual women when tested by a male, but not a female, experimenter. Biol. Psychol. 70, 197–204.

Mallet, A. I., Holland, K. T., Rennie, P. J., Watkins, W. J. and Gower, D. B. (1991) Applications of gas chromatography–mass spectrometry in the study of androgen and odorous 16-androstene metabolism by human axillary bacteria. J. Chromatogr-Biomed. 562, 647–658.

McClintock, M. K., Jacob, S., Zelano, B. and Hayreh, D. J. (2001) Pheromones and vasanas: The functions of social chemosignals. Nebraska Symposium on Motivation 47, 75–112.

McCollough, P. A., Owen, J. W. and Pollak, E. I. (1981) Does androstenol affect emotion? Ethol. Sociobiol. 2, 85–88.

Monti-Bloch, L. and Grosser, B. I. (1991) Effect of putative pheromones on the electrical activity of the human vomeronasal organ and olfactory epithelium. J. Steroid Biochem. 39, 573–582.

Moran, D. T., Jafek, B. W. and Rowley, J. C. I. (1991) The vomeronasal (Jacob's) organ in man: Ultrastructure and frequency of occurrence. J. Steroid Biochem. 39, 545–552.

Nixon, A., Mallet, A.I. and Gower, D.B. (1988) Simultaneous quantification of five odorous steroids (16-androstenes) in the axillary hair of men. J. Steroid Biochem. 29, 505–510.

Pause, B. M. (2004) Are androgen steroids acting as pheromones in humans? Physiol. Behav. 83, 21–29.

Penton-Voak, I. S., Little, A. C., Jones, B. C., Burt, D. M., Tiddeman, B. P. and Perrett, D. I. (2003) Female condition influences preferences for sexual dimorphism in faces of male humans (Homo sapiens). J. Comp. Psychol. 117, 264–271.

Penton-Voak, I. S., Perrett, D. I., Castles, D. L., Kobayashi, T., Burt, D. M., Murray, L. K. and Minamisawa, R. (1999) Menstrual cycle alters face preference. Nature 399, 741–742.

Restrepo, D., Arellano, J., Oliva, A. M., Schaefer, M. L. and Lin, W. (2004) Emerging views on the distinct but related roles of the main and accessory olfactory systems in responsiveness to chemosensory signals in mice. Horm. Behav. 46, 247–256.

Rhodes, G. (2006) The evolutionary psychology of facial beauty. Ann. Rev. Psychol. 57, 199–226.

Roberts, S. C., Little, A. C., Gosling, L. M., Jones, B. C., Perrett, D. I., Carter, V., and Petrie, M. (2005) MHC-assortative facial preferences in humans. Biol. Lett. 1, 400–403.

Rowe, C. (1999) Receiver psychology and the evolution of multicomponent signals. Anim. Behav. 58, 921–931.

Shinohara, K., Morofushi, M., Funabashi, T., Mitsushima, D. and Kimura, F. (2000) Effects of 5-androst-16-en-3-ol on the pulsatile secretion of luteinizing hormone in human females. Chem. Senses 25, 456–467.

Signoret, J. P. and du Mesnil du Buisson, F. (1961) Etude du comportement de la truie en oestrus. Proceedings of the 4th International Congress on Animal Reproduction, 2, 171–175.

Stensaas, L. J., Lakver, R. M., Monti-Bloch, L., Grosser, B. I. and Berliner, D. L. (1991) Ultrastructure of the human vomeronasal organ. J. Steroid Biochem. 39, 553–560.

Tirindelli, R., Mucignat-Caretta, C. and Ryba, N. J. (1998) Molecular aspects of pheromonal communication via the vomeronasal organ of mammals. Trends Neurosci. 21, 482–486.

Trotier, D., Eloit, C., Wassef, M., Talmain, G., Bensimon, J. L., Doving, K. B. and Ferrand, J. (2000) The vomeronasal cavity in adult humans. Chem. Senses 25, 369–380.

Vandenbergh, J. G. (Ed.) (1983) Pheromones and reproduction in mammals. Academic Press, New York.

Wyatt, T. D. (2003) Pheromones and Animal Behaviour. Cambridge University Press, Cambridge.

Wysocki, C.J. and Preti, G. (2004) Facts, fallacies, fears and frustrations with human pheromones. Anat. Rec. 281A, 1201–1211.

Chapter 11
The Influence of Sexual Orientation on Human Olfactory Function

Mark J.T. Sergeant, Jennifer Louie and Charles J. Wysocki

Abstract Sexual orientation influences human olfactory function. Following a brief review of the biological basis of homosexuality, this chapter explores exactly how olfactory function varies as a result of sexual orientation. Three separate areas of research are considered: recent studies on the neural processing of social odorants by heterosexuals and homosexuals; the influence of sexual orientation on the production and perception of body odours; and the influence of female sexual orientation on menstrual synchrony.

11.1 Introduction

An individual's sexual orientation is usually described in terms of one of two distinct and mutually exclusive groups, *heterosexual* (sexual interest in members of the opposite-sex) and *homosexual* (sexual interest in members of the same-sex). Some academics have questioned this form of classification, arguing that such groups reflect arbitrary and socially constructed categories (e.g., Muehlenhard 2000). Initial taxometric studies of male sexual orientation, however, strongly suggest that classifying individuals as either heterosexual or homosexual does reflect an objective system of categorization (Gangestad, Bailey and Martin 2000), as does evidence of a similar manifestation and etiology of male homosexuality across cultures with disparate views on sexuality (Whitam and Zent 1984). Although sexual orientation also appears to be biomodally distributed among females, research into this area is somewhat more complicated due to the greater degree of plasticity in female sexual expression (Chivers, Rieger, Latty and Bailey 2004). This, coupled with the lower incidence of female homosexuality (roughly 1% of the female population compared to 2% to 5% of the male population; Rahman and Wilson 2003a), has resulted in most empirical research focusing on male rather than female sexual orientation.

Several lines of research suggest a biological basis to homosexuality. Behavioural genetics studies have suggested a strong coinheritance of homosexuality among monozygotic twins compared to dizygotic twins and genetic and adopted

Mark J.T. Sergeant
Nottingham Trent University, Division of Psychology
mark.sergeant@ntu.ac.uk

siblings (e.g., Bailey, Dunne and Martin 2000; Kendler, Thornton, Gilman and Kessler 2000). A number of researchers have also noted a specific association between several genetic regions and the development of homosexuality among male siblings (Hamer, Hu, Magnuson, Hu and Pattatucci 1993; Mustanski, DuPree, Nievergelt, Bocklandt, Schork and Hamer 2005). Several studies have also documented differences between sexual orientation groups in several sexually dimorphic neural regions such as the supra-chiasmatic nucleus (Swaab and Hofman 1990), the interstitial nucleus of the anterior hypothalamus (LeVay 1991) and the midsagittal plane of the anterior-commissure (Allen and Gorski 1992). Readers seeking a more detailed overview of literature on the biological basis to homosexuality are referred to the review by Rahman and Wilson (2003a).

Given the evidence for sex differences in olfactory function (e.g., Brand and Millot 2001; Garcia-Falgueras, Junque, Gimenez, Caldu, Segovia and Guillamon 2006), and the sex-atypicality of homosexual individuals on tasks eliciting sexually dimorphic reactions (e.g., spatial rotation tasks; Rahman and Wilson 2003b), it is not surprising that several authors have investigated the influence of sexual orientation on olfactory function. The remainder of this chapter will review the literature on this topic, focusing on three separate areas of research:, recent studies on the neural processing of social odorants by heterosexuals and homosexuals, the influence of sexual orientation on the production and perception of body odours and the influence of female sexual orientation on menstrual synchrony

11.2 The Neural Processing of Social Odorants

Two substances frequently, though perhaps incorrectly, labeled as human pheromones are the progesterone derivative 4, 16-androstadien-3-one (androstadienone) and the estrogen-like estra-1, 3, 5(10), 16-tetraen-3-ol (estratetraenol). Both substances have been previously reported to influence mood state and physiological arousal in males and females, though the specific pattern of findings varies considerably between studies (Bensafi, Tsutsui, Levenson and Sobel 2004; Jacob and McClintock 2000). Bensafi et al. (2004) suggest that some of these inconsistencies may be due to the concentrations of odorants utilized in each study, as androstadienone at least appears to function in a dose-dependent fashion.

A number of recent studies have examined neuropsychological reactions to androstadienone and estratetraenol. Using functional magnetic resonance imaging (fMRI), Sobel, Prabhakaran, Hartley, Desmond, Glover, Sullivan and Gabrieli (1999) noted that male participants exposed to estratetraenol showed significant activation of the inferior frontal gyrus (primarily involved in language processing/production and face recognition) and anterior medial thalamus (a relay station to and from the cerebral cortex involved in arousal and the integration of sensory information). Furthermore, research by Savic, Berglund, Gulyas and Roland (2001) suggests that this form of neural activation is sexually dimorphic. Androstadienone was found to activate the preoptic and ventromedial nucleus in females, while estratetraenol was found to activate the paraventricular and dorsomedial nuclei in males. Thus there was a sexually dimorphic pattern of activation, with androstadienone activating the hypothalamic region in females and, to a lesser degree, the olfactory region in males, while estratetraenol activated the hypothalamic region in males and the olfactory region in females.

Using positron emission tomography (PET) imaging, Savic, Berglund and Lindström (2005) examined the reactions of 12 homosexual males to both androstadienone and estratetraenol, with 12 heterosexual males and 12 heterosexual females acting as control participants. The homosexual male participants demonstrated sex-atypical reactions to the compounds, showing hypothalamic activation in response to androstadienone but not estratetraenol. A similar study by Berglund, Savic and Lindström (2006) examined reactions to these compounds by 12 homosexual females, with 12 heterosexual males and 12 heterosexual females again acting as controls. Berglund et al. (2006) also report that homosexual females demonstrated sex-atypical reactions, showing hypothalamic activation in response to estratetraenol but not to androstadienone. The differences between the two female groups were not, however, as marked as those of the two male groups. However, these findings suggest that reactions to androstadienone and estratetraenol are not simply dependent upon an individual's biological sex, but also upon their sexual orientation (i.e., the target of their sexual and romantic attractions). These findings are also commensurate with two recent reports that sexual orientation has a strong influence on neural responses to both facial images (Kranz and Ishai 2005) and sexually arousing stimuli (Ponseti Bosinski, Wolff, Peller, Jansen, Mehdorn, Büchel, and Siebner 2006), with homosexual individuals again showing reactions more consistent with members of the opposite biological sex.

11.3 The Production and Perception of Body Odour

A growing body of research suggests body odour is an effective method of communication able to convey information on, among other things, an individual's immune system characteristics (Wedekind and Füri 1997), current fertility status (Kuukasjärvi, Eriksson, Koskela, Mappes, Nissinen and Rantala 2004) and the degree of fluctuating asymmetry they demonstrate (Gangestad and Thornhill 1998). Thus far, however, virtually all research in this area has focused on the characteristics and reactions of heterosexual participants, with no focus on how body odour is either produced or perceived by homosexual individuals. To date, only one published study has examined this area. Martins, Preti, Crabtree, Runyan, Vainius and Wysocki (2005) collected body odour samples from 24 heterosexual and homosexual males and females (6 in each group) which were then rated by a sample of 80 heterosexual and homosexual males and females (20 in each group). As with other studies of body odour, the samples were collected under strictly controlled conditions, in an attempt to control for numerous potentially confounding variables associated with individual dietary, hygiene and behavioural practices.

The pattern of preferences for body odour differed noticeably between orientation groups. Homosexual males displayed a strong preference for the odour of other homosexual males and, to a lesser degree, heterosexual females. Conversely, heterosexual males and females and homosexual females found the odour of homosexual males to be the least preferred odour that was presented. Homosexual and heterosexual females provided an equivalent pattern of responses, preferring odours from heterosexual rather than homosexual individuals. Heterosexual males showed no significant preferences towards the odour of any group, except for the strong dislike of odour from homosexual males. Intriguingly these findings do not appear to be based on quantitative differences between groups, as there were no significant differences in the intensity of body odour from any of the four orientation groups. This suggests that sex and sexual orientation influence body odour in a qualitative rather than quantitative manner.

Taken comprehensively these findings suggest there are clear differences in the production and perception of body odour for sexual orientation groups. In particular, the differences between heterosexual and homosexual males were more distinct than those between heterosexual and homosexual females. Differences in the production and perception of body odour may result from a combination of environmental or biological factors, for example, differences in axillary flora and gland function, lifestyle factors such as frequency of exercising or differences in HLA (human leukocyte antigen) alleles or other genetic regions. Given the wealth of information that can be conveyed through body odour, it is also possible that the behavioural effects of this information may vary between sexual orientation groups.

11.4 Menstrual Synchrony among Homosexual Females

Menstrual synchrony is the purported tendency of females who either co-habit or interact over an extended period of time to have co-occurring menses (Wyatt 2003). This process was first documented by McClintock (1971), but has since been corroborated by several other researchers (e.g., Graham and McGrew, 1980; Weller and Weller 1997). Furthermore, experimental evidence suggests this phenomenon relies on a semiochemical mechanism; axillary secretions collected from females at different stages of their menstrual cycle have the potential to accelerate or delay the onset of menses in other females (Preti, Cutler, Garcia, Huggins and Lawley 1986; Stern and McClintock 1998). However, other researchers have found no evidence for synchrony between heterosexual females (Strassman 1997; Wilson, Hildebrandt Kiefhaber and Gravel 1991), while other authors have highlighted that research on menstrual synchrony is often flawed due to a failure to statistically control for female participants with an irregular menstrual cycle or the convergence of menses by chance (Schank 2001).

Given inconsistent findings, the use of homosexual females as participants provides three distinct advantages compared to heterosexual females (Weller and

Weller 1992). Firstly, as synchrony is based on the interaction between closely affiliated females, the high degree of intimacy within a homosexual female couple should maximize the degree of synchrony that occurs. Secondly, studying homosexual female couples effectively eliminates the likelihood of female participants interacting sexually with males, which also influences synchrony (Cutler, Garcia and Krieger 1980; Veith, Buck, Getzlaf, Van Dalfsen and Slade 1983). Finally, homosexual females are less likely to be using hormone-based contraceptives to prevent pregnancy, though they may be used by homosexual females for other purposes such as to control acne or heavy menstrual bleeding (Cochran et al. 2001). This is important as hormone-based contraceptives have been demonstrated to influence olfactory function (Caruso, Grillo, Agnellio, Maolino, Intelisano and Serra 2001).

To date, no experimental study has assessed whether female axillary secretions can accelerate or delay the onset of menses in homosexual females. As a result it is difficult to comment on whether there is an analogous mechanism of synchronization among heterosexual and homosexual females. However, three studies have assessed whether homosexual female couples demonstrate the behavioral consequence of synchronization; a gradual convergence of menstrual cycle timing. As part of a larger study of sexual responsiveness, Matteo and Rissman (1984) examined the menstrual cycle timing of 7 homosexual female couples over a period of 14 consecutive weeks. Due to the small sample size involved the researchers were unable to statistically analyze menstrual cycle data, but they did report that none of the homosexual female couples were perfectly synchronized and that there was considerable variance in menstrual cycle timing. In a more detailed study, Weller and Weller (1992) examined the degree of synchrony among 20 homosexual female couples. The researchers noted that synchrony occurred at more than chance levels, with approximately half of the couples menstruating within 2 days of each other. Furthermore, synchrony was more noticeable among couples with regular menstrual cycles and a high degree of emotional intimacy and shared activities. These findings stand in stark contrast to those of Trevathan, Burleson and Gregory (1993), who found no evidence for menstrual synchrony among 29 homosexual female couples. Over the course of three consecutive menstrual cycles, there was no evidence for synchrony among couples, with the timing of menstruation appearing to diverge rather than converge in many cases. Furthermore, these findings did not differ when irregularity in menstrual cycle length or possible interactions with other roommates were taken into account.

Given the inconsistent findings for menstrual synchrony among heterosexual females, it is difficult to interpret the above findings for homosexual females. While one study (Weller and Weller 1992) appears to demonstrate that homosexual females do synchronize their cycles, two other studies detected no evidence of synchronization among homosexuals (Matteo and Rissman 1984; Trevathan et al. 1993). It is apparent that further research is required to interpret menstrual synchrony among homosexual females. In particular, there is a need for experimental studies to address whether a potentially analogous mechanism of synchronization described among heterosexual females exists in homosexual females.

11.5 Conclusions

Several avenues of research now suggest that sexual orientation has a noticeable impact on human olfaction. In particular, it has been demonstrated that homosexual individuals show neural reactions to social odorants that are similar to heterosexual members of the opposite biological sex (Berglund et al. 2006; Savic et al. 2005) and that sexual orientation influences the production and perception of body odour (Martins et al. 2005). With regard to menstrual synchrony among females, the influence of sexual orientation is less clear and is complicated by numerous conflicting studies based on heterosexual females. This may be further complicated by the comparative differences between sexual orientation groups for males and females. The reactions of homosexual females were comparatively similar to those of heterosexual females with regard to both the neural processing of odorants and perceptions of body odour. The reactions of homosexual males differed noticeably from those of heterosexual males in both of these areas. Such findings are consistent with the greater differences documented for other characteristics based on sexual orientation for males rather than females (Chivers et al. 2004).

Acknowledgments This work was supported by a European Chemoreception Research Organisation (ECRO) Traveling Fellowship to Mark Sergeant and by institutional funds from the Monell Chemical Senses Center. The authors would like to thank Deborah Lodge and Tom Dickins for their comments on earlier drafts of this work.

References

Allen, L.S. and Gorski, R.A. (1992) Sexual orientation and the size of the anterior commissure in the human brain. P. Natl. Acad. Sci. USA 89, 7911–7202.

Bailey, J.M., Dunne, M.P. and Martin, N.G. (2000) Genetic and environmental influences on sexual orientation and its correlates in an Australian twin sample. J. Pers. Soc. Psychol. 78, 524–536.

Bensafi, M., Tsutsui, T., Levenson, R.W. and Sobel, N. (2004) Sniffing a human sex-steroid derived compound affects mood and autonomic arousal in a dose-dependent manner. Psychoneuroendocrinol. 29, 1290–1299.

Berglund, H., Savic, I. and Lindström, P. (2006) Brain response to putative pheromones in lesbian women. P. Natl. Acad. Sci. USA 103, 8269–8274.

Brand, G. and Millot, J.L. (2001) Sex differences in human olfaction: beyond evidence and enigma. Q. J. Exp. Psychol-B 54, 259–270.

Caruso, S., Grillo, C., Agnellio, C., Maolino, L., Intelisano, G. and Serra, A. (2001) A prospective study evidencing rhinomanometric and olfactometric outcomes in women taking oral contraceptives. Hum. Reprod. 16, 2288–2294.

Chivers, M.L., Rieger, G., Latty, E. and Bailey, J.M. (2004) A Sex Difference in the Specificity of Sexual Arousal. Psychol. Sci. 15, 736–744.

Cochran, S.D., Mays, V.M., Bowen, D., Gage, S., Bybee, D., Roberts, S.J., Goldstein, R.S., Robison, A., Rankow, E.J. and White, J. (2001). Cancer-related risk indicators and preventive screening behaviors among lesbians and bisexual women. Am. J. Public Health 91, 591–597.

Cutler, W.B., Garcia, C.R. and Krieger, A.M. (1980) Sporadic sexual behavior and menstrual cycle length in women. Horm. Behav. 14, 163–172.

Gangestad, S.W., Bailey, J.M. and Martin, N.G. (2000) Taxometric analyses of sexual oriention and gender identity. J. Pers. Soc. Psychol. 78, 1109–1121.

Gangestad, S.W. and Thornhill, R. (1998) Menstrual cycle variation in women's preference for the scent of symmetrical men. P. Roy. Soc. Lond. B 265, 927–933.

Garcia-Falgueras, A., Junque, C., Gimenez, M., Caldu, X., Segovia, S. and Guillamon, A. (2006) Sex differences in the human olfactory system. Brain Res. 1116, 103–111.

Graham, C.A. and McGrew, W.C. (1980) Menstrual synchrony in female undergraduates living on a co-educational campus. Psychoneuroendocrinol. 5, 245–252.

Hamer, D.H., Hu, S., Magnuson, V.L., Hu, N. and Pattatucci, A.M.L. (1993) A linkage between DNA markers on the X chromosome and male sexual orientation. Science 261, 321–327.

Jacob, S. and McClintock, M.K. (2000) Psychological state and mood effects of steroidal chemosignals in women and men. Horm. Behav. 37, 57–78.

Kendler, K.S., Thornton, L.M., Gilman, S.E. and Kessler, R.C. (2000) Sexual orientation in a U.S. national sample of twin and non-twin sibling pairs. Am. J. Psychiat. 157, 1843–1846.

Kranz, F. and Ishai, A. (2006) Face perception is modulated by sexual preference. Curr. Biol. 16, 63–68.

Kuukasjarvi, S., Eriksson, C.J.P., Koskela, E., Mappes, T., Nissinen, K. and Rantala, M.J. (2004) Attractiveness of women's body odors over the course of the menstrual cycle: The role of oral contraceptives and receiver sex. Behav. Ecol. 15, 579–584.

LeVay, S. (1991) A difference in hypothalamic structure between heterosexual and homosexual men. Science 253, 1034–1037.

Martins, Y., Preti, G., Crabtree, C.R., Runyan, T., Vainius, A.A. and Wysocki, C.J. (2005) Preference for human body odors is influenced by gender and sexual orientation. Psych. Sci. 16, 694–701.

Matteo, S. and Rissman, E.F. (1984).Increased sexual activity during the midcycle portion of the human menstrual cycle. Horm. Behav. 18, 249–255.

McClintock, M.K. (1971) Menstrual synchrony and suppression. Nature 229, 244–245.

Muehlenhard, C.L. (2000) Categories and sexuality. J. Sex Res. 37, 101–107.

Mustanski, B.S., DuPree, M.G., Nievergelt, C.M., Bocklandt, S., Schork, N.J. and Hamer, D. (2005) A genomewide scan of male sexual orientation. Hum. Genet. 116, 272–278.

Ponseti, J., Bosinski, H.A., Wolff, S., Peller, M., Jansen, O., Mehdorn, H.M., Büchel, C. and Siebner, H.R (2006) A functional endophenotype for sexual orientation in humans. NeuroImage 33, 825–833.

Preti, G., Cutler, W.B., Garcia, C-R., Huggins, G.R. and Lawley, H.J. (1986) Human axillary secretions influence women's menstrual cycles: the role of donor extract of female. Horm. Behav. 20, 474–482.

Rahman, Q. and Wilson, G.D. (2003a) Born gay? The psychobiology of human sexual orientation. Pers. Indiv. Differ. 34, 1337–1382.

Rahman, Q. and Wilson, G.D. (2003b) Large sexual orientation related differences in performance on mental rotation and judgment of line orientation. Neuropsychol. 17, 25–31.

Savic, I., Berglund, H., Gulyas, B. and Roland, P. (2001) Smelling of odourous sex hormone-like compounds causes sex-differentiated hypothalamic activation in humans. Neuron 31, 661–668.

Savic, I., Berglund, H. and Lindström, P. (2005) Brain response to putative pheromones in homosexual men. P. Natl. Acad. Sci. USA 102, 7356–7361.

Schank, J.C. (2001) Menstrual-cycle synchrony: problems and new directions for research. J. Comp. Psychol. 115, 3–15.

Sobel, N., Prabhakaran, V., Hartley, C.A., Desmond, J.E., Glover, G.H., Sullivan, E.V. and Gabrieli, J.D.E. (1999) Blind smell: brain activation induced by an undetected air-borne chemical. Brain 122, 209–217.

Stern, K. and McClintock, M.K. (1998) Regulation of ovulation by human pheromones. Nature 392, 177–179.

Strassmann, B.I. (1997) The biology of menstruation in homo sapiens: total lifetime menses, fecundity, and nonsynchrony in a natural fertility population. Curr. Anthropol., 38 123–129.

Swaab, D.F. and Hofman, M.A. (1990) An enlarged suprachiasmatic nucleus in homosexual men. Brain Res. 537, 141–148.

Trevathan, W.R., Burleson, M.H. and Gregory, W.L. (1993) No evidence for menstrual synchrony in lesbian couples. Psychoneuroendocrinol. 18, 425–435.

Veith, J.L., Buck, M., Getzlaf, S., Van Dalfsen, P. and Slade, S. (1983) Exposure to men influences occurrence of ovulation in women. Physiol. Behav. 31, 313–315.

Wedekind, C. and Füri, S. (1997) Body odour preferences in men and women: do they aim for specific MHC combinations or simply heterozygosity? P. Roy. Soc. Lond. B 264, 1471–1479.

Weller, A. and Weller, L. (1992) Menstrual synchrony in female couples. Psychoneuroendocrinol. 17, 171–177.

Weller, L. and Weller, A. (1997) Menstrual variability and the measurement of menstrual synchrony. Psychoneuroendocrinol. 22, 115–128.

Whitam, F.L. and Zent, M. (1984) A cross-cultural assessment of early cross-gender behaviour and familial factors in male homosexuality. Arch. Sex Behav. 13, 427–439.

Wilson, H.C., Hildebrandt Kiefhaber, S. and Gravel, V. (1991) Two studies of menstrual synchrony: negative results. Psychoneuroendocrinol. 17, 565–569.

Wyatt, T.D. (2003) *Pheromones and Animal Behavior: Communication by Smell and Taste.* Cambridge: Cambridge University Press.

Part III
Recognition within Species: Individual, Sex, Group

Chapter 12
MHC-Associated Chemosignals and Individual Identity

Peter A. Brennan

Abstract The ability of animals to recognise and discriminate individual conspecifics is a vital feature of mammalian social systems. Genes of the major histocompatibility complex (MHC) have long been recognised to play an important role in influencing chemosensory cues of individual identity. In particular, the profile of urinary volatiles of mice has been related to MHC type, although a mechanism to explain this link has remained obscure. This article aims to review recent developments, which have revealed a new class of MHC-associated chemosignals. These are nine-amino acid peptide ligands bound by MHC class I molecules, which are presented at the cell surface for immune surveillance. In addition to this immune function, these peptides have been found to elicit highly sensitive and specific responses in sensory neurons of both the main olfactory and vomeronasal systems. They have also been shown to convey information about strain identity in biologically relevant contexts. Hence it now appears that there are multiple systems for signalling MHC identity, with distinct features that are likely to be adapted for use in different behavioural contexts.

12.1 The Importance of Individual and Kin Recognition

The recognition of individual identity and the relatedness of individuals play vital roles in mammalian social behaviour. They enable nepotistic behaviour towards kin, including mother-offspring interactions; the recognition of group membership and the maintenance of dominance hierarchies and territories; as well as influencing choice of mate in order to maintain immunological diversity and avoid inbreeding. Many species rely on visual or auditory cues to recognise individuals, but for most vertebrates, individual and kin recognition depend on being able to detect and discriminate differences in genetically determined chemosensory signals. In identifying these chemosensory signals of individual identity, most attention has been focused on families of polymorphic genes that differ in their expression among

Peter A. Brennan
University of Bristol, Department of Physiology
p.brennan@bristol.ac.uk

individuals. The best known of these are the genes of the major histocompatibility complex (MHC), which determine the recognition of self from non-self by the immune system, as a defence against pathogens.

One context in which individual recognition is influenced by MHC type is mate recognition, which is essential for preventing pregnancy block (the Bruce effect) in mice (Bruce 1959). The pregnancy block effect occurs when a recently mated female mouse has direct contact with the urinary chemosignals of an unfamiliar male. These chemosignals activate a neuroendocrine reflex, resulting in a fall in progesterone levels and failure of embryo implantation. Although the mating male also produces pregnancy-blocking chemosignals in his urine, they do not block his mate's pregnancy, as the female learns to recognise them during a sensitive period for memory formation at mating (Brennan, Kaba and Keverne 1990). Yamazaki and co-workers (Yamazaki, Beauchamp, Wysocki, Bard, Thomas and Boyse 1983) found that congenic mice, differing from the mating male only at the H2 region of the MHC, were not recognised by the female and were effective in blocking her pregnancy.

Mate choice in mice has also been found to be influenced by MHC genotype, as first reported by Boyse, and then investigated in a series of further studies by Yamazaki and Beauchamp (Boyse, Beauchamp and Yamazaki 1987). They studied mating preferences of male mice housed with a female of the same inbred strain and a congenic female that only differed only in their MHC genotype. They found a disassortative pattern of mating, with the male preferring to mate with the female of dissimilar MHC type, thus avoiding inbreeding. The difficulty in studying mate choice in the laboratory environment is perhaps evident from the confused picture to emerge from subsequent studies by other investigators (Jordan and Bruford 1998). Nevertheless, evidence for disassortative mate preference has been observed in semi-natural enclosures, in which colonies of mice produced fewer MHC homozygous offspring than expected from random matings (Potts, Manning and Wakeland 1991). The influence of MHC type on mate choice appears to depend on learning of kin odours in the nest environment, as it can be largely reversed by cross-fostering mouse pups onto MHC-dissimilar mothers (Yamazaki, Beauchamp, Kupniewski, Bard, Thomas and Boyse 1988; Penn and Potts 1998a).

MHC genotype has also been found to influence kin recognition, in the context of maternal behaviour in mice. Female mice are more likely to form communal nests with kin of MHC-similar genotype, minimising the delivery of maternal resources to genetically unrelated individuals (Manning, Wakeland and Potts 1992). Furthermore, if mouse pups have become scattered from the nest, females preferentially retrieve pups of similar MHC type to themselves (Yamazaki, Beauchamp, Curran and Boyse 2000). The pups themselves appear to be able to use MHC-related cues to learn the odour of their mother and siblings, as revealed by a preference for odours of maternal MHC-type in a choice test (Yamazaki et al. 2000).

MHC influences on behaviour are not restricted to mice. Female sticklebacks show a preference for the odour of males that is consistent with them attaining an optimum level of MHC diversity in their offspring (Reusch, Häberli, Aeschlimann and Milinski 2001). There is also good evidence that humans can discriminate and recognise the odours from individuals on the basis of MHC type. On average, humans perceive the odours of other individuals as being more pleasant if they share fewer MHC alleles with the perceiver than either no matches or a high degree of similarity (Wedekind and Furi 1997; Jacob, McClintock, Zelano and Ober 2002). Whether such MHC-related odour preferences play a role in the complexities of modern human society is an open question. However, it is notable that fathers, grandmothers and aunts are able to identify the odour of a related infant independently of prior experience, which could point to a role in parental or nepotistic behaviour (Porter, Balogh, Cernoch and Franchi 1986).

12.2 Volatile MHC-Associated Chemosignals

The first evidence that MHC genotype could influence odour identity came from the Y-maze tests of Yamazaki and Beauchamp (Yamazaki, Yamaguchi, Baranoski, Bard, Boyse and Thomas 1979; Yamazaki, Beauchamp, Imai, Bard, Phelan, Thomas and Boyse 1990). They were able to train mice to discriminate the urine odours of congenic mice that differed genetically only at the H2 region of their MHC. Similarly, untrained mice have also been found to be capable of discriminating the MHC type of urine odours in a habituation/dishabituation test (Penn and Potts 1998b). In both of these experimental designs, mice were able to discriminate between urine odours without direct contact, on the basis of their volatile components. Analysis of the volatile constituents of urine by gas chromatography has revealed that urine from MHC-congenic mice differ in the relative proportions of volatile carboxylic acids (Singer, Beauchamp and Yamazaki 1997). Moreover, these differences are sufficient to distinguish between MHC types and elicit significantly different patterns of neural activity in the main olfactory system (Schaefer, Yamazaki, Osada, Restrepo and Beauchamp 2002). It is not surprising that the main olfactory system should be capable of distinguishing MHC-related differences in the profile of urinary volatiles. After all, the main olfactory system is adapted to learn to recognise complex mixtures of odorants that make up odours. However, the role of the individual profile of urinary volatiles in conveying individuality in biologically-relevant contexts remains to be firmly established.

12.2.1 How Does MHC Genotype Influence Urine Odour?

The H2 region of mouse chromosome 17 codes for MHC proteins of classical class I type, which are expressed on the cell membrane of nearly all nucleated cells in the body. Their immunological role is to bind peptides resulting from the proteosomal degradation of endogenous and foreign proteins (Paulsson 2004). They then present the peptides at the cell surface for immune surveillance, providing a constantly updated indication of intracellular protein composition. This enables the immune system to recognise cells that have been invaded by pathogens, which triggers cell destruction. MHC class I proteins belong to a highly polymorphic gene family, with structurally diverse peptide binding grooves. Unrelated individuals within a population are highly unlikely to share the same MHC genotype.

Several hypotheses have been put forward to explain how MHC class I molecules could affect the profile of urinary volatiles (Penn and Potts 1998c). One of the most popular of these is the carrier hypothesis (Singh 2001). According to this hypothesis, the cleavage of MHC class I molecules from the cell membrane releases the peptide from the binding groove, which would then become available for binding small volatile molecules. Fragments of MHC class I proteins have been found in rat urine, at low concentrations (Singh, Brown and Roser 1987). And although there is no direct evidence that MHC class I proteins can bind urinary volatiles, the ability of mice to discriminate urine odours of MHC-congenic strains does appear to be related to polymorphism in their peptide-binding groove (Carroll, Penn and Potts 2002).

12.3 MHC Peptide Ligands as Individuality Chemosignals

Although most of the early work focused on the ability of mice to discriminate MHC-dependent urine odours, it does not mean that all individuality chemosignals are volatile. Indeed, lesions of the main olfactory epithelium do not prevent the pregnancy block effect, or affect the ability of female mice to recognise their mate (Lloyd-Thomas and Keverne 1982; Ma, Allen, Van Bergen, Jones, Baum, Keverne and Brennan 2002). Both pregnancy block and mate recognition are mediated by the vomeronasal system (Bellringer, Pratt and Keverne 1980), which has traditionally been associated with the detection of non-volatile stimuli following physical contact. This ability of the vomeronasal system to detect individuality is reinforced by the finding that neurons in the accessory olfactory bulb, respond highly selectively to the strain identity of an anaesthetized stimulus animal (Luo, Fee and Katz 2003). This puzzling situation was resolved by Boehm's hypothesis that the peptides bound by MHC class I molecules could convey information about MHC type, and represent a novel class of non-volatile chemosignal that could be detected by the vomeronasal system (Boehm and Zufall 2006).

Peptides bound by mouse MHC class I proteins are typically nine amino acids in length, a size determined by proteosomal processing. The major factor determining the specificity of their binding is the presence of large hydrophobic side chains known as anchor residues, which occupy pockets in the MHC binding groove. The position and shape of these anchor pockets varies among different MHC class I proteins. Therefore individuals with different MHC genotypes will possess a different combinations of peptides bound by their MHC class I proteins. For example, the H-2D^b haplotype, found in C57BL/6 mice, codes for the MHC class I D^b protein. This preferentially binds peptides that have asparagine (N) at position 5, such as AAPDNRETF. Whereas the H-2K^d haplotype, found in BALB/c mice, codes for the MHC class I K^d protein, which binds to peptides with tyrosine (Y) at position 2, such as **SYFPEITHI**. The position and nature of the anchor residues along the peptide chain reflects the binding characteristics of the MHC class I molecule, and therefore conveys information about MHC type. A chemosensory receptor with binding characteristics similar to a particular MHC class I protein will therefore be sensitive to peptides associated with that MHC type (Boehm and Zufall 2006).

12.3.1 Vomeronasal Responses to MHC Peptide Ligands

Sensory responses to synthetic peptides possessing the characteristic features of MHC peptide ligands have been recorded from slices of vomeronasal epithelium at concentrations down to 10^{-13} M (Leinders-Zufall, Brennan, Widmayer, Chandramani, Maul-Pavicic, Jäger, Li, Breer, Zufall and Boehm 2004). Moreover, the response characteristics showed a dependence on the presence of anchor residues similar to that determining binding to MHC class I proteins. Synthetic peptides of BALB/c-type (SYFPEITHI) or C57BL/6-type (AAPDNRETF) elicited selective responses from largely separate sub-populations of vomeronasal sensory neurons (VSNs). These responses were abolished when the anchor residues were substituted by alanines, which lack a side chain, or when the amino acid sequence was scrambled to change the position of the anchor residues. Moreover, responses were unaffected by changes in the amino acid residues between the anchor residues, implying that the position of the anchor residues determine the specificity of the VSN response.

MHC peptide ligands convey information about strain identity that influences mate recognition in the pregnancy block effect. The addition of synthetic peptides of C57BL/6 type to BALB/c urine was effective in blocking pregnancy of females that had mated with a BALB/c male, whereas the addition of peptides of BALB/c type was ineffective. Conversely, the addition of BALB/c type peptides to C57BL/6 urine altered its strain identity and caused it to block the pregnancy of a female that had mated with a C57BL/6 male (Leinders-Zufall et al. 2004). Both the responses of peptide-sensitive VSNs and their effectiveness in conveying individuality in the pregnancy block effect, were unaffected in knockout mice lacking functional TRPC2 channels (Kelliher, Spehr, Li, Zufall and Leinders-Zufall 2006). This implies that the peptide-sensitive VSNs utilise a different transduction mechanism from the TRPC2-dependent vomeronasal transduction mechanism operating in other VSNs.

Calcium imaging of peptide-sensitive VSNs revealed them to be located in the basal layer of the vomeronasal epithelium, and co-localised with V2R2 immunostaining (Leinders-Zufall et al. 2004). Although most of them responded selectively to synthetic peptides of either BALB/c-type or C57/BL6-type, a significant proportion of VSNs responded to both peptides (Leinders-Zufall et al. 2004). This could be explained by the co-expression of more than one V2R type per VSN (Martini, Silvotti, Shirazi, Ryba and Tirindelli 2001). However, it remains to be seen whether individual VSNs respond to particular combinations of MHC peptides, and the extent to which they could respond to peptides associated with specific combinations of MHC alleles.

Peptide-sensitive VSNs express receptors of the V2R class, which possess a large extracellular N-terminal domain, possibly involved in binding MHC peptide ligands. But intriguingly, V2Rs are co-expressed with non-classical MHC Ib proteins, which are not known to be expressed in any tissue other than the VNO (Ishii, Hirota and Mombaerts 2003; Loconto, Papes, Chang, Stowers, Jones, Takada, Kumanovics, Fischer-Lindahl and Dulac 2003), suggesting that they might have a specific chemosensory function. Moreover, V2Rs bind to MHC Ib proteins and β-microglobulin, suggesting that they might form a receptor complex (Loconto et al. 2003). Structural considerations suggest that the peptide-binding groove of the MHC Ib is empty and unlikely to bind peptides (Olson, Huey-Tubman, Dulac and Bjorkman 2005). Nevertheless, sequence variability among the nine members of the non-classical MHC Ib family is located in the peptide binding groove, and certain combinations of MHC Ib proteins are co-expressed with particular V2Rs (Ishii et al. 2003). It is possible that the combinatorial co-expression of MHC Ib proteins with V2Rs affects the peptide specificity of VSN responses, but their role remains enigmatic.

12.3.2 Main Olfactory Responses to MHC Peptide Ligands

According to the traditional view, the main olfactory system detects airborne volatile molecules, while the vomeronasal system detects non-volatile chemosignals following physical contact. However, it is becoming increasingly apparent that there is considerable overlap between the systems (Brennan and Zufall 2006). For example, both the main olfactory and vomeronasal systems respond to putative male mouse pheromones at high sensitivity. More surprisingly, olfactory sensory neurons (OSNs) from the main olfactory system also respond to MHC peptide ligands (Spehr, Kelliher, Li, Boehm, Leinders-Zufall and Zufall 2006). It may seem unlikely that the main olfactory system would respond to non-volatile stimuli such as MHC peptide ligands. However, when male mice investigated females whose anogenital region had been painted with rhodamine, which is a non-volatile fluorescent dye, fluorescence was found across all the nasal endoturbinates (Spehr et al. 2006). This suggests that non-volatile molecules are able to gain access to the main olfactory epithelium of mice following direct investigation of a stimulus.

OSN responses to peptides were found to be highly sensitive, with thresholds down to 10^{-11} M, although they were around two orders of magnitude less sensitive than VSN responses (Spehr et al. 2006). OSN responses to MHC peptide ligands also displayed a different dependence on anchor residues. Whereas replacement of anchor residues with alanines abolished the responses of VSNs, it shifted the stimulus response curve of around two thirds of the OSNs, making them less sensitive by approximately two orders of magnitude. However, OSNs still failed to respond to the scrambled version of the peptide or a mixture of its constituent amino acids (Spehr et al. 2006). This suggests that the response of OSNs to MHC peptide ligands may be more dependent on the overall sequence of amino acids in the peptide chain, rather being solely dependent on the position of anchor residues. This is highly interesting, as the ability of OSNs to recognise specific MHC peptide ligands could theoretically confer the ability to detect peptides of pathogenic origin, and therefore convey information about the health status of a conspecific.

Peptide-sensitive OSNs have a ciliated morphology, which is characteristic of sensory neurons in the main olfactory epithelium. They also appear to use a similar transduction mechanism based on cAMP and the canonical olfactory cyclic nucleotide gated ion channel (Spehr et al. 2006). This contrasts with peptide-sensitive VSNs, in which transduction appears to involve diacylglycerol gated ion channels (Leinders-Zufall et al. 2004). Calcium imaging of the main olfactory epithelium revealed that the responses of peptide-sensitive OSNs were highly specific. OSNs were never found to respond to both C57/BL6 and BALB/c MHC peptide ligands, as was the case in VSNs (Spehr et al. 2006), suggesting that they may use a different coding strategy.

Consistent with effectiveness as stimuli for OSNs, MHC peptide ligands have been shown to have behavioural effects that are mediated by the main olfactory system. Male mice normally spend more time investigating urine from females of a dissimilar strain to themselves, and the addition of dissimilar MHC peptide ligands to urine from females of the same strain as the mating male resulted in a similar increase in investigation (Spehr et al. 2006). This behavioural response required direct contact with the stimulus and was unaffected by removal of the VNO. Moreover, the increased investigation time in response to dissimilar MHC peptide ligands was abolished in CNGA2 knockout mice that lacked the classical main olfactory transduction mechanism.

12.4 Conclusions and Future Directions

In addition to their immunological role, MHC peptide ligands have recently been discovered to function as chemosignals, linking individuality at the immunological and behavioural levels. The discovery of peptide-sensitive OSNs in the main olfactory system extends their possible influence to species that lack a VNO, such as humans. Given the ubiquity of class I MHC proteins amongst vertebrates, the role of MHC peptide ligands in signalling individual identity is likely to be widespread, and not just limited to mice. Indeed, the addition of synthetic MHC peptide ligands has been shown to bias the mate preference of female sticklebacks, in a way consistent with attaining optimum MHC diversity in their offspring (Milinski, Griffiths, Wegner, Reusch, Haas-Assenbaum and Boehm 2005). Furthermore, there is a plausible argument that the role of peptides in signalling individuality actually preceded their role in immune recognition (Boehm 2006).

There is clear evidence that highly sensitive receptors for MHC peptide ligands can be found in both the main olfactory and vomeronasal systems of mice.

Furthermore these peptide ligands can elicit behavioural and physiological effects, when presented in biologically relevant contexts. And yet there remains a prominent gap in this picture, as MHC peptide ligands have not been identified in biological secretions. This is somewhat unusual in the field of pheromonal research in which research normally proceeds from finding a biological effect; via identification of a potential chemosignal released by an animal; to its synthesis and testing in a bioassay. Filling this gap in our understanding will not be a trivial, as the concentrations of the MHC peptide ligands are likely to be low, and difficult to identify among a forest of other small peptides. However, it at least serves as a reminder that there is more than one route to the identification of potential pheromones. Studies on MHC peptide ligands have so far concentrated on using urine as a stimulus. However, strain-specific responses of neurons in the accessory olfactory bulb were recorded during investigation of an anaesthetised animal, and were elicited equally effectively by investigation of the head and the anogenital area (Luo et al. 2003). Future investigations need to test for the presence of MHC peptide ligands in a wider range of biological secretions, from a variety of species, including saliva, skin and vaginal secretions, and numerous poorly characterised secretions from specialised scent glands.

Many interesting questions remain in the study of the receptor and transduction mechanisms involved in the detection of MHC peptide ligands. The role of the MHC Ib molecules in ligand binding, and in determining the specificity of the potential receptor complex with V2Rs, will be of particular interest. As will a comparison of the receptor proteins between peptide sensing OSNs and VSNs. Further investigations into the co-expression of V2Rs, and the receptor coding strategy of the peptide-sensing VSNs compared to OSNs, will be of great importance in elucidating how the information is handled by the neural systems that mediate the physiological and behavioural effects. There is also much to be gained from a greater understanding of the different transduction mechanisms used by the ever-expanding variety of chemosensory sub-systems (Brennan and Zufall 2006). This will enable them to be targeted more selectively using genetic technology, which will in turn allow a greater understanding of the role of MHC peptide-ligands in natural behavioural contexts.

Our understanding of the chemosensory role of the MHC has been revolutionised by the discovery of responses to MHC peptide ligands. However, it should not be forgotten that a wealth of evidence has accumulated over the last 20 years, linking MHC type with the profile of volatile constituents of mouse urine. Moreover, other genetic differences between individuals can also signal individuality, such as the major urinary proteins of mice (Hurst and Beynon 2004). Therefore a picture emerges of a variety of chemosensory signals of individuality that can be sensed by both the main olfactory and vomeronasal systems. Different chemosignals and chemosensory systems are likely to be used in different behavioural contexts. In particular, the main olfactory system's capacity to learn and recognise complex mixtures of airborne volatiles, provides a way of linking volatile and non-volatile signals into a multimodal sensory representation of an individual or social group. It additionally provides a signal that can be sensed at a distance, avoiding the potential dangers inherent in investigating non-volatile stimuli directly (Hurst and Beynon 2004). Teasing apart the relative contributions of innate and learnt responses to the different cues, in different species and behavioural contexts, will provide continuing interest to this field for some time to come.

References

Bellringer, J.F., Pratt, H.P.M. and Keverne, E.B. (1980) Involvement of the vomeronasal organ and prolactin in pheromonal induction of delayed implantation in mice. J. Reprod. Fert. 59, 223–228.

Boehm, T. (2006) Co-evolution of a primordial peptide-presentation system and cellular immunity. Nature Rev. Immunol. 6, 79–84.

Boehm, T. and Zufall, F. (2006) MHC peptides and the sensory evaluation of genotype. Trends Neurosci. 29, 100–107.

Boyse, E.A., Beauchamp, G.K. and Yamazaki, K. (1987) The genetics of body scent. Trends Genet. 3, 97–102.

Brennan, P., Kaba, H. and Keverne, E.B. (1990) Olfactory Recognition: a simple memory system. Science 250, 1223–1226.

Brennan, P. and Zufall, F. (2006) Pheromonal communication in vertebrates. Nature, in press.

Bruce, H. (1959) An exteroceptive block to pregnancy in the mouse. Nature 184, 105.

Carroll, L.S., Penn, D.J. and Potts, W.K. (2002) Discrimination of MHC-derived odors by untrained mice is consistent with divergence in peptide-binding region residues. Proc. Natl. Acad. Sci. USA 99, 2187–2192.

Hurst, J. and Beynon, R. (2004) Scent wars: the chemobiology of competitive signalling in mice. Bioessays 26, 1288–1298.

Ishii, T., Hirota, J. and Mombaerts, P. (2003) Combinational coexpression of neural and immune multigene families in mouse vomeronasal sensory systems. Curr. Biol. 13, 394–400.

Jacob, S., McClintock, M.K., Zelano, B. and Ober, C. (2002) Paternally inherited HLA alleles are associated with women's choice of male odor. Nat. Genet. 30, 175–179.

Jordan, W.C. and Bruford, M.W. (1998) New perspectives on mate choice and the MHC. Heredity 81, 127–133.

Kelliher, K., Spehr, M., Li, X.-H., Zufall, F. and Leinders-Zufall, T. (2006) Pheromonal recognition memory induced by TRPC2-independent vomeronasal sensing. Eur. J. Neurosci. 23, 3385–3390.

Leinders-Zufall, T., Brennan, P., Widmayer, P., Chandramani, P.S., Maul-Pavicic, A., Jäger, M., Li, X.-H., Breer, H., Zufall, F. and Boehm, T. (2004) MHC class I peptides as chemosensory signals in the vomeronasal organ. Science 306, 1033–1037.

Lloyd-Thomas, A. and Keverne, E.B. (1982) Role of the brain and accessory olfactory system in the block to pregnancy in mice. Neuroscience 7, 907–913.

Loconto, J., Papes, F., Chang, E., Stowers, L., Jones, E.P., Takada, T., Kumanovics, A., Fischer-Lindahl, K. and Dulac, C. (2003) Functional expression of murine V2R pheromone receptors involves selective association with the M10 and M1 families of MHC class 1b molecules. Cell 112, 607–618.

Luo, M.M., Fee, M.S. and Katz, L.C. (2003) Encoding pheromonal signals in the accessory olfactory bulb of behaving mice. Science 299, 1196–1201.

Ma, D., Allen, N.D., Van Bergen, Y.C.H., Jones, C.M.E., Baum, M.J., Keverne, E.B. and Brennan, P.A. (2002) Selective ablation of olfactory receptor neurons without functional impairment of vomeronasal receptor neurons in OMP-ntr transgenic mice. Eur. J. Neurosci. 16, 2317–2323.

Manning, C.J., Wakeland, E.K. and Potts, W.K. (1992) Communal nesting patterns in mice implicate MHC genes in kin recognition. Nature 360, 581–583.

Martini, S., Silvotti, L., Shirazi, A., Ryba, J.P. and Tirindelli, R. (2001) Co-expression of puta-
    tive pheromone receptors in the sensory neurons of the vomeronasal organ. J. Neurosci. 21,
    843–848.
Milinski, M., Griffiths, S., Wegner, K., Reusch, T., Haas-Assenbaum, A. and Boehm, T. (2005)
    Mate choice decisions of stickleback females predictably modified by MHC peptide ligands.
    Proc. Natl. Acad. Sci. USA 102, 4414–4418.
Olson, R., Huey-Tubman, K., Dulac, C. and Bjorkman, P. (2005) Structure of a pheromone
    receptor-associated MHC molecule with an open and empty groove. PLOS 3, e257.
Paulsson, K. (2004) Evolutionary and functional perspectives of the major histocompatibility com-
    plex class I antigen-processing machinery. Cell. Mol. Life Sci. 61, 2446–2460.
Penn, D. and Potts, W.K. (1998a) MHC-disassortative mating preferences reversed by cross-
    fostering. Proc. R. Soc. Lond. B 265, 1299–1306.
Penn, D. and Potts, W.K. (1998b) Untrained mice discriminate MHC-determined odors. Physiol.
    Behav. 63, 235–243.
Penn, D. and Potts, W.K.. (1998c) How do major histocompatibility complex genes influence odor
    and mating preferences? Adv. Immunol. 69, 411–436.
Porter, R.H., Balogh, R.D., Cernoch, J.M. and Franchi, C. (1986) Recognition of kin through char-
    acteristic body odors. Chem. Sens. 11, 389–395.
Potts, W.K., Manning, C.J. and Wakeland, E.K. (1991) Mating patterns in seminatural populations
    of mice influenced by MHC genotype. Nature 352, 619–621.
Reusch, T., Häberli, M., Aeschlimann, P. and Milinski, M. (2001) Female sticklebacks count alleles
    in a strategy of sexual selection explaining MHC polymorphism. Nature 414, 300–302.
Schaefer, M.L., Yamazaki, K., Osada, K., Restrepo, D. and Beauchamp, G.K. (2002) Olfactory fin-
    gerprints for major histocompatibility complex-determined body odors II: relationship among
    odor maps, genetics, odor composition, and behavior. J. Neurosci. 22, 9513–9521.
Singer, A.G., Beauchamp, G.K. and Yamazaki, K. (1997) Volatile signals of the major histocom-
    patibility complex in male mouse urine. Proc. Natl. Acad. Sci. USA 94, 2210–2214.
Singh, P.B. (2001) Chemosensation and genetic individuality. Reproduction 121, 529–539.
Singh, P.B., Brown, R.E. and Roser, B. (1987) MHC antigens in urine as olfactory recognition
    cues. Nature 327, 161–164.
Spehr, M., Kelliher, K., Li, X.-H., Boehm, T., Leinders-Zufall, T. and Zufall, F. (2006) Essential
    role of the main olfactory system in social recognition of major histocompatibility complex
    peptide ligands. J. Neurosci 26, 1961–1970.
Wedekind, C. and Furi, S. (1997) Body odour preferences in men and women: do they aim for
    specific MHC combinations or simply heterozygosity? Proc. R. Soc. Lond. B 264, 1471–1479.
Yamazaki, K., Beauchamp, G.K., Curran, M. and Boyse, E.A. (2000) Parent-progeny recognition
    as a function of MHC odotype identity. Proc. R. Soc. Lond. B 97, 10500–10502.
Yamazaki, K., Beauchamp, G.K., Kupniewski, D., Bard, J., Thomas, L. and Boyse E.A. (1988)
    Familial imprinting determines H-2 selective mating preferences. Science 240, 1331–1332.
Yamazaki, K., Beauchamp, G.K., Imai, Y., Bard, J., Phelan, S.P., Thomas, L. and Boyse, E.A.
    (1990) Odortypes determined by the major histocompatibility complex in germfree mice. Proc.
    Natl. Acad. Sci. 87, 8413–8416.
Yamazaki, K., Beauchamp, G.K., Wysocki, C.J., Bard, J., Thomas, L. and Boyse, E.A. (1983)
    Recognition of H-2 Types in relation to the blocking of pregnancy in mice. Science 221,
    186–188.
Yamazaki, K., Yamaguchi, M., Baranoski, L., Bard, J., Boyse, E.A. and Thomas, L. (1979) Recog-
    nition among mice: Evidence from the use of a Y maze differentially scented by congenic mice
    of different major histocompatibility types. J. Exp. Med. 150, 755–760.

Chapter 13
Pregnancy Block from a Female Perspective

Stuart D. Becker and Jane L. Hurst

Abstract Within a limited time after mating, exposure of female rodents to the scent of an unfamiliar conspecific male results in pregnancy termination. Since its discovery in mice, pregnancy block (or the 'Bruce Effect') has been confirmed in several other murine and microtine rodent species. Adaptive explanations for this behaviour have traditionally focused on advantages to the blocking male, but the suggested benefits to females remain controversial. Consideration of potential female benefits and the implications of female advantage in pregnancy block suggest that this behaviour could evolve with little or no reference to male advantage, and may represent a potential reproductive cost to stud males.

13.1 The Mechanism of Pregnancy Block

Following mating, exposure of female laboratory mice to the urinary scent of an unfamiliar male causes pregnancy disruption and return to oestrus (Parkes and Bruce 1961). The timing of exposure is critical. Around oestrus, female rodents show daily prolactin surges, increasing to twice daily after mating and peaking approximately one hour before the change to light and dark periods (Barkley, Bradford and Geschwind 1978; Ryan and Schwartz 1980). Pregnancy block occurs only if females are exposed to male scent coincident with two prolactin peaks, at least one during the light phase, while exposure outside these peaks fails to cause pregnancy block (Rosser, Remfry and Keverne 1989).

Pregnancy disruption is mediated through activation of a specific vomeronasal neuroendocrine pathway that inhibits prolactin release (Brennan and Binns 2005). As prolactin is essential for maintaining luteal function during early pregnancy in rodents (Stormshak, Zelinski-Wooten and Abdelgadir 1987), this inhibitory pathway causes luteolysis and hence pregnancy failure. The duration of sensitivity to pregnancy blocking signals varies between species, ranging from 4–5 days

Stuart D. Becker
University of Liverpool, Department of Veterinary Preclinical Science
stuart.becker@liv.ac.uk

J.L. Hurst et al., *Chemical Signals in Vertebrates 11.*
© Springer 2008

post-mating (pre-implantation) in *Mus* (Parkes and Bruce 1961) up to 17 days post-mating (pre- and post-implantation) in microtine species (Stehn and Jannett 1981).

During the period 4–6 h after mating, females learn the scent signature of the stud male, enabling them to recognise a different male scent as unfamiliar (Brennan, Kaba and Keverne 1990). Studies commonly define familiar and unfamiliar males according to whether they are from the same or different inbred strain of genetically identical individuals, such as C57BL/6, CBA or BALB/c (e.g. Yamazaki, Beauchamp and Wysocki 1983; Rosser et al. 1989; Peele, Salazar, Mimmack, Keverne and Brennan 2003). Other differences such as social status (Labov 1981a; Huck 1982), specific aspects of genotype (Coopersmith and Lenington 1998; Brennan and Peele 2003), or individual differences within an outbred strain (Bruce 1960) have also been investigated. However, exposure to unfamiliar individuals does not necessarily cause pregnancy block in wild mice (Coopersmith and Lenington 1998).

Exposure to the stud male's scent after mating fails to disrupt pregnancy (Bruce 1960) and may reduce the likelihood of pregnancy block if females are concurrently exposed to unfamiliar male scent (Thomas and Dominic 1987). Pregnancy blocking signals are thought to be androgen-dependent, and male scent gradually loses its efficacy to trigger pregnancy block over 6 weeks post-castration (Spironello-Vella and deCatanzaro 2001). However, in early studies castrated males were efficacious in blocking pregnancy throughout the 18 week post-operative period of study (Bruce 1960). Pregnancy disruption was also seen after exposure to the scent of females differing from the stud male's strain only at the major histocompatibility complex (MHC) (Yamazaki et al. 1983). The androgen-dependence of cues triggering pregnancy block is thus not entirely clear.

Memory formation has been shown to be contingent on mating (Kaba, Rosser and Keverne 1989), although some evidence suggests that prior familiarity gained through longer-term exposure to a particular male scent without mating may also reduce the efficacy of that scent to produce pregnancy block (Bloch 1974). Memorising the stud male's scent is thought to be mediated through disruption of the mating male's pregnancy-blocking signal by selectively enhanced inhibition in the accessory olfactory bulb (Binns and Brennan 2005). In the context of pregnancy block, females determine male familiarity via the vomeronasal/accessory olfactory system, through recognition of a combination of male-specific pheromones and MHC peptides (Brennan and Peele 2003; Leinders-Zufall, Brennan, Widmayer, Chandramani, Maul-Pavicic, Jäger, Li, Breer, Zufall and Boehm 2004). Fractionation studies of male urine demonstrate that major urinary proteins (MUPs), known to underlie individual recognition in mice (Hurst, Payne, Nevison, Marie, Humphries, Robertson, Cavaggioni and Beynon 2001), are not involved in the recognition of unfamiliarity in this context (Peele et al. 2003). MHC class I proteins bind a wide variety of peptide ligands (Engelhard 1994; Brennan and Zufall 2006) that stimulate directly both the vomeronasal organ (Leinders-Zufall et al. 2004) and main olfactory epithelium (Spehr, Kelliher, Li, Boehm, Leinders-Zufall and Zufall 2006). As the binding specificities of MHC molecules, and thus the range of bound peptides, vary between alleles (Engelhard 1994), MHC-associated urinary scents reflect the MHC alleles carried by the donor.

13.2 Functional Significance

Despite extensive investigation of pregnancy block since its discovery, a convincing explanation for its functional significance and evolutionary development has remained elusive. The postponement of reproduction inevitably impairs reproductive success, but in order to evolve the Bruce effect must offer an overall benefit. The costs and benefits may be very different for males and females.

13.2.1 Male Advantages

Many early studies into the Bruce effect appear to assume it evolved for male advantage (reviewed by Schwagmeyer 1979). Individual males that cause pregnancy disruption could potentially accrue selective advantages including siring of offspring at the expense of male competitors and avoiding the provision of paternal care to unrelated offspring post-partum (Rogers and Beauchamp 1976; Schwagmeyer 1979). However, male advantage alone cannot explain the evolution of a mechanism that relies on female response. Central to many of the arguments for male reproductive advantage in pregnancy block is the assumption that females will re-mate with the blocking male after terminating their current gestation, but this behaviour has not been demonstrated except in situations of enforced cohabitation (Labov 1981b). Indeed females able to evade such male induced reproductive costs are likely to be at a significant evolutionary advantage, and the adaptive advantages of a passive female response to male scent have been queried repeatedly (e.g. Bronson and Coquelin 1980; Brennan and Peele 2003).

13.2.2 Female Advantages

To address concerns over how apparently passive female responses to pregnancy blocking cues could have evolved, several hypothetical female advantages have been suggested. Pregnancy disruption in response to desertion by the original stud male would enable a female to re-mate and so potentially increase the likelihood of paternal investment in the offspring (Dawkins 1976). However, multiple paternity is common in litters of house mice (Dean, Ardlie and Nachman 2006), and males assist with communal nursing of offspring within their territory without evidence of bias (Manning, Dewsbury, Wakeland and Potts 1995; Lonstein and De Vries 2000).

It has been suggested that in order to avoid male infanticide (and hence wasted investment in gestation), females may terminate pregnancy resulting from an earlier mating and then re-mate with the infanticidal male (Labov 1981b; Storey 1986). However, in free-ranging tests of this hypothesis, artificial replacement of stud males did not alter inter-litter interval, suggesting that females did not block pregnancies when risk of infanticide was apparently increased (Mahady and Wolff 2002).

Others have suggested that females may terminate pregnancy, regardless of infanticide risk, to exert post-copulatory mate choice. Prospective drivers of female choice include competitive ability (Labov 1981a; Huck 1982), advantageous genetic combinations in offspring (Rülicke, Guncz and Wedekind 2006), avoidance of deleterious recessive alleles (Coopersmith and Lenington 1998) or phenotypic (and hence genotypic) rarity (Schwagmeyer 1979). Laboratory-based experiments that have manipulated likely aspects of male attractiveness show contradictory results. Behavioural observations show that inbred laboratory-strain females that mate with a different-strain stud male will block pregnancy if exposed to the scent of a male genetically identical to themselves (Rülicke et al. 2006). However, females are known to prefer less closely related mates (Penn 2002), and inbred offspring suffer poorer competitive success (Meagher, Penn and Potts 2000; Tregenza and Wedell 2000).

In almost all experiments examining potential behavioural and ecological mechanisms (e.g. Bruce 1963; Labov 1981a; Huck 1982), the presence of a male or his scent are assumed to result inevitably in female exposure. Thus most experimental designs have used small cages that prevent females from expressing a choice to avoid or approach male scent, or have applied the stimulus directly to the female nares or vomeronasal organ. Very few studies examine the role of female behaviour in controlling exposure.

The issue of timing has frequently been overlooked, but may be of critical importance in the interaction between behaviour and pregnancy block. As previously described, pregnancy block occurs only if females are exposed to male scent coincident with two prolactin peaks, at least one during the light phase, while exposure outside these peaks fails to block pregnancy (Rosser et al. 1989). By altering their exposure to male scent during these brief periods of sensitivity, females could choose to maintain or terminate pregnancy in the presence of unfamiliar male scent with minimal impact on normal behaviour at other times. Published studies have recorded female behaviour outside the critical period, 3–7 h after the expected dark phase prolactin peak, thus complicating behavioural interpretation. Drickamer (1989) tested female preference by presenting wild-derived females with paired samples of soiled male bedding, and found a general avoidance of unfamiliar male scent during the early stages of gestation. Conversely, an attraction to unfamiliar scent reported by deCatanzaro & Murji (2004) occurred when inbred CF1 females simultaneously chose between two CF1 inbred males, and one outbred laboratory strain male. In this case the increased investigation directed towards the novel strain male may be due to information gathering rather than preference (Hurst, Thom, Nevison, Humphries and Beynon 2005). Neither test corresponded to the sensitive period for the Bruce effect, rendering meaningful interpretation with regard to pregnancy block extremely difficult.

Female ability to control exposure to male scent at critical times may help to explain why similar pregnancy-blocking stimuli have produced conflicting results in different experiments. For example in one study manipulating male social status (Labov 1981a), females were housed directly below males, while a similar study (Huck 1982) housed females adjacent to males, separated by mesh. The pregnancy blocking ability of dominant males was equal to subordinate males in the former study, but more efficacious in the latter. As the former study enforced female proximity to male scent while the latter did not, female attraction to dominant males may account for their greater efficacy in pregnancy block rather than any intrinsic difference in potency between dominant and subordinate male scents. In another study designed to examine the effect of carrying the deleterious t-complex genotype on male pregnancy blocking efficacy (Coopersmith and Lenington 1998), the apparatus ensured that the female's environment was saturated with male scent, suggesting that t-complex carriers were inherently less able to trigger pregnancy block than unaffected males. Interestingly in this study of genetically heterogeneous wild-derived mice, unfamiliar t-complex carriers induced no more pregnancy block than seen in control females that were not exposed to unfamiliar males, although the unfamiliar males must have differed genetically from the stud male, including MHC type. Thus pregnancy block did not occur in response to individual recognition or MHC differences between the stud versus an unfamiliar male in this study. However, high control blocking rates and behavioural restrictions imposed by the experimental apparatus make functional interpretation of female benefit impossible.

13.2.3 Maximising Female Reproductive Success

Successful reproduction in females demands substantial investment in gestation and lactation (Johnson, Thomson and Speakman 2001). While the reproductive success of males may be determined by the number of mates he can fertilise, females are limited by the number of young they can produce (Andersson 1994). The survival of young is the single most important factor in determining lifetime reproductive success in female mice and other species (Clutton-Brock 1988; König 1994). Optimal timing of reproduction is critical to offspring survival, and may be delayed according to the social environment through pheromonally-mediated mechanisms including puberty delay and oestrus inhibition (Bronson and Coquelin 1980). Female control of the Bruce effect may represent an additional method to avoid suboptimal reproductive timing according to social conditions. As females' home ranges may overlap more than one male's territory (Hurst 1987; Manning 1995), it seems likely that they would have the ability to control their exposure to male scent. Further, limiting sensitivity to a short period of the day considerably reduces opportunities for males to manipulate the Bruce effect to their own advantage (for example, by scent marking resources that females cannot afford to avoid such as food sources, Hurst and Nevison 1994).

The main factor affecting offspring survival in mice is thought to be social disruption of maternal behaviour (Peripato, de Brito, Vaughn, Pletscher, Matioli and Cheverud 2002), although other factors have been implicated including infanticide especially by non-stud males (Huck 1984), infection (e.g. Parker and Richter 1982), and predation (Millar, Havelka and Sharma 2004). The importance of the former effect can be seen in the sharp decrease in pup survival, and hence female reproductive success, in nest sites that cannot be defended effectively, particularly those used by a large number of animals including non-stud males (Southwick 1955). Indeed, overcrowding has driven the evolution of pheromonally-primed reproductive suppression (Bronson and Coquelin 1980). Together with the timing of sensitivity to the Bruce effect, this suggests a novel functional explanation for pregnancy block in an ecological context—alteration of reproductive investment based on nest stability and the associated likelihood of offspring survival.

The sensitive period, occurring approximately 1 h before dark and up to 4 days post-mating (Parkes and Bruce 1961; Rosser et al. 1989), coincides with the time females are most likely to be in sheltered nest sites (Refinetti 2004). If females remain within the nest during this sensitive period, their exposure will be restricted to other animals that share their nest through the light phase. Pregnant females strongly defend their nest sites (Vom Saal, Franks, Boechler, Palanza and Parmigiani 1995) but their ability to do so depends on the physical protection afforded by the site and social pressure to use limited sites of shelter (Wolff 1985; Hurst 1987). The presence of fresh scents from other males, particularly from outside a familiar stable group, would indicate a nest site not defended effectively. Avoidance of novel male scents would allow pregnant females to avoid settling in such sites and, since pregnancy block occurs only in response to fresh scent (Peele et al. 2003), by the end of the light phase females will have had ample opportunity to exclude males or to leave the nest for an alternative. However, where this is not possible (e.g. because defendable nest sites are limited), females that terminate pregnancy until they can find a more suitable nest will avoid wasted investment, particularly prior to implantation.

Thus, rather than providing a reproductive benefit to males as traditionally assumed, the Bruce effect may have evolved solely to female advantage. Notably, this response also increases selective pressure on stud males to increase their investment in the territorial defence of nest sites that are preferred by females (Ims 1987). Females may improve their own reproductive success through threat of pregnancy block, compelling stud males to invest more heavily in nest defence.

13.3 Future Work: Challenging Assumptions

The hypothesis that nest stability alters the probability of pregnancy block needs to be tested using naturalistic enclosures where animals have the opportunity to exhibit normal choices that are restricted by the laboratory environment. Altering the apparent stability of nest sites artificially by manipulating scent cues and/or their occupation by different males will then allow female nesting decisions to be related to the outcome for the maintenance or blocking of pregnancy. Analysis of re-mating strategies following pregnancy block, including paternity and the subsequent willingness of females to re-mate, is also essential to evaluate advantages from both a female and male perspective.

Male scent is typically used as the pregnancy-blocking stimulus during investigation of the Bruce effect. However experiments addressing the androgen-dependency of pregnancy block have used only laboratory strains, many of which are thought to be very closely-related intersubspecific hybrids (Yoshiki and Moriwaki 2006). These inbred laboratory mouse strains lack the context of the complex genetic background variation between individuals found in wild populations. Examining androgen-dependent scent characteristics, while ignoring the potential relevance of other scent cues to females in the wild, risks artificially exaggerating the importance of androgens in defining pregnancy blocking scents. The hypothesis that scent unfamiliarity may be additive and multifactorial needs to be tested using genetically disparate mice (e.g. wild-derived) of both sexes as scent donors This would help to evaluate the significance of androgens and individual recognition in pregnancy blocking signals, and the extent to which other aspects of conspecifics' scents are also relevant to the Bruce effect.

Memorising MHC-associated scent enables females to discriminate between stud and unfamiliar males, but may also allow a female to monitor scent changes in the stud male (e.g. during disease). Pregnancy could be disrupted to avoid infection of the offspring, particularly where transplacental infection could result in foetal death (e.g. Fenner 1982). The sensitivity of females to unfamiliar MHC peptide ligands in the context of pregnancy block (Leinders-Zufall et al. 2004) provides a potential mechanism for detecting changes in familiar stud male infection status since MHC molecules bind foreign peptides from pathogens. Future work could include observations of female behaviour towards infected and uninfected males during the critical period for the Bruce effect, and examine the efficacy of such scents in inducing pregnancy block using different pathogens of varying life cycle and virulence.

Lastly, the Bruce effect may be part of a general response to stressful circumstances where reproductive investment may be threatened. Pregnancy block is controlled through selectively enhanced inhibition at the level of the accessory olfactory bulb and medial amygdala (Binns and Brennan 2005). Acute stress has been shown to increase activity in the medial amygdala (Gammie and Stevenson 2006), and wild mice are known to block pregnancy in response to apparently minor stressors such as handling and cage cleaning (Chipman and Fox 1966). The potential for generalising the pregnancy blocking response suggests that detection of unfamiliar scent may be only one aspect of a more complex stress response, and that females may use this behaviour to optimise reproductive investment, taking into account multiple risks present in the natural environment.

The challenges remain to show whether altering the likelihood of offspring survival alters the behaviour of females to maintain or terminate pregnancy, to examine whether males are advantaged or disadvantaged by the Bruce effect, and to understand the importance of this intriguing behavioural and neurophysiological mechanism in house mouse ecology.

References

Andersson, M. (1994) *Sexual selection*. Princeton University Press, Princeton, New Jersey.
Barkley, M.S., Bradford, G.E. and Geschwind, II. (1978) Pattern of plasma prolactin concentration during first half of mouse gestation. Biol. Reprod. 19, 291–296.

Binns, K.E. and Brennan, P. (2005) Changes in electrophysiological activity in the accessory olfactory bulb and medial amygdala associated with mate recognition in mice. Eur. J. Neurosci. 21, 2529–2537.

Bloch, S. (1974) Observations on the ability of the stud male to block pregnancy in the mouse. J. Reprod. Fertil. 38, 469–471.

Brennan, P. and Binns, K.E. (2005) Vomeronasal mechanisms of mate recognition in mice. Chem. Senses. 30 (suppl 1), i148–i149.

Brennan, P., Kaba, H. and Keverne, E.B. (1990) Olfactory recognition: a simple memory system. Science. 250, 1223–1226.

Brennan, P. and Zufall, F. (2006) Pheromonal communication in vertebrates. Nature. 444, 308–315.

Brennan, P.A. and Peele, P. (2003) Towards an understanding of the pregnancy-blocking urinary chemosignals of mice. Biochem. Soc. Trans. 31, 152–155.

Bronson, F.H. and Coquelin, A. (1980) The modulation of reproduction by priming pheromones in house mice: speculations on adaptive function. In: D. Müller-Schwartze and R.M. Silverstein (Eds.), *Chemical Signals: Vertebrates and Aquatic Invertebrates.* Plenum, New York, pp. 243–265.

Bruce, H.M. (1960) A block to pregnancy in the mouse caused by proximity of strange males. J. Reprod. Fertil. 1, 96–103.

Bruce, H.M. (1961) Time Relations in Pregnancy-Block Induced in Mice by Strange Males. Journal of Reproduction and Fertility. 2, 138-&.

Bruce, H.M. (1963) Olfactory block to pregnancy among grouped mice. J. Reprod. Fertil. 6, 451–460.

Chipman, R.K. and Fox, K.A. (1966) Oestrus synchronization and pregnancy blocking in wild house mice (*Mus musculus*). J. Reprod. Fertil. 12, 233–236.

Clutton-Brock, T.H. (1988) Reproductive Success. In: T.H. Clutton-Brock (Eds.), *Reproductive success: studies of individual variation in contrasting breeding systems.* Chicago University Press, Chicago, pp. 472–485.

Coopersmith, C.B. and Lenington, S. (1998) Pregnancy block in house mice (*Mus domesticus*) as a function of t-complex genotype: Examination of the mate choice and male infanticide hypotheses. J. Comp. Psychol. 112, 82–91.

Dawkins, R. (1976) *The Selfish Gene.* Oxford University Press, Oxford.

Dean, M.D., Ardlie, K.G. and Nachman, M.W. (2006) The frequency of multiple paternity suggests that sperm competition is common in house mice (*Mus domesticus*). Mol. Ecol. 15, 4141–4151.

deCatanzaro, D. and Murji, T. (2004) Inseminated female mice (*Mus musculus*) investigate rather than avoid novel males that disrupt pregnancy, but sires protect pregnancy. J. Comp. Psychol. 118, 251–257.

Drickamer, L.C. (1989) Pregnancy block in wild stock house mice, *Mus domesticus*—olfactory preferences of females during gestation. Anim. Behav. 37, 690–692.

Engelhard, V.H. (1994) Structure of peptides associated with class I and class II MHC molecules. Annu. Rev. Immunol. 12, 181–207.

Fenner, F. (1982) Mousepox. In: H.L. Foster, J.D. Small and J.G. Fox (Eds.), *The Mouse in Biomedical Research.* Academic Press, New York, pp. 209–230.

Gammie, S.C. and Stevenson, S.A. (2006) Effect of daily and acute restraint stress during lactation on maternal aggression and behavior in mice. Stress. 9, 171–180.

Huck, U.W. (1982) Pregnancy block in laboratory mice as a function of male social status. J. Reprod. Fertil. 66, 181–184.

Huck, U.W. (1984) Infanticide and the evolution of pregnancy block in rodents. In: G. Hausfater and S.B. Hrdy (Eds.), *Infanticide: comparative and evolutionary perspectives.* Aldine, New York, pp. 349–365.

Hurst, J.L. (1987) Behavioral variation in wild house mice *Mus domesticus* rutty - a quantitative assessment of female social organization. Anim. Behav. 35, 1846–1857.

Hurst, J.L. and Nevison, C. (1994) Do female house mice (*Mus musculus domesticus*) regulate their exposure to reproductive priming pheromones ? Anim. Behav. 48, 945–959.

Hurst, J.L., Payne, C.E., Nevison, C.M., Marie, A.D., Humphries, R.E., Robertson, D.H., Cavaggioni, A. and Beynon, R.J. (2001) Individual recognition in mice mediated by major urinary proteins. Nature. 414, 631–4.

Hurst, J.L., Thom, M.D., Nevison, C.M., Humphries, R.E. and Beynon, R.J. (2005) MHC odours are not required or sufficient for recognition of individual scent owners. Proc. R. Soc. Lond., B, Biol. Sci. 272, 715–724.

Ims, R.A. (1987) Male spacing systems in microtine rodents. Am. Nat. 130, 475–484.

Johnson, M.S., Thomson, S.C. and Speakman, J.R. (2001) Limits to sustained energy intake III. Effects of concurrent pregnancy and lactation in *Mus musculus*. J. Exp. Biol. 204, 1947–1956.

Kaba, H., Rosser, A. and Keverne, B. (1989) Neural basis of olfactory memory in the context of pregnancy block. Neuroscience. 32, 657–662.

König, B. (1994) Components of lifetime reproductive success in communally and solitarily nursing house mice - a laboratory study. Behav. Ecol. Sociobiol. 34, 275–283.

Labov, J.B. (1981a) Male social status, physiology, and ability to block pregnancies in female house mice (*Mus musculus*). Behav. Ecol. Sociobiol. 8, 287–291.

Labov, J.B. (1981b) Pregnancy blocking in rodents: adaptive advantages for females. Am. Nat. 118, 361–371.

Leinders-Zufall, T., Brennan, P., Widmayer, P., Chandramani S., P., Maul-Pavicic, A., Jäger, M., Li, X.-H., Breer, H., Zufall, F. and Boehm, T. (2004) MHC Class I peptides as chemosensory signals in the vomeronasal organ. Science. 306, 1033–1037.

Lonstein, J.S. and De Vries, G.J. (2000) Sex differences in the parental behavior of rodents. Neuroscience and Biobehavioral Reviews. 24, 669–686.

Mahady, S.J. and Wolff, J.O. (2002) A field test of the Bruce effect in the monogamous prairie vole (*Microtus ochrogaster*). Behav. Ecol. Sociobiol. 52, 31–37.

Manning, C.J., Dewsbury, D.A., Wakeland, E.K. and Potts, W.K. (1995) Communal nesting and communal nursing in house mice, *Mus musculus* domesticus. Anim. Behav. 50, 741–751.

Meagher, S., Penn, D.J. and Potts, W.K. (2000) Male–male competition magnifies inbreeding depression in wild house mice. Proc. Natl. Acad. Sci. U.S.A. 97, 3324–3329.

Millar, J.S., Havelka, M.A. and Sharma, S. (2004) Nest mortality in a population of small mammals. Acta Theriol. 49, 269–273.

Parker, J.C. and Richter, C.B. (1982) Viral diseases of the respiratory system. In: H.L. Foster, J.D. Small and J.G. Fox (Eds.), *The Mouse in Biomedical Research*. Academic Press, New York, pp. 109–158.

Parkes, A.S. and Bruce, H.M. (1961) Olfactory stimuli in mammalian reproduction. Science. 134, 1049–1054.

Peele, P., Salazar, I., Mimmack, M., Keverne, E.B. and Brennan, P.A. (2003) Low molecular weight constituents of male mouse urine mediate the pregnancy block effect and convey information about the identity of the mating male. Eur. J. Neurosci. 18, 622–628.

Penn, D.J. (2002) The scent of genetic compatibility: sexual selection and the Major Histocompatibility complex. Ethology. 108, 1–21.

Peripato, A.C., de Brito, R.A., Vaughn, T.T., Pletscher, L.S., Matioli, S.R. and Cheverud, J.M. (2002) Quantitative trait loci for maternal performance for offspring survival in mice. Genetics. 162, 1341–1353.

Refinetti, R. (2004) Daily activity patterns of a nocturnal and a diurnal rodent in a seminatural environment. Physiol Behav. 82, 285–294.

Rogers, J.G. and Beauchamp, G.K. (1976) Influence of stimuli from populations of *Peromyscus leucopus* on maturation of young. J. Mammal. 57, 320–330.

Rosser, A.E., Remfry, C.J. and Keverne, E.B. (1989) Restricted exposure of mice to primer pheromones coincident with prolactin surges blocks pregnancy by changing hypothalamic dopamine release. J. Reprod. Fert. 87, 553–559.

Rulicke, T., Guncz, N. and Wedekind, C. (2006) Early maternal investment in mice: no evidence for compatible-genes sexual selection despite hybrid vigor. J. Evol. Biol. 19, 922–928.

Ryan, K.D. and Schwartz, N.B. (1980) Changes in serum hormone levels associated with male-induced ovulation in group-housed adult female mice. Endocrinology. 106, 959–966.

Schwagmeyer, P.L. (1979) Bruce effect - evaluation of male-female advantages. Am. Nat. 114, 932–938.

Southwick, C.H. (1955) Regulatory mechanisms of house mouse populations: social behavior affecting litter survival. Ecology. 36, 627–634.

Spehr, M., Kelliher, K.R., Li, X.-H., Boehm, T., Leinders-Zufall, T. and Zufall, F. (2006) Essential role of the main olfactory system in social recognition of major histocompatability complex peptide ligands. J. Neurosci. 26, 1961–1970.

Spironello-Vella, E. and deCatanzaro, D. (2001) Novel male mice show gradual decline in the capacity to disrupt early pregnancy and in urinary excretion of testosterone and 17 beta-estradiol during the weeks immediately following castration. Horm. Metab. Res. 33, 681–686.

Stehn, R.A. and Jannett, F.J. (1981) Male-induced abortion in various microtine rodents. J. Mammal. 62, 369–372.

Storey, A.E. (1986) Influence of sires on male-induced pregnancy disruptions in meadow voles (Microtus pennsylvanicus) differs with stage of pregnancy. J. Comp. Psychol. 100, 15–20.

Stormshak, F., Zelinski-Wooten, M.B. and Abdelgadir, S.E. (1987) Comparative aspects of the regulation of corpus luteum function in various species. Adv. Exp. Med. Biol. 219, 327–360.

Thomas, K.J. and Dominic, C.J. (1987) Evaluation of the role of the stud male in preventing male-induced implantation failure (the Bruce effect) in laboratory mice. Anim. Behav. 35, 1257–1259.

Tregenza, T. and Wedell, N. (2000) Genetic compatability, mate choice and patterns of parentage. Mol. Ecol. 9, 1013–1027.

Vom Saal, F.S., Franks, P., Boechler, M., Palanza, P. and Parmigiani, S. (1995) Nest defence and survival of offspring in highly aggressive wild canadian female house mice. Physiol. Behav. 58, 669–678.

Wolff, R.J. (1985) Mating behaviour and female choice: their relation to social structure in wild caught house mice (Mus musculus) housed in a semi-natural environment. J. Zool. 207, 43–51.

Yamazaki, K., Beauchamp, G.K. and Wysocki, C.J. (1983) Recognition of H-2 types in relation to the blocking of pregnancy in mice. Science. 221, 186–188.

Yoshiki, A. and Moriwaki, K. (2006) Mouse phenome research: implications of genetic background. Ilar J. 47, 94–102.

Chapter 14
The Wing-Sac Odour of Male Greater Sac-Winged Bats *Saccopteryx bilineata* (Emballonuridae) as a Composite Trait: Seasonal and Individual Differences

Barbara Caspers, Stephan Franke, and Christian C. Voigt

Abstract Male *Saccopteryx bilineata* possess a sac-like organ for the storage and display of odoriferous secretion in their front wing membrane. Since males use the scent in agonistic and courtship activities, and compose it from different secretions of distinct sources, like saliva, urine and gland secretion, we hypothesized that multiple information is encoded in the male scent-profile. We expected that the odour profile of males varies seasonally, giving information on male reproductive status. In addition, the odour profile ought to vary between individuals, thus providing the possibility for individual recognition. We repeatedly collected samples from wing-sac liquids of 20 male *S. bilineata* in five Costa Rican colonies during the mating and non-mating season. Samples were analysed by gas chromatography coupled to mass spectrometry to compare wing-sac contents. Wing-sac odours included various substances such as carboxylic acids, terpenoids and aromatic compounds. Male scent profiles varied (1) between seasons in the relative amount of tetradecanoic and octadecanoic acid, and (2) between individuals in the relative amount of two species-specific substances. These results suggest that the wing-sac liquid of male *S. bilineata* is indeed a composite trait and may be useful for the simultaneous transfer of multiple information.

14.1 Introduction

Selection, natural as well as sexual, may shape signals in various ways when communication signals enhance or reduce the signallers' fitness. Although signals are often associated with a particular message, they often convey several messages. This has been demonstrated for acoustical signals (Runkle, Wells, Robb and Lance 1994), visual signals (Badyaev, Hill, Dunn and Glenn 2001; Basolo and Trainor 2002;

Barbara Caspers

Leibniz-Institute for Zoo and Wildlife Research, Berlin,

caspers@izw-berlin.de

J.L. Hurst et al., *Chemical Signals in Vertebrates 11.*

© Springer 2008

Grether, Kolluru and Nersissian 2004; Robson, Goldizen and Green 2005) as well as chemical signals (Buesching, Waterhouse and Macdonald 2002a, b). Signals, especially in long living species, in which reproduction is not limited to a single lifetime event and social groups are formed for a long time are often under multiple selective forces. To facilitate individual recognition, signals should carry a predictable individual signature. At the same time, potential mates should prefer signals that present a snapshot of the momentary health and social status of the signaller, selecting for ultimately honest signals (e.g. scents: Penn and Potts 1998). Lastly, hormonal changes during different reproductive status should be encoded in the signal if potential mates aim to assess the reproductive status of a potential mate (Marler 1961). Thus, either distinct signals encoding each type of information separately, or composite traits encoding simultaneously for more than one type of information, are required.

Composite traits have been found in different modalities among several organisms across the animal kingdom, e.g. in green swordtails, *Xiphophorus helleri* (Basolo and Trainor 2002), or the European badger, *Meles meles* (Buesching et al. 2002a, b). Studies on house finches, *Carpodacus mexicanus*, have shown that the plumage colour acts as a composite trait (Badyaev et al. 2001). Several independent components of the ornament, like coloration, pigment asymmetry or patch area asymmetry had different fitness consequences (Badyaev et al. 2001).

In mammals, olfactory cues play an important role in sexual selection, both during matechoice and male contests (Andersson 1994) and as olfactory cues are almost always multi-component, they provide a multitude of possibilities to encode information (Albone 1984) . Chemical signals may provide information about health (Penn and Potts 1998; Kavaliers, Colwell, Braun and Choleris 2003; Zala, Potts and Penn 2004), dominance status (Gosling and Roberts 2001), sex and group membership (Hofer, East, Sämmang and Dehnhard 2001; Safi and Kerth 2003) respectively population affiliation (Hayes, Richardson, Claus and Wyllie 2002).

Greater sac-winged bats, *Saccopteryx bilineata*, have a large behavioural repertoire that involves among others the use of odour (Bradbury and Emmons 1974; Bradbury and Vehrencamp 1976). Males possess, in contrast to females, a pouch in their front wing membrane, which is used to store and display odour (Voigt and von Helversen 1999). The wing sacs are free of secreting cells (Starck 1958; Scully, Fenton and Saleuddin 2000) but contain an odoriferous liquid (Bradbury and Emmons 1974). Male greater sac-winged bats show a unique behaviour during which they clean and actively refill the wing-sacs daily during a stereotypic time-consuming two-stage process with various odoriferous secretions from different body regions (Voigt and von Helversen 1999; Voigt 2002, 2005). Microbial analysis of the wing pouches showed that males have on average fewer microbes in the wing-sac than females in their rudimental pouch, and that every male has on average two out of 40 different microbes in their wing-pouch, possibly indicating an individual microbial flora (Voigt, Caspers and Speck 2005). The mating system of greater sac-winged bats has been described as harem-polygynous (Bradbury and Emmons 1974; Bradbury and Vehrencamp 1976), since the basic social unit within a colony comprises of a single adult male and a varying number of females. In the remainder of the text we refer to these social units as harems and to the territorial male as the harem male. Harem males display their wing-sac odour throughout the year during aerial displays and also when roosting next to a colony member or in agonistic male-male encounters. The most frequent odour display is a hovering flight during which males remain airborne in front of a female for a few seconds (Bradbury and Emmons 1974; Bradbury and Vehrencamp 1976; Voigt and von Helversen 1999; Voigt, von Helversen, Michener and Kunz 2001; Voigt 2005). This behaviour can be observed throughout the whole year, although it is more often performed by males prior and during the mating season (Voigt and Schwarzenberger 2007). Voigt and von Helversen (1999) argued that the potential for female choice in *S. bilineata* is large, because females are larger than males and have control over copulations (Tannenbaum 1975). This is also reflected in a high percentage (70%) of extra harem paternities (Heckel, Voigt, Mayer and von Helversen 1999). Thus, it is very likely that the chemical signals of male greater sac-winged bats may be shaped by sexual selection.

The wing-sac odour of male greater sac-winged bats is involved in several behaviours, is a mixture of several secretions from different body regions, and filled into a specialized storage organ, which makes *S. bilineata* a suitable subject for the study of compositional chemical traits. We hypothesized that the wing-sac odour of male *S. bilineata* is a compositional trait. We collected wing-sac secretion of 20 male individuals during different reproductive seasons, analysed the volatiles using a gas chromatograph linked to a mass spectrometer (GC-MS) and predicted that scents ought to carry information about (1) reproductive status and (2) individuality in male greater sac-winged bats.

14.2 Material and Methods

14.2.1 Study Population

All samples were collected from a population of greater sac-winged bats at La Selva Biological Station administered by the Organization for Tropical Studies (OTS) in Costa Rica. La Selva Biological Station (10°25' N, 84°00' W) is surrounded by tropical lowland rainforest at an elevation of approximately 100 m. Odour samples were collected during three field seasons (September 2003 to January 2004, August 2004, and December 2005 to March 2005) from individuals of five different colonies, all located in abandoned cottages surrounded by primary and secondary growth forest. Individuals were mist-netted in the morning between 0500-0600 hours, when they returned to their day-time roost or alternatively in the daytime roost, close after they returned early in the morning (0600-0830 hours) using nylon mist-nets (Avinet, CH2, Mist Net, 38 mm mesh in 50 denier, 2-ply nylon, 4 shelves, net size 2.6 m). We collected odour samples from all caught males, weighed them, and noted sex and reproductive condition. For individual recognition all individuals were marked with coloured plastic bands (Hughes, size XCL) on the forearm. Immediately afterwards all bats were released at the site of capture. None of the captured individuals disappeared from the daytime roost after data and sample collection.

14.2.2 Sample Collection

Odour samples were collected from *S. bilineata* by wiping out the wing-sac with a piece of cotton (Hartmann DIN 61640-CO, 100% cotton). Female odour samples were taken by wiping out the rudimental part of the wing sac (for pictures of the wing-sac and the wing-sac rudiment see Voigt et al. 2005). The cotton was washed with dichlormethane and dried prior to taking odour samples. We wiped out only one wing-sac each time we took a sample and alternated the wings between subsequent sampling events to ensure that the impact of sample collection was as low as possible and had minimal influence on the subsequent sample of the same individual. Samples were stored in Teflon-capped glass vials (2 ml) and 100 µl of dichlormethane were added to each sample for preservation. In total 41 samples from 20 different males in five colonies were analysed.

To test for seasonal variation of wing-sac odours, we collected odour samples from seven different males. Each of the seven males was caught once during the non-mating season (August 2004) and once during the mating season (December 2003 – January 2004 or December 2004 – January 2005). To test for the individuality of wing-sac odours, samples from adult males caught at least three times were used.

14.2.3 Chemical Analysis

Odour samples were analysed by GC and GC-MS using a Hewlett-Packard 5890 gas-chromatograph equipped with a 30 m J & W (J&W Scientific, Folsom, CA) DB5-coated capillary column and a linked Hewlett-Packard mass selective detector (MSD; 70eV EI). Dichlormethane extracts of the cotton samples were concentrated to approximately 5µl prior to analysis. Data were collected under the following GC conditions: 1.5µl split-less injection, helium as carrier gas, 60°C inlet temperature, 3 min initial time, 10°C/min rate, 280°C final temperature, 20 min final time. In addition, GC-MS chemical ionisation and high resolution experiments were performed with a VG 70-250 SE mass spectrometer on 30 m DB5MS and 50 m CPSil88CB capillary columns. For analyses the proportion of each peak area to the total peak area was calculated in percent (%). The following criteria were applied to choose substances for more detailed analysis: a) male specificity, b) fatty acids, and c) steroids. Substances fulfilling these criteria were compared using the relative contribution of their peak area in relation to the total peak area of all selected substances.

14.2.4 Statistical Analysis

Statistical analyses were performed with SPSS 8.0. We used a Wilcoxon signed ranks test to test for seasonality. To test for individuality we used a general linear model (GLM) with either individual or colony as a fixed factor. All tests were two-sided. For the parametric tests we used only homogenous and normally distributed data, transforming data using natural logarithms (1+ln) prior to statistical analysis if necessary.

14.3 Results

In total, 185 substances were found in the wing-sac liquid of male *S. bilineata* from a Costa Rican population. For a more detailed analysis the relative peak area of thirteen focus compounds was compared (Table 14.1). Of these nine were male-specific substances (indole, indol-3-carboxaldehyde, indole-3-carboxylic acid, 2-aminoacetophenon, anthranilic acid, 5H,10H-dipyrrolo[1,2-a:1′,2′-d]pyrazine-5,10-dione (pyrocoll), indolo[2,1-b]quinazoline-6,12-dione (tryptanthrin), 2,6,10-trimethyl-3-oxo-6,10-dodecadienolide, and a compound $C_{15}H_{24}O_2$ of unknown structure), three were fatty acids (tetradecanoic acid, hexadecanoic acid, and octadecanoic acid), and one a steroid (cholesterol). On average, the cumulative peak area of these substances made up $62.5 \pm 20.7\%$ of the whole chromatogram area.

14.3.1 Seasonality

Odour samples of seven males were compared between the non-mating season (August) and the mating season (December). None of the thirteen analysed substances was absent during either season. However, there was a significant differ-

Table 14.1 The thirteen focus scent compounds found in the wing-sac secretion of male *S. bilineata* sorted by their mean relative contribution (%) to the cumulative peak area of the focus substances. The table also shows the results of the GLM (fixed factor: individual) we used to compare the variances within the individuals and between individuals.

Substance	Mean±Std	Min-Max	P
Cholesterol	24.9 ± 12.0	0 - 62.6	0.02
Hexadecanoic acid	23.9 ± 14.7	6.9 - 67.4	0.19
$C_{15}H_{24}O_2$	17.1 ± 8.0	0 - 30.0	0.001***
Tryptanthrin	9.3 ± 4.7	1.9 - 18.5	0.52
Pyrocoll	6.4 ± 9.1	0 - 42.2	0.015
Octadecanoic acid	6.0 ± 4.7	0 - 21.8	0.34
2,6,10-trimethyl-3-oxo-6,10-dodecadienolide	4.9 ± 2.4	0 - 8.6	< 0.001***
Anthranilic acid	2.8 ± 4.0	0 - 20.1	0.40
Indole-3-carboxylic acid	1.3 ± 1.8	0 - 4.1	0.15
Indole	1.3 ± 1.1	0 - 5.1	0.65
Indole-3-carboxaldehyde	1.2 ± 1.1	0 - 4.2	0.16
Tetradecanoic acid	0.8 ± 0.9	0 - 3.7	0.48
2-Aminoacetophenon	0.1 ± 0.2	0 - 0.7	0.06

***significant after Bonferoni correction

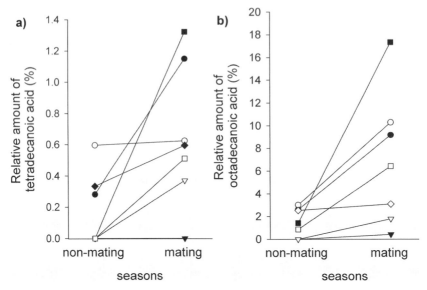

Fig. 14.1 The percentage peak area of tetradecanoic acid (a) and octadecanoic acid (b) in relation to the other twelve focus substances in the wing-sac liquid of seven male *S. bilineata*. Samples of the same individual from the non-mating season (July-August) and the mating season (December) are connected by a line. Individual symbols indicate samples from individual males

ence in the relative peak area of tetradecanoic acid (Wilcoxon signed ranks test: $n = 7, Z = 2.2, p = 0.028$) and octadecanoic acid (Wilcoxon signed ranks test: $n = 7, Z = 2.4, p = 0.018$) between seasons. The relative amount of both substances was significantly higher during the mating season than during the non-mating season (Fig. 14.1). For all other substances we found no seasonal differences in the relative contribution to individual odour profile.

14.3.2 Individuality

We compared the odour samples of seven adult male *S. bilineata* that were caught at least three times (four individuals three times, two individuals four times and one individual eight times). None of the substances was detected only in a single individual. The average relative portion of 2,6,10-trimethyl-3-oxo-6,10-dodecadienolide (GLM: $F_{6,21} = 14.99$, $p < 0.001$), and $C_{15}H_{24}O_2$ (GLM: $F_{6,21} = 5.95$, $p = 0.001$) differed significantly between individuals (Table 14.1). On two occasions, we had two males of the same colony in our analyses. Males from the same colony had different, unique, non-overlapping levels of 2,6,10-trimethyl-3-oxo-6,10-dodecadienolide (minimum – maximum values for Colony 1: male A 5.5 - 6.7% and male B 1.6 - 4.4%; Colony 2: male A 2.6 - 6.3% and male B 6.2 - 8.1%). Levels of $C_{15}H_{24}O_2$ showed much overlap between males from the same colony (Colony 1: male A 18.2

- 24.4% and male B 13.3 - 25.9%; Colony 2: male A 0 -20.2% and male B 15.8 - 30.0%). No significant differences were found for the relative peak areas of the other eleven substances.

14.4 Discussion

Male greater sac-winged bats have a specialized storage organ, the wing-sac in their front wing membrane, which they fill with substances from various body regions during a unique perfume-blending behaviour (Voigt and von Helversen 1999; Voigt 2002). The wing-sac odour of male greater sac-winged bats contains substances, such as fatty acids, terpenoids and aromatic compounds. This is consistent with the hypothesis that the wing-sac liquid of male *S. bilineata* might combine information: 1) on reproductive status and 2) on individuality.

14.4.1 Reproductive status/Seasonality

Analysis of hormonal metabolites in the faeces of *S. bilineata* showed most females were in oestrous in early December, suggesting that *S. bilineata* has a mating season of only approximately a few weeks each year (Voigt and Schwarzenberger 2007). In *S. bilineata,* odour is involved in several courtship or greeting displays, such as hovering flights (Bradbury and Emmons 1974; Voigt and von Helversen 1999; Voigt et al. 2001; Voigt 2005). The hovering flights can be observed most frequently in the morning after the bats have returned to the daytime roost and in the evening shortly before the bats emerge from the roost (Bradbury and Emmons 1974; Voigt and von Helversen 1999; Voigt 2002). Aerial greeting displays can be seen during the whole year in the daytime roost, although males hover more frequently prior to and during the mating season, indicating that the hovering flight and therewith the wing-sac content is part of the courtship behaviour (Voigt and Schwarzenberger 2007).

  Our study shows that the composition of the wing-sac odour changes between non-mating and mating seasons. The relative amount of tetradecanoic acid (C14) and octadecanoic acid (C18) is higher during the mating season, suggesting that females could use fatty acids for assessing the reproductive status of a male. The importance of fatty acids for olfactory communication has already been shown in several mammals, such as spotted hyenas, *Crocuta crocuta* (Hofer et al. 2001) and the Indian mongoose, *Herpestes auropunctatus* (Gorman 1976).

  A possible explanation for seasonal variation in odour profiles might be temporal changes in the bats' insect diet. Possibly males feed on different insect species during the mating season than during the non-mating season. Alternatively odour profiles may vary seasonally according to fluctuating hormone levels. Since hormones such as testosterone are known to vary with reproductive activity (bats: Hosken, Blackberry, Stewart and Stucki 1998) and since plasma androgen levels influence glandular secretion in mammals (Ebling 1977), it is possible that androgen-controlled glandular secretion causes seasonal changes in the odour profile of male greater sac-winged bats.

14.4.2 Individuality

None of the 185 substances detected in the male wing-sac odour occurred only in a single individual. Although most substances were shared by all males, we detected inter-individual difference in the composition of the wing-sac liquid. The relative contribution of 2,6,10-trimethyl-3-oxo-6,10-dodecadienolide and $C_{15}H_{24}O_2$ (species-specific substances, unpublished data) was constant throughout season, and differed between individuals.

Harems of *S. bilineata* are stable over several years (personal observations; Voigt 2005). Each morning, males "greet" returning females by hovering in front of them (Bradbury and Emmons 1974; Voigt and von Helversen 1999). We suggest that individual recognition based on odour might be important for the group maintenance in this social system. Individual recognition based on olfactory cues is a well-documented phenomenon in mammals (Halpin 1980, 1986; Johnston, Derzie, Chiang, Jernigan and Lee 1993), although the exact mechanisms are poorly understood. Buesching et al. (2001) analysed subcaudal gland secretions of badgers, *Meles meles,* and compared the variation between samples collected repeatedly from individuals. In badgers individual variation of subcaudal gland secretions was lower than seasonal variation, suggesting that the subcaudal gland secretion in badgers encodes information about badger individuality. To facilitate recognition, each individual ought to exhibit unique identity information in its odour. Thus, for individual recognition males ought to differ in the relative peak area of at least one odour compound. Individuals that can be mistaken, e.g. individuals of the same colony should have individual distinct, non-overlapping ranges of the relative amount of this substance. In our study, the relative amount of 2,6,10-trimethyl-3-oxo-6,10-dodecadienolide and $C_{15}H_{24}O_2$ differed between individuals. Individuals of the same colony had unequal levels of 2,6,10-trimethyl-3-oxo-6,10-dodecadienolide, making individual recognition based on the concentration of one substance possible.

14.4.3 The Wing-Sac Odour as a Composite Signal

A storage and display organ, in which a scent bouquet is actively blended from secretions of various body areas, has so far only been described for the greater sac-winged bat in mammals (Voigt and von Helversen 1999). The combination of information in such a specialized organ may be an adaptation to the specific needs of the social system of *S. bilineata. S. bilineata* is polygynous and females maintain a minimum distance to their neighbours within the harem territory. Aerial displays are the only possibility for males to bring females in contact with their scent from a distance since females respond aggressively to approaching males. Male greater sac-winged bats use a single morphological structure, the wing-sac, as the organ to display the scent. Here we showed that multiple information is potentially encoded in the odoriferous liquid, namely seasonality and individuality. Whether this information is used in intraspecific communication remains to be proven.

Acknowledgments We thank Frank C. Schroeder and Jerrold Meinwald for their help in analysing the odour samples and identifying the substances. Special thanks goes to Jürgen Streich for help in statistics. We thank the Costa Rican authorities (SINAC and MINAE) and especially Javier Guevara for his help with the collecting and export permits. We also thank OTS for supporting our research at "La Selva" biological station. This research was funded by a grant from the German science community (DFG) to C.C. Voigt (VO890/3) and a grant from the Humboldt University (Berliner Chancengleichheitsprogramm für Frauen) Berlin, to B. Caspers.

References

Andersson, M. (1994) *Sexual Selection*. Princeton University Press, Princeton, New Jersey.

Albone, E.S. (1984) *Mammalian Semiochemistry*. John Wiley & Sons limited, New York

Badyaev, A.V., Hill, G.E., Dunn, P.O. and Glenn, J.C. (2001) Plumage color as a composite trait: Developmental and functional integration of sexual ornamentation. Amer. Nat. 158, 221–235

Basolo, A.L. and Trainor, B.C. (2002) The conformation of a female preference for a composite male trait in green swordtails. Anim. Behav. 63, 469–474

Bradbury, J.W. and Emmons, L. (1974) Social organisation of some Trinidad bats I. Emballonuridae. Z. Tierpsychol. 36, 137–183

Bradbury, J.W and Vehrencamp, S.L. (1976) Social organization and foraging in emballonurid bats I. Field studies. Behav. Ecol. Sociobiol. 1, 337–381

Buesching, C.D., Waterhouse, J.S. and Macdonald, D.W. (2002 a) Gas-chromatographic analysis of the subcaudal gland secretion of the european badger (*meles meles*) part I: chemical differences related to individual parameters. J. Chem. Ecol. 28, 41–56

Buesching, C.D., Waterhouse, J.S. and Macdonald, D.W. (2002 b) Gas-chromatographic analysis of the subcaudal gland secretion of the european badger (*meles meles*) part II: time-related variation in the individual-specific composition. J. Chem. Ecol. 28, 57–69.

Ebling, F.J. (1977) Hormonal control of mammalian skin glands. In *Chemical Signals of Vertebrates*, 1, 17–33

Gosling, L.M. and Roberts, S.C. (2001) Scent-marking by male mammals: Cheat-proof signals to competitors and mates. Adv. Stud. Behav. 30, 169–217

Gorman, M.L. (1976) A mechanism for individual recognition by odour in *Herpestes auropunctatus* (carnivora: viverridae). Anim. Behav. 24, 141–145

Grether, G.F., Kolluru, G.R. and Nersissian, K. (2004) Individual colour patches as multicomponent signals. Biological Review 79, 583–610

Halpin, Z.T. (1980) Individual odors and individual recognition: Review and commentary. Biol. Behav. 5, 233–243

Halpin, Z.T. (1986) Individual odors among mammals: Origins and functions. Adv. Stud. Behav. 16, 39–70

Hayes, R.A., Richardson, B.J., Claus, S.C. and Wyllie, S.G. (2002) Semiochemicals and social signalling in the wild European rabbit in Australia: II. Variations in chemical composition of chin gland secretion across sampling sites. J. Chem. Ecol. 28, 2613–2625

Heckel, G., Voigt, C.C., Mayer, F. and von Helversen, O. (1999) Extra-harem paternity in the white-lined bat *Saccopteryx bilineata*. Behaviour 136, 1173–1185

Hofer, H., East, M., Sämmang, I. and Dehnhard, M. (2001) Analysis of volatile compounds in scent-marks of spotted hyenas (*Crocuta crocuta*) and their possible function in olfactory communication. In: A.Marchlewska-Koj, D. Muller-Schwarze and J. Lepri (Eds.), *Chemical Signals in Vertebrates* 9. Plenum Press, New York, pp. 141–148.

Hosken, D.J., Blackberry, M.A. Stewart, T.G. and Stucki, A.F. (1998) The male reproductive cycle of three species of Australien vespertilionid bats. J. Zool. Lond. 245, 261–270.

Johnston, R.E., Derzie, A., Chiang, G., Jernigan, P. and Lee, H.-C. (1993) Individual scent signatures in golden hamsters: evidence for specialization of function. Anim. Behav. 45, 1061–1070

Kavaliers, M., Colwell, D.D., Braun, W.J. and Choleris, E. (2003) Brief exposure to the odour of a parasitized male alters the subsequent mate odour responses of female mice. Anim. Behav. 65, 59–68

Marler, P. (1961) The logical analysis of animal communication. J.Theor. Biol. 1, 295–317

Penn, D. and Potts, W.K. (1998) Chemical signals and parasite-mediated sexual selection. Trends Ecol. Evolut. 13, 391–396

Robson, T.E., Goldizen, A.W. and Green, D.J. (2005) The multiple signals assessed by female satin bowerbirds: could they be used to narrow down females' choices of mates? Biol. Lett. 1, 264–267.

Runkle, L.S., Wells, K.D., Robb, C.C and Lance, S.L. (1994) Individual, nightly, and seasonal variation in calling behavior of the gray tree frog, *Hyla versicolor*: implications for energy expenditure. Behav. Ecol. 5, 318–325

Safi, K. and Kerth, G. (2003) Secretions of the interaural gland contain information about individuality and colony membership in the Bechstein's bat. Anim. Behav. 65, 363–369

Scully, W.M.R., Fenton, M.B. and Saleuddin, A.S.M. (2000) A histological examination of the holding sacs and glandular scent organs of some bat species (Emballonuridae, Hipposideridae, Phyllostomidae, Vespertilionidae and Molossidae). Can. J. Zool. 78, 613–623

Starck, D. (1958) Beitrag zur Kenntnis der Armtaschen und anderer Hautdrüsenorgane von *Saccopteryx bilineata* Temminck 1838 (Chiroptera, Emballonuridae). Gegenbaur morphologisches Jahrbuch 99, 3–25

Tannenbaum, R. (1975) Reproductive strategies in the white-lined bat. PhD Thesis, Cornell University

Voigt, C.C. (2002) Individual variation of perfume blending in male sac-winged bats. Anim. Behav. 63, 31–36

Voigt, C.C. (2005) The evolution of perfume blending and wing sacs in emballonurid bats. In: R.T. Mason, M.P. LeMaster, D. Müller-Schwarze (Eds.), *Chemical Signals in Vertebrates 10*. Springer Press, New York, pp. 93–100

Voigt, C.C. and von Helversen, O. (1999) Storage and display of odor in male S*accopteryx bilineata* Emballonuridea. Behav. Ecol. Sociobiol. 47, 29–40

Voigt, C.C., von Helversen, O., Michener, R and Kunz, T.H. (2001) The economics of harem maintenance in the sac-winged bat, *Saccopteryx bilineata* (Emballonuridae). Behav. Ecol. Sociobiol. 50, 31–36

Voigt, C.C., Caspers, B. and Speck, S. (2005) Bats, bacteria, and bat smell: Sex-specific diversity of microbes in a sexually selected scent organ. J. Mammal. 86, 745–749

Voigt, C.C. and Schwarzenberger, F. (unpublished manuscript) Female reproductive endocrinology of a small tropical bat (*Saccopteryx bilineata*; Emballonuridae) monitored by fecal hormone metabolites.

Zala, S.M., Potts, W.K. and Penn, D.J. (2004) Scent-marking displays provide honest signals of health and infection. Behav. Ecol. 15,338–344

Chapter 15
Gender Specific Expression of Volatiles in Captive Fossas (*Cryptoprocta ferox*) During the Mating Season

Barbara Renate Vogler, Frank Goeritz, Thomas Bernd Hildebrandt and Martin Dehnhard

15.1 Introduction

Madagascar is known for its high number of endemic species: 80% of the flora and fauna is unique to this island (USAID Environmental Program, Madagascar Mission). All eight Malagasy carnivore species, including the fossa as the main predator, are restricted to the island. Recent DNA-analyses (Yoder, Burns, Zehr, Delefosse, Veron, Goodmann and Flynn 2003) allowed the classification of the fossa as a mongoose-like animal (family *Herpestidae*).

Fossas are the only natural enemies of large lemur species and thus play an important role in Madagascars ecosystem. Strictly territorial (Hawkins 1998), these animals live in the rain and dry forests throughout Madagascar with most individuals staying in coastal forest areas. The wild fossa population is estimated at less than 2500 individuals (IUCN 2006). Due to the ongoing habitat destruction and hunting pressure of local farmers, a large decline in fossa numbers is noted, and the species is today considered to be endangered (IUCN 2006). Additionally, the fossa is listed in CITES Appendix II.

On Madagascar, the fossa is protected from human interference in different national parks. The total population of captive fossas was calculated as 85 individuals (49 males, 36 females) worldwide with 35 males and 23 females housed in 16 European zoos (Winkler 2002).

Studies of free-ranging fossas described an orange staining of belly and breast hair, with an increased intensity in male individuals during the breeding season (Hawkins 1998). It was assumed that this visual effect is supplemented by olfactory compounds. More recently, a difference in scent-marking behaviour of male and

Barbara Renate Vogler
Leibniz Institute for Zoo and Wildlife Research
vogler@izw-berlin.de

female fossas was described (Dickie 2005) including a whole body rub to mark objects within their enclosure.

Fossas have an unusual mating system. During the breeding season several females take residence on high branches of a "traditional mating tree" (Hawkins 1998) situated within the territory of one female. Male fossas meanwhile gather underneath this tree or on neighbouring trees waiting to mate with a female. It was proposed that the mating behaviour is dominated by the female fossa, being more selective than the male. A female may avoid mating by moving to the outer branches where the male cannot follow (Hawkins 1998).

From recent research on colouration, scent marking, solitary life style and mating strategy it was assumed that fossas use scent marks as a means of olfactory communication during the breeding season. Therefore it was hypothesized that (1) volatile substances are expressed in fossa hairs including the tail region, (2) the volatiles in fossa hairs might have a pheromonal function and (3) the composition of volatiles is different between the sexes and the seasons (mating vs. non-mating season). The present study aimed to characterise gender and season specific volatile patterns emanating from fossa hair.

15.2 Material and Methods

15.2.1 Animal and Sample Collection

Manipulations of any kind in untrained carnivores require general anaesthesia. Sample collection was therefore performed on anaesthetised individuals. Depending on the preferences of the veterinarian, the fossas were either caught with nets or anaesthetised by a combination of injectable narcotics (Xylazine: 2.5–5.0 mg/kg, Ketaminhydrochloride: 10.5–20 mg/kg and Diazepam: 0.5–1.0 mg/kg) prior to anaesthesia by inhalation gases (Isoflurane: 1.5–3 Vol. % and Oxygen: 1–2 L/min) applied by facial mask.

The present study included the volatile analysis of 16 male and 15 female individuals from 10 different European Zoos. All animals were mature (aged 3–15 years). From these, 13 males and 10 females were sampled during both seasons (breeding and non-breeding season).

Of all animals hair samples of four different body areas (throat region, sternal region, lateral abdomen, ventral of the base of the tail) (Fig. 15.1) were taken using an electric hair-clipping device. Additionally, during one breeding season, swabs of the anal gland secretion were collected from 8 male and 7 female fossas.

Sample material was obtained wearing single-use protective gloves. Both hair samples and anal swabs were collected into glass vials, thoroughly sealed and stored at −80 °C until analysis.

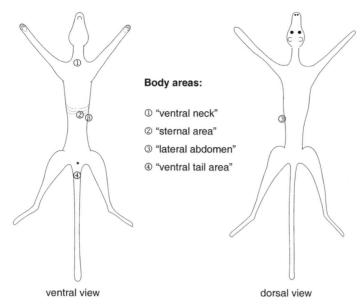

Body areas:

① "ventral neck"

② "sternal area"

③ "lateral abdomen"

④ "ventral tail area"

ventral view   dorsal view

Fig. 15.1 Areas of hair sampling

15.2.2 Solid-Phase Micro extraction (SPME) Sampling

The volatile substances were extracted from portions of 0.1g hair using solid-phase micro extraction (SPME). The method uses a fibre coated with an adsorbent that can extract organic compounds from the headspace above the sample. Extracted compounds are desorbed upon exposure of the SPME fibre in the heated injector port of a gas chromatograph (GC).

SPME was carried out in 20 ml headspace vials (Shimadzu) with a CTC Combi Pal system auto injector at 70 °C for 60 min using a StableFlex fibre with an 85 μm Carboxen / PDMS coating (Sigma). Prior to analyses 0.8 μg -camphor solved in 4 μl isopropanol were added as internal standard.

GC-MS determinations were conducted with a Shimadzu GCMS-QP 5050. MS acquisition was performed in TIC. The samples were analysed using a 50 m SE-54-CB1 capillary column (0.32 mm i.d. and 1 μm film thickness, CS, Langerwehe, Germany). Ultra pure helium was used as carrier gas, with a column head pressure setting of 41 kPa generating a column flow of 1 ml/min. Injector temperature was 300 °C; the interface temperature was maintained at 300 °C and ionising voltage was at 1.2 kV. A splitless injection mode was used, the purge valve was turned on 15 min after injection, with a split flow of 8 ml/min during the GC run. The GC oven was kept at 45 °C for 2 min, increased to 105 °C at 15 °C min^{-1}, from 105 to 165 °C at 10 °C min^{-1} and then from 165 to 290 °C at 4 °C min^{-1}. The mass spectra were identified by computer MS library research (NIST) and compared with those of the authentic standards. Standard chemicals were tested using the same protocols.

15.2.3 Data Evaluation

To calculate the total amount of volatiles included in the hair samples from different body regions the area beneath all peaks was calculated using the GCM-SPostrunAnalysis ® software including all peaks above a signal-to-noise ratio of 3:1 (exceeding an area of 200.000 units). From those, 20 distinct peaks were selected for peak comparison and the area beneath the peak was used to evaluate the differences between sex and season.

For each compound, means ± standard error (SEM) were calculated and differences were assessed by ANOVA. Calculated p values < 0.05 were considered to be significantly different. The statistical procedures were performed with the software programme Instat Version 3 (Graphpad Software Inc.).

15.3 Results

15.3.1 Comparison between Body Regions

An orange staining of the coat of abdomen and breast during the breeding season was not detected in all animals. When staining occurred, the throat and the ster-

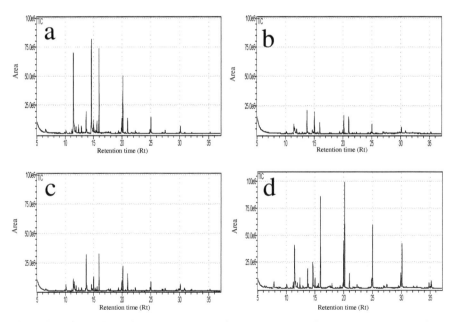

Fig. 15.2 Patterns of volatiles emitted from hairs during the breeding season. The comparison was carried out using SPME and GC-MS showing the total ion chromatograms (TIC) of four hair samples from one male including the throat region (a), the sternal region (b), the region of the lateral abdomen (c), and the tail region (d)

nal region showed the most intensive colouration. There was no relation between intensity of orange staining and measured quantity of volatile substances.

The composition of volatiles obtained from the different body areas was similar within the genders, but varied regarding their quantities (Fig. 15.2). The pattern of coat volatiles consisted of approximately 30 detectable compounds (with retention times between 9.8 and 30.2 minutes) whose highest concentrations were expressed in the tail region (Fig. 15.2). This difference was obvious both in the breeding and non-breeding season. Therefore, the tail region was chosen for further comparison of volatiles regarding gender and season specific differences, even though in females slightly higher amounts of volatiles were measured in the throat area during the breeding season.

15.3.2 Gender Specific Differences

During the breeding season significantly higher amounts of volatiles were obtained in males compared to females ($p < 0.025$) whereas this difference was not apparent during the non-breeding season. The peak-pattern, however, differed between the genders and the seasons. Figure 15.3 shows two selected samples from a male and a female individual.

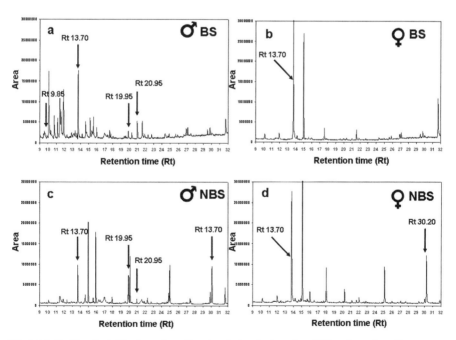

Fig. 15.3 Profile of volatiles from the tail region during the breeding (BS) and non-breeding season (NBS) of a male (a, c) and a female individual (b, d). Arrows indicate gender specific peaks

Table 15.1 Male (top) and female (bottom) specific substances (BS = breeding season, NBS = non-breeding season, M = male, F = female, n.d. = not detectable, n.s. = not significant)

Rt	BS	NBS	Identification
9.85	M > F (n.d.)	n.s.	unknown
19.95	M > F; p = 0.01	n.s.	similar to 2-nonen-1-ol
20.95	M > F; p = 0.002	M > F (n.d.)	6,10-dimethyl-5,9-undecadien-2-one
Rt	BS	NBS	Identification
13.70	F > M; p = 0.03	F > M; p = 0.005	unknown.
30.20	n.s.	F > M (n.s.)	isopropyl myristate (IPM)

Three male and two female specific peaks were detected. Two of the male specific peaks (Rt 19.95, Rt 20.95) showed significantly ($p = 0.01$ and $p = 0.0002$, respectively) higher levels during the breeding season compared to females (Table 15.1). In case of substance Rt 19.95 this difference was not recorded during the non-breeding season, as the substance did not occur in significantly higher concentrations in males compared to females. The substance Rt 20.95 retained its gender specificity ($p = 0.013$) during the non-breeding season, as it was not detectable in females during that time. The third male specific peak (Rt. 9.85) exclusively occurred in low amounts in males during the breeding season but was detectable in only one out of 10 males during the non-breeding season. This substance was not detectable in females at any time. Additionally, the last two substances (Rt 20.95 and Rt 9.85) show a seasonal specificity in male fossas: both were measured in significantly higher concentrations during the breeding compared to the non-breeding season ($p = 0.007$ and $p = 0.001$, respectively).

Initially the substance at Rt 19.95 was identified as 2-nonen-1-ol based on mass spectrum library search. The comparison with a commercial 2-nonen-1-ol standard indeed revealed a high degree of similarity between the mass spectra, but a distinct deviation regarding the retention time suggesting a similar molecule with a chain length greater than 2-nonen-1-ol. The substance Rt 20.95 was tentatively identified as 6,10-dimethyl-5,9-undecadien-2-one which corresponds with the authentic standard regarding mass spectra and retention time.

While the male specific substances occur in significantly higher concentrations during the breeding season, the female specific substance detected at Rt 13.70 showed significantly higher levels during breeding ($p = 0.03$) and non-breeding ($p = 0.005$) season compared to males. Mass spectra analysis revealed a typical benzaldehyde ion but this substance has yet to be identified. Another substance (Rt 30.20) showed higher concentrations in females during the non-breeding season; however, the difference between the genders was not significant (Table 15.1). Based on library search substance Rt 30.20 was identified as isopropyl myristate (IM), which was subsequently confirmed by comparison with the commercial standard.

15.4 Conclusions

Comparative GCMS analysis of fossa hair samples from the breeding and the non-breeding season revealed a distinct gender and season specific pattern of volatiles. It was possible to detect five gender specific compounds: three substances specific for males and two for females. One male specific compound was identified as 6,10-dimethyl-5,9-undecadien-2-one (geranyl acetone), another one shows a high similarity to 2-nonen-1-ol, but could not be confirmed by comparison with a commercial standard. These compounds were previously described as pheromone in several insect species (www.pherobase.com), whereas in mammals, 6,10-dimethyl-5,9-undecadien-2-one was present in red fox urine during mating season (Jorgensen, Novotny, Carmac, Copland and Wilson 1978). A third so far unidentified male specific substance occurs only in minor concentrations, predominantly during the breeding season. One female specific peak showed a typical benzaldehyde pattern, but was not identified yet. A second substance was identified as isopropyl myristate (IM), although the difference compared to males was not significant due to the high inter-female variation of concentration.

In addition seasonal differences were detected. Two specific substances were expressed by male fossas in higher concentrations during the breeding than during the non-breeding season.

The composition of volatiles in fossa hairs was different from those in anal gland secretions. It was therefore assumed that the anal gland did not contribute to the volatile patterns of hairs particularly from the tail region and that the described volatiles from hairs are secreted from scent glands within the skin. We hypothesise a close relation between the expression of volatiles in fossa hairs and scent marking behaviour observed during the mating season. Moreover, the gender and season specific expression of volatiles in a solitary animal with a brief mating season strongly suggests a pheromonal function of these substances.

Future work should identify further gender and season specific volatiles, to conduct a comparative study on free ranging fossas, and to perform bioassays to prove a pheromonal function of the identified substances.

Acknowledgments We thank the zoos of Amsterdam (NL), Berlin (West) (D), Bratislava (SK), Chemnitz (D), Colchester (UK), Duisburg (D), Dvur Králové (CZ), Frankfurt (D), Olomouc (CZ) and Ústí (CZ) for granting access to their fossas and for supporting the study. We are grateful to Gunnar Weibchen, University of Hamburg, Germany, for helping us to identify specific volatile substances. The attendance of the CSiV conference by the author BRV was supported by a grant of the Academy for Animal Health (AFT), Germany and GlaxoSmithKline, Germany.

References

Dickie, L. A. (2005) The reproductive physiology of the fossa in captivity. PhD Thesis, Queen Mary University of London.

Hawkins, C. E. (1998) Behaviour and ecology of the fossa, *Crytoprocta ferox* (Carnivora: Viverridae) in a dry deciduous forest, western Madagascar. Unpublished PhD thesis, University of Aberdeen, Schottland.

IUCN (2006) *2006 Red List of Threatened Species.* www.iucnredlist.org. Downloaded on 15 September 2006.

Jorgensen, J.W., Novotny, M., Carmac, M., Copland, G.B. and Wilson, S.R. (1978) Chemical scent constituents in the urine of the red fox (*Vulpes vulpes L.*) during the winter season. Science 199, 796–798.

Winkler, A. (2002) Fossa (*Cryptoprocta ferox*) - International Studbook (unpublished).

Yoder, A. D., Burns, M. M., Zehr, S., Delefosse, T., Veron, G., Goodmann, S. M. and Flynn, J. J. (2003) Single origin of Malagasy Carnivora from an African ancestor. Nature 421, 734–737.

Chapter 16
Do Spotted Hyena Scent Marks Code for Clan Membership?

Nicole Burgener, Marion L East, Heribert Hofer and Martin Dehnhard

Abstract The spotted hyena (*Crocuta crocuta*) is a territorial carnivore that lives in highly structured social groups called clans. Individuals of both sexes produce scent in a prominent anal scent gland. Gas-chromatographic analysis of 13 fatty acids and esters in scent profiles from 45 individuals belonging to three social groups demonstrated sufficient variation to suggest that odour may permit individual olfactory recognition. Further, anal scent secretions from members of the same clan are more similar to each other than those from different clans, consistent with the idea of a social group odour. We describe a mechanism involving both scent pasting and dry-pasting behaviour to explain how a group odour label may be concocted from individual scent secretions and how this group label is spread among members of a clan.

16.1 Introduction

Animals that live in societies need effective means of communication, and selection has favoured a wide variety of visual, acoustic and olfactory signals that convey information between members of social groups (Eisenberg and Kleiman 1972; Halpin 1986; Bradbury and Vehrencamp 1998). Social coherence of a group, at a basic level, requires group members to discriminate between individuals that are members of their group and those that are not, and an unambiguous mechanism to achieve this would be a group membership label. Evidence for such group identity labels have been found in a variety of taxa (e.g. mammals: Mykytowycz 1968; O'Riain and Jarvis 1997; Sun and Müller-Schwarze 1998; Bloss, Acree, Bloss, Hood and Kunz 2002; Safi and Kerth 2003; social insects: Gamboa, Reeve, Ferguson and Wacker 1986; Venkataraman, Swarnalatha, Nair and Gadagkar 1992; Breed, Diaz and Lucero 2004). Scent membership labels are likely to be honest and reliable signals because group odours are thought to be acquired through the transfer of

Nicole Burgener

Leibniz Institute for Zoo and Wildlife Research, Berlin, Germany and Freie Universität, Berlin, Germany

burgener@izw-berlin.de

J.L. Hurst et al., *Chemical Signals in Vertebrates 11.*
© Springer 2008

scent among members of a group (Halpin 1986). For example, in the naked mole-rat (*Heterocephalus glaber*), a distinctive colony odour is composed of scent that is distributed among and learnt by colony members (O'Riain and Jarvis 1997).

The spotted hyena is a crepuscular/nocturnal social carnivore that lives in fission-fusion groups termed clans (Kruuk 1972; Mills and Hofer 1998). Clan members use a wide spectrum of visual and acoustic signals for both short and long distance communication with both group and non-group members (Kruuk 1972; East and Hofer 1991; East, Hofer and Wickler 1993; Holekamp, Boydston, Szykman, Graham, Nutt, Birch and Piskiel 1999). Nocturnal activity and a fission-fusion social structure would be expected to also favour olfactory communication and spotted hyenas have both a prominent anal scent gland and inter-digital scent glands (Kruuk 1972; Mills and Gorman 1987).

Although clans defend communal territories, clan members may undertake short distance excursions (Höner, Wachter, East, Runyoro and Hofer 2005) or long distance foraging trips up to 70 km from their territory boundaries that necessitate movement through or foraging within territories belonging to other clans (Hofer and East 1993a,b,c). Therefore, spotted hyenas not only need a reliable means to distinguish members of their social group from those of other groups, they also need to identify owners of the territories into which they intrude so that they can respond appropriately when challenged (Hofer and East 1993b).

When clan members meet another group member of the same sex, they normally perform a ritualized greeting ceremony in which greeting partners stand head to tail, raise their hind legs and tail while investigating the anal genital region of their greeting partner (Kruuk 1972; East et al. 1993). The format of greetings suggests that odour plays a central role in reaffirming associations among clan members. Furthermore, during aggressive encounters subordinate individuals often protrude their anal scent gland (Kruuk 1972; East et al. 1993), suggesting that scent within the anal gland may confirm an individual's membership of the group thereby decreasing the probability of escalated aggression.

Scent from the anal gland is deposited on vegetation in a process known as pasting (Kruuk 1972). Pasting may serve as an olfactory means to communicate ownership of a particular area and the resources the area contains (Gorman and Mills 1984; Mills and Gorman 1987). It has also been suggested that pasted scent may signal the recent presence of a group member in an area (Hofer, East, Sämmang and Dehnhard 2001; Drea, Vignieri, Kim, Weldele and Glickman 2002).

In a preliminary analysis of scent from the anal gland of individually known spotted hyenas from several clans in the Serengeti National Park (NP), Tanzania, Hofer et al. (2001) presented evidence of a group odour based on gas-chromatography (GC) of fatty acids in the scent. In this study, we examine samples from the anal scent gland of a larger sample of individuals from three clans using solid-phase microextraction (SPME) of volatile compounds followed by gas-chromatography—mass spectrometry (GCMS). We outline details of scent pasting behaviour in the Serengeti NP that may explain how a group odour label is concocted from individual scent secretions from group members and how this group label is spread among members of a clan.

16.2 Methods and Materials

16.2.1 Study Animals and Collection of Scent

Between 2000 and 2003, a total of 45 anal gland scent samples were collected from spotted hyenas that were members of three clans in the Serengeti NP, Tanzania. Individuals were recognized by their spot patterns (East and Hofer 1991). Females were philopatric and males normally dispersed from their natal clan (East and Hofer 2001). Animals were classified as adults ($>$ 24 mo old), yearlings (12 — 24 mo), or cubs ($<$ 12 mo). The three study clans held territories at the woodland / plain boundary in the centre of the Serengeti NP. Scent marking behaviour was monitored during focal follows of individuals, and during dawn and dusk observations at the communal den (East and Hofer 1991), food resources and resting sites. Focal follows were conducted both within and outside the territory of the clan to which the individual belonged. All study clans were habituated to the presence of a vehicle (East and Hofer 1991).

Scent was collected immediately after an individual pasted on an individual post (see results), following the procedure described by Hofer et al. (2001). All samples were stored in 20 ml glass headspace vials (Shimadzu) at −20 °C until analysed.

16.2.2 Chemical and Statistical Analysis

Samples were analysed using a Shimadzu GCMS-QP 5050 fitted with a 50 m SE-54 capillary column (0.32 i.d. and 1 µm film thickness, CS, Langerwehe, Germany). Volatile compounds were desorbed upon exposure of the SPME fibre in the heated injector port of the GC. SPME was carried out with a CTC combi pal system autoinjector at 70 °C for 60 min using a fibre with 85 mm polyacrylate coating. SPME sampling was done in the headspace above the collected scent sample. GC analyses were conducted using ultrapure helium as carrier gas, with a column head pressure setting of 41.2 kPa. Injector temperature was 300 °C, the interface temperature was maintained at 300 °C and ionizing voltage was at 1.2 kV. Splitless injection mode was used and the purge valve was turned on 15 min after the injection with a split flow of 8 ml min^{-1} during the GC run. The GC oven was kept at 45 °C for 2 min, then increased to 105 °C at 15 °C min^{-1}, from 105 °C to 165 °C at 10 °C min^{-1} and then from 165 °C to 290 °C at 4 °C min^{-1}. Mass spectrometer (MS) acquisition was performed in a total ion chromatogram (TIC). Compounds were identified by their retention times and comparison of mass spectra to those in the NIST 1998 mass spectral library.

To compare different profiles, peaks were matched by their retention times and mass spectra. The contribution of each peak to the overall area of the profile was calculated as a percentage (% area) and used for analyses. For the current study we focussed our analyses on fatty acids and esters (Poddar-Sarkar and Brahmachary 1999; Hofer et al. 2001).

Statistical analyses were carried out using PRIMER 5.2.9 and SYSTAT 10. Because data were not normally distributed we applied a non-parametric test. The similarity of 13 key substances was compared using the Bray-Curtis-Similarity index and the statistical significance was tested using analysis of similarities (ANOSIM) with clan membership as a factor (Clarke and Green 1988). A 3-D multidimensional scaling plot of the similarity indices was used to display the greater level of similarity between individuals within a clan in comparison to individuals between clans. "Stress" is a measure of "goodness of fit" that evaluates how well a particular configuration reproduces the observed distance matrix.

16.3 Results

16.3.1 Chemical Composition

In the 45 scent profiles we analysed, 13 compounds that were either fatty acids or esters were identified, and these were used as the key components in our analysis. There was sufficient variation in these 13 components between the scent profiles of different individual members of a social group to suggest that each group member may have a unique profile that would permit individual odour recognition (Fig. 16.1).

Scent profiles of different individuals within a clan were more similar to each other than to the profiles of individuals from other clans (N = 8, N = 23, N = 14, Global R = 0.177, P = 0.007, Fig. 16.2).

16.3.2 Pasting and Dry-pasting Behaviour

Both adults and yearlings of both sexes pasted mostly in the vicinity (within 50 m) of the communal den, at sites along tracks frequently used by spotted hyenas, and at communal latrines, where clan members also deposited faeces. Individuals either scent marked on individual scent posts or on communal scent posts. An individual scent post was created when a spotted hyena pasted on a stalk of grass or herb that had not been pasted with scent by another spotted hyena and thus the scent post only presented the odour from one individual. A communal scent post was created either when individuals pasted on a stalk of vegetation that already had scent pasted on it by one or several other clan members, or when several spotted hyenas pasted on stalks that were spatially close together (within less than 0.5 m). Communal scent posts thus presented scent from several individuals. Both individual and communal scent posts were investigated (sniffed) by clan members that passed by the post. Some scent posts elicited either pasting by the investigating individual, or dry-pasting in which the animal performed the normal behaviour associated with pasting except that the anal gland was not protruded, and scent was not deposited.

Communal scent posts were established at the communal den, at communal latrines, and close to frequently used hyena paths and resting sites. Individuals that

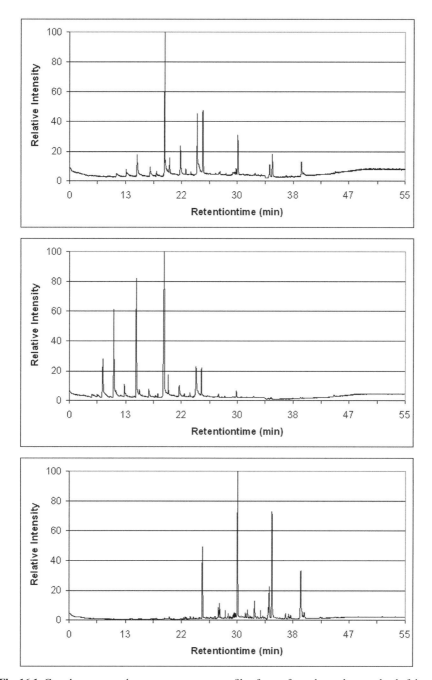

Fig. 16.1 Gas-chromatograph mass spectrometer profile of scent from the anal scent gland of three adult female spotted hyenas from the same social group

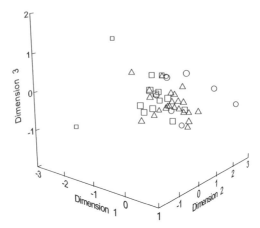

Fig. 16.2 A multidimensional scaling (MDS) plot in three dimensions for scent profiles of individuals from three spotted hyena clans (○ Isiaka clan, N = 8; □ Pool clan, N = 14; Δ Mamba clan, N = 23). Similarities between samples were calculated with the Bray-Curtis coefficient. MDS Stress: 0.08

pasted on communal scent posts not only deposited their own scent secretions on the post, but also rubbed the scent secretions previously deposited by other individuals into their scent gland and onto the fur of their ventral side.

16.4 Discussion

We found considerable individual variation between the scent profiles of individual clan members in spotted hyenas, but we also demonstrated a general similarity of odour among individual members of a clan, suggesting the existence of a clan scent signature. This result is consistent with that of Hofer et al. (2001) who also found evidence for a clan odour when comparing, by means of SPME and GC, the fatty acid content of anal scent secretion from members of the same three clans. Group odour has been proposed as a mechanism by which members of a group can distinguish between group and non-group members (e.g. Gorman, Nedwell and Smith 1974; O'Riain and Jarvis 1997; Safi and Kerth 2003). In honey bees (*Apis mellifera*), colony guards are thought to use a colony template, rather than individual templates, when discriminating colony members from intruders (Breed et al. 2004).

   Our behavioural observations suggested that spotted hyenas actively acquired scent from other clan members when they pasted over stalks of vegetation that held scent deposited by another clan member. During this process the scent of other clan members may be taken into the protruded scent gland as well as being rubbed onto the pasting individual's fur. As clan membership is not static over time, there is probably a need for clan members to continuously anoint themselves with the scent of other clan members to retain the current clan odour.

Although an intruder in a clan territory could potentially use communal scent posts to acquire the group odour of the territory owning clan, we have no evidence that intruders do so. As scent is relatively persistent, there may be costs to such behaviour if acquisition of an alien clan odour results in increased aggression from group members when an intruder returns "home". For example, in house mice (*Mus musculus domesticus*), the absence of the group odour led to aggressive behaviour (Hurst, Fang and Barnard 1993).

It could be expected that a persistent honest signal of group membership would be favoured in a species, such as the spotted hyena, that lives in a fission-fusion society in which individuals may not meet for long periods. In the Serengeti NP, spotted hyenas undertake long distance foraging trips, and as these commuting trips are not coordinated between clan members there is the potential for clan members not to meet for several weeks or more (Hofer and East 1993b).

In addition to scent pasting as described by Kruuk (1972), we have observed dry-pasting when individuals do not protrude their gland or deposit scent. We suggest that dry-pasting over individual or communal scent posts is an important mechanism by which spotted hyenas acquire a general clan odour. Dry-pasting is most notice-able among cubs and yearlings, and it is likely that the acquisition of the group odour by these young age classes is important for their social integration into the clan.

Bacteria have been suggested to be external vectors that produce volatiles in scent glands (bacterial fermentation hypothesis: e.g. Gorman et al. 1974; Albone and Perry 1975; Voigt, Caspers and Speck 2005) even though this idea has been questioned (e.g. Svendsen and Jollick 1978). In several mammalian species, bacteria are active in scent glands (e.g. *Herpestes auropunctatus*: Gorman et al. 1974, *Vulpes vulpes*: Gosden and Ware 1976; *Castor canadensis*: Svendsen and Jollick 1978; *Saccopteryx billineata*: Voigt et al. 2005). Currently it is not known whether bacteria play any role in the production of group odour in spotted hyenas, but if so, we suggest that the use of communal scent posts would serve as a mechanism for the transfer of a common bacterial mix among members of a clan.

Acknowledgments We are grateful to the Tanzanian Commission for Science and Technology (COSTECH) for permission to conduct the study and the Director Generals of the Tanzania Wildlife Research Institute and Tanzania National Parks for cooperation. We acknowledge M. Hilker for ideas and discussion and we thank B. Caspers, O. Höner, B. Kostka, K. Paschmionka, J. Streich, D. Thierer, B. Wachter and K. Wilhelm for assistance. The Leibniz Institute for Zoo and Wildlife Research funded the study.

References

Albone, E.S. and Perry, G.C. (1975) Anal sac secretion of the red fox, *Vulpes vulpes*: volatile fatty acids and diamines: Implications for a fermentation hypothesis of chemical recognition. J. Chem. Ecol. 2, 101–111.

Bloss J., Acree, T.E., Bloss, J.M., Hood, W.R. and Kunz, T.H. (2002) Potential use of chemical cues for colony-mate recognition in the big brown bat, *Eptesicus fuscus*. J. Chem. Ecol. 28, 819–834.

Bradbury, J.W. and Vehrencamp, S.L. (1998) *Principles of Animal Communication*. Sinauer Associates, Sunderland, Massachusetts.

Breed, M.D., Diaz, P.H. and Lucero, K.D. (2004) Olfactory information processing in honey bee, *Apis mellifera*, nestmate recognition. Anim. Behav. 68, 921–928.

Clarke, K.R and Green, R.H. (1988) Statistical design and analysis for a "biological effects" study. Mar. Ecol. Trog. Ser. 46, 213–226.

Drea, C.M., Vignieri, S.N., Kim, H.S., Weldele, M.L. and Glickman, S.E. (2002) Responses to olfactory stimuli in spotted hyenas (*Crocuta crocuta*): II. Discrimination of conspecific scent. J. Comp. Psychol. 116, 342–349.

East, M.L. and Hofer, H. (1991) Loud calling in a female-dominated mammalian society: I. Structure and composition of whooping bouts of spotted hyaenas (*Crocuta crocuta*). Anim. Behav. 42, 637–649.

East, M.L., Hofer, H. and Wickler, W. (1993) The erect "penis" is a flag of submission in a female-dominated society: greetings in Serengeti spotted hyenas. Behav. Ecol. Sociobiol. 33, 355–370.

East, M.L. and Hofer H. (2001) Male spotted hyenas (*Crocuta crocuta*) queue for status in social groups dominated by females. Behav. Ecol. 12, 558–568.

Eisenberg, J.F. and Kleiman, D.G. (1972) Olfactory communication in mammals. Annu. Rev. Ecol. Syst. 3, 1–32.

Gamboa, J.G., Reeve, H.K., Ferguson, I.D. and Wacker, T.L. (1986) Nestmate recognition in social wasps: the origin and acquisition of recognition odours. Anim. Behav. 34, 685–695.

Gorman, M.L., Nedwell, D.B. and Smith, R.M. (1974) An analysis of the contents of the anal scent pockets of *Herpestes auropunctatus*. J. Zool. 172, 389–399.

Gorman, M.L. and Mills, M.G.L. (1984) Scent marking strategies in hyaenas (Mammalia). J. Zool. 202, 535–547.

Gosden, P.E. and Ware, G.C. (1976) The aerobic baterial flora of the anal sac of the red fox. J. Appl. Bacteriol. 41, 271–275.

Halpin, T.Z. (1986) Individual odors among mammals: origins and functions. Adv. Stud. Behav. 16, 40–70.

Hofer, H. and East, M.L. (1993a) The commuting system of Serengeti spotted hyaenas: how a predator copes with migratory prey. I. Social organization. Anim. Behav. 46, 547–557.

Hofer, H. and East, M.L. (1993b) The commuting system of Serengeti spotted hyaenas: how a predator copes with migratory prey. II. Intrusion pressure and commuters' space use. Anim. Behav. 46, 559–574.

Hofer, H. and East, M.L. (1993c) The commuting system of Serengeti spotted hyaena: how a predator copes with migratory prey. III. Attendance and maternal care. Anim. Behav. 46, 575–589.

Hofer, H., East, M.L., Sämmang, I. and Dehnhard, M. (2001) Analysis of volatile compounds in scent-marks of spotted hyenas (*Crocuta crocuta*) and their possible function in olfactory communication. In: A. Marchlewska-Koj, J.J. Lepri and D. Müller-Schwarze (Eds.), *Chemical Signals in Vertebrates 9*. Kluwer Academic / Plenum Publishers, New York, pp. 141–148.

Holekamp, K.E., Boydston, E.E., Szykman, M., Graham, I., Nutt, K.J., Birch, S., Piskiel, A. and Singh, M. (1999) Vocal recognition in the spotted hyaena and its possible implications regarding the evolution of intelligence. Anim. Behav. 58, 383–395.

Höner, O.P., Wachter, B., East, M.L., Runyoro, V.A. and Hofer, H. (2005) The effect of prey abundance and foraging tactics on the population dynamics of a social, territorial carnivore, the spotted hyena. Oikos 108, 544–554.

Hurst, J.L., Fang, J. and Barnard, C. (1993) The role of substrate odours in maintaining social tolerance between male house mice, *Mus musculus domesticus*: relatedness, incidental kinship effects and the establishment of social status. Anim. Behav. 48, 157–167.

Kruuk, H. (1972) *The Spotted Hyena*. University of Chicago Press, Chicago.

Mills, M.G.L. and Gorman, M.L. (1987) The scent-marking behaviour of the spotted hyaena *Crocuta crocuta* in the southern Kalahari. J. Zool. 212, 483–497.

Mills, G. and Hofer, H. (1998) *Hyaenas – Status Survey and Conservation Action Plan*. IUCN/SSC Hyaena Specialist Group. IUCN, Gland and Cambridge.

Mykytowycz, R. (1968) Territorial marking in rabbits. Sci. Am. 218, 116–126.

O'Riain, M.J. and Jarvis, J.U.M. (1997) Colony member recognition and xenophobia in the naked mole-rat. Anim. Behav. 53, 187–198.

Poddar-Sarkar, M. and Brahmachary, R.L. (1999) Can free fatty acids in the tiger pheromone act as an individual finger print? Curr. Sci. India 76, 141–142.

Safi, K. and Kerth, G. (2003) Secretions of the interaural gland contain information about individuality and colony membership in the Bechstein's bat. Anim. Behav. 65, 363–369.

Sun, L. and Müller-Schwarze, D. (1998) Anal gland secretion codes for family membership in the beaver. Behav. Ecol. Sociobiol. 44, 199–208.

Svendsen, G.E. and Jollick, J.D. (1978) Bacterial contents of the anal and castor glands of beaver (Castor canadensis). J. Chem. Ecol. 4, 563–569.

Venkataraman, A.B., Swarnalatha, V.B., Nair, P. and Gadagkar, R. (1992) The mechanism of nest-mate discrimination in the tropical social wasp Ropalidia marginata and its implications for the evolution of sociality. Anim. Behav. 43, 95–102.

Voigt, C.C., Caspers, B. and Speck, S. (2005) Bats, bacteria, and bat smell: sex-specific diversity of microbes in a sexually selected scent organ. J. Mammal. 86, 745–749.

Chapter 17
The Ontogeny of Pasting Behavior in Free-living Spotted Hyenas, *Crocuta crocuta*

Kevin R. Theis, Anna L. Heckla, Joseph R. Verge and Kay E. Holekamp

Abstract The function of scent marking varies not only between species, but also among different age/sex classes within the same species. We conducted a longitudinal study of 26 free-living spotted hyenas to describe the ontogeny of anal gland scent marking ('pasting') in this species. Males increased the proportional abundance of overmarking between cub and subadult periods, while female pasting remained consistent throughout early development. Pasting was sexually dimorphic in that male cubs pasted more frequently than female cubs, and male subadults countermarked more often than their female peers. By examining the anal pouches of 113 anaesthetized individuals, we also determined that spotted hyenas do not consistently produce paste until their third year. In light of these findings, the potential functions of pasting by juvenile hyenas were discussed.

17.1 Introduction

An efficient communication system is a critical component of any animal society. Communication between conspecifics can function to coordinate group activities, modulate intragroup aggression, and facilitate complex social interactions. In mammals, communication is often achieved via chemical signals that may convey information about the caller's reproductive state, social status, or individual, species and group identity (Wyatt 2003). Mammals often communicate chemically through scent marking, which is the deliberate deposition of glandular secretions, urine or feces in the environment. For many mammals, the temporal and spatial patterns of scent marking facilitate territorial maintenance and mate attraction via advertisement of resource holding potential (Rich and Hurst 1998; Gosling and Roberts 2001). However, it is becoming increasingly clear that the function of scent marking varies not only between species, but also among different age/sex classes within the same species (e.g. Drea, Vignieri, Kim, Weldele

Kevin R. Theis
Michigan State University, Department of Zoology
theiskev@msu.edu

and Glickman 2002; White, Swaisgood and Zhang 2003). To date, few studies have addressed the ontogeny of scent marking in mammals (but see French and Cleveland 1984; Woodmansee, Zabel, Glickman, Frank and Keppel 1991). The current study describes the ontogeny of anal gland scent marking ('pasting') in free-living spotted hyenas. Although some earlier studies have considered pasting in spotted hyenas (Kruuk 1972; Mills 1990; Woodmansee et al. 1991; Boydston, Morelli and Holekamp 2001; Drea et al. 2002), we still do not have a complete understanding of its potential functions, particularly among juveniles.

Spotted hyenas are large, gregarious carnivores found throughout sub-Saharan Africa, living in clans of up to 90 individuals (Kruuk 1972). They are unique in that their societies are female-dominated, with adult immigrant males being lower-ranking socially than all natal animals. Natal animals inherit their mothers' rank positions within the clan's linear dominance hierarchy (Holekamp and Smale 1993), while immigrant males queue for social status (East and Hofer 2001). In addition to their unique biology, spotted hyenas are intriguing subjects for communication studies because they emit a rich array of visual, vocal and chemical signals. Their most conspicuous chemical signaling behavior is pasting, wherein a hyena deposits anal gland secretions on grass stalks. Pasting is performed by all members of the Family Hyaenidae (Kruuk 1972; Mills 1990). In the spotted hyena, all age/sex classes frequently paste, although juvenile spotted hyenas may do so ineffectively because they may not produce paste within their anal glands until they are at least one year old (Kruuk 1972). Juvenile pasting therefore presents a Darwinian puzzle.

The objectives of this study were: 1) to systematically determine the age at which spotted hyenas begin producing paste in their anal glands, 2) to determine how rates of juvenile pasting and overmarking vary during development, 3) to evaluate the effect of social rank on pasting behavior among juveniles, and 4) to identify the behavioral contexts in which pasting occurs among juvenile spotted hyenas.

17.2 Methods

This study was conducted in the Talek region of the Masai Mara National Reserve, Kenya. The subject population was a single *Crocuta* clan, whose members were identified by their unique spot patterns and other conspicuous characteristics such as ear notches. Sex was determined from the dimorphic glans morphology of the erect phallus (Frank, Glickman and Powch 1990). Cubs were assigned birthdates by estimating their ages ($\pm$ 7d) when first observed, based on their pelage and size. In this study, hyenas were considered cubs until they were one year old, and subadults from 13 to 30 months old. We further defined early, middle and late subadulthood as 13–18, 19–24 and 25–30 months respectively. Since juvenile hyenas inherit their social ranks directly from their mothers (Engh, Esch, Smale and Holekamp 2000), each subject was assigned its mother's relative rank within the clan at the time of the subject's birth.

Between 1994 and 2003, 113 different Talek hyenas were anaesthetized with Telazol (W.A. Butler Co.; 6.5 mg/kg) delivered from a CO_2-powered rifle (Telinject Inc.). To determine the age at which hyenas begin to produce paste, we examined the anal pouch of each hyena for the presence of secretions. If a hyena was darted multiple times during this period, we only used data from the first darting. The ages of immigrant males were estimated from patterns of tooth wear using methods described by Van Horn, McElhinny and Holekamp (2003).

The Talek clan has been monitored continuously since 1988. The current study utilized archived behavioral data gathered between 1988 and 1996. We collected repeated measures from 14 male and 12 female hyenas, from birth through 30 months. Observations were made at dens, kills and elsewhere. During observations, agonistic interactions, scent marking, sniffing of other hyenas, and sniffing of grass stalks were recorded via all occurrence sampling (Altmann 1974). Rates of pasting and overmarking were calculated as the number of events per individual per hour observed (Woodmansee et al. 1991).

In considering the behavioral contexts eliciting pasting behavior, we only used data from hyenas who were observed pasting at least 20 times. Each pasting event was assigned a context based upon the events that most immediately preceded the pasting. Pasting by juvenile hyenas occurred following pasting by another hyena, investigation of grass stalks, sniffing of other hyenas, agonistic interactions, and play. Pasting atop marks previously deposited in the same observation session were labeled as 'overmarks.' Pastings that were preceded by more rare stimuli, such as border patrols, inter-clan conflicts and allomarking of cubs, were considered together as 'other.' Pasting events that occurred within two minutes of pasting by another hyena, and which were not immediately preceded by another event, were considered to be 'socially facilitated.' Pasting events not preceded within two minutes by evident stimuli were considered to be spontaneous. For each individual, we calculated the proportion of pastings that occurred in each of these contexts.

To evaluate the effects of age and sex on the likelihood of paste production we employed multiple logistic regression (SAS v8.2). Since behavioral data were not normally distributed, comparisons were made using nonparametric Friedman ANOVA, Mann-Whitney U (Statistica v6), and Wilcoxon signed-rank tests (SAS v8.2). Results were considered statistically significant when $P < 0.05$ (two-sided). When considering the contexts that elicited pasting, we utilized the Holm's sequentially-rejective Bonferroni method to evaluate the statistical significance of multiple comparisons (Shaffer 1995). Descriptive statistics have been presented as mean ± standard error.

17.3 Results

Anaesthetized hyenas ranged from 7 to 143 months of age. Investigation of their anal pouch contents revealed that there was a highly significant effect of age (Wald $\chi^2 = 16.85$, $P < 0.0001$), but not of sex (Wald $\chi^2 = 0.007$, $P = 0.93$), on the

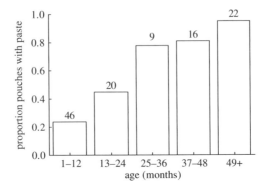

Fig. 17.1 Effect of age on paste production in spotted hyenas

likelihood of paste production. It appears that spotted hyenas do not consistently produce paste until their third year (Fig. 17.1). Nine months was the earliest age at which secretions were observed within an anal pouch. Paste production at such a young age is rare, as only three of the twelve cubs darted in their ninth month had visible secretions in their anal pouch. Eight cubs were darted prior to nine months, and all their pouches appeared devoid of secretion.

For the behavioral portion of this study, we observed the 26 juvenile hyenas for 98 ± 7 hours each (range: 37-165). Collectively, they exhibited 1491 paste events, for an average of 57 ± 7 pastings per hyena (range: 8–138). For both males and females, cub and subadult pasting rates were similar ($N_m = 14$, $S = -16.5$, $P = 0.33$; $N_f = 12$, $S = 14$, $P = 0.30$). Additionally, neither males nor females varied their pasting rates between early, middle and late subadult periods ($N_m = 8$, Friedman ANOVA $\chi^2 = 1.75$, $P = 0.42$; $N_f = 12$, $\chi^2 = 3.87$, $P = 0.14$; minimum 3 hr. observed in each period). Although there was no effect of age, males pasted more frequently than females when cubs but not when subadults (Fig. 17.2). Pasting rates were not

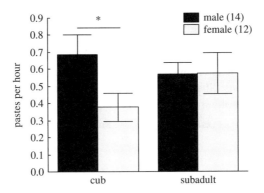

Fig. 17.2 Pasting rates of male and female spotted hyenas by age. Male cubs pasted more frequently than female cubs (Mann-Whitney U test, $U = 42$, $P = 0.03$), however, no sex difference was apparent when the hyenas were subadults ($U = 76$, $P = 0.71$)

influenced by social rank in any sex/age class examined here (male cub: $R^2 = 0.133$, $F_{1,12} = 1.85$, $P = 0.2$; female cub: $R^2 < 0.001$, $F_{1,10} < 0.01$, $P = 0.98$; male subadult: $R^2 = 0.082$, $F_{1,12} = 1.07$, $P = 0.32$; female subadult: $R^2 = 0.073$, $F_{1,10} = 0.79$, $P = 0.4$).

Males overmarked more when they were subadults than cubs (Fig. 17.3). Males did not, however, vary the proportional abundance of overmarking across early, middle and late subadult periods (N = 6, Friedman ANOVA $\chi^2 = 3.0$, P = 0.22). Female overmarking did not vary with age (Fig. 17.3). Although cubs did not overmark disproportionately by sex, subadults males overmarked more than females (Fig. 17.3). Social rank did not influence the extent to which hyenas overmarked in any age/sex class (male cub: $R^2 = 0.21$, $F_{1,12} = 3.14$, P = 0.1; female cub: $R^2 < 0.001$, $F_{1,10} < 0.01$, $P = 0.97$; male subadult: $R^2 < 0.001$, $F_{1,12} < 0.01$, $P = 1.0$; female subadult: $R^2 = 0.07$, $F_{1,10} = 0.71$, $P = 0.42$).

The proportion of pastings occurring in each behavioral context did not vary between cub and subadult periods for either males or females (Fig. 17.4). Additionally, among cubs there were no significant differences between males and females (Fig. 17.4a). Among subadults, however, more female than male pastings were spontaneous and more male than female pastings were socially facilitated (Fig. 17.4b). The proportions of pasting events that followed investigation of the environment, agonistic interactions, and play did not differ between male and female subadults. Lastly, in this smaller dataset than the one presented above in Fig. 17.3, male subadults did not overmark the pastings of conspecifics significantly more than female subadults (U = 35, P = 0.19) but the difference was in the same direction.

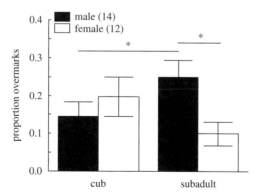

Fig. 17.3 Proportion of spotted hyena pastings that were overmarks by age and sex. Males increased their overmarking activity with age (Wilcoxon signed-rank test, S = 32.5, P = 0.04), however, females did not (cub vs. subadult: S = − 15.5, P = 0.24; across subadult periods: N = 6, Friedman ANOVA $\chi^2 = 3.5$, P = 0.17). Although male and female cubs did not differ in their frequency of overmarking (Mann-Whitney U test, U = 75, P = 0.67), a sex difference was apparent among subadults (U = 37, P = 0.02)

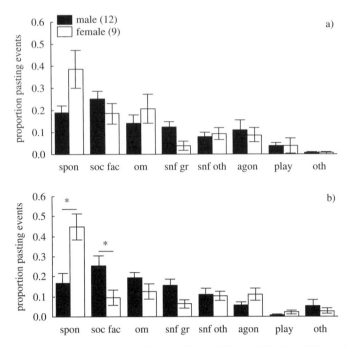

Fig. 17.4 Behavioral contexts of pasting by (a) cubs and (b) subadults. Sex did not significantly affect pasting context among cubs (Mann-Whitney U test; spontaneous (spon): U = 31, P = 0.11; socially facilitated (soc fac): U = 35, P = 0.19; overmark (om): U = 49, P = 0.75; sniff grass (snf gr): U = 26, P = 0.05; sniff another hyena (snf oth): U = 50.5, P = 0.81; agonistic interaction (agon): U = 49, P = 0.75; play: U = 32, P = 0.13). Female subadults were more likely than males to paste spontaneously (U = 12, P < 0.002), while males exhibited more socially facilitated marking (U = 21.5, P = 0.02). Other contexts did not vary by sex among subadults (snf gr: U = 29.5, P = 0.08; snf oth: U = 53.5, P = 0.97; agon: U = 38.5, P = 0.28; play: U = 41, P = 0.38). Neither males nor females varied pasting context with age (Wilcoxon signed-rank test; male: spon: S = −8, P = 0.52; soc fac: S = 0, P = 1.0; om: S = 19, P = 0.15; snf gr: S = 7, P = 0.58; snf oth: S = 6, P = 0.68; agon: S = − 5, P = 0.7; play: S = − 15.5, P = 0.07; female: spon: S = 3.5, P = 0.73; soc fac: S = − 8, P = 0.31; om: S = − 6.5, P = 0.5; snf gr: S = 7, P = 0.28; snf oth: S = 1.5, P = 0.91; agon: S = 3.5, P = 0.73; play: S = 1.5, P = 0.81).

17.4 Discussion

Although a few individuals in our study population produced paste as early as nine months, it appears that spotted hyenas do not consistently generate anal gland secretions until their third year. This is when both males and females mature sexually (Matthews 1939), with males beginning to disperse from their natal clans (Smale, Nunes and Holekamp 1997), and many females conceiving their first litters (Holekamp, Smale and Szykman 1996). It is also during this time that juveniles in our study began to participate in territorial defense. Therefore, it appears that paste production in spotted hyenas is similar to excretory gland activity in many mammals, in that it seemingly coincides with the appearance of adult typical behaviors (Yahr 1983). Interestingly, although puberty appeared to coincide with consistent paste production, this did not lead to increased marking activity. Woodmansee et al. (1991) found that captive spotted hyenas increased their frequency of pasting as they approached puberty. The difference between their study and ours reaffirms the influence of social and environmental contexts on mammalian scent marking behavior (French and Cleveland 1984; Müller-Schwarze 2001). Yet the puzzle remains. Why do juveniles who are not producing paste exhibit pasting behavior? The behavior may not be adaptive, but given the frequency at which young hyenas pasted in this study, functional hypotheses should certainly be considered.

First, rather than depositing scents in the environment, cubs may instead be acquiring group-specific odors that facilitate conspecific recognition of them as members of the group before they are widely known as individuals. Rasa (1973) proposed an analogous hypothesis for communal allomarking of newborn African dwarf mongooses (Helogale undulate rufula). In our study, we observed multiple instances of older subadults and adult females allomarking young cubs (unpublished data). Overmarking may be a way for cubs to proactively don group-specific scents. It has been suggested that European badger (*Meles meles*) cubs accomplish the same objective by rubbing themselves against the scent glands of adult conspecifics (Buesching, Stopka and Macdonald 2003). A second functional hypothesis for pasting by juvenile hyenas is that it exposes them to symbiotic bacteria that are potentially necessary for paste production in general, and for generation of group-specific odors in particular. In bobtail squid (*Euprymna scolopes*), bacterial colonization is a prerequisite for normal development of the luminescent organ in juveniles (McFall-Ngai and Ruby 1991). The rudimentary organs of juvenile squid have ciliated projections that facilitate bacterial inoculation. Once inoculation has occurred, the ciliated structures regress (McFall-Ngai and Ruby 1991). We see a rough analogy between these ciliated structures and overmarking by hyena cubs. In mammals, bacteria may be responsible for generating many of the semiochemicals emanating from excretory glands. One hypothesis for mammalian group-specific odors is that members of a social group, through bodily contact and overmarking, share a common microbial population in their scent-glands (Albone 1984). Pasting may therefore provide both short and long-term group-recognition benefits to juveniles, by facilitating their acquisition of both group-specific odors and bacteria.

We found that male cubs pasted more frequently than female peers, suggesting that this sexually dimorphic behavior might be mediated by differential exposure of males and females to specific gonadal steroid hormones during development (Yahr 1983). Although male and female subadults did not differ with respect to their frequency of pasting (see also Mills and Gorman 1987; Woodmansee et al. 1991), there were sex differences in the contexts that elicited subadult pasting behavior. Subadult males pasted more in response to scent marking activity by clanmates than did females, who most often pasted spontaneously. These sex differences suggest that the functions of pasting may differ between subadult males and females. Countermarking potentially advertises the signaler's competitive ability to both competitors and prospective mates (Rich and Hurst 1998). Even for subadult male hyenas who have begun to produce both paste in their anal glands and sperm in their testes, pasting in response to the scent marking activity of clanmates is unlikely to have a reproductive function because males seldom sire offspring in their natal clans (Engh, Funk, Van Horn, Scribner, Bruford, Libants, Szykman, Smale and Holekamp 2002). In the current study, we did not find a correlation between subadult male social rank and countermarking behavior (data not presented), suggesting that pasting by subadult males is also unlikely to serve a competitive function. To further evaluate this hypothesis, however, data must be collected on the targets of countermarking. The subject of countermarking by spotted hyenas will be addressed in an additional manuscript, as will the potential functions of pasting behavior by adult spotted hyenas.

Acknowledgments We thank the Office of the President of Kenya, the Kenya Wildlife Service, the Narok County Council and the Senior Warden of the Masai Mara National Reserve for their cooperation in completion of this research. E. Brown, L. Cisneros, A. Kuczynski and K. Pyle provided valuable assistance in data collection. J. Kolowski and J. Tanner provided many helpful suggestions. The methods in this study were approved by the Michigan State University Institutional Animal Care and Use Committee, approval number 05/05-064-00. This research was supported by funding from the National Science Foundation (IBN0343381 and IOB0618022).

References

Albone, E.S. (1984) *Mammalian Semiochemistry*. John Wiley, New York.

Altmann, J. (1974) Observational study of behavior: sampling methods. Behaviour 49, 227–265.

Boydston, E.E., Morelli, T.L. and Holekamp, K.E. (2001) Sex differences in territorial behavior exhibited by the spotted hyena. Ethology 107, 369–385.

Buesching, C.D., Stopka, P. and Macdonald, D.W. (2003) The social function of allo-marking in the European badger (*Meles meles*). Behaviour 140, 965–980.

Drea, C.M., Vignieri, S.N., Kim, H.S., Weldele, M.L. and Glickman, S.E. (2002) Responses to olfactory stimuli in spotted hyenas (*Crocuta crocuta*): II. Discrimination of conspecific scent. J. Comp. Psychol. 116, 342–349.

East, M.L. and Hofer, H. (2001) Male spotted hyaenas *Crocuta crocuta* queue for status in social groups dominated by females. Behav. Ecol. 12, 558–568.

Engh, A.L., Esch, K., Smale, L. and Holekamp, K.E. (2000) Mechanisms of maternal rank 'inheritance' in the spotted hyena, *Crocuta crocuta*. Anim. Behav. 60, 323–332.

Engh, A.L., Funk, S.M., Van Horn, R.C., Scribner, K.T., Bruford, M.W., Libants, S., Szykman, M., Smale, L. and Holekamp, K.E. (2002) Reproductive skew among males in a female-dominated mammalian society. Behav. Ecol. 13, 193–200.

Frank, L.G., Glickman, S.E. and Powch, I. (1990) Sexual dimorphism in the spotted hyaena (*Crocuta crocuta*). J. Zool. Lond. 221, 308–313.

French, J.A. and Cleveland, J. (1984) Scent-marking in the tamarin, *Saguinus oedipus*: sex differences and ontogeny. Anim. Behav. 32, 615–623.

Gosling, L.M. and Roberts, S.C. (2001) Scent-marking by male mammals: Cheat-proof signals to competitors and mates. Adv. Stud. Behav. 30, 169–217.

Holekamp, K.E. and Smale, L. (1993) Ontogeny of dominance in free-living spotted hyaenas: juvenile rank relations with other immature individuals. Anim. Behav. 46, 451–466.

Holekamp, K.E., Smale, L. and Szykman, M. (1996) Rank and reproduction in the female spotted hyaena. J. Reprod. Fertil. 108, 229–237.

Kruuk, H. (1972) *The Spotted Hyena*. University of Chicago Press, Chicago.

Matthews, L.H. (1939) Reproduction in the spotted hyaena, *Crocuta crocuta*. Philos. T. Roy. Soc. B. 230, 1–78.

McFall-Ngai, M.J. and Ruby, E.G. (1991) Symbiont recognition and subsequent morphogenesis as early events in an animal-bacterial mutualism. Science 254, 1491–1494.

Mills, M.G.L. (1990) *Kalahari Hyaenas*. Unwin-Hyman, London.

Mills, M.G.L. and Gorman, M.L. (1987) The scent-marking behaviour of the spotted hyaena *Crocuta crocuta* in the southern Kalahari. J. Zool. 212, 483–497.

Müller-Schwarze, D. (2001) From individuals to populations: Field studies as proving grounds for the role of chemical signals. In: A. Marchlewska-Koj, J.L. Lepri and D. Muller-schwarze (Eds.), *Chemical Signals in Vertebrates* 9. Kluwer Academic / Plenum Publishers, New York, pp. 1–10.

Rasa, O.A.E. (1973) Marking behaviour and its social significance in the African dwarf mongoose, *Helogale undulata rufula*. J. Tierpsychol. 32, 293–318.

Rich, T.J. and Hurst, J.L. (1998) Scent marks as reliable signals of the competitive ability of mates. Anim. Behav. 56, 727–735.

Shaffer, J.P. (1995) Multiple hypothesis testing. Annu. Rev. Psychol. 46, 561–584.

Smale, L., Nunes, S. and Holekamp, K.E. (1997) Sexually dimorphic dispersal in mammals: patterns, causes and consequences. Adv. Stud. Behav. 26, 180–250.

Van Horn, R.C., McElhinny, T.L. and Holekamp, K.E. (2003) Age-estimation and dispersal in spotted hyenas (*Crocuta crocuta*). J. Mammal. 84, 1019–1030.

White, A.M., Swaisgood, R.R. and Zhang, H. (2003) Chemical communication in the giant panda (*Ailuropoda melanoleuca*): the role of age in the signaller and assessor. J. Zool. Lond. 259, 171–178.

Woodmansee, K.B., Zabel, C.J., Glickman, S.E., Frank, L.G. and Keppel, G. (1991) Scent marking (pasting) in a colony of immature spotted hyenas (*Crocuta crocuta*): A developmental study. J. Comp. Psychol. 105, 10–14.

Wyatt, T.D. (2003) *Pheromones and Animal Behaviour*. Cambridge University Press, Cambridge.

Yahr, P. (1983) Hormonal influences on territorial marking behavior. In: B.B. Svare (Ed.), *Hormones and Aggressive Behavior*, New York, pp. 145–175.

Chapter 18
Human Body Odour Individuality

Pavlina Lenochova and Jan Havlicek

Abstract Humans produce temporarily stable, genetically mediated odour signatures and possess the ability to recognise, discriminate and identify other people through the sense of smell. The capability of self, gender, kin and non-kin odour recognition plays a role in social interactions. It seems that despite the stability of olfactory cues, the hedonic quality of body odour may vary over time.

18.1 Introduction

> *"It is the odour of a body what invites for love feast, not its beauty. . ."*
> *(Asklepiades from Erithrea, Greek philosopher and rhetorician, 124-60 BC)*

Human skin is covered with abundant apocrine, eccrine and sebaceous glands. The latter produces a fatty sebum over the whole body surface, especially on the chest, back, forehead, cheeks, scalp, breast areolae and external genitals. The highest incidence of eccrine glands occurs in axillae, palm, sole and forehead areas. In contrast, apocrine glands are only bound to tertiary hair and are to be found in the regions of axillae, external genitals, eyelids, nipples, around the umbilicus and in the ears and nose. Fresh secretion (fat sebum and sweat) of these glands is virtually odourless but the metabolic activity of skin bacteria transforms it to odourous compounds (Leyden, McGinley, Holzle, Labows and Kligman 1981). The characteristic odour of every human, called the "body odour signature", is an essential source of information about the odour producer.

Axillary area is the pivotal place of personal odour origin. Due to the upright posture in recent humans it is especially suitable for chemical communication. Axillary substances are potential cues of individuality, gender, reproductive status, age, health, diet, smoking habits, hygiene and so on. The main aim of this paper is to review studies dealing with different aspects of body odour individuality. We mainly focus on odour recognition of (1) gender specific cues, (2) self and sexual partner

Pavlina Lenochova
Charles University, Department of Anthropology and Human Genetics
p.lenoska@seznam.cz

J.L. Hurst et al., *Chemical Signals in Vertebrates 11.*
© Springer 2008

cues, (3) kin cues and (4) mother-infant cues. Subsequently, we discuss possible proximal causes of odour individuality.

18.2 Sex Differences

It is probable that besides individual characteristics, some general olfactory features differentiate the two sexes. Resulting from these differences, people can distinguish between males and females through smell alone. After experiencing both male and female hand odours, a group of eight women and eight men, blindfolded, went through thirty trials trying to identify the source of odour as one of those intro-duced previously. During the session, the raters were informed of the correctness of their judgment, thus systematic learning of the target odours could occur (Wal-lace 1997). In another study a three-choice discrimination test was used while the subject's own T-shirt, a strange male's and strange female's T-shirt was evaluated. A significant ability to discriminate the sexes on the basis of olfactory cues was found. Male odours were in most cases described as musky and female odours as sweet (Russell 1976). The capability of discrimination between male and female axillary odours was further shown in a cross-cultural study in 20% of Italian, 30% of German and 60% of Japanese raters. T-shirts worn for 7 consecutive nights were laid out in a separate room, then taken by the subject and evaluated for olfactory sex recognition and hedonic qualities, confirming the quality difference in the two sexes' odour (Schleidt, Hold and Attili 1981). In most studies the female odours were evaluated as more pleasant and male odours as unpleasant (Doty, Green, Ram and Yankell 1982; Schleidt et al. 1981; Schleidt 1980). These ratings often negatively correlated with the odour intensity (Havlicek, Bartos, Dvorakova and Flegr 2006; Weisfeld, Czilli, Phillips, Gall and Lichtman 2003; Doty, Kligman, Leyden and Orndorff 1978). An inverse relationship between the perceived intensity and pleasantness of body odour has also been shown in a study where people tried to identify gender from oral odours. After sniffing breath scents from glass tubes (the donors were hidden behind plywood barriers), the judges assessed the male odours as more intense and less pleasant (Doty et al. 1982). Even if people only suppose some odour to be male, they tend to judge it as rather unpleasant (Schleidt 1980).

The subjectively perceived qualitative variance between the two sexes' odours may be attributed to different levels of sex steroid homornes. However, analytical studies did not discover any qualitative differences in male and female odours (Zeng, Leyden, Spielman and Preti 1996). Only variable levels of steroid substances, like androstenes, have been found (Gower et al. 1985).

It has been shown that women's body odour changes across their menstrual cycle (Havlicek et al. 2006; Singh and Bronstad 2001). Odours of women in the fertile phase were rated as the least intense and the most attractive. This may result in misidentification of women's gender in some phases of their cycle (e.g. in men-strual phase). Another possible explanation is the fact that women have on average smaller apocrine glands than men and therefore produce less intense body odour

(Doty 1977). To sum up, men and women differ in their body odour, thus the two sexes can be discriminated but often the commonly perceived divergence in hedonic qualities of male and female body odour are likely to be a result of different intensity.

18.3 Self-Recognition and Sexual Partner's Odour

The ability to recognize one's own body odour was tested for the first time by Russell (1976). The participants had to pick out their own T-shirt from a group of three placed in wax-coated cardboard buckets with a small hole for sniffing. Thirteen of the sixteen males and nine of the thirteen females answered correctly. Raters in a different study had a much harder choice. Only one-third of the subjects identified their own odour correctly, but this time their own odour was one among another nine T-shirts (Hold and Schleidt 1977). These results were however still significantly better from chance. The shirts were laid out randomly in a row in a separate experimental room without any external covering and the raters were tested simultaneously, thus the actual proportion of subjects able to recognize their own sample could be underestimated as a consequence of touching the samples and interfusion of odours.

The quality of own scents was evaluated more often as being pleasant in women and more often unpleasant in men (Schleidt et al. 1981; Hold and Schleidt 1977). Elsewhere, a comparison of self-odour hedonic evaluation was made between a group of people who had a strictly prescribed hygienic treatment and participants without any restrictions, who followed their usual personal hygienic routine (i.e. using perfumes). In the former, men rated their odour as unpleasant, unlike the latter group, where men assessed their odour as predominantly pleasant. Women in both groups described their scents as mainly pleasant (Schleidt 1980). Different results were found when after three hours of sampling axillary secretions the participants tried to find their own gauze in a group of five vials and evaluated all samples for the hedonic quality. 59.4% of females, but only 5.6 % of males were able to identify their self-odour. At the same time, women tended to rate their own secretions as significantly less pleasant than males (Platek, Burch and Gallup 2001).

Apart from the self-odour, another scent which one would expect to be relatively familiar with—due to frequent contact with naked skin—is body odour of a sexual partner. Twenty-four couples were recruited for the study of Hold and Schleidt (1977). Their T-shirts, worn for seven consecutive nights, were used for the experiment with nine other samples. Correct identification of the partner's T-shirt on the basis of individual odour qualities were found in one third of the subjects. The odour of the partner was judged as predominantly pleasant (Hold and Schleidt 1977). In another study, no hygienic restrictions were demanded in the group of participants while odour sampling. Even though they could use any perfumes and deodorants, the partner's body odour recognition was comparable to that in the previous study (Schleidt 1980). From the cross-cultural point of view, the sexual partners' body odours were evaluated more often as pleasant in German females than in Italian and Japanese females, while in men no cross-cultural differences were found (Schleidt et al. 1981).

On the whole, it seems that humans are able to recognize their own body odour and women mostly tend to assess it as pleasant and men mostly as unpleasant. The significant results from experiments dealing with the question of the sexual partner's body odour recognition indicate that humans can learn their partner's olfactory cues, probably through their mutual intimacy.

18.4 Kin Recognition

Most likely, specific olfactory cues (genetically based similarities in body odours) are perceivable among all family members. During human evolution, the ability of distinguishing kin from non-kin might have played an important role in kin altruism and inbreeding avoidance. Individuals with the advantage of discerning a kin member could have had higher levels of survival, successful reproduction and consequently higher fitness. Olfactory kin recognition was tested for the first time by Porter and Moore (1981). A T-shirt of one child was sniffed in a two-choice test, together with a T-shirt worn by an unfamiliar, age-matched individual, by the child's mother and by a sibling. Mothers and siblings alike were able to discriminate their relative's scent (Porter and Moore 1981). The possibility that the sibling's odour is discernable simply due to shared environmental cues, was dismissed by Porter et al. (1986). Siblings who had lived a long time apart rated the odours of their brother or sister and a control person of the same sex and age. Twenty-seven of the forty subjects correctly identified the shirt worn for three nights by their sibling. This significant result supports the probable existence of family olfactory cues (Porter, Balogh, Cernoch and Franchi 1986).

Recognition of kin is expected to fluctuate according to the level of consanguinity. The closer the family members are the more similar olfactory signatures with more family cues they could have. Varying degrees of consanguinity were observed in a study of Porter et al. (1986), where newborn babies' fathers, grandmothers and aunts, who were exposed to its odour before testing, identified correctly the odour of donor infant. Later on, the odour was described by fathers as similar to their own or the newborn mother's body odour. Positive results in relatively distant relatives (i.e. aunts and grandmothers) can be explained by the fact of a prior knowledge about the aim of study and consequent possibility of self-training. Thus, level of consanguinity did not show an effect on ability to recognise relative newborn baby. Different levels of consanguinity were the main topic of another study in which preadolescent children identified their full siblings (first-degree relatives) but not half siblings or step-siblings (further relatives) by the odour (Weisfeld et al. 2003). In the same study, twenty-two Canadian university students were tested for the ability to discriminate the odours of their mother, father, a brother, a sister, a familiar but unrelated person, a stranger, and themselves. Kin members were rarely confused with non-kin and the participants remarked that even though the scent very often resembled a familiar one, they were not sure to which target person it belonged to. Close relatives of one species seem to have a similar but individually distinguishable odour signature.

Discrimination between kin and non-kin is also connected with the phenomenon of inbreeding avoidance and incest taboo. The combination of related genes might mean a lower probability of viability of the offspring. Hedonic evaluations of kin members' odours provide good support for these suggestions. People recognized their relatives' odour but described it as less pleasant than the odour of a non-kin person in general. A mutual father-daughter and brother-sister aversion was found as most significant in tests of preferences of odours in the nuclear family (Weisfeld et al. 2003). These relationships are the examples of the greatest danger of incest. The exact mechanism of kin recognition is not yet fully explained, but there are two main theories. Firstly, odour association may be an important factor in individual odour discrimination based on olfactory imprinting during the sensitive period of life ("sensitive period phenomenon"). Thus, the recognition of sibling's odour could be strengthened through exposure to a sibling during childhood (e.g. Porter et al. 1986). Secondly, the family member cue provides support for the mechanism of "phenotypic matching". On the basis of similar features in the odour signature, the individuals may discriminate between the stimuli of known kin and those of possible kin. For example, T-shirts of mothers and their children were correctly matched by non-kin raters on the basis of perceivable similarities (Porter, Cernoch and Balogh 1985).

18.5 Mother-Infant Recognition

The special case in kin recognition studies is mutual recognition of mothers and newborn infants. Presumably, olfaction plays a crucial role in the mother-infant first contact and the linking-up of their mutual relationship.

In a study testing the olfactory phenotype matching theory (see above), a child's garment together with garments worn by unfamiliar children were presented in cardboard buckets and mothers were asked to decide which one belonged to their own baby. It was found that newly-parturient mothers can identify their own infants based on body odour within the first six days postpartum (Porter, Cernoch and McLaughlin 1983). To test whether mothers are especially sensitive for infant recognition, a group of non-mothers was also asked to participate. After 45 minutes of cradling a baby, women tested its undershirt. Unexpectedly, it was found that there was no difference between mothers and non-mothers in this task (Kaitz and Eidelman 1991). The authors concluded that both mothers and non-mothers can learn an infant's odour features and they do so without intention during the course of routine interactions. Another experiment using pre-exposure to the target odours came to similar conclusions: no improved ability of infant odour identification was found amongst mothers than non-mothers (Fleming, Corter, Surbey, Franks and Steiner 1995). These authors suggest that mothers are not uniquely primed to recognize the infant odours. In the second part of the study, the authors focused on factors associated with the successful recognition. They discovered that the amount of time spent with the child in the first day after the delivery positively influences the mother's success in odour recognition. In addition, mothers who were able to recognize their infant's odour differed from mothers who were not in having a shorter interval to first breast-feeding, more nurturing attitudes and closer nasal interactions with the child.

A newborn's ability to recognise its mother is the other side of the coin. First contact with maternal olfactory stimuli occurs *in utero*. The child is in an environment poor in external visual and auditory stimuli and olfaction might have an outstanding role in perception of the surroundings. A rich range of chemical information probably stimulates fetal chemoreceptors via the flow of the amniotic fluid through the nasal passages of the baby (Davis and Porter 1991). The odour of amniotic fluid itself was a main interest in several previous studies. It is supposed that there are characteristic familiar cues in the odour of the amniotic fluid. The study of Schaal and Marlier (1998) showed that both mothers and fathers are able to discriminate and accurately identify the odour of amniotic fluid from their own infant at a better-than-chance level. The scent of amniotic fluid was described as similar to the body odour of the mother, especially at the end of gestation (Schaal and Marlier 1998).

The individually unique quality of amniotic fluid is confirmed by the newborn infant's own responsiveness to its odour. Own amniotic fluid and a control stimulus were presented to 2 day old newborns. Most frequently, infants turned their head to a gauze pad scented with their own amniotic fluid odour (Schaal, Marlier and Soussignan 1995). Neither differences between breast-fed and bottle-fed neonates nor between the two sexes were found. The authors interpret the amniotic fluid preference as a consequence of its odour similarity to lacteal secretions, which might help while breast-feeding. The uniqueness of olfactory cues in amniotic fluid has also been shown in several other studies. Immediately after parturition, the newborn was washed and one nipple of the mother was moistened with a small amount of amniotic fluid. The baby was then laid between the mother's breasts and, during the initial attempt to locate the mother's nipple, spontaneous choice of the nipple scented with amniotic fluid occurred (Varendi, Porter and Winberg 1996). In the study of Marlier et al. (1997), however, 2 day old breast-fed children did not show preferential reaction to amniotic fluid or to lacteal secretions. At four days of age, they preferred odour of milk (Marlier, Schaal and Soussignan 1997). On the contrary, one year later the same research team found somewhat different results. Two and four day old bottle-fed neonates were exposed to the odours of their amniotic fluid, formula milk and distilled water as a control sample. Both 2 and 4 day old infants preferred the amniotic fluid to the odour of formula milk (Marlier, Schaal and Soussignan 1998a). A discrepancy between these two studies can be explained by different development of preferences in breast fed and bottle fed infants. Positive attraction to amniotic fluid was also found in the work of Schaal et al. (1998). At the age of three days, both breast-fed and bottle-fed neonates preferred their familiar amniotic fluid, when presented together with a non-familiar one, and a familiar or non-familiar amniotic fluid when presented with the control stimulus (Schaal, Marlier and Soussignan 1998). Another study by Marlier et al. (1998b) convincingly documents the fluent transition of preferences from amniotic fluid to lacteal odours. When the reactions of 2 day old breast-fed neonates to presented pairs of stimuli (amniotic fluid, colostrum or control) were compared, the amniotic fluid was favoured at this age. On the other hand, 4 day old children preferred milk to amniotic fluid and the control. It seemed that with increasing age and exposure to feeding cues, infants display greater attraction to their mothers' lacteal odour than to their own amniotic fluid odour (Marlier, Schaal and Soussignan 1998b). The last study to be mentioned here examined whether the odours of amniotic fluid, mother's breast or a control stimulus influence the crying of the newborn infant. After the first breast-feed, at the moment of the first distance of crying, the infants were exposed to one of the odourous samples. Infants exposed to amniotic fluid cried significantly less than other exposed to the other stimuli (Varendi, Christensson, Porter and Winberg 1998). Despite the fact that the opportunity for breast odour sampling was no more than two hours, which could explain the lack of ameliorative effect of breast odour, the importance of familiar amniotic fluid smell is obvious. To sum up, the attraction to amniotic fluid odour may reflect foetal exposure to that substance. The human foetus can detect and store the unique chemosensory information available in the prenatal environment and thus prenatal experience might influence the earliest odour preferences.

In the last three decades, the question of discernible uniqueness of mother's odour for its newborn baby has been tested repeatedly. It is thought that, early in life, the key olfactory cues are produced by the mother's nipple and areola region, which may support the first successful suckling (Schaal, Doucet, Sagot, Hertling and Soussignan 2006). In the first, second and sixth week after delivery, the mostly sleeping children smelled cotton pads impregnated with an odourless substance, then pads worn for three hours in the breast area of their mother and another lactating mother (Russell 1976). There was no difference between the one and 2 week old newborns' responses. At the age of six weeks, 6/10 of infants identified their mother's odour from the non-mother's. The author suggests the possibility that this effect could be explained by the infant's own scent contaminating the odour of the mother's skin during earlier contacts (Russell 1976). However, the value of odours produced by the lactating woman's breast was confirmed in the Makin and Porter study (1989). A group of full-term, 2 week old, healthy but bottle-fed neonates was exposed to three samples: a breast sample of an unrelated non-parturient female, a breast sample of an unrelated lactating mother and an axillary odour sample of the same unrelated female. In a series of two-choice tests, children oriented preferentially to the breast sample of the lactating woman without any previous exposure to her odour, indicating the attractiveness of the areolar odour for newborns in general (Makin and Porter 1989). Varendi et al. (1994) provided another supportive study confirming the extraordinary importance of breast odour. They washed one randomly chosen breast of mothers immediately after delivery, placed the newborn between her breasts and observed that the most children spontaneously chose the unwashed one, grasped a nipple and started to suckle (Varendi, Porter and Winberg 1994). Additionally, preference for familiar axillary odours has been tested. Two week-old breast-fed newborns sniffed their own mother's axillary odour for a longer time than samples of control women (Cernoch and Porter 1985). No similar preference for the father's odour was found. A different reaction was found in bottle-fed newborns of the same age. They were not able to recognize the odour of their mother, which indicated the lack of familiarization of the mother's unique olfactory signature, presumably because of reduced contact with her naked skin. This same effect could also explain the lack of preference for father's odours.

18.6 Stability of Odour Quality

The origin of odour individuality could be at least partly explained by genetic factors. This is supported by several twin studies. As monozygotic twins possess the same genotype, it is supposed that their olfactory signatures would resemble each other. The hand odours of two female monozygotic twins, having been on the same diet for three days, and two female twins on a different diet, were judged by nine women. The number of correct discriminations between the two twins was better than chance in both groups. The hand odours of twins on different diets were significantly easier to recognize (Wallace 1977). In another study, cotton pads worn for one night in the axillae of twins living apart were gathered and presented to subjects who tried to match the odours of the twins together (Roberts, Gosling, Spector, Miller, Penn and Petrie 2005). Scents of identical twins were matched by raters better than could be expected by chance and with higher frequency than odours of dizygotic twins. Duplicate odour samples of the same individual were correctly matched with the same frequency as matching the identical twins scents. This suggests high odour similarity in pairs of twins which can be perceived by human nose, even if the siblings live apart.

Influence of genetic factors on the odour signature predicts its relative stability throughout life. On the other hand, body odour can be shaped by a number of environmental factors as well, including diet and reproductive or health status (for review see Havlicek and Lenochova in this issue). It was shown by Porter et al. (1985) that environmental influences alone cannot produce sufficiently similar body odours to enable matching of two unrelated individuals. They asked spouses who lived together to wear T-shirts, and then asked unfamiliar raters to match the spouses according to their smell. Raters were not able to match odours of spouses at rates different from chance. Thus, genetic factors seem to be crucial for individual or kin recognition

18.7 Conclusions

The main aim of this paper was to document in which social contexts human body odour individuality may be functionally important. There is clear evidence that humans can recognize other people and discriminate or identify their body odour. Contrary to traditional thinking, smell therefore has the potential to be an integral part of nonverbal behaviour and plays a role in social interactions and establishment of relationships. Olfactory communication appears highly important in very early human development, potentially facilitating the mutual mother-infant relationship. The majority of studies are based on conscious and verbally declared recognition. We suggest that this approach might in fact underestimate the significance of odour cues. It is expected that odour perception partly runs on the subconscious level. Presumably, humans perceive the unique body odour of other individuals and evaluate its hedonic quality being not always fully aware of it. This is supported by some studies based on EEG (Electroencephalography) (e.g. Pause, Krauel, Sojka and Ferstl 1998). However, it still presents a field to be explored. Although the olfactory signature seems to be genetically mediated and consequently stable over the time, our data indicate that its quality could oscillate, possibly due to environmental factors.

Acknowledgments We were funded by the Josef, Marie and Zdenka Hlavka Foundation, grants no. GAUK 393/2005 and GACR 406/06/P377. We thank Barbara Husarova, Jitka Lindova, Nassima Boulkroune and Tom Hadrava for many helpful comments on the manuscript and language corrections.

References

Cernoch, J.M. and Porter, R.H. (1985) Recognition of maternal axillary odors by infants. Child. Dev. 56, 1593–1598.

Davis, L.B. and Porter, R.H. (1991) Persistent effect of early odor exposure on human neonates. Chem. Senses 16, 169–174.

Doty, R.L., Green, P.A., Ram, C. and Yankell, S.L. (1982) Communication of gender from human breath odors: relationship to perceived intensity and pleasantness. Horm. Behav. 16, 13–22.

Doty, R.L., Kligman, A., Leyden, J. and Orndorff, M.M. (1978) Communication of gender from human axillary odors: Relationship to perceived intensity and hedonicity. Behav. Biol. 23, 373–380.

Doty, R.L. (1977) A review of recent psychophysical studies examining the possibility of chemical communication of sex and reproductive status in humans. In: Müller-Schwarz D and Mozell, M.M., *Chemical Signals in Vertebrates*. Plenum, New York, pp. 273–286.

Fleming, A., Corter, C., Surbey, M., Franks, P. and Steiner, M. (1995) Postpartum factors related to mother's recognition of newborn infant odours. J. Reprod. Infant. Psyc. 13, 197–210.

Gower, D.B., Bird, S., Sharma, P. and House, F.R. (1985) Axillary 5 alpha-androst-16-en-3-one in men and women: relationship with olfactory acuity to odorous 16-androstenes. Experientia 41, 1134–1136.

Havlicek, J., Bartos, L., Dvorakova, R. and Flegr, J. (2006) Non-advertised does not mean concealed. Body odour changes across the human menstrual cycle. Ethology 112, 81–90.

Hold, B. and Schleidt, M. (1977) The importance of human odor in nonverbal communication. Z. Tierpsychol. 43, 225–238.

Kaitz, M. and Eidelman, A.I. (1991) Smell-recognition of newborns by women who are not mothers. Chem. Senses 17, 225–229.

Leyden, J.J., McGinley, K.J., Holzle, E., Labows, J.N. and Kligman, A.M. (1981) The microbiology of the human axilla and its relationship to axillary odor. J. Invest. Dermatol. 77, 413–416.

Makin, J.W. and Porter, R.H. (1989) Attractiveness of lactating females' breast odors to neonates. Child. Dev. 60, 803–810.

Marlier, L., Schaal, B. and Soussignan, R. (1997) Orientation responses to biological odours in the human newborn. Initial pattern and postnatal plasticity. C. R. Acad. Sci. Paris, Science de la vie. 320, 999–1005.

Marlier, L., Schaal, B. and Soussignan, R. (1998a) Bottle-fed neonates prefer an odor experienced in utero to an odor experienced postnatally in the feeding context. Dev. Psychobiol. 33, 133–145.

Marlier, L., Schaal, B. and Soussignan, R. (1998b) Neonatal responsiveness to the odor of amniotic and lacteal fluids: a test of perinatal chemosensory continuity. Child. Dev. 69, 611–623.

Pause, B.M., Krauel, K., Sojka, B. and Ferstl, R. (1998) Body odor evoked potentials: a new method to study the chemosensory perception of self and non-self in humans. Genetica 104, 285–294.

Platek, S.M., Burch, R.L. and Gallup, G.G. (2001) Sex differences in olfactory self-recognition. Physiol. Behav. 73, 635–640.

Porter, R.H. and Moore, J.D. (1981) Human kin recognition by olfactory cues. Physiol. Behav. 27, 493–495.

Porter, R.H., Cernoch, J.M. and McLaughlin, F.J. (1983) Maternal recognition of neonates through olfactory cues. Physiol. Behav. 30, 151–154.

Porter, R.H., Cernoch, J.M. and Balogh, R.D. (1985) Odor signatures and kin recognition. Physiol. Behav. 34, 445–448.

Porter, R.H., Balogh, R.D., Cernoch, J.M. and Franchi, C. (1986) Recognition of kin through characteristic body odors. Chem Senses 11, 389–395.

Roberts, S.C., Gosling, L.M., Spector, T.D., Miller, P., Penn, D.J. and Petrie, M. (2005) Body odor similarity in noncohabiting twins. Chem. Senses 30: 1–6.

Russell, M.J. (1976) Human olfactory communication. Nature 260, 520–522.

Schaal, B., Doucet, S., Sagot, P.,Hertling, E. and Soussignan, R. (2006) Human breast areolae as scent organs: morphological data and possible involvement in maternal-neonatal coadaptation. Dev. Psychobiol. 48, 100–110.

Schaal, B. and Marlier. L. (1998) Maternal and paternal perception of individual odor signatures in human amniotic fluid – potential role in early bonding? Biol. Neonate 74, 266–273.

Schaal, B., Marlier, L. and Soussignan, R. (1998) Olfactory function in the human fetus: evidence from selective neonatal responsiveness to the odor of amniotic fluid. Behav. Neurosci. 6, 1438–1449.

Schaal, B., Marlier, L. and Soussignan, R. (1995) Responsiveness to the odor of amniotic fluid in the human neonate. Biol. Neonate 67, 397–406.

Schleidt, M., Hold, B. and Attili, G. (1981) A cross-cultural study on the attitude towards personal odors. J. Chem. Ecol. 7, 19–31.

Schleidt, M. (1980) Personal odor and nonverbal communication. Ethol. Sociobiol. 1, 225–231.

Singh, D. and Bronstad, M. (2001) Female body odor is a potential cue to ovulation. Proc. R. Soc. Lond. B 268, 797–801.

Varendi, H., Christensson, K., Porter, R.H. and Winberg, J. (1998) Soothing effect of amniotic fluid smell in newborn infants. Early Hum. Dev. 51, 47–55.

Varendi, H., Porter, R.H. and Winberg, J. (1996) Attractiveness of amniotic fluid odor: evidence of prenatal olfactory learning? Acta Paediatr. 85, 1223–1227.

Varendi, H., Porter, R.H. and Winberg, J. (1994) Does the newborn baby find the nipple by smell? Lancet 344, 989–990.

Wallace, P. (1977) Individual discrimination of humans by odor. Physiol. Behav. 19, 577–579.

Weisfeld, G.E., Czilli, T., Phillips, K.A., Gall, J.A. and Lichtman, C.M. (2003) Possible olfaction-based mechanisms in human kin recognition and inbreeding avoidance. J. Exp. Child Psycho. 85, 279–295.

Zeng. X.N., Leyden, J.J., Spielman, A.I. and Preti, G. (1996) Analysis of characteristic human female axillary odors: Qualitative comparison to males. J. Chem. Ecol. 22, 237–257.

Chapter 19
Environmental Effects on Human Body Odour

Jan Havlicek and Pavlina Lenochova

Abstract Human body odour is individually specific and several lines of evidence suggest that to some extent it is under genetic control. There are however numerous other sources of variation, commonly labelled as environmental factors, which are the main aim of this paper. These include: 1) reproductive status, 2) emotional state, 3) diet and 4) diseases. We primarily focus on axillary and genital odours as they have been proposed to have communicative function. We prelusively conclude that a specific diet and some diseases have major impact on variations in human body odour.

19.1 Introduction

In common with most other creatures, humans emit numerous odours from a variety of sources. The father of sexology, Havelock Ellis, assorted the sources as (1) the general skin odour, a faint, but agreeable fragrance often to be detected on the skin even immediately after washing; (2) the smell of the hair and scalp; (3) the odour of the breath; (4) the odour of the armpit; (5) the odour of the feet; (6) the perineal odour; (7) in men, the odour of the preputial smegma; (8) in women the odour of the mons veneris, that of vulvar smegma, of vaginal mucus and the menstrual odour (Ellis 1927). Odours of all above mentioned sites participate to a different extent on our body odour individuality, which was labelled an "odour signature" (Porter, Cernoch and Balogh 1985). In particular it is the axillary odour, odour of skin, scalp and genital region which are the most distinctive in healthy humans. It was speculated by different theorists that apocrine glands in the axilla (e.g. Comfort 1971) and in the genital region (e.g. Michael, Bonsall and Kutner 1975) developed primarily for communicative purposes. As the main aim of this paper is communication (in its broad sense), we will mostly focus on odours emitted from these body sites. On the other hand, we believe that even odours produced as by-products of an organism's

Jan Havlicek
Charles University, Department of Anthropology
jan.havlicek@fhs.cuni.cz

J.L. Hurst et al., *Chemical Signals in Vertebrates 11.*
© Springer 2008

metabolic activity may have informational value for a perceiving individual. Thus, odours emitted from other sites will be mentioned as well.

The significance of genetic control on body odour individuality has been shown previously in several twin studies. Wallace (1977) found that hand odour was more easily distinguishable between dizygotic than between monozygotic twins. A more recent paper by Roberts, Gosling, Spector, Miller, Penn and Petrie (2005) found that odours of monozygotic twins were correctly matched by human subjects higher than chance but this was not true for odours of dizygotic twins. Twins lived apart in this study, which excluded the possibility that their body odour was similar due to shared environmental factors. As with humans, dogs can match odours of twins but not of genetically unrelated individuals (Sommerville, McCormick and Broom 1994).

Heritability of the odour signature is further supported by a study which showed that subjects not acquainted with odour donors were able to match odour of parents and their offspring but not odours of the spouses (Porter et al. 1985). The latter result again excluded the possibility that correct matching was due to shared environment.

On the other hand, there are also numerous influences on body odour from the environment. In this paper we describe the effects of environmental factors in a broad sense and *sensu stricto* we mean non-genetic factors. We consider this division a major help for our working purposes. However, one should not forget that in reality it is extremely difficult to disentangle numerous interactions between influences of the genes and the environment. The main aim of this paper is to review current knowledge about environmental factors, including psycho-physiological effects, and to propose possible avenues for future research. Environmental factors supposedly having a particularly profound impact on body odour are: (1) reproductive status, (2) emotional status, (3) diet, and (4) infections and diseases.

19.2 Reproductive Status

It is conventionally thought that ovulation in humans is concealed. This is in contrast with some primate species having conspicuous visual cues to ovulation. That is why a number of theorists have speculated on the evolution of concealed ovulation (for review, see Pawlowski 1999). However, several recent studies showed that hedonicity of body odour fluctuates across the menstrual cycle, being most attractive around the time of ovulation. Singh and Bronstad (2001) asked women to wear T-shirts for 24 h when in the follicular phase and again when in the luteal phase of their cycle. Male raters assessed odour of the T-shirts worn in the follicular phase significantly more attractive. Using a similar approach to odour collection (i.e. T-shirts) but worn by each woman only once, a Finnish team found a quadratic correlation of body odour pleasantness and day of the cycle, peaking in mid cycle (Kuukasjärvi, Eriksson, Koskela, Mappes, Nissinen and Rantala 2004). This pattern was significant in male raters and also a similar non-significant trend was found in female raters. They found no changes in subjectively perceived quality of body odour in women taking hormonal contraception. Two further studies, carried out by one team, used one-shot data. There was a correlation between the probability of conception and body odour attractiveness in one study (Thornhill, Gangestad, Miller, Scheyd, McCollough and Franklin 2003) but not in the other (Thornhill and Gangestad 1999).

Most recently, Havlicek, Bartos, Dvorakova and Flegr (2006) tested women's body odour repeatedly across the whole menstrual cycle and focused specifically on axillary scents. Male subjects rated the smell of the samples as least attractive during menstrual bleeding and most attractive during the follicular (i.e. fertile) period of the cycle.

Axillary odour might not be the only olfactory cue to women's reproductive status. Due to the cyclic fluctuations in composition of vaginal secretion one may expect similar changes in vaginal odour. This was demonstrated thirty years ago by Doty, Ford, Preti and Huggins (1975). However, as we have suggested previously, we believe that due to bipedality the axillary odour is of higher importance in humans (Havlicek et al. 2006).

Current evidence strongly suggests the existence of olfactory cues to ovulation in humans. Such cues could be used by men for monitoring their current or potential partner. It should be noted, however, that men cannot determine ovulation exactly but rather in a probabilistic manner. Therefore we believe that cyclic odour cues are rather a byproduct of physiological changes than a specific signal of ovulation.

Apart from cyclic changes we may expect pre- and post-reproductive odour cues. Empirical evidence is however highly limited. In a study, primarily focused on mood changes, it was found that odour of postmenopausal women was misidentified as the odour of men (Chen and Haviland-Jones 1999). This phenomenon was more poetically described by Gould and Pyle (1896): children smell of butyric acid and odour of older people resembles dry leaves. This suggests that there might be cues related to reproductive life period. Development of active apocrine glands during puberty also supports this view (Stoddart 1990). As far as we know there has been no study on changes in body odour during puberty.

19.3 Emotional State

For a socially living individual it is of high relevance to be aware of the emotional state of one's conspecifics. It is a fairly well-known phenomenon that animals of different species react to odour emitted by another animal which is under stressful conditions. This led several researchers to the question of whether emotional state in humans can influence their body odour. This possibility was first tested by Chen and Haviland-Jones (2000), who asked their subjects to wear armpit pads while watching 13 min excerpts from a comedy or a fearful movie. Most of them watched both excerpts. Subsequently three and six choice tests were presented to raters who were instructed to pick the sample that smelled of people when they are happy or afraid. Women correctly identified "happy odour" collected from both sexes and "fearful odour" from men but not of women. Men correctly identified "happy odour" collected from women and "fearful odour" of men. Authors speculate that failure to identify women's "fearful odour" is due to the lower intensity of their body odour compared to men. However, the authors did not check for the effect of shaving of axillary hair which is more prevalent in women and we suppose it might have biased the gender comparison.

The second study used a similar method to induce specific mood in odour donors (Ackerl, Atzmueller and Grammer 2002). In this case, subjects watched a horror movie or a neutral movie. From three presented odours, female raters were asked to choose the sample different from the other two and say which one "smells like fear". The different sample was picked significantly more often, but not judged correctly as "fearful odour". However, "fearful odour" was rated as less pleasant and more intense.

Finally, the issue of "fearful and happy odours" was intensively explored by Cantafio (2003). He carried out a series of four methodologically similar studies. In the first, odour donors watched a 12 min excerpt from a scary movie twice over (i.e. 24 min). Axillary samples were collected after 6, 12 and 24 min. Female raters found odour of men taken after 24 min. more likely to smell like fear. No similar effect was observed in female donors. In the second study, donors watched first a comedy, a horror movie a week after and odour in a neutral context was collected after energetic exercise. Data analysis unexpectedly showed that 12 min samples smelled more like fear than 24 min samples; samples collected during a comedy were judged to smell like fear and samples collected while watching a horror were judged to smell like happiness. The author explains these counter-intuitive results by suggesting that raters can differentiate between the smell of the emotions even though they give them wrong labels. He therefore addressed this question in other two experiments. It was tested whether raters judge samples collected under particular emotional conditions the same during a second test, irrespective of whether they were labelled correctly or not. In one experiment axillary samples were used, in the other samples were collected from the forehead. In both experiments, the raters labelled the samples significantly more often the same in the second session as in the first. Moreover, male axillary samples collected during the fearful movie were judged correctly. Neither women's axillary fear odour, nor happy odour of both sexes, nor forehead odours, were identified correctly. Even though we found Cantafio's work highly innovative, we suggest it should be interpreted cautiously due to several methodological shortcomings. For instance, sample collection was not balanced and different length of storage may have affected the results. Using the same excerpt repeatedly during one session to induce a particular mood is also disputable. Moreover, he did not control for menstrual cycle effects or using hormonal contraception in either women raters or donors (see above). All these variables have been shown to have an impact on olfactory abilities and body odour, respectively.

Some 19th century writers reported about cases of distinctive body odour emanating after sexual excitement or coitus (Gould and Pyle 1896). According to our knowledge this question has not been addressed in any empirical study yet.

In conclusion, research on odours and emotions is a potentially fruitful field. The reviewed studies suggest that at least some emotional states may result in changes in body odour. We feel however, that the method of verbal recognition of mood according to odour is not the most appropriate one. Even though odours emitted in specific emotional states may have informative value they do not have to be perceived under full conscious control. Thus using hedonic ratings or more objective physiological measures such as changes in skin conductance, heart beat or hormonal levels seems more suitable to us.

19.4 Diet

The effect of diet on body odour was previously observed in several animal species. In a series of simple choice experiments it was found that guinea pigs fed on diet usually used in laboratories for rats were judged less attractive by the opposite sex than those fed on diet designed for guinea pigs (Beauchamp 1976). Ferkin, Sorokin, Johnston and Lee (1997) tested how diet containing different amounts of protein influences odour of different sources in male meadow voles. The more protein the males consumed, the longer females spent investigating their scents. This was particularly true for marks from the anogenital region and faeces and, to a lesser extent, for urine. A follow-up study focusing on the effect of food deprivation (Pierce and Ferkin 2005) showed that odours of meadow vole females deprived of food for 24 h were less attractive to male conspecifics. However, this effect disappeared 48h after they were re-fed. Food deprivation also affected the females' odour preferences. When deprived, they did not show preferences for male odours but this returned 72 h after being re-fed. Contrary to humans, the main source of odour in meadow voles and other rodents is urine. As metabolites of ingested food are excreted partly in urine, one may expect significant changes in its composition due to diet. A pathway to axillary glands in humans is much less straightforward.

Most of the researchers testing genetic components of the body odour signature mention diet as the most significant source of environmental variability (e.g. Roberts et al. 2005). This is supported by a study which found that hand odour of twins on the same diet was not matched correctly by human subjects (Wallace 1977). Also dogs who were able to recognize monozygotic adult twins on a different diet failed in the same task when smelling monozygotic suckling twins on the same diet (part breast-fed and part bottle-fed; Hepper 1988). However, when we look closer to find out which diet components shape our body odour in fact we realize that we know very little.

To fill this gap we specifically focused on the effect of red meat consumption (Havlicek and Lenochova 2006). We asked 8 male odour donors to adopt for two weeks a diet containing daily portions of meat, and 9 subjects to adopt a similar diet with a non-meat alternative. For the first ten days they were given a list of meals they should eat. The list of dishes for the two groups differed only by inclusion of meat. For the last four days pre-testing, all meals were provided to the subjects to further standardize dietary intake. Odour donors were asked not to use deodorants or perfumes, and to avoid spicy food, smoking and drinking alcohol. Every day they also wrote down their mood, fatigue and level of stress on seven point scales. During the last 24 h, they wore pads in their armpits under controlled conditions and immediately after the collection of samples, 30 female raters assessed all samples on their pleasantness, attractiveness, intensity and masculinity using seven point scales. A month later, we performed the same procedure with odour donors on the opposite diet. To analyze the data we used repeated measures ANOVA with raters as the unit of the analysis. In other words, we compared mean women's ratings when they assessed donors on the meat diet and when on the non-meat diet. We found that odour of donors when on the non-meat diet was rated more pleasant, attractive and less intense. As we did not find differences in subjectively-rated mood, stress and fatigue between meat and non-meat condition we believe that the observed effect is likely to be due to the diet effect.

At this point we do not know how long one must consume meat to elicit a discernible difference in body odour, neither do we know about the amount of meat already perceptible in body odour. This study specifically focused on the effect of red meat. It is unknown whether a similar effect would be found for consumption of fish. Some ethnologic remarks suggest so. For instance, Stoddard (1990) mentions that people whose main diet is fish exude a fishy, trimethylamine-like scent. However, it is not clear what the source of this note is. From an evolutionary perspective, our results on the decrease of odour attractiveness due to red meat consumption are rather surprising. It is conventional wisdom that meat consumption played a substantial role in social aspects of human evolution as meat sharing is widespread in hunter gatherer societies (e.g. Lovejoy 1981). Therefore one would rather expect an increase of body odour attractiveness of a person eating meat, which could signify good hunting skills (i.e. a good quality mate). However, it should be noted that consumption of a relatively large amount of meat on a daily basis, which is now widespread in so-called developed countries, is a recent phenomenon. It may have an opposite effect, perhaps resembling some sort of metabolic disorder.

It is a common phenomenon to speak about individuals of other cultures as having an unpleasant smell. This view has not only been expressed by lay people but also by anthropologists of the late 19th and early 20th century, who often stressed unpleasant body odour of savant or lower races (Largey and Watson 1972). In many cases this is used to support their ethnocentric (and racist) views. However, in other cases it is difficult to judge whether it is merely prejudice. More recently, ethnic differences in body odour are commonly interpreted as the results of different eating habits (Classen, Howes and Synnott 1994). We concur to some extent with such an interpretation. However, without experimental evidence it remains on the level of speculation only. On the other hand, there is no reason to exclude a genetic influence *a priori* as well. Stated simply: why should people whose appearance is different not also differ in their body odour?

In most odour studies researchers try to control the diet of their odour donors. In particular, consumption of garlic, onion, chilies, pepper, vinegar, blue cheese, cabbage, radish, fermented milk products and marinated fish are recommended to be avoided (e.g. Thornhill and Gangestad 1999; Singh and Bronstad 2001; Roberts et al. 2005). This list is, with minor changes, copied between the studies without any reference to its relevance. Our daily experience says that consumption of food containing plants of the family *Alliacea* (e.g onion, garlic) has a distinct impact on breath odour. Secondary metabolites (i.e. essential oils) of these plants contain numerous sulphur compounds and may have an influence on axillary odour (Sastry, Buck, Janak, Dressler and Preti 1980). However, no such studies have been conducted yet. Another notable group of plants commonly used in European cuisine is the *Brasicacea* family (i.e cauliflower, cabbage, broccoli). These plants emit high amounts of dimethyl trisulphide and other sulphur compounds during cooking (Buttery, Guadagni, Ling, Seifert and Lipton 1976) and again might have an influence on axillary odour. Clearly we need more studies to progress from speculation to more solid knowledge.

Some specific parts of food eaten by mothers can influence the body odour of newborn babies, presumably through placental passage. One case study in Turkey showed that the body and urine of a healthy newborn baby smelled suspiciously of maple syrup, which is the diagnostic sign of maple syrup urine disease (Yalcin, Tekinalp and Ozalp 1999). However, the diagnosis was not confirmed by laboratory tests. Instead it was found that the baby's mother had eaten a dish containing fenugreek seeds just prior to delivery. This spice, commonly used in traditional Turkish and other Middle East cuisines, contains 3-hydroxy-4,5-dimethyl-2(5H) furanone which is responsible for its distinctive odour (Podebrad, Heil, Reichert, Mosandl, Sewell and Bohles 1999). We found the studies from Ferkin's lab (see above) on the deteriorating effect of starvation on odour attractiveness in voles highly thought-provoking. It would be interesting to test whether dieting in humans may have a similar effect. However, as already noted above, for rodents the main source of odour is urine which may reflect diet more easily than axillary products.

To sum up, diet appears to have a significant impact on human body odour and may contribute to differences between populations. We suppose that its hedonic value would depend on familiarity with such odours. As was already noted above, we know very little about effects of specific parts of the diet ingested, and even less about their interactions. This is of importance not only for the methodology of odour experiments, but it also may influence olfactory based mate choice, individual or kin recognition and possibly other odour cues. Effect of diet may mask or interact with individually (gender, species etc.) specific olfactory cues. Such an effect was observed in spiny-mouse pups which preferred odour cues of heterospecific females fed on a familiar diet over odours of conspecific females fed on an unfamiliar diet (Porter and Doane 1977). Concerning dietary effects in mate choice, we suggest that outbreeding avoidance may to some extent rely on unattractiveness of highly unfamiliar odours. This may result, for instance, from interactions between the odour of genetically distant individuals and odours due to different eating habits of such individuals.

19.5 Diseases

Current evolutionary theorists mostly agree upon the role of pathogens as a significant driving force in evolution. In social animals it is of particular relevance to monitor health status of potential social or sexual partners. Closer contact with infected conspecifics may promote pathogen transmission and, in the case of sexual partners, susceptibility to infection or an inherited deficiency that can deteriorate offspring viability. It is well documented in many animal species that olfaction is used to monitor health status in mate choice. For instance mice can discriminate between parasitized and non-infected individuals (Kavaliers and Colwell 1995).

It is therefore not surprising that some human pathologies are also related to changes in body odour. In fact, using smell as a diagnostic tool dates back to Avicenna's and Galen's treatises (Stoddart 1990). Current medicine, relying mostly on more advanced technologies, sees such a method as too subjective. However, in some fields (e.g. dermatology) it is one of the most effective diagnostic tools. To review all diseases and infections that impact on body odour is beyond the scope of this paper; instead, we will focus on several model cases.

From the point of etiology and of possible communicative value, it is wise to discriminate between the effect of metabolic or structural disorders and infectious diseases. Metabolic and structural disorders have a high level of inheritance (often simple Mendelian). Strictly speaking, they should not be included as environmental effects. Metabolic disorders are characterized by missing enzymes, resulting in production of different metabolites which may influence body odour of the affected individual. Similarly, in structural disorders an organism may not express structural proteins, which results in pathogenic changes of affected tissue. Examples of structural disorders influencing body odour are some forms of ichtyosis, a particularly bullous congenital ichtyosiform erythroderma. It is caused by mutation in the gene coding for keratin, which in turn dramatically affects cutaneous structure. Skin of affected individuals forms quills and often is covered with blisters. The disease is followed by distantly noticeable malodour, presumably because of metabolic activity of bacteria colonizing exfoliating scales and blisters (Shwayder 2004).

Examples of metabolic disorders influencing body odour are more widespread. The most common is diabetes, particularly the type I which is characterized by insufficient secretion of insulin. This may lead to ketoacidosis and production of acetone which is responsible for the typical sweet breath odour of diabetics in this metabolic state (Laffel 1999). Another example of such a metabolic disorder is trimethylaminuria. This relatively rare disorder is caused by an inability to metabolize trimethylamine. Foods containing large amounts of choline (e.g. eggs, legumes, brassicas and offal) or trimethyl N-oxide (fish) are metabolized by gut bacteria to trimethylamine. Patients suffering from trimethylaminuria are unable to transform this compound to odourless trimethyl N-oxide, resulting in distinctive odour in their urine, breath and sweat reminiscent of decaying fish (Chalmers, Bain, Michelakakis, Zschocke and Iles 2006).

The last example of metabolic disorders we are going to present here is quite unusual as it is connected with pleasant odour. The disorder is termed Maple syrup urine disease and is characterized by a metabolic defect of branched chain amino acids (leucine, valine and isoleucine) which concentrate in body fluids, and of branched chain 2-oxo acids which are excreted in urine and give this a pleasant maple syrup-like odour (Menkes 1959).

The other group of diseases changing personal body odour is caused by activity of bacteria and other pathogens. These may invade different parts of the human body such as skin, oral cavity, genital and urinary tract etc. Most human body surfaces are inhabited by indigenous microflora which may in some specific conditions turn to pathogenic agents. However, in most cases, indigenous microflora have a protective effect. Obviously not all pathogenic activity produces odorous changes. In some infections odour is rather non-specific. This may be partly caused by the incidence of super infections. Below, we briefly review typical odoriferous infections of some main body sites.

Malodour of the oral cavity is commonly labeled halitosis and in about 80% of cases it is caused by bacterial activity. Oral malodour is not always connected with pathogenic processes (for instance, it is not involved in causing temporary morning breath or bad odour due to hunger). Major sites producing the odour are the periodontal pockets and posterior part of the tongue. The tongue coating, especially, is often related to bad odour. However, the majority of cases are of pathogenic origin. The most common malodour related oral disease is periodontitis. Main contributors of oral malodour are volatile sulphur compounds (e.g. hydrogen sulphide, methyl mercaptans and dimethyl sulphide) which are products of bacterial metabolic activity on sulphur containing amino acids (e.g. cysteine) (Morita and Wang 2001). Volatile sulphur compounds are toxic to epithelial tissue and thus can facilitate infection of the tissue. The most widespread odourous disease of the female genital tract is vaginosis. This is caused by changes in resident microflora and characterized by discharge and fishy odour due to production of trimethylamine (Wolrath, Stahlbom, Hallen and Forsum 2005). Odoriferous infections of skin are numerous, including wounds, ulcers etc. They often include multi-infections and produce strong malodour. For instance, cutaneous ulcers are frequently accompanied by foul smell (Finlay, Bowszyc, Ramlau and Gwiezdzinski 1996).

As far as we know there are no human studies or reports on changes in body odour due to intestinal parasites (not counting smell of faeces in infections causing diarrhoea). For instance, it was found in mice that females spend longer investigating the urine of non-infected males compared to those that were infected with the nematode *Heligmosomoides polygyrus* (Kavaliers and Colwell 1995). It is important to note that infection by this nematode produces no clear signs of disease. The choosiness for non-infected mates depends on health status, sex, reproductive and other factors (for review see Kavaliers, Choleris, Agmo and Pfaff, 2004).

Up to now odoriferous changes due to disease have been investigated in humans almost entirely from the medical perspective without any evolutionary perspective. This could be due to the fact that most researchers live in developed countries where numerous life threatening infections have been almost eradicated and thus can be supposed to be something unusual. However, this is definitely not true in the rest of the world and similarly over human history. Infections and parasites were the main cause of death, both in childhood and at reproductive age. Thus, avoiding contact with infected companions or mates might have had a crucial value and smell could have been one of the main cues. There is vast historical evidence from plague, typhus and other pandemias connected with their distinctive smell.

19.6 Conclusions

In this paper we have focused on non genetic factors shaping human body odour. As stressed above, the interaction of genetic and environmental effects is often highly complicated. For instance, we can find individual differences in metabolic products of certain diets and simultaneously these odorous dietary components interact with genetically controlled components resulting in complicated individual body odour. Although current evidence is rather sparse and we need more empirical work, we may prelusively conclude that diet and infections are the most influential environmental factor on body odour. One should not forget that humans in the vast majority of cultures tend to alter their body odour by use of different fragrances, perfumes, and deodorants. However, this is another story.

Acknowledgments We would like to express our gratitude to Jaroslav Flegr, Jindra Havlickova, Barbara Husarova, Anna Kotrcova and Jitka Lindova, for many helpful comments on the manuscript and language corrections. The study was supported by the grants GAUK 393/2005 and GACR 406/06/P377.

References

Ackerl, K., Atzmueller, M. and Grammer, K. (2002) The scent of fear. Neuro Endocrinol. Lett. 23, 79–84.

Beauchamp, G.K. (1976) Diet influences attractiveness of urine in guinea-pigs. Nature 263, 587–588.

Buttery, R., Guadagni, D., Ling, L., Seifert, R. & Lipton, W. (1976) Additional volatile components of cabbage, broccoli, and cauliflower. J. Agric. Food Chem. 24, 829–832.

Cantafio, L.J. (2003) Human olfactory communication of alarm and safety. *PhD. Thesis*. Rutgers University, New Jersey.

Chalmers, R.A., Bain, M.D., Michelakakis, H., Zschocke, J. & Iles, R.A. (2006) Diagnosis and management of trimethylaminuria (FMO3 deficiency) in children. J. Inherit. Metab. Dis. 29, 162–172.

Chen, D. & Haviland-Jones, J. (1999) Rapid mood change and human odors. Physiol Behav 68, 241–250.

Chen, D. and Haviland-Jones, J. (2000) Human olfactory communication of emotion. Percept. Motor Skill. 91, 771–781.

Classen, C., Howes, D. and Synnott, A. (1994) *Aroma - The Cultural History of Smell.* Routledge, London.

Comfort, A. (1971) Likelihood of human pheromones. Nature 230**,** 432–433.

Doty, R.L., Ford, M., Preti, G. and Huggins, G.R. (1975) Changes in the intensity and pleasantness of human vaginal odors during the menstrual cycle. Science 190, 1316–1317.

Ellis, H. (1927) *Studies in the Psychology of Sex IV. Sexual Selection in Man.* Random House, New York.

Ferkin, M.H., Sorokin, E.S., Johnston, R.E. and Lee, C.J. (1997) Attractiveness of scents varies with protein content of the diet in meadow voles. Anim. Behav. 53, 133–141.

Finlay, I.G., Bowszyc, J., Ramlau, C. and Gwiezdzinski,Z. (1996) The effect of topical 0.75% metronidazole gel on malodorous cutaneous ulcers. J. Pain Symptom Manag. 11, 158–162.

Gould, G.M. & Pyle, W.L. (1996) Anomalies and curiosities of medicine. New York: Bell Publ. Comp.

Havlicek, J. and Lenochova, P. (2006) The effect of meat consumption on body odour attractiveness. Chem. Senses 31, 753–759.

Havlicek, J., Bartos, L., Dvorakova, R. and Flegr, J. (2006) Non-advertised does not mean concealed. Body odour changes across the human menstrual cycle. Ethology 112, 81–90.

Hepper, P.G. (1988) The discrimination of human odor by the dog. Perception 17, 549–554.

Kavaliers, M., Choleris, E., Agmo, A. and Pfaff, D.W. (2004) Olfactory-mediated parasite recognition and avoidance: linking genes to behavior. Horm. Behav. 46, 272–283.

Kavaliers, M. and Colwell, D.D. (1995) Odors of parasitized males induce aversive responses in female mice. Anim. Behav. 50, 1161–1169.

Kuukasjärvi, S., Eriksson, C.J.P., Koskela, E., Mappes, T., Nissinen, K. and Rantala, M.J. (2004) Attractiveness of women's body odors over the menstrual cycle: the role of oral contraceptives and receiver sex. Behav. Ecol. 15, 579–584.

Laffel, L. (1999) Ketone bodies: a review of physiology, pathophysiology and application of monitoring to diabetes. Diabetes Metab. Res. 15, 412–426.

Largey, G.P. and Watson, D.R. (1972) The sociology of odors. Am. J. Sociol. 77, 1021–1034.

Lovejoy, C.O. (1981) The origin of man. Science 211, 314–349.

Menkes, J. (1959) Maple syrup disease: isolation and identification of organic acids in the urine. Pediatrics 23, 348–353.

Michael, R.P., Bonsall, R.W. and Kutner, M. (1975) Volatile fatty acids, "copulins", in human vaginal secretions. Psychoneuroendocrino. 1, 153–163.

Morita, M. & Wang, H. L. (2001) Association between oral malodor and adult periodontitis: a review. J. Clin. Periodontol. 28, 813–819.

Pawlowski, B. (1999) Loss of oestrus and concealed ovulation in human evolution - the case against the sexual-selection hypothesis. Curr. Anthropol. 40, 257–275.

Pierce, A.A. and Ferkin, M.H. (2005) Re-feeding and the restoration of odor attractivity, odor preference, and sexual receptivity in food-deprived female meadow voles. Physiol. Behav. 84, 553–561.

Podebrad, F., Heil, M., Reichert, S., Mosandl, A., Sewell, A.C. and Bohles, H. (1999) 4,5-dimethyl-3-hydroxy-2[5H]-furanone (sotolone) - The odour of maple syrup urine disease. J. Inher. Metabol. Dis.22, 107–114.

Porter, R.H. and Doane, H.M. (1977) Dietary-dependent cross-species similarities in maternal chemical cues. Physiol. Behav. 19, 129–131.

Porter, R.H., Cernoch, J.M. and Balogh, R.D. (1985) Odor signatures and kin recognition. Physiol. Behav. 34, 445–448.

Roberts, S.C., Gosling, L.M., Spector, T.D., Miller, P., Penn, D.J. and Petrie, M. (2005) Body odor similarity in noncohabiting twins. Chem. Senses 30, 651–656.

Sastry, S.D., Buck, K.T., Janak, J., Dressler, M. & Preti, G. (1980) Volatiles emitted by humans. In: G.R.Waller and O.C.Dermer (Eds.), *Biochemical Applications of Mass Spectrometry, First Supplementary Volume.* John Wiley, New York, pp. 1086–1129.

Shwayder, T. (2004) Disorders of keratinization diagnosis and management. Am. J. Clin. Dermatol. 5, 17–29.

Singh, D. and Bronstad, P.M. (2001) Female body odour is a potential cue to ovulation. Proc. R. Soc. Lond. B. 268, 797–801.

Sommerville, B.A., McCormick, J.P. and Broom D.M. (1994) Analysis of human sweat volatiles - an example of pattern-recognition in the analysis and interpretation of gas chromatograms. Pestic. Sci. 41, 365–368.

Stoddart, D.M. (1990) *The Scented Ape - The Biology and Culture of Human Odour.* Cambridge: Cambridge University Press.

Thornhill, R. and Gangestad, S.W. (1999) The scent of symmetry: a human sex pheromone that signals fitness? Evol. Hum. Behav. 20, 175–201.

Thornhill, R., Gangestad, S.W., Miller, R., Scheyd, G., McCollough, J.K. and Franklin, M. (2003) Major histocompatibility complex genes, symmetry, and body scent attractiveness in men and women. Behav. Ecol. 14, 668–678.

Wallace, P. (1977) Individual discrimination of human by odor. Physiol. Behav. 19, 577–579.

Wolrath, H., Stahlbom, B., Hallen, A. and Forsum, U. (2005) Trimethylamine and trimethylamine oxide levels in normal women and women with bacterial vaginosis reflect a local metabolism in vaginal secretion as compared to urine. APMIS 113, 513–516.

Yalcin, S.S., Tekinalp, G. and Ozalp, I. (1999) Peculiar odor of traditional food and maple syrup urine disease. Pediatr. Int. 41, 108–109.

Part IV
Sexual Communication

Chapter 20
A Candidate Vertebrate Pheromone, SPF, Increases Female Receptivity in a Salamander

Lynne D. Houck, Richard A. Watts, Louise M. Mead, Catherine A. Palmer, Stevan J. Arnold, Pamela W. Feldhoff and Richard C. Feldhoff

Abstract Plethodontid (lungless) salamanders have evolved an unusual pheromone delivery system in which the male courtship pheromone is applied to the skin of the female, apparently diffusing through the mucosal-rich epithelia into her superficial capillary system. In *Desmognathus ocoee*, a plethodontid salamander that uses the diffusion mode of pheromone delivery, we conducted a behavioural bioassay to test a 20–25 kDa molecular weight fraction of the male courtship pheromone: this fraction was effective in increasing female receptivity. The principal component of the *D. ocoee* pheromone fraction was identified as a 25 kDa protein that had significant sequence similarity with the precursor of a newt reproductive pheromone (a decapeptide termed sodefrin). We termed the principal protein component in the *D. ocoee* pheromone "Sodefrin Precursor-like Factor" (SPF). SPF also occurs in other plethodontid salamanders, including species of *Plethodon*, *Aneides* and *Eurycea*. Across these species, SPF is a highly variable protein that bears the signature of positive selection. The presence of SPF in distantly related genera suggests that the sodefrin precursor gene has been retained as a courtship signal throughout the evolutionary radiation of plethodontid salamanders.

20.1 Introduction

Social communication in vertebrates often is mediated by chemical signaling via olfactory pheromones (Karlson and Lüscher 1959, Johnston 2003, Wyatt 2003). In particular, vertebrate sexual behaviour can be powerfully influenced when a courting partner delivers chemical signals. These pheromone signals typically are transduced into behavioural changes via a pathway beginning with direct olfactory stimulation. Receptors of the main olfactory or accessory olfactory epithelia respond to pheromones, with subsequent reception and processing by the central nervous system (Halpern and Martinez-Marcos 2003; Luo, Fee and Katz 2003; Rodriguez 2004).

Lynne D. Houck
Oregon State University, Zoology,
houckl@science.oregonstate.edu, wattsri@science.oregonstate.edu, arnoldst@onid.orst.edu

J.L. Hurst et al., *Chemical Signals in Vertebrates 11.*
© Springer 2008

Reproductive chemical signals in vertebrates also may be delivered by non-olfactory pathways. Although such pathways generally are uncommon across vertebrates, non-olfactory courtship pheromone delivery is widespread in plethodontid (lungless) salamanders (Arnold 1977; Fig. 20.1). About 100 million years ago, these salamanders evolved a highly stylised courtship that includes the delivery of phero-mones from the male to the female (Arnold 1977; Houck and Sever 1994; Houck and Arnold 2003). During the courtship season, a male develops glandular tissue on the chin (the mental gland). In most plethodontid species, the male's premaxillary teeth also hypertrophy at this time. When courting a female, the male delivers mental gland secretions by "scratching" the female's dorsal skin with his hypertrophied teeth and by rubbing his gland over the abraded site (Fig. 20.2). This mode of pheromone delivery presumably results in the diffusion of the pheromone into the female's superficial circulatory system (Houck 1986).

The function of the male courtship pheromone in plethodontids was first documented in a study of *Desmognathus ocoee*: Houck and Reagan (1990) showed that the diffusion delivery of a crude extract from the male pheromone gland increased

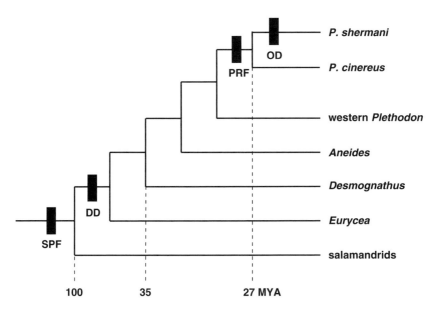

Fig. 20.1 Cladogram showing the relationships of salamandrids (e.g., newts) and various clades of plethodontid salamanders, the evolutionary origin of courtship pheromones and modes of delivery. Relationships at the generic level are concordant across studies using both morphological and molecular characters (Chippindale, Bonett, Baldwin and Wiens 2004; Macey 2005; Min, Yang, Bonnett, Vleltes, Brandon and Wake 2005). Relationships of species groups within the genus *Plethodon* are based on allozyme data (Larson and Highton 1978; Highton and Peabody 2000). Approximate divergence times (shown at bottom) are based on albumin immunology (Larson, Weisrock and Kozak 2003). Rectangular boxes show the point of origin of Sodefrin-like Precursor Factor (SPF), diffusion delivery of courtship pheromones (DD), Plethodon Receptivity Factor (PRF), and olfactory delivery of courtship pheromones (OD)

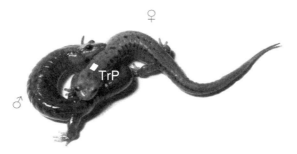

Fig. 20.2 Pheromone delivery in *Desmognathus ocoee* salamanders and the method of pheromone delivery used during behavioural trials. A receptive female places her chin on the tail base of the male and typically straddles his tail. The male turns back towards the female and places his submandibular mental gland on her dorsum. The male then uses his premaxillary teeth to scratch the site on her dorsum that he has swabbed with his mental gland secretions. To mimic pheromone delivery in behavioural trials, each male was deglanded and a treatment solution was delivered to each female in a treatment patch (TrP) placed on her dorsum just posterior to the head. Photograph by Stevan J. Arnold

female receptivity to the male, as measured by a significant reduction in courtship duration. Subsequent studies of another plethodontid salamander (*Plethodon shermani*) showed that a single pheromone protein (PRF) could act alone to elicit change in female courtship behaviour (Feldhoff, Rollmann and Houck 1999). Thus, we tested whether this would be the case for the *D. ocoee* courtship pheromone. We conducted a behavioural bioassay of *D. ocoee* females in response to receiving a 20–25 kDa pheromone fraction that contained a primary protein presumed to be the pheromone signal.

20.2 Materials and Methods

20.2.1 Animal Collection and Care

During May 2001 and 2002, we collected 100 adult males and 100 adult females from a population of *Desmognathus ocoee* in Macon County, North Carolina, USA (35° 02'20" N, 083° 33'08" W). Collecting permits were obtained. Only animals in courting condition were taken, as judged by females being gravid and males having enlarged premaxillary teeth. Each male was sedated and had his mental gland surgically removed (cf Houck and Reagan 1990) at least 14 days in advance of courtship trials so that pheromone delivery could be restricted to experimental presentation. Animals were sent to Oregon State University and maintained at 16–18 °C on a late-May North Carolina photoperiod.

20.2.2 Behavioural Experiments

Female receptivity was assessed in response to the application of either pheromone extract (0.7 mg/ml in 1/2x Phosphate Buffered Saline, PBS) or a saline control (1/2x PBS) to the dorsal skin of a female that subsequently engaged in courtship with a deglanded male. On a given trial night, we placed each female in a separate courtship box. We then applied to her dorsum a treatment patch (Fig. 20.2) consisting of a 1×2 mm rectangular piece of filter paper with low-protein binding capacity (Glass microfibre GF/A: Whatman) moistened with 5μl of either pheromone extract or saline.

After 30 min, a deglanded male was added to each courtship box, and a combination of scan and focal sampling was used to record all occurrences of courtship behaviours. These behaviours included the time of the first occurrence of the male orienting to and physically contacting the female, and the time when the male completed spermatophore deposition. The same male-female pair was observed on a subsequent night when the female received a treatment (saline control or pheromone extract) that was different from the treatment she received on an earlier trial. In this way, each male-female pair was its own control. On each trial night, equal numbers of females were treated with each solution.

Courtship duration was defined as the time from first tactile contact (typically the male contacting the female) until spermatophore deposition was completed. We selected spermatophore deposition as the end of courtship because (a) this is a distinct behaviour that any observer can score, (b) the time between spermatophore deposition and insemination typically is only 1–2 min, and (c) some females do not get inseminated (e.g., walk past the spermatophore) and so the end of courtship would be ambiguous for these females. In addition, treatment (pheromone or saline) was not correlated with insemination success for *D. ocoee*, given that a pair reached the spermatophore deposition stage (Houck, unpublished data).

Courtship encounters were staged during 4 trial nights (May 30, June 3, 7 and 12) so that each male-female pair had 3 days off between trials. As earlier work showed that a crude homogenate of the pheromone-producing mental gland was effective in reducing courtship duration (Houck and Reagan 1990), we used a one-tailed paired *t*-test to analyse the data.

20.2.3 Preparation of Pheromone Extracts

Males were anesthetised and mental glands ($N \approx 200$) were surgically removed. Secreted components were extracted into 0.8 mM acetylcholine chloride in 1/2x PBS (cf Rollmann, Houck and Feldhoff 1999). Gland extracts were centrifuged for 10 min at 14,000 g and the supernatant was removed. The supernatant was filtered (0.2 μm) and loaded as aliquots onto a Sephadex Superfine G-75 gel filtration column (1.6 cm x 15.5 cm; Pharmacia, Piscataway, NJ) on a Waters HPLC system (Millipore, Milford, MA). The column had previously been equilibrated with one-half strength Dulbecco's phosphate buffered saline (1/2x PBS). The column was eluted

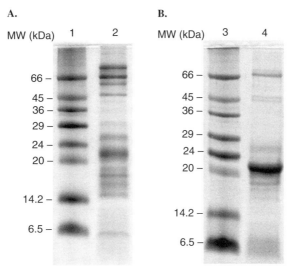

Fig. 20.3 Preparation of pheromone extract for behavioural trials. Secreted proteins from *D. ocoee* pheromone were size fractionated on Sephadex-G75 gel filtration resin prior to behavioural testing. AU (absorption units) = wavelength = 220 nm). Inset SDS-PAGE (15% Tris-Tricine) lanes are A. Standards, B. Whole extract, C-G. pooled and concentrated fractions #12–16, #17–20, #21–24, #25–28 and #25–39 respectively. The experimental pheromone fraction was D (#17—-20)

at 1 ml/min and individual 4 ml protein fractions were collected. Fractions were collected as shown in Fig. 20.3 and concentrated by ultrafiltration (3,000 MW cut-off). Pooled pheromone fraction D (see inset in Fig. 20.3) was used in behavioural trials. Prior to use, the final protein concentration of this pooled fraction was adjusted with 1/2x PBS to 0.7 mg/ml, which is the approximate protein concentration of whole pheromone extract (Houck, Bell, Reagan-Wallin and Feldhoff 1998).

20.3 Results

20.3.1 Male D. ocoee Pheromone Increased Female Receptivity

Twenty male-female pairs each mated on two different trial nights, once with a pheromone treatment (the 20–25 kDa fraction of the male courtship pheromone) and once with a saline treatment (n = 20 paired values). The duration of courtship was reduced significantly when pairs were treated with the pheromone fraction vs the saline control ($t = 1.73$, df = 19, $P < 0.02$). For these 20 pairs, the average courtship duration was 54 min when treated with the pheromone fraction, and 70 min when treated with the saline control. Thus, the control treatment resulted in courtships that were approximately 30% longer.

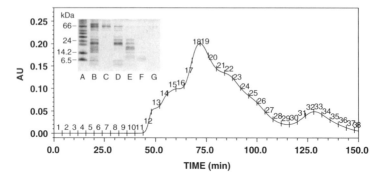

Fig. 20.4 Comparison of secreted proteins in pheromone extracted (A) from mental glands of *Desmognathus ocoee* (which has non-olfactory pheromone delivery) with (B) pheromone extracted from *Plethodon shermani* (which has olfactory delivery). Lanes 1 and 3 are MW standards. SDS-PAGE (15% Tris-Tricine)

20.3.2 Characteristics of D. ocoee SPF and Related Proteins

Compared with the protein content of mental gland secretions of *P. shermani* males, the protein profile for *D. ocoee* showed a much more complex pattern of protein staining (Fig. 20.4). In *D. ocoee*, no single dominant protein occurred in this "non-olfactory" pheromone profile.

20.4 Discussion

In the plethodontid salamander *D. ocoee*, courtship duration was reduced for male-female pairs in which the female received a protein signal from the 20–25 kDa fraction of the male courtship pheromone. We interpret this reduction in courtship duration as an increase in receptivity for females receiving the pheromone.

The demonstration of this behavioural response to a male pheromone signal is significant because of the manner in which the pheromone was delivered. Most other vertebrate examples of reproductive pheromones involve reception via the olfactory system(s). In contrast, *D. ocoee* females received the pheromone via diffusion through the dorsal skin. We assume that the well developed superficial capillary system of these lungless salamanders is the route by which the male pheromone was transported to whatever target tissue(s) initiated responses that affected female reproductive behaviour.

Multiple lines of evidence suggest that SPF, a single protein within the 20–25 kDa pheromone fraction, is responsible for female behavioural response. First, this major protein component within the *D. ocoee* fraction was genetically very similar to the precursor of sodefrin (Palmer, Watts, Houck, Picard and Arnold 2007), and sodefrin is a known reproductive pheromone in newts (Kikuyama, Toyoda, Ohmiya, Matsuda, Tanaka and Hayashi 1995; Kikuyama and Toyoda 1999). Second, a separate study showed that the cDNA library of proteins expressed in male *D. ocoee* mental glands contained a high proportion (25%) of transcripts for SPF (Palmer et al. 2007). This pattern of expression, production and secretion is similar to that described for PRF, a courtship pheromone identified for the plethodontid *P. shermani* (Rollmann et al. 1999; Palmer, Watts, Gregg, McCall, Houck, Highton and Arnold 2005). Third, a gene tree for SPF varieties showed that SPF in *D. ocoee* is highly diverse: there are five distinct types of SPF, in addition to isoforms for each type (Palmer et al. 2007). This diversification greatly exceeds that expected under processes of drift and purifying selection: instead, this level of diversity reflects a history of strong positive selection.

The extensive, rapid diversification documented for SPF is unusual as most proteins evolve more slowly. Recent work, however, shows that rapid evolution driven by positive selection is characteristic of certain reproductive proteins that are involved in interactions between the sexes (Swanson and Vacquier 2002; Panhuis, Clark and Swanson 2006). Diagnosis of this evolutionary mode in both SPF and PRF suggests that diversification in both of these pheromones is a consequence of a dynamic evolutionary interaction between male signals and female receptors (Watts, Palmer, Feldhoff, Feldhoff, Houck, Jones, Pfrender, Rollmann and Arnold 2004; Palmer et al. 2005; Palmer et al. 2007). Thus, in addition to the behavioural bioassay results reported here, a strong case is made for SPF being an active component in *D. ocoee* courtship pheromones due to (a) its similarity with the newt pheromone precursor, (b) the mental gland abundance of SPF mirroring that of the courtship pheromone PRF, and (c) the extraordinary molecular evolution experienced by both SPF and PRF.

The delivery of male courtship pheromones is widespread among plethodontid salamanders (Houck and Arnold 2003), and other courtship pheromones are being discovered for this group (Houck, Palmer, Watts, Arnold, Feldhoff and Feldhoff 2007). The mode by which these pheromones are transferred to the female apparently has been modified from delivery via diffusion into the circulatory system to delivery that directly stimulates vomeronasal receptors (Fig. 20.1; Houck and Sever 1994; Watts et al. 2004; Palmer et al. 2005; Palmer et al. 2007). The behavior patterns and morphologies associated with these two delivery modes often remain static for millions of years. In contrast, evolution at the level of pheromone signals is apparently an incessant process that continuously alters the protein sequence and composition of pheromones both within and among species (Watts et al. 2004; Palmer et al. 2005; Palmer et al. 2007).

Acknowledgments We thank Martha Baugh and Amy Picard for help with behavioural observations. Laboratory facilities and support were provided by the staff at the Highlands Biological Station in Highlands, North Carolina, USA. This research was supported by National Science Foundation grants 0110666 and 0416724 to LDH, and 0416834 to RCF.

References

Arnold, S.J. (1977) The courtship behavior of North American salamanders with some comments on Old World salamandrids. In: D. Taylor and S. Guttman (Eds.), *The Reproductive Biology of Amphibians*. Plenum Press, New York, pp. 141–183.

Chippindale, P.R., Bonett, R.M., Baldwin, A.S. and Wiens, J.J. (2004) Phylogenetic evidence for a major reversal of life-history evolution in plethodontid salamanders. Evol. 58, 2809–2822.

Feldhoff, R.C., Rollmann, S.M. and Houck, L.D. (1999) Chemical analyses of courtship pheromones in a plethodontid salamander. In: R.E. Johnston, D. Müller, Schwarze and P. Sorensen (Eds.), *Advances in Chemical Signals in Vertebrates*. Kluwer Academic/Plenum Publishers, New York, pp. 117–125.

Halpern, M. and Martinez-Marcos, A. (2003) Structure and function of the vomeronasal system: and update. Prog. Neurobiol. 70, 245–318.

Highton, R. and Peabody, R.B. (2000) Geographic protein variation and speciation in salamanders of the *Plethodon jordani* and *Plethodon glutinosus* complexes in the southern Appalachian Mountains with the descriptions of four new species. In: R.C. Bruce, R.G. Jaeger, and L.D. Houck. (Eds.) *The biology of plethodontid salamanders*. Plenum, New York, pp. 31–94.

Houck, L.D. 1986. The evolution of salamander courtship pheromones. In: D. Duvall, D. Müller-Schwarze and R.M. Silverstein. (Eds.) *Chemical Signals in Vertebrates*, Vol. IV: *Ecology, Evolution, and Comparative Biology*. Plenum Press, New York. Pp. 173–190.

Houck, L.D. and Arnold, S.J. (2003) Courtship and mating. In: D.M. Sever (Ed.), *Phylogeny and Reproductive Biology of Urodela (Amphibia)*. Science Publishers Inc., Enfield, New Hampshire, 383–424.

Houck, L.D., Bell, A.M., Reagan-Wallin, N.L. and Feldhoff, R.C. (1998) Effects of experimental delivery of male courtship pheromones on the timing of courtship in a terrestrial salamander, *Plethodon jordani* (Caudata: Plethodontidae). Copeia 1998, 214–219.

Houck, L.D, Palmer, C.A., Watts, R.A., Arnold, S.J., Feldhoff, P.W. and Feldhoff, R.C. (2007) A new vertebrate courtship pheromone, PMF, affects female receptivity in a terrestrial salamander. Anim. Behav. 73, 315–320.

Houck, L.D. and Reagan, N.L. (1990) male courtship pheromones increase female receptivity in a plethodontid salamander. Anim. Behav. 39, 729–734.

Houck, L.D. and Sever, D.M. (1994) Role of the skin in reproduction and behavior. In: H. Heatwole and G.T. Barthalmus (Eds.), *Amphibian Biology Volume 1: The Integument*. Surrey Beatty & Sons, Chipping Norton, Australia, pp. 351–381.

Johnston, R.E. (2003) Chemical communication in rodents: From pheromones to individual recognition. J. Mammal. 84, 1141–1162.

Karlson, P. and Lüscher, M. (1959) Pheromones: a new term for a class of biologically active substances. Nature 183, 55–56.

Kikuyama, S. and Toyoda, F. (1999) Sodefrin: a novel sex pheromone in a newt. Rev. Reprod. 4, 1–4.

Kikuyama, S., Toyoda, F., Ohmiya, Y., Matsuda, K., Tanaka, S. and Hayashi, H. (1995) Sodefrin: a female-attracting peptide pheromone in newt cloacal glands. Science 267, 1643–5.

Larson, A. and Highton, R. (1978) Geographic protein variation and divergence of the salamanders in the *Plethodon welleri* group (Amphibia: Plethodontidae). Syst. Zool. 27, 431–448.

Larson, A., Weisrock, D.W. and Kozak, K. H. (2003) Phylogenetic systematics of salamanders (Amphibia: Caudata), a review. In: D.M. Sever (Ed.), *Phylogeny and Reproductive Biology of Urodela (Amphibia)*. Science Publishers Inc., Enfield, New Hampshire, pp. 31–108.

Luo, M., Fee, M.S. and Katz, L.C. (2003) Encoding pheromonal signals in the accessory olfactory bulb of behaving mice. Science 299, 1196–201.

Macey, J.R. (2005) Plethodontid salamander mitochondrial genomics: a parsimony evaluation of character conflict and implications for historical biogeography. Cladistics 21, 194–202.

Min, M.S., Yang, S.Y, Bonnett, R.M., Vleltes, D.R., Brandon, R.A. and Wake, D.B. (2005) Discovery of the first Asian plethodontid salamander. Nature 435, 87–90.

Palmer, C., Watts, R.A., Gregg, R., McCall, M., Houck, L.D., Highton, R. and Arnold, S.J. (2005) Lineage-specific differences in evolutionary mode in a salamander courtship pheromone. Mol. Biol. Evol. 22, 2243–2256.

Palmer, C., Watts, R.A., Houck, L.D., Picard, A.L. and Arnold, S.J. (2007) Evolutionary replacement of pheromone components in a salamander signaling complex: more evidence for phenotypic-molecular decoupling. Evolution 61, 202–215.

Panhuis, T.M., Clark, N.L. and Swanson, W.J. (2006). Rapid evolution of reproductive proteins in abalone and *Drosophila*. Phil. Trans. Roy. Soc. B 361, 261–268.

Rodriguez, I. (2004) Pheromone receptors I mammals. Horm. Behav. 46, 219–30.

Rollmann, S.M., Houck, L.D. and Feldhoff, R.C. (1999) Proteinaceous pheromone affecting female receptivity in a terrestrial salamander. Science 285, 1907–9.

Rollmann, S.M., Houck, L.D. and Feldhoff, R.C. (2000) Population variation in salamander courtship pheromones. J. Chem. Ecol. 26, 2713–2724.

Swanson, W.J. and Vacquier, V.D. (2002) The rapid evolution of reproductive proteins. Nat. Rev. Genet. 3, 137–144.

Watts, R.A., Palmer, C.A., Feldhoff, R.C., Feldhoff, P.W., Houck, L.D., Jones, A.G., Pfrender, M.E., Rollmann, S.M. and Arnold, S.J. (2004) Stabilizing selection on behavior and morphology masks positive selection on the signal in a salamander pheromone signaling complex. Mol. Biol. Evol. 21, 1032–41.

Wyatt, T..D. (2003) Pheromones and animal behaviour: Communication by smell and taste. Cambridge University Press, Cambridge.

Chapter 21
Cross-dressing in Chemical Cues: Exploring 'She-maleness' in Newly-emerged Male Garter Snakes

Michael P. LeMaster, Amber Stefani, Richard Shine and Robert T. Mason

Abstract She-males are male garter snakes that elicit courtship behavior from other males during the breeding season. Initially thought to consist of a small sub-set of males which retained their attractive nature throughout the breeding season, recent behavioral data suggests that most, if not all, males undergo a period of 'she-maleness' upon first emerging from winter hibernation before losing their attractive nature shortly after emergence. Utilizing behavioral experiments and chemical analyses, we sought to discern whether newly-emerged male red-sided garter snakes (*Thamnophis sirtalis parietalis*) display a pheromone profile similar to the female sexual attractiveness pheromone. Sequestered in the skin lipids of females and responsible for triggering male courtship behavior, this pheromone has been previously linked with long-term she-maleness in this species. Results from courtship trials demonstrated that newly-emerged males are attractive to other males, although not to the same degree as females. Subsequent chemical analyses of skin lipids from females and newly-emerged males showed no quantitative or qualitative difference in the components constituting the sexual attractiveness pheromone. Thus, it appears that the majority of males in this species emerge with a female-like pheromone profile and subsequent physiological changes, yet to be identified, are responsible for the short- vs. long-term nature of this phenomenon.

21.1 Introduction

Snakes are excellent models for vertebrate pheromone research because the initiation of many behaviors in snakes is dependent on pheromone production and expression (reviewed in Carpenter and Ferguson 1977; Burghardt 1980; Mason 1992). In particular, the reproductive success of snakes depends on the production and perception (via the vomeronasal organ—Halpern 1987) of specific sex pheromones. These pheromones coordinate multiple activities in the reproductive process, including locating potential mates over long distances (e.g., Gehlback, Watkins and Kroll 1971; Ford 1981), eliciting courtship behavior once mates come in contact

Michael P. LeMaster
Western Oregon University, Department of Biology,
lemastm@wou.edu

(e.g., Noble 1937; Andrén 1986), and terminating courtship once mating is complete (e.g., Ross and Crews 1977; Shine, Olsson and Mason 2000b).

Of all snakes species studied to date, the chemical ecology of the red-sided garter snake, *Thamnophis sirtalis parietalis*, is the best understood (reviewed in Mason 1993). Indeed, the female sexual attractiveness pheromone for this species represents the first reptilian pheromone characterized (Mason, Fales, Jones, Pannell, Chinn and Crews 1989). Composed of a suite of homologous long-chain saturated and monounsaturated methyl ketones sequestered in the skin lipids of females, this pheromone is responsible for triggering male courtship behavior during the breeding season (Mason et al. 1989; Mason, Jones, Fales, Pannell and Crews 1990). Subsequent investigations have demonstrated that this pheromone not only triggers male courtship behavior, but is also responsible for mediating male trailing behavior during the breeding season (LeMaster and Mason 2001). In addition, qualitative variation in the methyl ketone profiles displayed by females has been linked to male mate choice (LeMaster and Mason 2002) and interpopulational sexual isolation (LeMaster and Mason 2003).

Interestingly, the female red-sided garter snake is not the only sex to express a methyl ketone profile attractive to males during the breeding season. It has been known for the past two decades that a small subset of males, termed she-males, exists in the populations which elicit long-term courtship from other males (Mason and Crews 1985). These males are morphologically and anatomically similar to other males except that they express a female-like sexual attractiveness pheromone on their dorsal surface (Mason 1993). Recent behavioral evidence, however, suggests that this 'she-maleness' is not limited to a select group of males. Instead, it appears that the majority of males in this species are attractive to other males upon first emergence from winter dormancy (start of breeding season) but then lose their attractive nature soon after (Shine, Harlow, LeMaster, Moore and Mason 2000a; Shine and Mason 2001).

The mechanism responsible for the attractive nature of newly-emerged males is unknown. Here we present a small study designed to investigate whether newly-emerged males express a female-like sexual attractiveness pheromone, similar to that observed in long-term she-males. If this hypothesis is correct, then we would expect newly-emerged males to display a methyl ketone profile that is quantitatively and qualitatively similar to that expressed by females during the breeding season.

21.2 Materials and Methods

Red-sided garter snakes utilized in this study were obtained from a field site near the community of Inwood in the Interlake region of Manitoba, Canada (50°31.58'N; 97°29.71W) during May of 2006. This site is located in an abandoned gravel quarry containing a single over-wintering hibernaculum possessing in excess of ten thousand garter snakes during the winter months and subsequent breeding season (M.P. LeMaster, personal observation).

21.2.1 Experimental Animals

Newly-emerged males (n = 10) and females (n = 10) were collected at the field site immediately upon emergence. Small females were chosen for this study (snout-vent length (SVL) < 55 cm) so that the average body length between the sexes was similar. This was done because previous work has demonstrated that the composition of the female sexual attractiveness pheromone is highly correlated with body size (LeMaster and Mason 2002). In addition, actively courting males were also collected (n > 200). Animals were segregated in their appropriate groups and held at ambient temperatures until testing, after which the majority of animals were returned to the hibernaculum and released.

21.2.2 Behavioral Experiments

Courtship trials were performed in outdoor areas measuring $1 \times 1 \times 1$ m and constructed of nylon fabric attached to metal posts (Moore, LeMaster and Mason 2000). Groups of ten actively courting male snakes were placed in an arena and allowed to acclimate. A female (n = 10) or newly-emerged male (n = 10) was then placed into each arena and the snakes were allowed to interact undisturbed for three minutes. The number of males actively courting the experimental animal at the end of the three-minute period was then recorded. Unique experimental animals and actively courting males were used for each trial.

21.2.3 Chemical Analyses

Following their use in the behavioral trials, sub-sets of the experimental females (n = 8) and experimental newly-emerged males (n = 8) were immediately sacrificed with an overdose of brevital sodium and their skin lipids were extracted with hexane overnight (LeMaster and Mason 2002). After removal of the animals, the extracts were returned to Western Oregon University where the methyl ketones composing the female sexual attractiveness pheromone were isolated using column chromatography (described in LeMaster and Mason 2002).

To examine variation in the quality of methyl ketones expressed by females and newly-emerged males, we determined the number of unique methyl ketones expressed by individual snakes and compared the relative concentrations of individual methyl ketones comprising the overall pheromone profiles for the two groups. The methyl ketones present in the pheromone extracts were identified by gas chromatography / mass spectrometry (Hewlett Packard 5890 Series II gas chromatograph coupled with a Hewlett Packard 5971 Series mass selective detector— see LeMaster and Mason 2003 for full description of the GC/MS platform and methods).

To examine variation in the quantity of methyl ketone expressed by females and newly-emerged males, we calculated the amount (micrograms) of methyl ketones expressed per unit (square centimeter) skin surface for individual snakes. This was accomplished by dividing the weight of the isolated methyl ketones extracted from an individual snake by the total surface area of the animal. Methyl ketone weights were determined using GC/MS by running each sample with an external standard of known weight (methyl stearate 10 mg / ml; 0.5 μl aliquot per sample). A general measure of skin surface area for each snake was determined by multiplying the SVL of a snake by its circumference at mid-body (Mason et al. 1990).

21.3 Results

21.3.1 Behavioral Experiments

The average SVL (± SD) of small females used in the courtship trials was 49.5 (± 5.0) cm whereas the average SVL (± SD) of newly-emerged males was 43.3 (± 5.7) cm. When presented with both experimental groups, male garter snakes responded with stereotypical courtship behaviors including increased tongue-flick rate, chin rubbing along the dorsum of the female, and body alignment with the female (Noble 1937). Overall, we observed a significant difference in the proportion of males actively courting the two groups with a higher proportion of males preferring to court the small females compared to the newly-emerged males (t-test, $t = 3.26$, $P = 0.004$; 65% of males courted small females versus 37% of males courting newly-emerged males; Fig. 21.1).

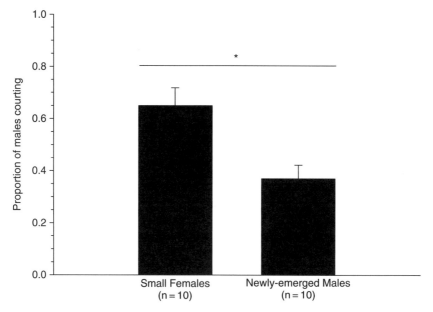

Fig. 21.1 Proportion (mean ± SD) of male red-sided garter snakes displaying courtship behavior to small females and newly-emerged males

Table 21.1 Quantitative and qualitative variation in expression of the methyl ketone profile among small female and newly emerged male red-sided garter snakes

Group (n = 8 per group)	Methyl ketone expression ($\mu g/cm^2$)	Unique methyl ketones observed	Relative (%) methyl ketone concentrations	
			Saturated	Unsaturated
Small Females	1.3 ($\pm$ 2.0)	16.0 ($\pm$ 1.1)	57.8 ($\pm$ 17.3)	42.2 ($\pm$ 17.3)
Newly-emerged Males	1.5 ($\pm$ 1.2)	15.8 ($\pm$ 1.3)	67.9 ($\pm$ 17.7)	32.1 ($\pm$ 17.7)

21.3.2 Chemical Analyses

There was no difference in the average SVL among the small females and newly-emerged males sampled for pheromone analysis (small females $= 50.4 \pm 5.0$ cm; newly-emerged males $= 45.0 \pm 5.1$ cm: t-test, $t = 2.12$, $P > 0.05$). The hexane extractions of individual snakes yielded an average ($\pm$ SD) of 17.0 ($\pm$ 4.4) mg of skin lipids per small female and 21.0 ($\pm$ 7.4) mg per newly-emerged male. Subsequent fractionation of the skin lipids yielded an average ($\pm$ SD) of 0.3 ($\pm$ 0.5) mg per small female and 0.2 ($\pm$ 0.2) mg per newly-emerged male. Overall, the methyl ketones accounted for an average ($\pm$ SD) of 1.4 ($\pm$ 2.2) % of the skin lipids collected from small females and 1.0 ($\pm$ 0.8) % of the skin lipids collected from newly-emerged males. After accounting for variation in skin surface area, we did not observe a significant difference between the small females and newly-emerged males with respect to the amount of methyl ketones extracted per unit of surface area (rank-sum test, $T = 59.0$, $P = 0.38$; Table 21.1).

Complete GC-MS analysis of the methyl ketone fractions revealed the presence of 18 unique long-chained methyl ketones—nine long-chain saturated methyl ketones and nine long-chain ω-9 cis-unsaturated methyl ketones. Small females and newly-emerged males varied in the number of methyl ketones expressed although the difference was not significant (t-test, $t = 0.42$, $P = 0.68$; Table 21.1). In addition, there was no significant difference in the relative contribution of saturated versus unsaturated methyl ketones to the pheromone profiles of females and newly-emerged males (t-test, $t = -1.15$, $P = 0.27$; Table 21.1), nor was there a significant difference in the relative contribution of individual methyl ketones between the two groups (randomization test, $P > 0.05$; Fig. 21.2).

21.4 Discussion

The results of this study demonstrate that upon emergence from winter hibernation, newly-emerged males are attractive to other males, albeit not to the same degree as females. Furthermore, it appears that the attractive nature of these males is due to the presence of a pheromone profile similar to the female sexual attractiveness pheromone. As hypothesized, complete GC/MS analysis showed no significant

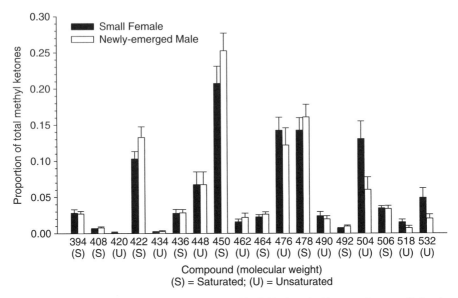

Fig. 21.2 Relative concentrations (mean ± SD) of individual methyl ketones from small female and newly-emerged male red-sided garter snakes

difference in the quantity or quality of the methyl ketone profiles between females and newly-emerged males.

The results from the courtship trials confirm previous studies demonstrating the attractive nature of most, if not all, newly-emerged males (Shine et al. 2000a; Shine et al. 2001). The adaptive advantage for newly-emerged males to appear attractive to other males appears to be related to survival: courting males transfer heat to the newly-emerged snakes and also are believed to provide cover from potential predators (Shine et al. 2001). This contrasts with the suggested advantage for long-term she-males (males which retain their attractive nature throughout the breeding season) where it is thought that the mimicry is utilized to 'confuse' other males, leading to increased reproductive success for the male expressing the female-like pheromone profile (Mason and Crews 1985).

The similarity between the methyl ketone profiles of females and newly-emerged males provides strong evidence that the expression of a female-like sexual attractiveness pheromone is primarily responsible for the attractive nature of newly-emerged males. This pheromone mimicry is similar to what was observed for she-males that retain their attractiveness for long periods during the breeding season (Mason 1993). Interestingly, the work with long-term she-males found that their methyl ketone profile was intermediate between non-attractive males and attractive females with respect to saturated and unsaturated methyl ketone concentrations (Mason 1993). We did not see such a difference in our study but instead found very similar methyl ketone concentrations between the newly-emerged males and females. This discrepancy is most likely due to differences in sampling technique: we only sampled small females ($< 55 \, \text{cm}$) whereas the previous work utilized females randomly selected from the population (average female size $= 60 \, \text{cm}$, Shine and Mason 2001). It is known that females of this species vary in saturated / unsaturated methyl ketone concentrations with respect to body size; larger females have higher concentrations of unsaturated methyl ketones whereas smaller females have higher concentrations of saturated methyl ketones (LeMaster and Mason 2002), similar to previous observations with long-term she-males.

It should be noted that while newly-emerged males were attractive in this study, small females garnered significantly more attention from courting males. This suggests that other cues beyond a female-like sexual attractiveness pheromone are involved in regulating male courtship to newly-emerged males. Visual and thermal cues are known to assist in the identification of potential mates in this species (Shine and Mason 2001) and thus might have played a role in this study. Additional chemical cues might also be involved. For example, squalene, a chemical found in high abundance in the skin lipids of male garter snakes, is known to reduce or eliminate male courtship behavior (Mason et al. 1989)—to what extent this chemical exists in the skin lipid of newly-emerged males is unknown. Thus, future studies are necessary to explore additional potential cues utilized by males to discriminate between females and newly-emerged males.

Acknowledgments We thank the Manitoba Department of Natural Resources, Dave Roberts, and Al and Gerry Johnson for assistance in the field. This research was supported by a Western Oregon University Faculty Development Award to M.P.L.

References

Andrén, C. (1986) Courtship, mating and agonistic behavior in the free-living population of adders, *Vipera berus* (L.). Amphibia-Reptilia 7, 353–383.

Burghardt, G.M. (1980) Behavioral and stimulus correlates of vomeronasal functioning in reptiles: feeding, grouping, sex, and tongue use. In: D. Müller-Schwarze and R.M. Silverstein (Eds.), *Chemical Signals—Vertebrates and Aquatic Invertebrates*. Plenum Press, New York, pp. 275–301.

Carpenter, C.C. and Ferguson, G.W. (1977) Variation and evolution of stereotyped behavior in repiles. In: C. Gans and D.W. Tinkly (Eds.), *Biology of the Reptilia*. Academic Press, New York, pp. 335–554.

Ford, N.B. (1981) Seasonality of pheromone trailing behavior in two species of garter snake, *Thamnophis* (Colubridae). Southwest. Nat. 26, 385–388.

Gehlback, F.R., Watkins, J.F. and Kroll, J.C. (1971) Pheromone trail-following studies of typhlopid, leptotyphlopid, and colubrid snakes. Behaviour 40, 282–294.

Halpern, M. (1987) The organization and function of the vomeronasal system. Annu. Rev. Neurosci. 10, 325–362.

LeMaster, M.P. and Mason, R.T. (2001) Evidence for a female sex pheromone mediating male trailing behavior in the red-sided garter snake, *Thamnophis sirtalis parietalis*. Chemoecology 11, 149–152.

LeMaster, M.P. and Mason, R.T. (2002) Variation in a female sexual attractiveness pheromone controls male mate choice in garter snakes. J. Chem. Ecol. 28, 1269–1285.

LeMaster, M.P. and Mason, R.T. (2003) Pheromonally mediated sexual isolation among denning populations of red-sided garter snakes, *Thamnophis sirtalis parietalis*. J. Chem. Ecol. 29, 1027–1043.

Mason, R.T. (1993) Chemical ecology of the red-sided garter snake, *Thamnophis sirtalis parietalis*. Brain Behav. Evol. 41, 261–268.

Mason, R.T. (1992) Reptilian Pheromones. In: C. Gans and D. Crews (Eds.), *Biology of the Reptilia, Vol. 18*. The University of Chicago Press, Chicago, pp. 115–216.

Mason, R.T. and Crews, D. (1985) Female mimicry in garter snakes. Nature 316, 59–60.

Mason, R.T., Fales, H.M., Jones, T.H., Pannell, L.K., Chinn, J.W. and Crews, D. (1989) Sex pheromones in garter snakes. Science 245, 290–293.

Mason, R.T., Jones, T.H., Fales, H.M., Pannell, L.K. and Crews, D. (1990) Characterization, synthesis, and behavioral response to sex pheromone in garter snakes. J. Chem. Ecol. 16, 27–36.

Moore, I.T., LeMaster, M.P. and Mason, R.T. (2000) Behavioural and hormonal responses to capture stress in the male red-sided garter snake, *Thamnophis sirtalis parietalis*. Anim. Behav. 59, 529–534.

Noble, G.K. (1937) The sense organs involved in the courtship of *Storeria*, *Thamnophis*, and other snakes. B. Am. Mus. Nat. Hist. 73, 673–725.

Ross, P., Jr. and Crews, D. (1977) Influence of the seminal plug on mating behavior in the garter snake. Nature 267, 344–345.

Shine, R. and Mason, R.T. (2001) Courting male garter snakes use multiple cues to identify potential mates. Behav. Ecol. Sociobiol. 49, 465–473.

Shine, R., Harlow, P., LeMaster, M.P., Moore, I.T. and Mason, R.T. (2000a) The transvestite serpent: Why do male garter snakes court (some) other males? Anim. Behav. 59, 349–359.

Shine, R., Olsson, M.M. and Mason, R.T. (2000b) Chastity belts in garter snakes: the functional significance of mating plugs. Biol. J. Linn. Soc. 70, 377–390.

Shine, R., Phillips, B., Waye, H., LeMaster, M.P. and Mason R.T. (2001) Advantage of female mimicry to snakes. Nature 414, 267.

Chapter 22
The Neurobiology of Sexual Solicitation: Vaginal Marking in Female Syrian Hamsters (*Mesocricetus auratus*)

Laura Been and Aras Petrulis

Abstract Vaginal marking is a reproductively-oriented scent marking response, in which female Syrian hamsters deposit vaginal secretions in response to odor cues from male conspecifics. Converging lines of evidence suggest that vaginal marking functions as a solicitational signal, or an advertisement of a female's impending sexual receptivity. Although vaginal marking is commonly used as an assay of proceptivity, the neural control of vaginal marking remains largely unknown. In this chapter, we will review the existing literature on vaginal marking, synthesizing evidence from behavioral, endocrine, and neuroanatomical studies that indicate targets for the neural control of vaginal marking. Lastly, we will describe preliminary data from our laboratory that suggests a possible neural circuit for the descending control of vaginal marking in female Syrian hamsters.

22.1 Introduction

In many species, exposure to social odors from conspecifics leads to scent marking behaviors, in which animals deposit their own chemical signals (Thiessen and Rice 1976). Scent marking is particularly important for solitary species for which odors are often the only means of synchronizing reproduction between the sexes. One such reproductively-oriented scent marking response to male odor cues is vaginal marking in Syrian hamsters. In this stereotyped, species-specific behavior, a female hamster lowers her pelvis, pressing her anogenital region to a substrate, and thrusts or moves forward while depositing a small amount of vaginal secretion (Johnston 1977). Typically, the female's tail will be deflected from the substrate, although the intensity of this behavior and the duration of anogenital contact with the substrate may vary across individuals (Johnston 1977). Based on localization of a putative pheromone, aphrodisin, within the vaginocervical tract, it is likely that vaginal secretion is produced by glandular tissue in the lower uterus, cervix, and

Laura Been
Georgia State University, Department of Psychology,
lbeen1@student.gsu.edu

upper vaginal tract and secreted into the vaginal lumen (Kruhoffer, Bub, Cieslak, Adermann, Kunstyr, Forssmann and Magert 1997).

While vaginal marking has been well-studied with regard to behavioral description, endocrine and chemosensory modulation, and some aspects of central nervous system involvement, the neurobiology underlying vaginal marking behavior is largely unknown. In this chapter, we will briefly review the known literature on vaginal marking, describe evidence that indicates targets for the central and peripheral control of vaginal marking, and, lastly, describe preliminary data from our laboratory that suggests a possible neural circuit for the descending control of vaginal marking in female Syrian hamsters.

22.2 Functional Significance

Converging lines of evidence suggest that vaginal marking functions as an advertisement of a female hamster's impending sexual receptivity. First, vaginal marking levels increase across the estrous cycle, peaking on the day prior to behavioral receptivity (Lisk and Nachtigall 1988). Second, females vaginal mark more to male conspecific odors than to the odors of other females, as demonstrated by the decreased latency and higher frequency of vaginal marking in male cages compared to female cages (Johnston 1977). Third, the secretion deposited by vaginal marking is both attractive to males and stimulates male copulatory behavior. Specifically, vaginal secretions can stimulate ultrasonic calling in males and attract males over short distances (Johnston and Kwan 1984). Furthermore, males are more attracted to intact females over vaginectomized females, which do not produce vaginal secretions, when the only cues available are odor cues (Kwan and Johnston 1980). Finally, when castrated males or ovariectomized females are scented with vaginal secretions, males' mounting frequency and time spent near the scented animals increases while their latency to mount the scented animals decreases (Johnston 1975). This evidence strongly suggests that the function of vaginal marking is sexual solicitation of male conspecifics.

22.3 Gonadal Steroid Hormones Modulate Vaginal Marking

22.3.1 Systemic Gonadal Steroid Hormones

One critical factor that modulates the expression of vaginal marking behavior is the cyclic production of gonadal steroid hormones, such as estradiol and progesterone, across the estrous cycle. It is known that vaginal marking frequency changes dramatically over the 4 day estrous cycle of Syrian hamsters. Marking frequency peaks during proestrous, the 24 h period before behavioral receptivity, when estradiol production is highest, and is markedly lower in frequency during the diestrous phase, when estradiol levels are low (Lisk and Nachtigall 1988). Furthermore, vaginal marking is almost never observed during estrus, when progesterone levels are highest (Johnston 1977; Takahashi and Lisk 1983). In addition to activational effects of hormones, the expression of vaginal marking is also determined by the neonatal hormone environment, as males gonadectomized as adults never show vaginal marking, whereas males gonadectomized neonatally show low levels of vaginal marking which increases significantly following estrogen treatment (Lisk and Nachtigall 1988). This evidence suggests that vaginal marking is stimulated by high levels of estrogen and inhibited by high levels of progesterone.

22.3.2 Gonadal Steroid Hormones in the Central Nervous System

Site-specific implantation of estradiol and/or progesterone within the brain has provided evidence for which steroid-sensitive forebrain areas are critical for vaginal marking behavior. Single implants of estradiol into the ventromedial hypothalamus (VMH) facilitate vaginal marking behavior, whereas single implants of estradiol into the anterior hypothalamus (AH) have no effect (Takahashi and Lisk 1985). Dual implants of estradiol, on the other hand, into both the medial preoptic area (MPO) and AH facilitate vaginal marking behavior (Takahashi, Lisk and Burnett 1985). Finally, progesterone implants into the VMH facilitate rapid display of lordosis, the species-specific receptive posture assumed by many rodents, whereas progesterone implants into the MPO and AH regulate the ability to prolong the expression of lordosis (Takahashi et al. 1985). This suggests that progesterone action in these areas may be required for the inhibition of vaginal marking, as lordosis and vaginal marking are never expressed simultaneously. Ultimately, data from the hormone implant studies suggest that specific hypothalamic structures mediate the hormone-dependent expression of vaginal marking behavior.

22.4 Chemosensory Cues Modulate Vaginal Marking

22.4.1 Peripheral Processing Chemosensory Cues

In addition to gonadal steroid hormones, vaginal marking is also regulated by odors, as female hamsters will preferentially direct their vaginal marks to male odors over female odors (Johnston 1977). Syrian hamsters detect odors through two olfactory systems: the main olfactory system, which mainly processes volatile odorants via the main olfactory epithelium (MOE) and main olfactory bulbs (MOB), and the accessory olfactory system, which mainly processes non-volatile cues via the vomeronasal organ (VNO) and accessory olfactory bulbs (AOB) (Wood 1997). Research has shown that vaginal marking is mediated by both the main and accessory olfactory systems. Specifically, damaging the main olfactory mucosa with zinc sulfate reduces the frequency of vaginal marking, suggesting the main olfactory system is necessary for normal levels of vaginal marking (Johnston 1992). In contrast, removal of the VNO does not affect overall frequency of vaginal marking, but does impair the ability of females to mark differentially to male and female odors, suggesting the accessory olfactory system is critical for appropriate expression of vaginal marking (Petrulis, Peng and Johnston 1999). Taken together, this evidence suggests that both the main and accessory olfactory systems are necessary for appropriately directed vaginal marking.

22.4.2 Central Processing of Chemosensory Cues

Lesion studies have further clarified which forebrain areas are important for the chemosensory modulation of vaginal marking. Importantly, these studies targeted the medial amygdala (MeA), which receives input from both the main and accessory olfactory systems, as well as the orbital/agranular insular cortex, which receives both direct and indirect input from the MOB and other main olfactory structures. Lesions of the orbital/agranular insular cortex do not affect vaginal marking behavior (Petrulis, DeSouza, Schiller and Johnston 1998), whereas lesions of the MeA significantly reduce vaginal marking (Takahashi and Gladstone 1988; Petrulis and Johnston 1999), suggesting that the MeA is critical for the chemosensory modulation of vaginal marking. Lesions of the hippocampal system provide additional information regarding which forebrain areas may be important for the chemosensory modulation of vaginal marking. Lesions of the fimbria-fornix, the source of modulatory input and subcortical connections to the hippocampus (Cassel, Duconseille, Jeltsch and Will 1997), decrease overall vaginal marking frequency and eliminate differential marking to male odors over females (Petrulis, Peng and Johnston 2000). However, this deficit could be due to unintended damage to the ventrolateral septum and not the fimbria-fornix, and thus is difficult to interpret. Significantly, lesions of the parahippocampal region including the entorhinal cortex, which receives MOB projections and provides the major source of cortical connections to the hippocampus (Witter 1993), have very little effect on vaginal marking (Petrulis et al. 2000). This suggests that the main olfactory system is not regulating vaginal marking through its connections to entorhinal cortex.

22.5 Hypothetical Forebrain Circuitry

Based on what we know about the hormonal and chemosensory regulation of vaginal marking, it is reasonable to suggest specific forebrain circuitry that may be responsible for the control of vaginal marking. First, it is likely that male social odor cues are detected by the female via both the MOE and the VNO, as both structures are necessary for the appropriate expression of vaginal marking. The MOE and VNO project to the MOB and AOB, respectively. From the MOB, information could be transmitted to a variety of cortical and subcortical structures, although likely not via the entorhinal connections to the orbital agranular/insular cortex. The MeA may be a particularly important structure for vaginal marking, as information from both the MOB and AOB converge on the MeA, which has been shown to be critical for the expression of vaginal marking. Furthermore, the MeA projects to the AH, MPOA, and the VMH, three structures in which estradiol implants stimulate marking (see above) and lesions impair vaginal marking (Malsbury, Kow and Pfaff 1977). However, we have little information as to how the MeA and hypothalamic areas interact with the midbrain, hindbrain and spinal circuitry that directly regulate the motor control of vaginal marking. We are addressing this gap in the literature by using neurotropic viral tract-tracing methods to elucidate the connections from muscles relevant to vaginal marking to the central nervous system.

22.6 Peripheral Anatomy and Musculature

In order to delineate the motoric circuits involved in vaginal marking, we first need to identify the relevant musculature involved in the behavior. To do this, it is instructive to think of vaginal marking as having two major, separable motor components: (1) the lowering and thrusting of the pelvis and (2) the deposition of vaginal secretion onto the substrate. Although the identity of these muscles is not known, muscle stimulation studies in female rats suggest the pelvic and pudendal nerves likely serve as regulators of the two main motor components of vaginal marking. For example, stimulation of the somatomotor branch of the pelvic nerve produces downward pelvic thrusting in female rats (Martinez-Gomez, Chirino, Beyer, Komisaruk and Pacheco 1992). This suggests that this nerve, through its connections to the iliococcygeus and pubococcygeus muscles (pelvic floor muscles), may regulate the lowering and thrusting of the pelvis observed during vaginal marking. In contrast, the vicerocutaneous branch of the pelvic nerve, through its connections to the smooth muscle of the vaginal wall, is a likely candidate for the control of the extrusion of vaginal secretion from the vagina onto the substrate. This hypothesis is supported by the finding that stimulation of this nerve increases vaginal wall pressure and contractions (Giuliano, Allard, Compagnie, Alexandre, Droupy and Bernabe 2001). It is also possible that the extrusion of vaginal secretion is not controlled by descending nerves, but rather by a local spinal reflex that responds to somatosensory feedback during anogenital contact with the substrate. If this is the case, it is likely that the pudendal nerve would control such a reflex, as stimulation of the anogenital region causes muscle movement that is mediated by the pudendal nerve (McKenna, Chung and McVary 1991). In order to determine whether sensory feedback is required for the extrusion of vaginal secretion, we plan to use topical anesthetic to render the anogenital region incapable of transducing sensory signals and quantify the effect on vaginal marking.

22.7 Neuroanatomical Tract Tracing from Relevant Muscles

22.7.1 Multi-synaptic Retrograde Tracing from the Rat Vagina/Clitoris

Recent neuroanatomical tract tracing studies have identified the neural circuits that control some of the same muscles that may be involved in vaginal marking. For example, Marson and colleagues used pseudorabies virus (PRV) to map the central nervous system neurons that innervate the vagina and clitoris in the rat (Marson and Murphy 2006). Neurotropic viruses, such as PRV, invade the nervous system of their hosts in a retrograde manner and spread through chains of synaptically connected neurons making it possible to define entire hierarchically connected circuits within an individual animal (Song, Enquist and Bartness 2005). Following injections of the virus into the vagina and clitoris, PRV-immunoreactive (PRV-IR) neurons were distributed widely throughout the spinal cord, specifically in the superficial, medial, and lateral regions of the dorsal horn, the lateral gray including the sacral parasympathetic nucleus, the dorsal gray commissure and intermediomedial nucleus, the intermediate gray, and the ventral horn. In the brainstem, PRV-IR neurons were found in rostroventral medulla (raphe magnus and ventral gigantocellular reticular formation) and ventral lateral medulla (nucleus paragiantocellularis). At the level of the midbrain, PRV-IR labeling was observed in the caudal ventrolateral periaqueductal gray (PAG), raphe pallidus, raphe obscurus, Barrington's nucleus, lateral lemniscus, and pontine reticular formation. Finally, at the level of the forebrain, PRV-labeled neurons were observed in the lateral hypothalamus (LH), fornix, paraventricular nucleus (PVN), MPO, medial preoptic nucleus (MPN), VMH, ventral bed nucleus of the stria terminalis (BNST), zona incerta, caudal MeA, dorsal hypothalamic area, and ventral hypothalamic area (Marson and Murphy 2006).

22.7.2 Multisynaptic Retrograde Tracing from the Hamster Vagina

To identify the neural control of vaginal secretion extrusion, we injected PRV into the smooth muscle of the vaginal wall. Following different survival times (48, 72, 96, 120, and 144 h post-injection), animals were sacrificed and brain tissue was processed for immunohistochemical localization of PRV. One hundred and twenty hours post-injection, PRV-immunoreactive neurons were distributed widely throughout the hindbrain, midbrain, and forebrain. Similarly to the pattern of labeling found in the rat, we observed PRV-labeled neurons in hindbrain areas including the rostroventral medulla (gigantocellular reticular nucleus, raphe obscurus nucleus) and the ventral lateral medulla (lateral paragigantocellular nucleus). In the midbrain, we observed PRV labeling in areas such as the lateral periaqueductal gray, rostral linear nucleus of raphe, and medial longitudinal fasciculus. Finally, in the forebrain, PRV-immunoreactive neurons were visible in areas including several thalamic areas (parafascicular nucleus, reticular nucleus, paraventricular nucleus), hypothalamic areas (ventromedial nucleus, medial preoptic area, medial preoptic nucleus, arcuate nucleus, anterior hypothalamus), as well as the posteromedial nucleus of the bed nucleus of the stria terminalis and the posterodorsal nucleus of the medial amygdala (Been and Petrulis, unpublished data). While this pattern of labeling is similar to what was found in the rat, some differences exist. For example, in the rat, PRV-IR labeling was found in areas such as Barrington's nucleus and the lateral lemniscus, whereas these areas were not labeled in the hamster. However, differences such as these are potentially due to the addition of clitoral injections in the rat or to actual interspecies differences.

22.7.3 Supporting Data from Related Models

The known neural circuitry for other female rodent sexual behaviors that share physiological characteristics with vaginal marking may also provide insight into which brain and spinal cord areas may be important for the neural control of vaginal marking. For example, many of the central and peripheral structures active during lordosis share similarities with the central and peripheral structures active during vaginal marking. Furthermore, lordosis and vaginal marking have a reciprocal relationship: lordosis is only exhibited during estrus, and never during the rest of the estrous cycle, whereas vaginal marking is exhibited during the diestrous and proestrous phases, but rarely during estrus. Thus, the already well-defined neural circuitry underlying lordosis in hamsters may also provide relevant information regarding the neural control of vaginal marking.

For example, neuroanatomical tract tracing studies in the hamster have shown that descending projections from the brainstem nucleus retroambiguus to the iliopsoas motoneuronal cell groups and the cutaneous trunci motor neurons may be responsible for the arching of the back and elevating of the tail seen during lordosis (Gerrits and Holstege 1999; Gerrits, Vodde and Holstege 2000). Given the similarities between some of the physical components of lordosis and vaginal marking behavior (i.e. tail elevation and postural changes) it is likely that these brain areas may also play a role in the central control of vaginal marking. Similarly, hormone-sensitive midbrain structures known to be involved in the lordosis circuit, such as the PAG, peripeduncular region, and mesencephalic reticular formation (Pfaff 1980) may also be involved in vaginal marking. Finally, at the forebrain level, our hypothesized forebrain circuitry for vaginal marking is corroborated in part by data from the lordosis circuit, in which estrogen-sensitive neurons in the VMH are a critical part of the ascending forebrain circuitry.

22.8 Conclusions and Future Directions

In the future, we plan to continue to use viral tract tracing to completely define the neural network underlying vaginal marking. First, we plan to co-localize PRV-labeled neurons resulting from injections of different strains of PRV into vaginal smooth muscle and the pelvic floor muscles. Next, as vaginal marking is strongly regulated by chemosensory cues, we plan to identify the location(s) where chemosensory information interacts with the descending control of vaginal marking by coupling anterograde tracer injections in the olfactory bulbs with PRV injections in muscles relevant to vaginal marking. Additionally, we plan to co-localize PRV injections in muscles relevant to vaginal marking with cells that express estrogen and progesterone receptors. Ultimately, this research program will yield a complete understanding of the neurobiology underlying vaginal marking, as well as provide a model for the neural control of proceptive sexual behavior.

References

Cassel, J. C., Duconseille, E., Jeltsch, H. and Will, B. (1997) The fimbria-fornix/cingular bundle pathways: a review of neurochemical and behavioural approaches using lesions and transplantation techniques. Prog. Neurobiol. 51, 663–716.

Gerrits, P. O. and Holstege, G. (1999) Descending projections from the nucleus retroambiguus to the iliopsoas motoneuronal cell groups in the female golden hamster: possible role in reproductive behavior. J. Comp. Neurol. 403, 219–28.

Gerrits, P. O., Vodde, C. and Holstege, G. (2000) Retroambiguus projections to the cutaneus trunci motoneurons may form a pathway in the central control of mating. J. Neurophysiol. 83, 3076–83.

Giuliano, F., Allard, J., Compagnie, S., Alexandre, L., Droupy, S. and Bernabe, J. (2001) Vaginal physiological changes in a model of sexual arousal in anesthetized rats. Am. J. Physiol. Regul. Integr. Comp. Physiol. 281, R140–9.

Johnston, R. E. (1975) Sexual excitation function of hamster vaginal secretion. Anim. Learn. Behav. 3, 161–6.

Johnston, R. E. (1977) The causation of two scent-marking behaviour patterns in female hamsters (Mesocricetus auratus). Anim. Behav. 25, 317–27.

Johnston, R. E. (1992) Vomeronasal and/or olfactory mediation of ultrasonic calling and scent marking by female golden hamsters. Physiol. Behav. 51, 437–48.

Johnston, R. E. and Kwan, M. (1984) Vaginal scent marking: effects on ultrasonic calling and attraction of male golden hamsters. Behav. Neural Biol. 42, 158–68.

Kruhoffer, M., Bub, A., Cieslak, A., Adermann, K., Kunstyr, I., Forssmann, W. and Magert, H. (1997) Gene expression of aphrodisin in female hamster genital tract segments. Cell. Tissue Res. 287, 153–60.

Kwan, M. and Johnston, R. E. (1980) The role of vaginal secretion in hamster sexual behavior: males' responses to normal and vaginectomized females and their odors. J. Comp. Physiol. Psychol. 94, 905–13.

Lisk, R. D. and Nachtigall, M. J. (1988) Estrogen regulation of agonistic and proceptive responses in the golden hamster. Horm. Behav. 22, 35–48.

Malsbury, C. W., Kow, L. M. and Pfaff, D. W. (1977) Effects of medial hypothalamic lesions on the lordosis response and other behaviors in female golden hamsters. Physiol. Behav. 19, 223–37.

Marson, L. and Murphy, A. Z. (2006) Identification of neural circuits involved in female genital responses in the rat: a dual virus and anterograde tracing study. Am. J. Physiol. Regul. Integr. Comp. Physiol. 291, R419–28.

Martinez-Gomez, M., Chirino, R., Beyer, C., Komisaruk, B. R. and Pacheco, P. (1992) Visceral and postural reflexes evoked by genital stimulation in urethane-anesthetized female rats. Brain Res. 575, 279–84.

McKenna, K. E., Chung, S. K. and McVary, K. T. (1991) A model for the study of sexual function in anesthetized male and female rats. Am. J. Physiol. 261, R1276–85.

Petrulis, A., DeSouza, I., Schiller, M. and Johnston, R. E. (1998) Role of frontal cortex in social odor discrimination and scent-marking in female golden hamsters (*Mesocricetus auratus*). Behav. Neurosci. 112, 199–212.

Petrulis, A. and Johnston, R. E. (1999) Lesions centered on the medial amygdala impair scent-marking and sex-odor recognition but spare discrimination of individual odors in female golden hamsters. Behav. Neurosci. 113, 345–57.

Petrulis, A., Peng, M. and Johnston, R. E. (1999) Effects of vomeronasal organ removal on individual odor discrimination, sex-odor preference, and scent marking by female hamsters. Physiol. Behav. 66, 73–83.

Petrulis, A., Peng, M. and Johnston, R. E. (2000) The role of the hippocampal system in social odor discrimination and scent-marking in female golden hamsters (*Mesocricetus auratus*). Behav. Neurosci. 114, 184–95.

Pfaff, D. (1980) *Estrogens and Brain Function*. New York, New York, Springer-Verlag.

Song, C. K., Enquist, L. W. and Bartness, T. J. (2005) New developments in tracing neural circuits with herpesviruses. Virus Res. 111, 235–49.

Takahashi, L. K. and Gladstone, C. D. (1988) Medial amygdaloid lesions and the regulation of sociosexual behavioral patterns across the estrous cycle in female golden hamsters. Behav. Neurosci. 102, 268–75.

Takahashi, L. K. and Lisk, R. D. (1983) Organization and expression of agonistic and socio-sexual behavior in golden hamsters over the estrous cycle and after ovariectomy. Physiol. Behav. 31, 477–82.

Takahashi, L. K. and Lisk, R. D. (1985) Estrogen action in anterior and ventromedial hypothalamus and the modulation of heterosexual behavior in female golden hamsters. Physiol. Behav. 34, 233–9.

Takahashi, L. K., Lisk, R. D. and Burnett, A. L., 2nd (1985) Dual estradiol action in diencephalon and the regulation of sociosexual behavior in female golden hamsters. Brain Res. 359, 194–207.

Thiessen, D. and Rice, M. (1976) Mammalian scent gland marking and social behavior. Psychol. Bull. 83, 505–39.

Witter, M. P. (1993) Organization of the entorhinal-hippocampal system: a review of current anatomical data. Hippocampus 3, 33–44.

Wood, R. I. (1997) Thinking about networks in the control of male hamster sexual behavior. Horm. Behav. 32, 40–5.

Chapter 23
Olfactory Control of Sex-Recognition and Sexual Behavior in Mice

Matthieu Keller, Michael J. Baum and Julie Bakker

Abstract In this chapter, we review recent data about the involvement of both the main and the accessory olfactory system in mate recognition and the control of sexual behavior in mice. Whereas the main olfactory system seems to play a central role in mate recognition in both male and female mice, clear sex differences emerge with regard to which olfactory system plays a more important role in the control of sexual behavior. Indeed, the main but not the accessory olfactory system seems to be more important in regulating sexual behavior in male mice, whereas in female mice, the accessory olfactory system seems to play a critical role in the control of mating.

Olfaction is of primary importance for social recognition in mammals, including mice. Thus mice use odors to distinguish sex, social or reproductive status of conspecifics (Brennan and Zufall 2006; Brown 1979). In addition, odors have been shown to facilitate the display of sexual behavior (e.g. Thompson and Edwards 1972) and to induce neuroendocrine responses (e.g. pregnancy block in female mice; Brennan and Keverne 1997).

23.1 Neural Processing of Odors: the Two Olfactory Systems

Two olfactory systems have evolved in terrestrial vertebrates which differ in both their peripheral anatomy and central projections. The main olfactory system is usually conceived as a general analyzer that detects and differentiates among complex chemosignals of the environment (Firestein 2001). Odors are detected by olfactory sensory neurons located in the main olfactory epithelium (MOE); these neurons project to glomeruli in the main olfactory bulb (MOB). The mitral and tufted neurons abutting these MOB glomeruli then transmit olfactory signals to various

Matthieu Keller
University of Liège, Center for Cellular and Molecular Neurobiology, Belgium and CNRS/INRA/University of Tours, Laboratory of Behavioral and Reproductive Physiology, France
keller@tours.inra.fr

forebrain targets including the piriform cortex, the entorhinal cortex or the anterior-cortical nucleus of the amygdala (Scalia and Winans 1975).

By contrast, the accessory olfactory system is thought to be involved in the detection of odors that influence a variety of reproductive and aggressive behaviors (Keverne 1999). Sensory neurons are located in the vomeronasal organ (VNO) and detect pheromones which gain access to the VNO by a pumping mechanism (Meredith and O'Connell, 1979). VNO neurons send projections to the accessory olfactory bulb (AOB). Mitral cells of the AOB project in turn to the medial nucleus of the amygdala; olfactory information is then dispatched to several hypothalamic regions such as the bed nucleus of the stria terminalis, the medial preoptic area and the ventromedial hypothalamus (Scalia and Winans 1975).

In mice, volatile urinary compounds activate a subset of mitral cells in the MOB (Lin, Zhang, Block and Katz 2005) whereas mitral cells in the AOB are only activated when mice make direct physical contact with conspecifics (Luo, Fee and Katz 2003). These observations suggest that the main olfactory system detects volatile odorants whereas the accessory olfactory system detects non-volatile body odors. However, the functional dichotomy between both systems is not so clear-cut. Indeed, the view that the VNO is only involved in the detection of non-volatile odors has recently been challenged. Using both electrophysiological and imaging methods, several groups have demonstrated that using in vitro preparations, VNO neurons can express very sensitive and specific responses to volatile compounds contained in male urine, such as farnesenes or brevicomin (Leinders-Zufall, Lane, Puche, Ma, Novotny, Shipley and Zufall 2000; Del Punta, Leinders-Zufall, Rodriguez, Jukam, Wysocki, Ogawa, Zufall and Mombaerts 2002). However, the general view is that these volatile ligands need to be transported into the VNO by transport proteins, such as major urinary proteins (MUPs; Brennan and Zufall 2006; Hurst and Beynon 2004). Another set of evidence for a complementary role of both olfactory systems comes from recent data about chemosignals derived from the major histocompatibility complex (MHC), a system which influences the odors produced by various species, including mice (Restrepo, Lin, Salcedo, Yamazaki and Beauchamp 2006). Indeed, MHC-class I peptide ligands activated sensory neurons in the VNO (Leinders-Zufall, Brennan, Widmayer, Prasanth Chandramani, Maul-Pavicic, Jager, Li, Breer, Zufall and Boehm 2004) as well as in the MOE (Spehr, Kelliher, Li, Boehm, Leinders-Zufall and Zufall 2006), suggesting that olfactory recognition of MHC peptides in mice occurs through both the main and accessory olfactory systems.

23.2 Main vs Accessory Olfactory System in Mate Recognition and Sexual Behavior

23.2.1 Mate Recognition

Although the case is not unequivocal, evidence suggests that the main olfactory system plays a more important role in sex discrimination than the accessory olfactory system in mice. Several studies (e.g., Lloyd-Thomas and Keverne 1982) reported that VNO removal had no effect on the preference of female mice to approach olfactory cues from an intact as opposed to a castrated male. Conversely, female mice in which the MOE was destroyed by intranasal application of zinc sulfate (ZnSO$_4$) no longer preferred to approach soiled bedding from intact as opposed to castrated males (Lloyd-Thomas and Keverne 1982).

Nonetheless, studies using different transgenic mouse models in which VNO function was manipulated by mutating a cluster of V1r receptor genes (Del Punta et al. 2002) or deleting the transient receptor potential 2 cation channel (TRP2, Leypold, Yu, Leinders-Zufall, Kim, Zufall and Axel 2002; Stowers, Holy, Meister, Dulac and Koentges 2002) suggested that the VNO is required for sex discrimination in male mice. Indeed, deletion of TRP2 results in a dramatic reduction of various electrophysiological responses in VNO sensory neurons after exposure to urinary odorants. At the behavioral level, TRP2 knock-out (KO) male mice mounted male and female subjects indiscriminately (Leypold et al. 2002; Stowers et al. 2002), leading to the conclusion that the VNO is required for sex discrimination. However neither study measured olfactory sex discrimination directly. It remains thus questionable whether TRP2-KO males can discriminate between the sexes on the basis of olfactory cues.

Indeed, a study using surgical lesions of the VNO could not confirm such a role for the VNO in sex discrimination (Pankevich, Baum and Cherry 2004). VNO-lesioned male mice were able to discriminate between various conspecific urinary odors during habituation/dishabituation tests. In addition, a recent study (Kelliher, Spehr, Li, Zufall and Leinders-Zufall 2006) investigated whether TRP2 is essential for memory formation in the context of the Bruce effect given the proposed key role of TRP2 in VNO signal transduction. Surprisingly, the loss of the TRP2 channels did not significantly influence memory formation in contrast to the effect of surgical lesions of the VNO, which prevented pregnancy block in these females. Electrophysiological recordings showed that TRP2 channels are not necessary for the transduction of MHC peptide ligands by sensory neurons in the basal zone of the VNO (Kelliher et al. 2006) thereby offering an explanation for some of the discrepancies observed between studies using surgical lesion of the VNO or genetically engineered mice.

23.2.2 Sexual Behavior

Which olfactory system mediates sexual behavior in mice has also been a matter of debate. Early studies (Thompson and Edwards 1972; Edwards and Burge 1973) suggested a role for the main olfactory system in the display of female sexual receptivity, since destructions of the MOE by intranasal infusion with a zinc sulfate solution (ZnSO$_4$) attenuated lordosis behavior in estrogen-progesterone–treated mice. However, the VNO may be also involved as a reduction in female sexual receptivity by VNO removal has been reported in various other rodent species (rats: Rajendren,

Dudley and Moss 1990; hamsters: Mackay-Sim and Rose 1986; voles: Curtis, Liu and Wang 2001).

In the male, the situation is even more confusing. Initial studies by Edwards and Burge (1973) suggested that the MOE is not involved in controlling male sexual behavior since intranasal $ZnSO_4$ infusion did not affect male copulatory behavior. However, this result was recently challenged by two studies, using either chemical lesion of the MOE with dichlobenil (Yoon, Enquist and Dulac 2005) or genetically engineered mice lacking CNGA2, a channel that is only expressed in MOE sensory neurons (Mandiyan, Coats and Shah 2005). Both studies showed a dramatic loss in sexual behavior after disruption of the MOE, which is in sharp contrast with the results of Edwards and Burge (1973). Finally, the VNO does not seem to be involved in male copulatory behavior since VNO lesioning had no effect (Pankevich et al. 2004).

23.3 A Re-examination of Olfactory function in Mice

Questions clearly remain about which olfactory system is more important in mate recognition and in sexual behavior in mice. To address these issues, we conducted a series of experiments to systematically assess the role of either olfactory system in mate recognition and sexual behavior. Our main results are summarized here.

23.3.1 General Methodology

To assess the contribution of the VNO and MOE to mate recognition and sexual behavior, we used C57Bl/6 mice from our breeding colony. Acknowledging the possible impact of early exposure to chemosignals from adult conspecifics of the opposite sex on later attractive properties of sexual pheromones (Moncho-Bogani, Lanuza, Hernandez, Novejarque and Martinez-Garcia 2002), we considered that our animals were not "chemically naïve" since they have been exposed to chemical signals of conspecifics (siblings, mother, other adult male and female mice housed in the same room).

Mice were subjected to either surgical VNO removal versus sham surgery, or to chemical destruction of the MOE with $ZnSO_4$ versus saline injection into the nostrils (see Keller, Douhard, Baum and Bakker 2006a, b; Keller, Pierman, Douhard, Baum and Bakker 2006c for all technical details). In contrast to the use of genetically engineered mouse models, these methods do not allow functional recovery processes during development as the animals are usually tested a few days after the lesions have been performed.

The efficacy of each type of lesion (either surgical or chemical) was always assessed by combining various behavioral and histological methods (see Keller et al. 2006a, b, c for further technical details) to exclude the possibility that lesioning one olfactory system affected the other one as well. Thus, we used habituation/dishabituation tests to determine the behavioral responsiveness of the animal to the presentation of a new odor to assess the efficacy of the $ZnSO_4$ treatment. At the neurobiological level, we combined soybean agglutinin (SBA) histochemistry and Fos immunocytochemistry to determine the efficacy of the lesion (either of the MOE or VNO). SBA is a morphological marker that stains the axons of VNO neurons projecting to the glomerular layer of the AOB and thus serves as a useful marker of VNO neurons that remain intact after VNO removal (Wysocki and Wysocki 1995). Fos labeling is used as a marker of neural activation to assess the response of each olfactory system after olfactory stimulation. Animals in which a lesion was not complete or not restricted to the olfactory system investigated were omitted from the experiments.

To investigate the effects of the lesions on mate recognition, subjects' odor preferences for a variety of different odor stimuli were assessed using a Y-maze (see Bakker, Honda, Harada and Balthazart 2002 for a full description of the maze and procedures). Thus animals were given a choice between either volatile or non-volatile odors derived from anesthetized conspecifics, urine or soiled bedding. To determine the effects of lesions on sexual behavior, males were paired with a sexually receptive female whereas females were paired with a sexually experienced male (see Keller et al. 2006a, b, c for details). With regard to male sexual behavior, the number of mounts, intromissions and ejaculations and their latencies were scored during several 30 min tests. With regard to female sexual behavior, the number of lordosis responses to the mount of the male was recorded. Females were tested either until they received 20 mounts from the stimulus male or 15 min elapsed. At the end of each test, a lordosis quotient was calculated by dividing the number of lordosis responses by the number of mounts received.

23.3.2 Sex-Recognition in VNO or MOE Lesioned Mice

When provided with volatile odor stimuli in the Y-maze, both male and female $ZnSO_4$-treated mice showed severe deficits in olfactory investigation. While saline-treated animals usually showed a clear odor preference for the volatile odors from the opposite sex, $ZnSO_4$ treated mice showed no preference and in general very low levels of olfactory investigation (Fig. 23.1; Keller et al. 2006b, c). These results suggest that the main olfactory system and, most likely, not the VNO is used to localize and identify the sex and endocrine status of conspecifics on the basis of their volatile odors. Animals were not able to rely on their VNO to compensate for the loss of main olfactory function in order to discriminate between the different odor stimuli. Another notable result is that the same pattern was observed after using non-volatile odor stimuli in the Y-maze: when provided with direct access to soiled bedding or urine samples, $ZnSO_4$-treated males and females showed no odor preferences as well as very low levels of olfactory investigation, probably because they could not detect the volatile odors released from these odor sources. This result is in line with earlier observations in OMPntr mice, where targeted destruction of

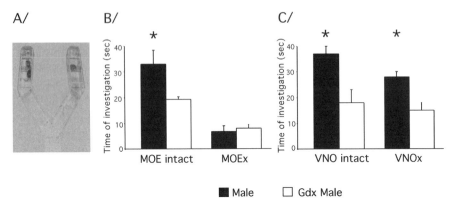

Fig. 23.1 Results of the odor preference test in a Y-maze (A). Female mice carrying lesions in either the MOE (B) or VNO (C) were given a choice between volatile odors derived from an intact or a gonadectomized male

the MOE made it impossible for female mice to locate male urine spots placed in their home cage (Ma, Allen, Van Bergen, Jones, Baum, Keverne and Brennan 2002). These data show that the MOE is necessary for localizing odor sources.

Following VNO removal, both male and female mice were still able to discriminate reliably between volatile odors derived from intact males and those from estrous females, thereby confirming our previous hypothesis suggesting that the VNO is not necessary for the detection and processing of volatile odors (Keller et al. 2006c; Pankevich et al. 2004). Thus, we confirmed that mice can use volatile odors for sex discrimination, as previously demonstrated (Pankevich et al. 2004; Moncho-Bogani et al. 2002; Moncho-Bogani, Lanuza, Lorente and Martinez-Garcia 2004). Similar results were found in other species such as hamsters (Petrulis, Peng and Johnston 1999) and ferrets (Woodley, Cloe, Waters and Baum 2004). In addition, VNO-lesioned animals showed clear deficits in olfactory investigation of various non-volatile odor stimuli, such as soiled bedding and urine, in the Y-maze (Keller et al. 2006a), suggesting that the VNO is involved in the detection and processing of non-volatile odors. Thus, our results confirm the older, widely held view of the VNO – accessory olfactory system being used to process, when in close contact, non-volatile components of body and urinary odorants that may trigger behavioral or endocrine responses (Luo et al. 2003; Powers, Fields and Winans 1979; O'Connell and Meredith 1984).

23.3.3 Olfactory Control of Sexual Behavior in VNO or MOE Lesioned Mice

In females, removal of the VNO clearly disrupted lordosis behavior in ovariectomized, hormonally primed, animals (Fig. 23.2). This is in line with results from other rodent species where a reduction in female sexual receptivity by VNO removal

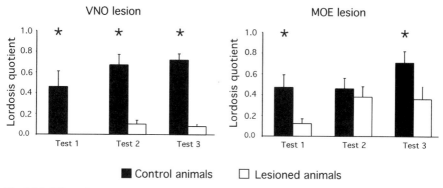

Fig. 23.2 Effect of VNO or MOE lesion on female sexual receptivity in mice

has been reported in rats (Rajendren et al. 1990), hamsters (Mackay-Sim and Rose 1986) and voles (Curtis et al. 2001). Intriguingly, studies on the role of the VNO in the Bruce effect showed that VNO ablation did not prevent female mice from becoming pregnant (Lloyd-Thomas and Keverne 1982; Kelliher et al. 2006), suggesting that the VNO may not mediate female sexual receptivity in mice. Because we only measured lordosis behavior during three short term tests, it is possible that, in the previous studies, VNO-lesioned females become sexually receptive after long-term exposure to the male. For example, it has been shown in rats that the lordosis quotient of VNO-lesioned females increased after prolonged exposure to the male (Rajendren et al. 1990). Whether females are sexually experienced or not prior to VNO removal may also play a role. It has been shown in male hamsters that once animals gain sexual experience, either olfactory system could sustain sexual behavior and only lesioning both olfactory systems disrupted sexual performance (Winans and Powers 1977; Meredith 1986).

Destruction of the MOE by intranasal $ZnSO_4$ treatment in female mice reduced lordosis quotients by approximately one half (Fig. 23.2, Keller et al. 2006a). These results confirm previous findings by Edwards and Burge (1973) who showed that peripheral anosmia induced by intranasal application of $ZnSO_4$ solution attenuated lordosis in sexually experienced and hormone-primed female mice, although not to the same extent as following lesion of the VNO (Keller et al. 2006c). This reduction in sexual receptivity may be directly related to the absence of any main olfactory inputs since sexual receptivity is determined by inputs from a range of different external, sensory as well as hormonal signals. Thus, deprivation of one sensory input (in this case from the main olfactory system) may induce less activation of the brain centers regulating lordosis and as a consequence impairs lordosis behavior. Alternatively, this reduction in lordosis behavior may be due indirectly to a deficit in VNO signaling. The fact that VNO-lesioned females fail to show lordosis behavior suggests that nonvolatile male odors are necessary to induce female sexual receptivity in this species (Keller et al. 2006c). Thus, if the main olfactory system is not functional, the VNO never has a chance to be activated because the animal never locates the non-volatile odor stimuli needed to activate it.

In the male, a recent experiment by Baum and collaborators (Pankevich et al. 2004) showed that lesion of the VNO was without any impact on male sexual behavior, suggesting a possible role for the main olfactory system. In line with this hypothesis, we showed that destruction of the MOE by intranasal application of zinc sulfate completely disrupted male sexual behavior, indicating that the MOE plays a central role. Our results also corroborate the recent observations of strong deficits in sexual behavior in CNGA2 mutant mice (Mandiyan et al. 2005) and following treatment with dichlobenil (Yoon et al. 2005). These convergent studies are in contrast however with an early report by Edwards and Burge (1973) that showed that lesioning the MOE with zinc sulfate solution did not affect sexual behavior in male mice. One possible explanation for this discrepancy could be differences in the amount of sexual experience of the subjects. Indeed, the mice used in the Edwards and Burge study (1973) were sexually experienced whereas Mandiyan et al. (2005) and Yoon et al. (2005) used sexually naïve mice, and studies performed in hamsters clearly demonstrate that sexual experience can compensate for the loss of information processed by one of the olfactory systems (e.g. Meredith 1986; Powers et al. 1979; Powers and Winans 1975; Winans and Powers 1977). However, we observed a similar loss of sexual behavior by intranasal infusion of zinc sulfate in sexually experienced male mice (Keller et al. 2006b). Therefore, in contrast with observations in male hamsters, sexual experience did not compensate for the loss of input from the main olfactory system in male mice.

23.3.4 Conclusions

Our work clearly demonstrates that the main as opposed to the accessory olfactory system plays a central role in mate recognition in both male and female mice. By contrast, clear sex differences emerge in the relative roles of these two olfactory systems in the regulation of mating. Thus, the main olfactory system seems to be more important in regulating sexual behavior in the male, whereas in the female, the accessory olfactory system seems to predominate.

Acknowledgments This work was supported by the Belgian Fonds National de la Recherche Scientifique (FNRS) and a National Institute for Child Health and Human Development grant N° HD 044897 to MJB and JB.

References

Bakker, J., Honda, S., Harada, N. and Balthazart, J. (2002) The aromatase knock-out mouse provides new evidence that estradiol is required during development in the female for the expression of sociosexual behaviors in adulthood. J. Neurosci. 22, 9104–9112.

Brennan, P.A. and Keverne, E.B. (1997) Neural mechanisms of mammalian olfactory learning. Prog. Neurobiol. 51, 457–481.

Brennan, P.A. and Zufall, F. (2006) Pheromonal communication in vertebrates. Nature 444, 308–315.

Brown, R.E. (1979) Mammalian social odors. Adv. Stud. Behav. 10, 107–161.

Curtis, J. T., Liu, Y. and Wang, Z. (2001) Lesions of the vomeronasal organ disrupt mating-induced pair bonding in female prairie voles (*Microtus ochrogaster*). Brain Res. 18, 167–174.

Del Punta, K., Leinders-Zufall, T., Rodriguez, I., Jukam, D., Wysocki, C. J., Ogawa, S., Zufall, F. and Mombaerts, P. (2002) Deficient pheromone responses in mice lacking a cluster of vomeronasal receptor genes. Nature 419, 70–74.

Edwards, D.A. and Burge, K.G. (1973) Olfactory control of sexual behavior of female and male mice. Physiol. Behav. 11, 867–872.

Firestein, S. (2001) How the olfactory system makes sense of scents. Nature 413, 211–218.

Hurst, J.L. and Beynon, R. (2004) Scent wars: the chemobiology of competitive signalling in mice. Bioessays 26, 1288–1298.

Keller, M., Douhard, Q., Baum, M.J. and Bakker, J. (2006a) Destruction of the main olfactory epithelium reduces female sexual behavior and olfactory investigation in female mice. Chem. Senses 31, 4, 315–323.

Keller, M., Douhard, Q., Baum, M.J. and Bakker, J. (2006b) Sexual experience does not modulate the detrimental effects of zinc sulfate–lesioning of the main olfactory epithelium on sexual behavior in male mice. Chem. Senses 31, 8, 753–762.

Keller, M., Pierman, S., Douhard, Q., Baum, M.J. and Bakker, J. (2006c) The vomeronasal organ is required for the expression of lordosis behahavior, but not sex discrimination in female mice. Eur. J. Neurosci. 23, 521–530.

Kelliher, K.R., Spehr, M., Li, X.H., Zufall, F. and Leinders-Zufall, T. (2006) Pheromonal recognition memory induced by TRPC2-independent vomeronasal sensing. Eur. J. Neurosci. 23, 3385–3390.

Keverne, E.B. (1999) The vomeronasal organ. Science 286, 716–720.

Leinders-Zufall, T., Brennan, P.A., Widmayer, Prasanth Chandramani, C., Maul-Pavicic, A., Jager, M., Li, X.H., Breer, H., Zufall, F. and Boehm, T. (2004) MHC class I peptides as chemosignals in the vomeronasal organ. Science 306, 1033–1037.

Leinders-Zufall, T., Lane, A.P., Puche, A.C., Ma, W., Novotny, M.V., Shipley, M.T. and Zufall, F. (2000) Ultrasensitive pheromone detection by mammalian vomeronasal neurons. Nature 405, 470–477.

Leypold, B.G., Yu, C.R., Leinders-Zufall, T., Kim, M.M., Zufall, F. and Axel, R. (2002) Altered sexual and social behaviors in trp2 mutant mice. Proc. Natl. Acad. Sci. USA 99, 6376–6381.

Lin, D.Y., Zhang, S.Z., Block, E. and Katz, L.C. (2005) Encoding social signals in the mouse main olfactory bulb. Nature 434, 470–477.

Lloyd-Thomas, A. and Keverne, E.B. (1982) Role of the brain and accessory olfactory system in the block of pregnancy in mice. Neuroscience 7, 907–913.

Luo, M., Fee, M.S. and Katz, L.C. (2003) Encoding pheromonal signal in the accessory olfactory bulb of behaving mice. Science 299, 1196–1201.

Ma, D., Allen, N.D., Van Bergen, Y.C., Jones, C.M., Baum, M.J., Keverne, E.B. and Brennan, P.A. (2002) Selective ablation of olfactory receptor neurons without functional impairment of vomeronasal receptor neurons in OMP-ntr transgenic mice. Eur. J. Neurosci. 16, 2317–2323.

Mackay-Sim, A. and Rose, J.D. (1986) Removal of vomeronasal organ impairs lordosis in female hamsters: effect is reversed by luteinising hormone-releasing hormone. Neuroendocrinology 42, 489–493.

Mandiyan, V.S., Coats, J.K. and Shah, N.M. (2005) Deficits in sexual and aggressive behaviors in Cnga2 mutant mice. Nat. Neurosci. 8, 1660–1662.

Meredith, M. (1986) Vomeronasal organ removal before sexual experience impairs male hamster mating behavior. Physiol. Behav. 36, 737–743.

Meredith, M. and O'Connell (1979) Efferents control of stimulus access to the hamster vomeronasal organ. J. Physiol. 286, 301–316.

Moncho-Bogani, J., Lanuza, E., Hernandez, A., Novejarque, A. and Martinez-Garcia, F. (2002) Attractive properties of sexual pheromones in mice: innate or learned? Physiol. Behav. 77, 167–176.

Moncho-Bogani, J., Lanuza, E., Lorente, M.J. and Martinez-Garcia F. (2004) Attraction to male pheromones and sexual behavior show different regulatory mechanisms in female mice. Physiol. Behav. 81, 427–434.

O'Connell, R.J. and Meredith, M. (1984) Effects of volatile and nonvolatile chemical signals on male sexual behaviors mediated by the main and accessory olfactory systems. Behav. Neurosci. 98, 1083–1093.

Pankevich, D.E., Baum, M.J. and Cherry, J.A. (2004) Olfactory sex-discrimination persists, whereas the preference for urinary odorants from estrous females disappears in male mice after vomeronasal organ removal. J. Neurosci. 24, 9451–9457.

Petrulis, A., Peng, M. and Johnston, R.E. (1999) Effects of vomeronasal organ removal on individual odor discrimination, sex-odor preference, and scent marking by female hamsters. Physiol. Behav. 66, 73–83.

Powers, J.B. and Winans, S.S. (1975) Vomeronasal organ: critical role in mediating sexual behavior of the male hamster. Science 187, 961–963.

Powers, J. B., Fields, R. B. and Winans, S. S. (1979) Olfactory and vomeronasal system articipation in male hamsters' attraction to female vaginal secretions. Physiol. Behav. 22, 77–84.

Rajendren, J.B., Dudley, C.A., and Moss, R.L. (1990) Role of the vomeronasal organ in the male-induced enhancement of sexual receptivity in female rats. Neuroendocrinology 52, 368–372.

Restrepo, D., Lin, W., Salcedo, E., Yamazaki, K. and Beauchamp, G. (2006) Odortypes and MHC peptides: complementary chemosignals of MHC haplotype? Trends Neurosci. 29, 604–609.

Scalia, F. and Winans, S.S. (1975) The differential projections of the olfactory bulb and accessory olfactory bulb in mammals. J. Comp. Neurol. 161, 31–35.

Spehr, M., Kelliher, K.R., Li, X.H., Boehm, T., Leinders-Zufall, T. and Zufall, F. (2006) Essential role of the main olfactory system in social recognition of major histocompatibility complex peptide ligands. J. Neurosci. 26, 1961–1970.

Stowers, L., Holy, T. E., Meister, M., Dulac, C. and Koentges, G. (2002). Loss of sex-discrimination and male-male aggression in mice deficient for TRP2. Science 295, 1493–1500.

Thompson, M.L. and Edwards, D.A. (1972) Olfactory bulb removal impairs the hormonal induction of sexual receptivity in spayed female mice. Physiol. Behav. 8, 1141–1146.

Winans, S. S. and Powers, J. B. (1977) Olfactory and vomeronasal deafferantation of male hamsters: Histological and behavioral analyses. Brain Res. 126, 325–344.

Woodley, S.K., Cloe, A.L., Waters, P. and Baum, M.J. (2004) Effects of vomeronasal organ removal on olfactory sex discrimination and odor preferences of female ferrets. Chem. Senses 29, 659–669.

Wysocki, C.J. and Wysocki, L.M. (1995) Surgical removal of the vomeronasal organ and its verification. In: Speilman, A.I. and Brands, J.G. (Eds.), Experimental Cell Biology of Taste and Olfaction. CRC Press, New York, pp. 49–57.

Yoon, H., Enquist, L.W. and Dulac, C. (2005) Olfactory inputs to hypothalamic neurons controlling reproduction and fertility. Cell 123, 669–682.

Chapter 24
The Role of Early Olfactory Experience in the Development of Adult Odor Preferences in Rodents

Pamela M Maras and Aras Petrulis

Abstract Mate recognition is an essential component of successful reproductive behavior, and in rodent species, is primarily guided by the perception of social odors in the environment. Importantly, there is substantial evidence that species or sexual odor preferences may be regulated by early olfactory experience, although considerable variability in the plasticity of these behaviors has been observed. The current chapter summarizes what is known regarding the role of early olfactory experience in the development of adult odor preferences, synthesizing data across species, sex, and behavioral paradigms.

24.1 Introduction

Successful reproductive behavior relies on the ability to identify and approach appropriate mating partners within the environment. Critically, mate recognition requires identifying species and sex characteristics of possible mates. As in many mammalian species, rodents use odor cues as the primary mechanism for mate recognition (Johnston 1983). Thus, sexually mature rodents typically display strong behavioral preferences for conspecific odors from opposite-sex individuals compared to odors from the same-sex or heterospecific individuals (Johnston 1983).

There is substantial evidence that these preferences may be shaped by prior experience rather than being fixed across the lifespan. Although experience during adulthood can alter various behavioral responses to social odors (Pfaus, Kippin and Centeno 2001), it appears that olfactory experience during early life plays a critical role in the development of species and sexual odor preferences in many rodent species (D'Udine 1983). Indeed, previous studies using cross-fostering or artificial odor exposure provide behavioral evidence that odors present in the rearing environment are preferred over unfamiliar odors in adulthood in a variety of contexts (D'Udine 1983; Nyby and Whitney 1980). More recently, Martinez-Garcia and colleagues have shown that early experience with opposite-sex odors is in

Pamela M Maras
Georgia State University, Department of Psychology
pmaras1@student.gsu.edu

fact necessary for the expression of sexual odor preferences in adult female mice (Moncho-Bogani, Lanuza, Hernandez, Novejarque and Martinez-Garcia 2002).

Together, these data suggest that associating specific odor cues with stimuli in the early environment may act as a mechanism to direct adult social or sexual behavior to opposite-sex conspecifics. Thus, the current review summarizes what is known regarding the role of early olfactory experience in the development of adult odor preferences, with a focus on the female Syrian hamster as a model for experience-dependent responses to sex-specific odor cues. We hypothesize that a classical conditioning model of olfactory learning explains the development of sexual and species odor preferences in rodents, and we will suggest possible neural mechanisms underlying this behavioral plasticity.

24.2 Functional Chemosensory Systems in Young Rodents

24.2.1 Early Chemosensory Processing

If adult odor preferences are in fact influenced by early olfactory experience, then chemosensory processing must be functional early in rodent development. Indeed, the chemosensory systems develop very early in rodents (Alberts 1976; Astic and Saucier 1981), and chemosensory processing appears to be functional both prenatally (Pedersen and Blass 1982; Stickrod, Kimble and Smotherman 1982) and perinatally, as evidenced by behavioral responses to odors (Devor and Schneider 1974; Gregory and Bishop 1975; Leon and Moltz 1971; Porter and Etscorn 1974). For example, Syrian hamster pups display behavioral preferences for different types of artificial odorants as early as postnatal day 4 (Devor and Schneider 1974). These data suggest that young rodents are able to both process and respond to odors present in their early environment.

24.2.2 Early Olfactory Learning

Importantly, many of the behavioral responses to odors expressed in rodent pups may reflect a conditioned attraction to familiar odors. In fact, preferences and aversions for olfactory cues can be conditioned *in utero* (Pedersen and Blass 1982; Stickrod et al. 1982), as well as during early postnatal life (Leon 1992; Wilson and Sullivan 1994). Perinatally, young rats display conditioned preferences for initially neutral odors paired with a variety or rewarding stimuli, including the nesting environment, milk presentation, tactile stimulation or contact with warm surfaces (Alberts and May 1984; Galef and Kaner 1980; Johanson and Hall 1982). Furthermore, these examples of early learning appear to follow the temporal constraints consistent with classical conditioning (Johanson and Hall 1982; Sullivan, Hofer and Brake 1986).

Research by Leon and colleagues described a possible neural mechanism underlying these conditioned odor preferences. By pairing artificial odor presentation with perineal tactile stimulation, they demonstrated that a behavioral preference for the conditioned odor is associated with changes in glomerular morphology in odor-specific regions of the olfactory bulb (Woo, Coopersmith and Leon 1987) as well as enhanced mitral/tufted cellular responses to the learned odors (Coopersmith and Leon 1984; Johnson, Woo, Duong, Nguyen and Leon 1995; Sullivan and Leon 1986). These physiological changes persist into adulthood (Coopersmith and Leon 1986) and are likely mediated by activation of the substantial noradrenergic inputs into the olfactory bulb (Rangel and Leon 1995; Sullivan, Wilson and Leon 1989). This research highlights the ability of the olfactory system to learn and remember olfactory cues early in development, and suggests one possible substrate for olfactory learning of adult odor preference.

24.3 Preferences for Species Odors

Much of the previous research on the role of early olfactory experience on adult odor preferences has used the approach of cross-fostering young pups to a lactating dam of a different species (D'Udine 1983). Thus, a shift in preference toward odor of the foster parent indicates that species-specific odors are learned via the early experience with the foster parent. Several important themes have emerged from this literature that shed light on the degree to which species preference is learned during early life.

First, biological constraints limit those behaviors that can be altered by early olfactory experience. Specifically, although changes in the expression of sexual behavior have been reported in the guinea pig (Beauchamp and Hess 1971), the vast majority of cross-fostering studies in rodents report more subtle shifts in social or olfactory preferences (D'Udine 1983), such as the time spent near stimulus animals or their odors. These data support the idea that different aspects of reproductive behavior may be differentially regulated by developmental factors (Beach 1976); consummatory aspects of reproduction, including the ability to copulate with conspecifics, may be relatively fixed, but appetitive aspects of reproduction, such as mate recognition and approach, are readily shaped by early experience (D'Udine 1983; Quadagno and Banks 1970).

Second, rather than a complete reversal of species preference, observed changes are often characterized by either increased attraction to the heterospecific foster species (Denenberg, Hudgens and Zarrow 1964; Lagerspetz and Heino 1970) or decreased attraction to the conspecific species (McCarty and Southwick 1977; McDonald and Forslund 1978; Quadagno and Banks 1970). For example, male Syrian hamsters fostered to Turkish hamster dams no longer display a species preference, primarily due to decreased investigation of conspecific females (Murphy 1980). Similarly, male Syrian hamsters reared by rats display decreased attraction to the soiled bedding of conspecifics compared to males reared normally (Surov,

Solovieva and Minaev 2001). These data demonstrate that, for male Syrian hamsters, early experience with heterospecific odors eliminates species preference primarily by altering behavioral responses to conspecific odors.

Third, although changes in odor preference are consistent throughout the literature, the direction and magnitude of these effects vary according to species and sex of the subjects. For example, Quadagno and Banks (1970) reciprocally cross-fostered two species of mice, *Mus musculus* and *Baiomys taylori ater*, and found that whereas both male and female cross-fostered *Mus musculus* display decreased attraction to conspecific odors, only the female cross-fostered *Baiomys taylori ater* differ from controls in their species preferences. A similar interaction between sex and genotype has been reported in other species of mice (McCarty and Southwick 1977) and in voles (McDonald and Forslund 1978). In a reciprocal cross-fostering design using Syrian and Turkish hamsters, Murphy (1980) found that cross-fostering affected more behavioral measures of species preference in Syrian compared to Turkish males. These variations may reflect ecological or reproductive factors, including parental care styles, sociality, and mating strategies, and require consideration when comparing the development of preference behavior across species or between sexes.

24.4 Preferences for Artificial Odors

Another common approach to manipulating the early olfactory environment has been to add an artificial odor either to the nesting environment or directly to the mother of young pups (Nyby and Whitney 1980). This approach provides a more direct method for manipulating early olfactory cues compared to cross-fostering, which is confounded by changes in parental care and social environment. Importantly, these studies highlight that rodents can impart social significance to initially neutral and biologically irrelevant odors. In addition, they suggest that many biological odors may initially be neutral and must acquire social relevance during development.

In general, rodents that have been exposed to an artificial odor during early life show an increased attraction to stimulus animals scented with that odor as adults. Increased odor preference following early exposure has been observed in female mice (Mainardi, Marsan and Pasquali 1965) and male rats (Marr and Gardner 1965). In some cases, this odor preference translates into increases in mating efficiency with females of the familiar scent. Specifically, male rats reared with citral-scented dams ejaculate faster when mating with citral-scented females compared to normal-scented females (Fillion and Blass 1986).

However, some studies report only marginal effects of early odor exposure on adult odor preference. These contradictory findings may be due to differences either in the behavioral measures or species tested. For example, although Moore, Jordan and Wong (1996) observed decreases in ejaculation latencies similar to those observed by Fillion and Blass (1986), the author found no effect of early odor exposure when male rats are given a simultaneous choice between females with the familiar odor and control females. In Syrian hamsters, males reared with vanillin-scented dams display modest increases in investigation of the vanillin odor, and these males still investigate female vaginal secretion substantially more than the familiar, vanillin odor (Macrides, Clancy, Singer and Agosta 1984). Furthermore, rearing condition does not affect sexual behavior with either the scented or control stimulus females in this species (Macrides et al. 1984).

One possible explanation for these apparently contradictory findings is due to the potential variation in how easily different artificial odors can be conditioned, such that a preference for some odors may be more difficult or impossible to learn. In fact, Nyby and Whitney (1980) report that both the chemical substance and the length of odor exposure affect whether certain artificial odors can acquire the ability to induce ultrasonic vocalizations in male mice. Furthermore, as with cross-fostering studies, the artificial odor approach adds a novel odor to a potentially over-saturated early olfactory environment, such that these additional odors may be masked by the more salient conspecific odors that are produced by the mother (Leon 1974), siblings (Hudgens, Denenberg and Zarrow 1968; Vasilieva 1994), and the pup itself (Mateo and Johnston 2000, 2003). Finally, odor saturation may also be a critical factor during preference or mating tests; sexual odors can stimulate robust and stereotyped behavioral responses (Johnston 1974) that may override attraction to artificial odors.

24.5 Preferences for Sexual Odors

24.5.1 Preference for Male Volatile Odors in Female Mice

More recently, an approach that removes odor cues from the postnatal environment has been used to ask whether certain types of olfactory experience are required for the development of sexual odor preference. Specifically, in several studies by Martinez-Garcia and colleagues, all male siblings have been removed from experimental litters of mice at 19 days after birth in an attempt to restrict the contact of female subjects with opposite-sex odors (Moncho-Bogani et al. 2002; Moncho-Bogani, Martinez-Garcia, Novejarque and Lanuza 2005). The authors hypothesized that attraction to the volatile components of sexual odors requires previous association with the non-volatile components of these odors. Therefore, control and chemically naïve female mice were tested for their investigation behavior under conditions that either prevented contact (providing volatile components only) or allowed contact (providing both volatile and non-volatile components) with the odor stimuli.

Unlike control females, chemically naïve females do not prefer the volatile components of male odors compared to female odors. When these females are allowed contact with the odors, however, they demonstrate robust preferences for the male odors. Critically, initially naïve females that receive contact experience with male soiled bedding as adults display sexual odor preferences when subsequently presented with only the volatile components of the odors. These data suggest that previous experience with non-volatile components of male odors is required for the attraction to the volatile components of these odors.

24.5.2 Preference for Male Volatile Odors in Female Syrian Hamsters

Our laboratory has begun to determine if this phenomenon of experience-dependent attraction to male volatile cues also occurs in female Syrian hamsters, as experiential effects on odor preference may vary according to species. Syrian hamsters provide an ideal model species for studying sexual preference as these behaviors are almost exclusively mediated by chemosensory cues (Johnston 1983). Furthermore, both males and females display robust preferences for opposite-sex volatile odors that are independent of adult sexual experience (Landauer, Banks and Carter 1977; Petrulis and Johnston 1999), suggesting that early olfactory experience may play a critical role in the development of these behaviors.

In addition to odor preferences, female hamsters also display a sexually-motivated scent marking behavior that is directed toward males or their odors (Johnston 1977, 1990). Vaginal marking is a stereotyped behavior in which females lower and thrust their perineal region onto the substrate and deposit vaginal secretion (Johnston 1977, 1990). The expression of this behavior requires sexual odor recognition and is likely more energetically expensive to produce compared to general investigatory behavior. Vaginal marking may therefore be under greater selective pressure to be expressed under restricted environmental conditions and may likely be shaped by prior olfactory experience. Thus, the goals of our lab are to determine if early contact experience with male odors is required for (a) the preference for male over female volatile odors and (b) the sex-specific direction of vaginal marking in adult female Syrian hamsters.

Using an experimental design similar to Moncho-Bogani et al. (2002), preliminary data in our lab suggest that female Syrian hamsters display an attraction to the volatile components of male odors that is independent of previous experience with male non-volatile odors. In fact, chemically naïve females prefer to investigate male odors over female odors, regardless of whether they are allowed contact with the odor stimuli. These results suggest that, unlike mice, female hamsters do not require previous contact experience with male chemosignals for the expression of opposite-sex odor preference. Our results are consistent, however, with the minimal effects of early artificial odor exposure on later attraction to these odors in male Syrian hamsters (Macrides et al. 1984).

Although chemically-naïve female hamsters prefer to investigate volatile male odors, early contact experience with male odors is required for the differential expression of vaginal marking in response to male and female volatile odors. Whereas control females vaginal mark more to male odors than to female odors, chemically naïve females mark equally to the volatile components of these two odor sources. This lack of differential marking by naïve females is primarily due to decreased levels of marking toward male odors. Furthermore, we have found that chemically naïve females born into litters with more male siblings vaginal mark more specifically to male odors compared to those females with fewer male siblings. Consequently, prenatal exposure to either steroid hormones or odor signals produced by male siblings may interact with postnatal olfactory experience to shape behavioral responses to sexual odors in adulthood.

24.5.3 Functional Interaction of Main and Accessory Olfactory Systems

As with many macrosmatic mammals, rodents have two separate chemosensory systems, the main olfactory system (MOS) and accessory olfactory system (AOS), which respond to social odors. Importantly, these sensory systems differ not only in their peripheral morphology and central projections, but also in the types of chemosignals that they process (Meredith 1991). Sensory neurons of the MOS, which are located in the main olfactory epithelium and project to the main olfactory bulbs, process volatile chemicals and can detect odors at a distance. In contrast, sensory neurons of the AOS, which are located in the vomeronasal organs (VNO) and project to the accessory olfactory bulbs, primarily process large, non-volatile chemicals and require contact for stimulation (Meredith 1991).

These distinctions between the MOS and AOS suggest that the experience-dependent responses to male volatile odors described in mice and hamsters may involve a functional interaction between the two sensory systems. Specifically, the ability of the MOS to mediate appropriate responses to volatile components of sexual odors may change as a function of previous chemosensory processing by the AOS. We and others (Halpern and Martinez-Marcos 2003; Moncho-Bogani et al. 2002) hypothesize that male non-volatile cues, presumably processed by the AOS, are intrinsically rewarding, whereas the volatile components of male odors, processed primarily by the MOS, acquire attractive properties through a form of classical conditioning. This conditioning can result from contact experience with male odors, during which non-volatile odor activation of the AOS conditions an attraction to the simultaneously-presented volatile cues processed by the MOS.

Although the initial projections of the MOS and AOS are anatomically distinct, these systems eventually converge within specific areas of the ventral forebrain known to regulate reproductive behavior in rodents. Specifically, the medial amygdala nucleus (MeA) and posteromedial cortical nucleus of the amygdala (PMCo) receive direct and indirect projections from the AOS and MOS, respectively. Functionally, chemosensory processing of social odors by either the MOS or AOS causes increases in c-fos expression, a marker for neuronal activity, within MeA neurons (Fewell and Meredith 2002; Swann, Rahaman, Bijak and Fiber 2001). Similarly, the activity of single neurons within the PMCo can be driven by electrical stimulation of either the main olfactory bulb or VNO (Licht and Meredith 1987). Finally, evidence suggests that sexual experience increases the response of MeA neurons to MOS processing of female odors (Fewell and Meredith 2002). These data suggest that the MeA and PMCo may be critical nuclei for the functional convergence of MOS and AOS processing and may therefore provide a neural substrate for experience-dependent interaction between the two systems.

24.6 Concluding Remarks

Taken together, this body of work demonstrates that adult behavioral responses to social odors are shaped by early olfactory experience. Indeed, heterospecific or artificial odor cues associated with the rearing environment acquire attractive properties that can last into adulthood in many rodent species. Furthermore, early experience with opposite-sex odors appears to be critical for the normal development of appropriate behavioral responses to sexual odors in mice and hamsters. Importantly, the behavioral plasticity observed using these different experimental approaches may all be mediated by a classical conditioning model of olfactory learning. The experience-dependent development of odor preference in rodents therefore provides a powerful model for understanding how the olfactory system recognizes and learns the salience of social odors, a function that is critical for the appropriate expression of reproductive behavior.

References

Alberts, J. R. (1976) Olfactory contributions to behavioral development in rodents. In: R. L. Doty (Ed.), *Mammalian Olfaction, Reproductive Processes and Behavior*. Academic Press, New York, San Francisco, London, pp. 67–91.

Alberts, J. R. and May, B. (1984) Nonnutritive, thermotactile induction of filial huddling in rat pups. Dev. Psychobiol. 17, 161–181.

Astic, L. and Saucier, D. (1981) Ontogenesis of the functional activity of rat olfactory bulb: autoradiographic study with the 2-deoxyglucose method. Brain Res. 254, 243–256.

Beach, F. A. (1976) Sexual attractivity, proceptivity and receptivity in female mammals. Horm. Behav. 7, 105–138.

Beauchamp, G. K. and Hess, E. H. (1971) The effects of cross-species rearing on the social and sexual preferences of guinea pigs. Z. Tierpsychol. 28, 69–76.

Coopersmith, R. and Leon, M. (1984) Enhanced neural response to familiar olfactory cues. Science 225, 849–851.

Coopersmith, R. and Leon, M. (1986) Enhanced neural response by adult rats to odors experienced early in life. Brain Res. 371, 400–403.

Denenberg, V. H., Hudgens, G. A. and Zarrow, M. X. (1964) Mice Reared with Rats: Modification of Behavior by Early Experience with Another Species. Science 143, 380–381.

Devor, M. and Schneider, G. E. (1974) Attraction to home-cage odor in hamster pups: Specificity and changes with age. Behav. Biol. 10, 211–221.

D'Udine, B. a. A., E. (1983) Early experience and sexual preferences in rodents. In: P. Bateson (Ed.), *Mate Choice*. Cambridge University Press, Cambridge, pp. 311–327.

Fewell, G. D. and Meredith, M. (2002) Experience facilitates vomeronasal and olfactory influence on Fos expression in medial preoptic area during pheromone exposure or mating in male hamsters. Brain Res. 941, 91–106.

Fillion, T. J. and Blass, E. M. (1986) Infantile experience with suckling odors determines adult sexual behavior in male rats. Science 231, 729–731.

Galef, B. G., Jr. and Kaner, H. C. (1980) Establishment and maintenance of preference for natural and artificial olfactory stimuli in juvenile rats. J. Comp. Physiol. Psychol. 94, 588–595.

Gregory, E. H. and Bishop, A. (1975) Development of olfactory-guided behavior in the golden hamster. Physiol. Behav. 15, 373–376.

Halpern, M. and Martinez-Marcos, A. (2003) Structure and function of the vomeronasal system: an update. Progress in Neurobiology 70, 245–318.

Hudgens, G. A., Denenberg, V. H. and Zarrow, M. X. (1968) Mice reared with rats: effects of preweaning and postweaning social interactions upon adult behaviour. Behaviour 30, 259–274.

Johanson, I. B. and Hall, W. G. (1982) Appetitive conditioning in neonatal rats: conditioned orientation to a novel odor. Dev. Psychobiol. 15, 379–397.

Johnson, B. A., Woo, C. C., Duong, H., Nguyen, V. and Leon, M. (1995) A learned odor evokes an enhanced Fos-like glomerular response in the olfactory bulb of young rats. Brain Res. 699, 192–200.

Johnston, R. E. (1974) Sexual attraction function of golden hamster vaginal secretion. Behav. Biol. 12, 111–117.

Johnston, R. E. (1977) The causation of two scent-marking behaviour patterns in female hamsters (Mesocricetus auratus). Anim. Behav. 25, 317–327.

Johnston, R. E. (1983) Chemical signals and reproductive behavior. In: J. G. Vandenbergh (Ed.), *Pheromones and Reproduction in Mammals*. Academic Press, New York, pp. 3–37.

Johnston, R. E. (1990) Chemical communication in golden hamsters: from behavior to molecules and neural mechanisms. In: D. A. Dewsbury (Ed.), *Contemporary Issues in Comparative Psychology*. Sinauer, Sunderland, MA, pp. 381–412.

Lagerspetz, K. and Heino, T. (1970) Changes in social reactions resulting from early experience with another species. Psychological Reports 27, 255–262.

Landauer, M. R., Banks, E. M. and Carter, C. S. (1977) Sexual preferences of male hamsters (Mesocricetus auratus) for conspecifics in different endocrine conditions. Horm. Behav. 9, 193–202.

Leon, M. (1974) Maternal pheromone. Physiol. Behav. 13, 441–453.

Leon, M. (1992) Neuroethology of olfactory preference development. J. Neurobiol. 23, 1557–1573.

Leon, M. and Moltz, H. (1971) Maternal pheromone: discrimination by pre-weanling albino rats. Physiol. Behav. 7, 265–267.

Licht, G. and Meredith, M. (1987) Convergence of main and accessory olfactory pathways onto single neurons in the hamster amygdala. Exp. Brain Res. 69, 7–18.

Macrides, F., Clancy, A. N., Singer, A. G. and Agosta, W. C. (1984) Male hamster investigatory and copulatory responses to vaginal discharge: an attempt to impart sexual significance to an arbitrary chemosensory stimulus. Physiol. Behav. 33, 627–632.

Mainardi, D. M., Marsan, M. and Pasquali, A. (1965) Causation of sexual preferences of the house mouse: The behaviour of mice reared by parents whose odour was artificially altered. Atti. Soc. Italiana Nat. Milano 104, 325–338.

Marr, J. N. and Gardner, L. E. (1965) Early olfactory experience and later social behavior in the rat: Preference, sexual responsiveness and care of young. J.Genetic Psych. 107, 167–174.

Mateo, J. M. and Johnston, R. E. (2000) Kin recognition and the 'armpit effect': evidence of self-referent phenotype matching. Proc. Biol. Sci. 267, 695–700.

Mateo, J. M. and Johnston, R. E. (2003) Kin recognition by self-referent phenotype matching: weighing the evidence. Anim. Cogn. 6, 73–76.

McCarty, R. and Southwick, C. H. (1977) Cross-species fostering: effects on the olfactory preference of Onychomys torridus and Peromyscus leucopus. Behav. Biol. 19, 255–260.

McDonald, D. L. and Forslund, L. G. (1978) The development of social preferences in the voles Microtus montanus and Microtus canicaudus: effects of cross-fostering. Behav. Biol. 22, 497–508.

Meredith, M. (1991) Sensory processing in the main and accessory olfactory systems: comparisons and contrasts. J. Steroid Biochem. Mol. Biol. 39, 601–614.

Moncho-Bogani, J., Lanuza, E., Hernandez, A., Novejarque, A. and Martinez-Garcia, F. (2002) Attractive properties of sexual pheromones in mice: innate or learned? Physiol. Behav. 77, 167–176.

Moncho-Bogani, J., Martinez-Garcia, F., Novejarque, A. and Lanuza, E. (2005) Attraction to sexual pheromones and associated odorants in female mice involves activation of the reward system and basolateral amygdala. Eur. J. Neurosci. 21, 2186–2198.

Moore, C. L., Jordan, L. and Wong, L. (1996) Early olfactory experience, novelty, and choice of sexual partner by male rats. Physiol. Behav. 60, 1361–1367.

Murphy, M. R. (1980) Sexual preferences of male hamsters: importance of preweaning and adult experience, vaginal secretion, and olfactory or vomeronasal sensation. Behav. Neural Biol. 30, 323–340.

Nyby, J. and Whitney, G. (1980) Experience affects behavioral responses to sex odors. In: D. Muller-Schwarze and R. M. Silverstein (Ed.), *Chemical Signals in Vertebrates and Aquatic Invertebrates*. Plenum Press, New York, pp. 173–190.

Pedersen, P. E. and Blass, E. M. (1982) Prenatal and postnatal determinants of the 1st suckling episode in albino rats. Dev. Psychobiol. 15, 349–355.

Petrulis, A. and Johnston, R. E. (1999) Lesions centered on the medial amygdala impair scent-marking and sex-odor recognition but spare discrimination of individual odors in female golden hamsters. Behav. Neurosci. 113, 345–357.

Pfaus, J. G., Kippin, T. E. and Centeno, S. (2001) Conditioning and sexual behavior: a review. Horm. Behav. 40, 291–321.

Porter, R. H. and Etscorn, F. (1974) Olfactory imprinting resulting from brief exposure in Acomys cahirinus. Nature 250, 732–733.

Quadagno, D. M. and Banks, E. M. (1970) The effect of reciprocal cross fostering on the behaviour of two sepcies of rodents, Mus musculus and Baiomys taylori ater. Anim. Behav. 18, 379–390.

Rangel, S. and Leon, M. (1995) Early odor preference training increases olfactory bulb norepinephrine. Brain Res. Dev. Brain Res. 85, 187–191.

Stickrod, G., Kimble, D. P. and Smotherman, W. P. (1982) In utero taste/odor aversion conditioning in the rat. Physiol. Behav. 28, 5–7.

Sullivan, R. M., Hofer, M. A. and Brake, S. C. (1986) Olfactory-guided orientation in neonatal rats is enhanced by a conditioned change in behavioral state. Dev. Psychobiol. 19, 615–623.

Sullivan, R. M. and Leon, M. (1986) Early olfactory learning induces an enhanced olfactory bulb response in young rats. Brain Res. 392, 278–282.

Sullivan, R. M., Wilson, D. A. and Leon, M. (1989) Norepinephrine and learning-induced plasticity in infant rat olfactory system. J. Neurosci. 9, 3998–4006.

Surov, A. V., Solovieva, A. V. and Minaev, A. N. (2001) The olfatory sexual preferences of golden hamster (Mescricetus auratus): the effects of early social and sexual experience. In: A. Marchlewska-Koj, J. L. Lepri and D. Muller-Schwarze (Ed.), *Chemical Signals in Vertebrates*. Kluwer Academic/Plenum Publishers, New York, pp.

Swann, Rahaman, F., Bijak, T. and Fiber, J. (2001) The main olfactory system mediates pheromone-induced fos expression in the extended amygdala and preoptic area of the male Syrian hamster. Neuroscience 105, 695–706.

Vasilieva, N. Y. (1994) Social cues influencing reproductive characteristics in Golden hamsters: the role of male flank gland secretion, vaginal discharge and litter composition. Advances in the Biosciences 93, 317–323.

Wilson, D. A. and Sullivan, R. M. (1994) Neurobiology of associative learning in the neonate: early olfactory learning. Behav. Neural Biol. 61, 1–18.

Woo, C. C., Coopersmith, R. and Leon, M. (1987) Localized changes in olfactory bulb morphology associated with early olfactory learning. J. Comp. Neurol. 263, 113–125.

Chapter 25
Have Sexual Pheromones Their Own Reward System in the Brain of Female Mice?

Fernando Martínez-García, Carmen Agustín-Pavón, Jose Martínez-Hernández, Joana Martínez-Ricós, Jose Moncho-Bogani, Amparo Novejarque and Enrique Lanuza

Abstract Even in rodents, there is no clear evidence of the existence of sexual pheromones mediating instinctive intersexual attraction. In this review we discuss previous results of our group indicating that female mice reared in the absence of male-derived chemosignals are 'attracted' by some components of male-soiled bedding, presumably detected by the vomeronasal organ. In contrast, male odors (olfactory stimuli) only acquire attractiveness by means of their association with the innately 'attractive' vomeronasal-detected pheromones. These 'attractive' male pheromones are rewarding to adult females, since they induce conditioned preference for a place where they are repeatedly presented to the females. Pheromone reward seems independent of the dopaminergic neurotransmission in the tegmento-striatal pathway, and uses mechanisms and circuits apparently different to those of other natural reinforcers.

25.1 Introduction: the Amygdala and Chemical Signals

In vertebrates, chemical signals are detected by the vomeronasal (VN) organ and olfactory epithelium, which give rise to parallel pathways that converge in the amygdala, where the secondary VN centers and important secondary olfactory nuclei are located (Shipley, Ennis and Puche 2004). Therefore, the amygdala is probably involved in sensory processing of chemical signals and, by means of its projections to key forebrain centers, might mediate adaptive responses to them. Functional studies of the amygdala suggest that it is implicated in the expression and acquisition of fear (LeDoux 2000) and of reward-related behaviors (Baxter and Murray 2002). In contrast, the chemosensory function of the amygdala and, more specifically, its role in detecting and responding to chemical signals from conspecific or allospecific individuals, has been largely neglected (but see Meredith, this volume). Our research

Fernando Martínez-García
Univ. València, Dept. Biologia Funcional
Fernando.mtnez-garcia@uv.es

J.L. Hurst et al., *Chemical Signals in Vertebrates 11.*
© Springer 2008

is aimed at testing the hypothesis that the amygdala mediates emotional responses to VN and olfactory stimuli, which include the attraction (a reward-related response) to sexual pheromones and emotional tagging of odors associated with them.

25.2 Sexual Pheromones Mediate Intersexual Attraction in Mice

Pheromones involved in intersexual attraction are usually called sexual pheromones. According to the classical definition by Karlson and Luscher (1959), pheromones are secreted substances that elicit *stereotyped* responses in conspecifics. Therefore, to be considered a sexual pheromone, a secreted substance should elicit intersexual attraction without the need for previous chemical or sexual experience. This is important since some reactions, even neuroendocrine, to chemical signals might result from a learning process. For instance, estrous induction by ram odors only occurs in sexually experienced ewes (Gelez, Archer, Chesneau, Campan and Fabre-Nys 2004). Although these odors constitute important chemical signals, they cannot be considered pheromones.

Against this background, we investigated whether intersexual attraction in mice is mediated by sexual pheromones, by checking if adult animals reared in the absence of chemosignals from adults of the other gender showed attraction for these chemicals. In mammals, this can only be done with females, since males have early experience with an adult female conspecific, their mother. Therefore, pregnant females were located in a room without males, and 19 days after delivery, males were removed and their sisters reared in a clean room until adulthood. These 'chemically naïve' adult females were used in two-choice tests to check if they were 'attracted' by the chemicals contained in male-soiled bedding. Our first experiments showed that chemically naïve female mice preferred male- to female-soiled bedding (Moncho-Bogani, Lanuza, Hernández, Novejarque and Martínez-García 2002). Females also show innate preference for male chemosignals in intact vs castrated male-soiled bedding preference tests (Fig. 25.1, left; see Martínez-Ricós, Agustín-Pavón, Lanuza and Martínez-García 2007). Although male chemical signals are new to chemically naïve females, their preference for male-soiled bedding is not due to novelty. Thus, females show a similar preference for male-soiled bedding at the beginning and at the end of the test (Moncho-Bogani et al. 2002) and preference persists through successive tests (Fig. 25.1 shows data for the first and fourth tests).

25.3 Mice Sexual Pheromones: Olfactory or Vomeronasal?

Whereas the olfactory epithelium only detects airborne volatiles, Wysocki, Wellington and Beauchamp (1980) demonstrated that the VN organ detects involatile chemicals, which are actively pumped into it (vomeronasal pumping; Meredith, Marques, O'Connell and Stern, 1980). Although *in vitro* some volatiles are able

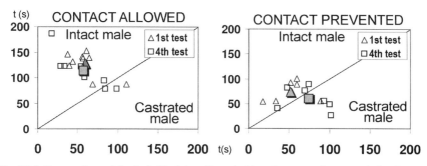

Fig. 25.1 Scatter plots of the individual (small symbols) and average (large symbols) time that female mice spent investigating intact vs castrated male-soiled bedding during 5 min preference tests. The diagonal line indicates no preference for either bedding. When contact with the bedding is allowed (left) females significantly prefer the bedding of intact males. If contact with the bedding is prevented but females can detect airborne male chemosignals (right), no preference for either bedding is observed (see Martínez-Ricós et al. 2007)

to activate VN neurons (Leinders-Zufall, Lane, Puche, Ma, Novotny, Shipley, Zufall 2000) through specific VN receptors (Boschat, Pelofi, Randin, Roppolo, Luscher, Broillet and Rodriguez 2002), for unknown reasons during exploratory behavior the mitral cells of the accessory olfactory bulb (the target of VN axons) are activated only when the nose of the subject contacts the source of stimuli (Luo, Fee and Katz, 2003).

Therefore, we tested whether contact with the source of stimulus was needed for the females to be attracted by male-soiled bedding. We used preference tests similar to the ones described above, but a perforated platform separated the females from the bedding some 3-4 cm. In these conditions, the chemically naïve females must have detected the volatiles that emanated from the bedding, but they did not prefer intact male- to castrated male- (Fig. 25.1, right) or to female-soiled bedding (Fig. 25.2, right). This further supports the view that the preferential chemoinvestigation of females towards male-soiled bedding is due to something else than novelty (male-derived volatiles are new to chemically naïve females). Moreover, since contact is required for the females to detect the attractive male sexual pheromone, the VN organ is very likely involved in its detection. In contrast, male odorants (stimuli detected by the olfactory epithelium) are not primarily attractive to females.

The expression of c-fos in the brain of females after chemoinvestigation of male-soiled bedding during two-choice preference tests (Moncho-Bogani, Martínez-García, Novejarque and Lanuza 2005) strongly supports this view. When contact with the bedding was allowed, the AOB, medial amygdala, posteromedial cortical amygdala and posteromedial bed nucleus of the stria terminalis were significantly activated as compared to controls (clean vs clean bedding). In contrast, olfactory (main olfactory bulb and anterior cortical amygdala), but not VN centers, were activated in the group of animals that explored only the airborne chemicals emanating from the bedding soiled by males.

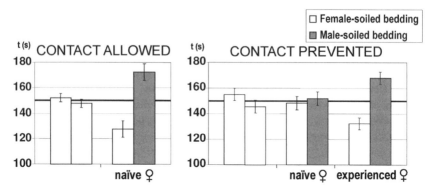

Fig. 25.2 Bar histograms showing the time (mean ± SEM) that females spent investigating the two compartments of a test cage that contained bedding soiled by males or females. Females significantly preferred male- to female-soiled bedding if contact with bedding was allowed (left). As in Fig. 25.1, chemically naïve females showed no preference for the airborne chemicals emanating from either bedding (right), but females that had contact experience with male-soiled bedding (experienced ♀) significantly preferred male-derived airborne chemosignals

25.4 Experience and the Attraction to Male-Derived Odorants

Beauchamp and collaborators reported that male guinea pigs whose VN organs had been removed showed an extinction-like decrease in chemoinvestigation of female urine (Beauchamp, Martin, Wysocki and Wellington 1982; Beauchamp, Wysocki and Wellington 1985). Thus, in the absence of a functional VN organ, the olfactory input temporarily sustains the ability to respond to female urine odors, but this olfactory-guided response requires the reinforcing input through the VN system. In other words, the VN-dependent response to female urine is unconditioned and the response to female odors is conditioned.

We tested whether this might also apply to the intersexual attraction in mice by comparing the chemoinvestigation of female mice to male-derived odorants prior to (chemically naïve females) and after repeated exposure (contact allowed) to male-soiled bedding (Moncho-Bogani et al. 2002, 2005). This exposure resulted in a Pavlovian association between the VN-dependent, attractive pheromone (unconditioned stimulus) and the primarily unattractive odorants, the latter becoming conditioned attractors for females (Fig. 25.2, right). Therefore, not only the sexual experience, but also the chemosensory history of the animals should be considered to identify sexual pheromones.

25.5 Male Sexual Pheromones are Rewarding to Female Mice

As suggested by Beauchamp's group, a female's unlearnt attraction to male-soiled bedding is probably reflecting the rewarding properties of sexual pheromones. To date, the only attempt to demonstrate this hypothesis rendered inconclusive

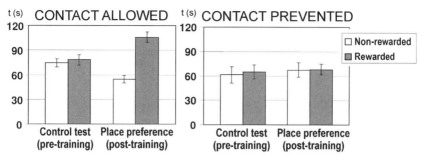

Fig. 25.3 Bar histogram showing the time spent by females in two particular locations of a test cage (mean ± SEM), both containing bedding soiled by castrated males in two tests, a control test prior to training, and a place-preference test, after four training sessions in which one of the locations was systematically rewarded with bedding soiled by intact males. Those animals that could contact the bedding during training (left) acquired a significant preference for the rewarded location. If contact was prevented (right), no place preference was acquired (see Martínez-Ricós et al. 2007)

results (Coppola and O'onnell 1988), probably due to an inappropriate experimental design. Therefore, we designed an experiment aimed at demonstrating the rewarding properties of male sexual pheromones to female mice. Since an operant paradigm is difficult to implement when the reward (pheromone) is not a consumable item, we decided to test the ability of male-soiled bedding to induce conditioned place preference (Tzschentke 1998). Therefore, we tried to demonstrate this hypothesis by assessing the ability of male-soiled bedding to induce conditioned place preference in females, using the so-called operant place conditioning (Crowder and Hutto 1992). Briefly, animals were introduced in a cage in which they could freely visit the reinforced location where they could obtain the putative reinforcer contained in male-soiled bedding and a second location, the non-reinforced one, where bedding soiled by castrated males was present (note that this looks like a regular two-choice test). After four 10-min (daily) training sessions in this situation, we tested the preference of the females for the place where the male-soiled bedding was previously available, by means of a two-choice test in which both locations (reinforced and non-reinforced) contained the same kind of bedding (soiled by castrated males). By comparing the results of this test with those of a similar pre-training control test, we concluded that conditioning occurred when contact with the bedding was allowed during training, but not when it was prevented. Therefore, the VN-detected sexual pheromone contained in male-soiled bedding is rewarding to female mice (Fig. 25.3; Martínez-Ricós et al. 2007).

25.6 Neural Basis of Pheromone-Induced Reward

Like other natural reinforcers (water, food, sex -somatosensory genital stimulation, and sweet taste), sexual pheromones should stimulate the neural circuits responsible for reward signaling. Current ideas on the neuroanatomy of reward are based on the pioneer work of cerebral autostimulation by Olds and Milner (1954), who observed that rats lever-pressed at huge rates to obtain electrical (auto)stimulation of the medial forebrain bundle. Three decades of intense pharmacological and behavioral studies lead to the 'dopamine hypothesis of reward' (Wise and Rompre 1989), according to which natural reinforcers and addictive drugs are rewarding because they elicit dopamine (DA) release in the nucleus accumbens, by activating the cells of the ventral tegmental area (VTA) that give rise to the DAergic tegmento-striatal pathway (TSP) (Spanagel and Weiss 1999). Since this hypothesis has been seriously challenged in the last decade (see Berridge and Robinson 1998; Wise 2004), the identification of sexual pheromones as new natural reinforcers constitutes a good opportunity to further explore the role of DA in reward.

The few available data on the afferent connections of the VTA in rats (Phillipson 1979; Geisler and Zahm 2005) and our unpublished data in mice indicate that some VN structures (bed nucleus of the stria terminalis and anterior medial amygdala) project weakly to the VTA. In contrast, the VN amygdala shows massive direct and indirect (through the basolateral amygdala) projections to the ventral striatum (Wright, Beijer and Groenewegen 1996; Novejarque, Lanuza and Martínez-Garcia 2005). This raises the possibility that pheromone reward signaling would be mediated by amygdalo-striatal projections bypassing the VTA. In fact, as suggested by c-fos expression (see Moncho-Bogani et al. 2005). Therefore, we decided to test the role of DA and of the DAergic TSP pathway in the reward induced by sexual pheromones by means of two kinds of experiments.

25.6.1 Effects of DAergic Drugs on 'Pheromone Seeking'

If DA were the causal factor of pheromone reward, the attraction to pheromones, DAergic drugs would decrease (antagonists) or increase (agonists) pheromone seeking in female mice. We have tested this possibility by analyzing the effects of systemic injections of DA drugs on the preferential chemoinvestigation of male-soiled bedding of chemically naïve females (Agustín-Pavón, Martínez-Ricós, Martínez-García and Lanuza 2007). As summarized in Table 25.1, the highest doses of D1 (SCH 23390) or D2 (Sulpiride) antagonists that did not impair locomotion nor significantly decrease exploratory behavior, rendered a preference for male-soiled bedding similar to saline-treated animals. In contrast, low doses (0.5 mg/kg) of a non-specific indirect DA agonist, amphetamine, completely abolished the preference for male chemosignals, without affecting locomotor activity or general exploratory behavior. This dose of amphetamine did not impair (Doty and Ferguson-Segall 1989) but even increased the olfactory function of the females, as revealed by habituation-dishabituation tests. Thus, the amphetamine-induced lack of preference for male-soiled bedding observed in female mice seems not to be a motor or a sensory effect of the drug, but reflects a decrease in the motivation of females to explore male pheromones.

Table 25.1 Effects of dopaminergic drugs on preference for male-soiled bedding.

		Treatment (mg / kg)	Locomotion	Exploratory Behavior	Preference Male vs Female	Olfactory function
Antagonists	D1	SCH 23390 (0.05)	=	=	YES	
	D2	Sulpiride (20)	=	=	YES	
Agonists	D1 + D2	Amph (0.5)	=	=	NO	↑
	D1	SKF 38393 (20)	↓	=	NO	↓
	D1	Quinpirole (0.2)	↓	=	YES	↓

Administration of D1 (SKF 38393) or D2 (quinpirole) agonists did not reproduce these results. Although at the doses used, both drugs significantly reduced the
locomotion of the females, quinpirole-treated animals displayed a clear preference
for male-soiled bedding whereas SKF treatments (D1 agonist) suppressed this preference. This differential effect of both drugs suggests that D1 receptors might be
responsible of the effects of amphetamine on pheromone reward.

This pharmacological approach suggests that DA is not the causal factor of
pheromone reward, since DA agonists do not enhance but abolish the motivation of
females to explore male-soiled bedding, whereas DA antagonists, at the doses used,
had no obvious effect on this behavior. Our finding that dopamine, probably acting
via D1 receptors, reduces pheromone reward, clearly contradicts the predictions of
the DA hypothesis of reward but supports modern views on the role of DA in reward
(Berridge and Robinson, 1998).

25.6.2 Pheromone Reward After Lesions of the Tegmento-Striatal Pathway

We also studied the role of the DAergic TSP in pheromone-induced reward by analyzing the effects of specific lesions of the VTA DAergic cells on the preferential
chemoinvestigation of male-soiled bedding by chemically naïve females (Martínez-
Hernández et al. 2006). Bilateral injections of a specific neurotoxic of DAergic cells
(6-hydroxydopamine) in the VTA of female mice resulted in a significant reduction in the number of dopaminergic cells in the VTA and, consequently, in the
DAergic innervation of the ventral striatum. Lesions were small enough to have
no significant effect on locomotor activity. After recovery, we tested the performance of these mice in two different appetitive behaviors, namely consumption of
sucrose-sweetened water (as compared to water alone) and the chemoinvestigation
of male-soiled bedding (as compared to clean bedding). Lesioned females showed a
reduced preference for sucrose. In fact, when considering both lesioned and sham-
operated animals, there was a moderate but highly significant statistical correlation
of the number of DAergic cells in the VTA and the preferential consumption of
sucrose in a 48 h period ($r = 0.547$, $p = 0.008$). In contrast, lesioned and sham-
operated females showed a similar preference for male-soiled bedding, and there

was no correlation of this preference with the number of DAergic cells in the VTA (r = 0.308, p = 0.163).

Therefore, in contrast to appetitive behaviors directed to 'classical' natural reinforcers (sweet taste or postingestive effects of sucrose), pheromone reward is independent of the DAergic TSP. Instead, pheromonal VN-detected stimuli might reach the reward circuitry of the brain by using direct and indirect (through the basolateral amygdala) pathways linking the VN amygdala with the ventral striatum.

25.7 The Vomeronasal-Olfactory Tandem

Taken together, these data suggest that male mice produce a testosterone-dependent pheromone that constitutes a VN-detected rewarding stimulus for females, thus mediating intersexual attraction in mice. Male odorants are not intrinsically attractive but acquire conditioned attractive properties after their association with the VN-detected rewarding sexual pheromone. This associative learning probably takes place in the basolateral amygdala (Moncho-Bogani et al. 2005), a hypothesis consistent with previously described functions of the basolateral amygdala in emotional learning. For instance, Schoenbaum, Chiba and Gallagher (1999) showed that cells in the basolateral amygdala encode the motivational significance of odors, after their association with sweet or bitter tasting fluids. Similarly, association of odors with VN-detected, innately rewarding sexual pheromones would change the motivational value of odorants (Moncho-Bogani et al. 2002), which would allow anticipatory (long-distance) responses, such as trailing the source of sexual pheromone, thus facilitating sexual encounter.

The VN system seems critical for intraspecific communication. Vomeronasal cells have ultrasensitive, highly specific receptors (Leinders-Zufall et al. 2000), which are appropriate to mediate innate responses to a limited repertoire of chemical signals, although the detection of these chemicals by the VN organ requires close proximity (Luo et al. 2003). In contrast, the olfactory system detects a myriad of airborne substances from a long distance and even allows animals to track their source. However, since olfactory receptors are generalist, encoding the identity of specific chemicals is too complex to be useful for innate responses (although exceptions exist; see Coureaud, this volume). The Pavlovian association of olfactory and vomeronasal stimuli takes advantage of the best of each system and overcomes their limitations.

Experimental data suggest that VN stimuli might also play a relevant role in prey-predator interactions by mediating affective responses to prey or predator chemical cues. For instance, one of the preferred prey for the snake *Thamnophis sirtalis* is earthworms. Halpern (1988) demonstrated that earthworm wash constitutes a VN stimulus that is rewarding for these snakes. On the other hand, it has been shown that rats display defensive reactions to a collar that has been worn by a cat, even if they have no previous experience with cats. For these defensive behavioral responses to occur, direct contact with the collar is needed (Dielenberg and McGregor 2001).

This suggests a role of the VN organ in detection of the fear-inducing cat's chemical cue, a hypothesis clearly supported by data on c-fos expression (Dielenberg and McGregor 2001).

Therefore, olfactory-vomeronasal associative learning might constitute a powerful tool for allowing long-distance, anticipatory responses to the presence of prey, predators and mates, elicited by conditioned olfactory stimuli, thus increasing survival and reproductive success. This might explain the remarkable evolutionary stability of the structure and connections of the amygdala in tetrapodian vertebrates (Martínez-García, Novejarque and Lanuza 2007).

Acknowledgments Supported by the Spanish MEC-FEDER (BFU2004-04272), the Valencian Government (Conselleria d'Empresa, Universitats i Ciència, ACOMP06/258) and the J.C. Castilla-La Mancha-FEDER (PAC-05-007-2).

References

Agustín-Pavón, C., Martínez-Ricós, J., Martínez-García, F. and Lanuza, E. (2007) The role of dopamine in the innate attraction towards male pheromones displayed by female mice. Behav Neurosci., in press.

Baxter, M.G. and Murray, EA. (2002) The amygdala and reward. Nat. Rev. Neurosci. 3, 563–573.

Beauchamp, G.K., Martin, I.G., Wysocki, C.J. and Wellington, J.L. (1982) Chemoinvestigatory and sexual behavior of male guinea pigs following vomeronasal organ removal. Physiol. Behav. 29, 329–236.

Beauchamp, G.K., Wysocki, C.J. and Wellington JL. (1985) Extinction of response to urine odor as a consequence of vomeronasal organ removal in male guinea pigs. Behav. Neurosci. 99, 950–955.

Berridge, K.C. and Robinson, T.E. (1998) What is the role of dopamine in reward: hedonic impact, reward learning, or incentive salience? Brain Res. Rev. 28, 309–69.

Boschat, C., Pelofi, C., Randin, O., Roppolo, D., Luscher, C., Broillet, M.C. and Rodriguez, I. (2002) Pheromone detection mediated by a V1r vomeronasal receptor. Nat. Neurosci. 5, 1261–1262.

Coppola, D.M. and O'Connell, R.J. (1988) Are pheromones their own reward? Physiol. Behav. 44, 811–816.

Crowder, W.F. and Hutto, C.W.Jr. (1992) Operant place conditioning measures examined using morphine reinforcement. Pharmacol Biochem Behav. 41, 825–835.

Dielenberg, R.A. and McGregor, I.S. (2001) Defensive behavior in rats towards predatory odors: a review. Neurosci. Biobehav. Rev. 25, 597–609.

Dielenberg, R.A., Hunt, G.E. and McGregor, I.S. (2001) "When a rat smells a cat": the distribution of Fos immunoreactivity in rat brain following exposure to a predatory odor. Neuroscience. 104,1085–1097.

Doty, R.L. and Ferguson-Segall, M. (1989) Influence of adult castration on the olfactory sensitivity of the male rat: a signal detection analysis. Behav. Neurosci. 103, 691–694.

Geisler, S. and Zahm, D.S. (2005) Afferents of the ventral tegmental area in the rat-anatomical substratum for integrative functions. J Comp Neurol. 490, 270–294.

Gelez, H., Archer, E., Chesneau, D., Campan, R. and Fabre-Nys, C. (2004) Importance of learning in the response of ewes to male odor. Chem. Senses. 29, 555–563.

Halpern, M. (1988) Vomeronasal system functions: Role in mediating the reinforcing properties of chemical stimuli. In: W.K. Schwerdtfeger and W.J.A.J. Smeets (Eds.), *The Forebrain of Reptiles. Current Concepts of Structure and Function.* Karger, Basel, pp. 142–150.

Karlson, P. and Luscher, M. (1959) Pheromones: a new term for a class of biologically active substances. Nature 183, 55–56.

Ledoux, J.E. (2000) Emotion circuits in the brain. Annu. Rev. Neurosci. 23, 155–84.

Leinders-Zufall, T., Lane, A.P., Puche, A.C., Ma, W., Novotny, M.V.; Shipley, M.T.; Zufall, F. (2000) Ultrasensitive pheromone detection by mammalian vomeronasal neurons. Nature. 405, 792–796.

Luo, A.H., Cannon, E.H., Wekesa, K.S., Lyman, R.F., Vandenbergh, J.G. and Anholt, R.R. (2002) Impaired olfactory behavior in mice deficient in the alpha subunit of G(o). Brain Res. 941, 62–71.

Luo, M., Fee, M.S. and Katz, L.C. (2003) Encoding pheromonal signals in the accessory olfactory bulb of behaving mice. Science. 299, 1196–1201.

Martínez-García, F., Novejarque, A. and Lanuza, E. (2007). Evolution of the amygala in vertebrates, In: J.H. Kaas (Ed.), *Evolution of the Nervous Systems*, Vol. 2. Academic Press, Oxford, pp. 255–334.

Martínez-Hernández, J., Lanuza, E., and Martínez-García, F. (2006) Selective dopaminergic lesions of the ventral tegmental area impair preference for sucrose but not for male sexual pheromones in female mice. Eur. J. Neurosci. 24, 885–893.

Martínez-Ricós, J., Agustín-Pavón, C., Lanuza, E., and Martínez-García, F. (2007) Intraspecific comunication through chemical signals in female mice: Reinforcing properties of involatile male sexual pheromones. Chem. Senses. 32, 139–148.

Meredith, M., Marques, D.M., O'Connell, R.O. and Stern, F.L. (1980) Vomeronasal pump: significance for male hamster sexual behavior. Science. 207, 1224–1226.

Moncho-Bogani, J., Lanuza, E., Hernández, A., Novejarque, A. and Martínez-García, F. (2002) Attractive properties of sexual pheromones in mice. Innate or learned? Physiol. Behav. 77, 167–176.

Moncho-Bogani, J., Martínez-García, F., Novejarque, A. and Lanuza, E. (2005) Attraction to sexual pheromones and associated odorants in female mice involves activation of the reward system and basolateral amygdala. Eur. J. Neurosci. 21, 2186–2198.

Olds, J. and Milner, P. (1954) Positive reinforcement produced by electrical stimulation of septal area and other regions of rat brain. J. Comp. Physiol. Psychol. 47, 419–427.

Phillipson, O.T. (1979) Afferent projections to the ventral tegmental area of tsai and interfascicular nucleus: A horseradish peroxidase study in the rat. J. Comp. Neurol. 187, 117–144.

Schoenbaum, G., Chiba, A.A. and Gallagher, M. (1999) Neural encoding in orbitofrontal cortex and basolateral amygdala during olfactory discrimination learning. J. Neurosci. 19, 1876–1884.

Shipley, M.T., Ennis, M. and Puche, A. (2004) Olfactory system. In: G. Paxinos (Ed.), *The Rat Nervous System*. Academic Press, San Diego, pp. 923–964.

Spanagel, R. and Weiss, F. (1999) The dopamine hypothesis of reward: past and current status. Trends Neurosci. 22, 521–527.

Tzschentke, T.M. (1998) Measuring reward with the conditioned place preference paradigm: a comprehensive review of drug effects, recent progress and new issues. Prog. Neurobiol. 56, 613–672.

Wise, R.A. (2004) Dopamine, learning and motivation. Nat. Rev. Neurosci. 5, 483–494.

Wise, R.A. and Rompre, P.P. (1989) Brain dopamine and reward. Annu. Rev. Psychol. 40, 191–225.

Wright, C.I., Beijer, A.V.J. and Groenewegen, H.J. (1996) Basal amygdaloid complex afferents to the rat nucleus accumbens are compartmentally organized. J. Neurosci. 16, 1877–1893.

Wysocki, C.J., Wellington, J.L. and Beauchamp, G.K. (1980) Access of urinary nonvolatiles to the mammalian vomeronasal organ. Science 207, 781–783.

Chapter 26
The Effect of Familiarity on Mate Choice

Sarah A. Cheetham, Michael D. Thom, Robert J. Beynon and Jane L. Hurst

Abstract The ability to recognize familiar conspecifics appears to be widespread among vertebrates and influences a variety of behavioural interactions including mate selection. Female choice of males has been shown to vary according to male familiarity, but interestingly in some species this favours familiar males, while in others unfamiliar males are preferred. Preference for unfamiliar partners might result from the attempt to minimise inbreeding costs by avoiding mating with individuals encountered during development, or with those sharing relatedness cues. Conspecifics that are familiar through prior mating experience might be avoided in species that benefit from a promiscuous mating system, again resulting in preference for unfamiliar mates. Conversely, familiar mates may be favoured in monogamous species where formation of a pair bond is important for parental investment, and when familiarity provides an opportunity for females to assess the quality and compatibility of potential mates. Thus different types of familiarity may have differing effects on mate choice, with the direction of preference being determined by other aspects of life history, such as the likelihood of inbreeding, the importance of polyandry, and the role of social dominance and territoriality in reproductive success.

26.1 Introduction

The ability to recognize familiar conspecifics has been found in many vertebrate species, and this capacity to remember previously-encountered individuals is likely to modify subsequent behavioural interactions. For example, familiarity has long been thought to play an important role in mediating competitive behaviour, allowing animals to learn about the abilities of their competitors and to establish recognised territory borders so that direct conflicts can be reduced (Gosling 1990). Animals may also gain advantages from prior familiarity with individuals when selecting mates. There is some evidence that females can recognize familiar males using

Sarah A. Cheetham
Mammalian Behaviour and Evolution Group, University of Liverpool
sacheet@liv.ac.uk

a variety of signals, including those in the acoustic (ungulates: Reby, Hewison, Izquierdo and Pepin 2001), visual (fish: Zajitschek, Evans and Brooks 2006) and olfactory (rodents: Hurst 1990) modalities. Females sometimes use the ability to discriminate familiar from unfamiliar males to adjust their mating preferences. Intriguingly, however, the direction of this effect is not consistent: while some authors have found that females are more likely to mate with familiar males (e.g. Fisher, Swaisgood and Fitch-Snyder 2003), others report female preference for unfamiliar males (e.g. Kelley, Graves and Magurran 1999). Here we consider the ways in which familiarity might affect mate choice, and how this might be influenced by social context.

26.1.1 What is Familiarity, and How is it Recognized?

Familiarisation through direct association is perhaps the most obvious means by which animals can discriminate familiar from unfamiliar conspecifics (Porter 1988). This describes the process whereby familiarity is established through direct association with an animal or its odours, mediated by the formation of a memory trace or template of the animal (Tang-Martinez 2001) which can subsequently be matched to the current phenotype of the familiar individual. Template formation can be facilitated through imprinting, habituation and associative learning (Mateo 2004), with learning at different stages of life potentially leading to divergent outcomes.

The process of becoming familiar with an individual can occur either through short-term or long-term contact, and both can influence mate choice. Long-term social contact can occur through early social experience, e.g. between those sharing the same nest, and appears to affect adult sexual preferences (discussed below). Short term familiarity can occur during mate assessment, or when an individual's scent is encountered in the environment and provides information about the owner before animals meet. To recognise and discriminate between individuals on the basis of their odours requires, at minimum, the ability to determine whether a new odour matches (familiarity) or does not match (novelty) a previously encountered odour. Such recognition by association involves the animal learning individually distinctive cues to be able to recognise specific individuals. The term "familiarity" is therefore sometimes used to describe individual recognition, but it is important to recognize that discrimination of familiarity is insufficient evidence for the ability to identify individuals. Animals may respond to familiarity without having the ability to identify the individual scent owner from a range of familiar conspecifics. When animals first meet an unfamiliar scent, they usually spend some time investigating and extracting information from the novel odour, but then reduce investigation on future encounters because most of the information contained in the cue has already been obtained. This reliable response to novelty and familiarity is the basis of the widely-used habituation—dishabituation test (Gregg and Thiessen 1981). More prolonged investigation when animals encounter a new scent indicates that animals have recognised some novelty in the new scent that requires further investigation, i.e. they can discriminate a difference between a familiar and unfamiliar scent (or animals), but this tells us nothing about the information, if any, gained on investigating the source. The novel scent could be from a previously familiar individual whose scent has changed (e.g. due to a change of status) or has been forgotten, or from an unfamiliar individual. To understand whether animals recognise individuals or other information from familiar scents requires a test of their subsequent response indicating that they have gained specific information from investigating the scent (Thom and Hurst 2004).

26.2 Avoidance of Familiar Individuals as Mates

In many vertebrate species, familiarity effects on adult sexual preferences begin early in life. During development, offspring may imprint on the odours of their parents and siblings and use this information as adults when choosing a mating partner. While immature animals may prefer the odours of their opposite sex siblings (Kruczek 2007), in many species the adults of both sexes tend to reject close kin as mates to avoid the substantial fitness costs associated with inbreeding (Thom, Stockley, Beynon and Hurst this volume). This behaviour is frequently modulated through odour cues. For example, female bank voles prefer the novel odour of unrelated over related males (Kruczek and Golas 2003), regardless of whether they were reared in the presence of the related subject (Kruczek 2007); mature pine voles of both sexes are more attracted to the chemical cues of unrelated opposite-sex conspecifics (Solomon and Rumbaugh 1997) and when given a choice between a sibling and non-sibling male, female prairie voles choose the non-sibling's odour (Smale, Pedersen, Block and Zucker 1990). This behaviour can be mediated by familial imprinting, as demonstrated by cross-fostering studies. For example, inbred laboratory strain male mice reverse their mate preference following cross-fostering (Yamazaki, Beauchamp, Kupniewski, Bard, Thomas and Boyse 1988). Familial imprinting can also influence the reproductive priming effects of scents. While puberty in juvenile female rodents can be accelerated by exposure to the scents of unfamiliar adult males, they do not show this response to the scents of familiar males with which they are reared (e.g. Berger, Negus and Day 1997). Penn and Potts (1998) have argued that familial imprinting may provide a more effective means of recognizing close relatives and avoid inbreeding than self-inspection alone because this mechanism allows animals to recognise a greater range of close relatives than just those bearing their own genotype. While imprinting on the genotypes of other animals during rearing would allow recognition of a much greater range of genotypes than self-referent matching, it is also likely to mean greater recognition and rejection of potential mates that may not be close kin. Imprinting is likely to be error-prone in species where multiple paternity is common or offspring are reared communally by females, when self-referent matching may be a much more reliable mechanism (Mateo 2004). However there is currently insufficient evidence to determine whether self-referent phenotype matching of odour cues is used to avoid inbreeding.

In many species, female choice does not involve simply mating with a single preferred male and rejecting all alternatives; instead, females often mate with multiple males in a single reproductive event. The many explanations for polyandry include: direct benefits associated with avoiding the costs of resisting mating or acquisition of material resources (Andersson 1994), or indirect genetic benefits resulting from genetic diversification of litters or facilitation of post-copulatory choice mechanisms to overcome females' inability to correctly identify the most genetically superior or genetically compatible male through pre-copulatory mate choice (Jennions and Petrie 2000). Under all these hypotheses except the first, females should improve their fitness by mating with multiple males. Similar behaviour is expected in males, where advantages of multiple mating are more straightforward: there is generally a positive relationship between the number of females mated and male reproductive success (Andersson 1994). We might thus expect males in general, together with females of polyandrous species, to actively discriminate against previous mates in favour of seeking matings with novel partners. This behaviour is indeed widespread. In males, discrimination against previous partners underlies the well known Coolidge effect, in which sexually exhausted males are able to resume copulation upon introduction to a novel receptive female (Wilson, Kuehn and Beach 1963). The decline in male sexual performance during repeated copulation with the same female may allow sperm to be conserved for when additional females are encountered (Wedell, Gage and Parker 2002). There is now evidence for male discrimination against familiar or previously mated females in a number of species (e.g. Olsson and Shine 1998; Kelley et al. 1999; Tokarz 2006). As predicted from the putative benefits of polyandry, females of many species also discriminate against previous partners. In pseudoscorpions (*Cordylochernes scorpioides*) females are nonreceptive towards males from which they have already received sperm in the first few hours after mating (Zeh, Newcomer and Zeh 1998). Female guppies, which use visual signals to choose mates, tend to prefer novel males (Zajitschek et al. 2006) and discriminate against previous mates and those that look like previous mates (Eakley and Houde 2004). Odour cues may play an important role in male avoidance of previous mates through mechanisms such as the Coolidge effect (Johnston and Rasmussen 1984), but there is currently little definitive evidence for any involvement of odour cues in the equivalent female behaviour.

Females may avoid males that are familiar because they are previous mates, but they may also prefer novel males because of the genetic benefits of mating with locally rare genotypes (negative frequency dependent selection). In guppies (*Poecilia reticulata*), females familiarized (but not mated) with males of a particular colour morph are significantly more likely to mate subsequently with a male bearing a novel colour pattern than with a familiar colour-type male (Hughes, Du, Rodd and Reznick 1999). Similarly polyandrous female house wrens (*Troglodytes aedon*) are more likely to accept extra-pair copulations from males carrying locally rare alleles (Masters, Hicks, Johnson and Erb 2003). Negative frequency dependent selection, which might be based on preference for unfamiliar MHC odours, is widely thought to be a key mechanism underlying the maintenance of MHC heterozygosity in vertebrates (Jordan and Bruford 1998), although evidence for this remains controversial (Thom et al. this volume).

Pregnancy block, or the Bruce effect, is a phenomenon whereby recently mated females exposed to the odours of an unfamiliar non-sire male experience pregnancy failure and return to oestrus (Bruce 1959). Exposure to the sire male's scent fails to disrupt pregnancy and, if this coincides with exposure to unfamiliar male odour, has a protective effect and reduces the likelihood of pregnancy block occurring (Parkes and Bruce 1961; Thomas and Dominic 1987). This mechanism relies on a female memorising the odours of the mating male (Kaba, Rosser and Keverne 1989), and distinguishing these from the unfamiliar odours of novel males. A number of hypotheses have been put forward to explain why females may abort their current litter in order to remate with the unfamiliar male, for example, to increase paternal investment from the unfamiliar male, or because of the threat of infanticide from an unfamiliar male towards another male's offspring (see review by Becker and Hurst this volume). Assuming that the female remates with the unfamiliar male that induces pregnancy block, the Bruce effect would result in preference for the unfamiliar male as a mate over the familiar stud male. However, so far the effect has only been observed in artificial laboratory situations and, as yet, there is no evidence that the unfamiliar blocking male goes on to successfully mate with the female. An alternative explanation for the Bruce effect is that unfamiliar male scent in the nest signifies a situation where the likelihood of offspring survival is low and females use this to abort a pregnancy and avoid wasted investment in gestation and lactation without choosing to re-mate with the unfamiliar male (Becker and Hurst this volume). It thus remains to be proven whether the Bruce effect represents a form of post-copulatory mate choice where unfamiliar males gain an advantage.

26.3 Preference for Familiar Individuals as Mates

While mate choice for unfamiliar partners might be explained by the fitness benefits of multiple mating, inbreeding avoidance, and favouring rare genotypes, preference for familiar males is also sometimes observed. Why would females choose relatively familiar males as mates, given the benefits outlined above of avoiding such males?

In many species, males defend territories, with relatively more competitive or dominant males holding higher quality territories. Territory holders may be preferred by females, either because of the better physical resources they can offer females, or because their ability to acquire and defend a territory is associated with positive genetic traits which will be inherited by the choosing female's offspring, particularly their sons. This system is usually mediated through deposition of scent marks in the environment by territorial males, while non-territory holders typically do not mark or have qualitatively different scents. Intruders into a territory may leave competing scent marks, but these are rapidly countermarked by the resident male, which either directly overmarks the intruder's scent (hamsters: Johnston, Chiang and Tung 1994; meadow voles: Johnston, Sorokin and Ferkin 1997b) or deposits a fresher scent nearby (Rich and Hurst 1999). Scent marks and countermarks provide a continuous record of competitive challenges between conspecifics providing a reliable advertisement of an individual's ability to dominate or defend an area to other competitors and potential mates (Hurst, Beynon, Humphries, Malone, Nevison, Payne, Robertson and Veggerby 2001). Females use these marks to assess male quality: for example, female house mice prefer males owning exclusively marked territories over males whose territories also contain a few fresh countermarks from another unfamiliar competitor male (Rich and Hurst 1998). The continued presence of one individual's marks in a specific area indicates that the owner is a successful territory owner because it has not been displaced. Females will thus tend to become more familiar with the odours of locally dominant territory holders, and because such males are usually preferred mating partners, familiarity could act as a surrogate measure of male quality in mate choice. This preference for familiarity may also apply for non-territorial species where male dominance is associated with higher levels of general scent marking of the environment. Like territoriality, male dominance is associated with higher mating and reproductive success across a range of species (see reviews in: Dewsbury 1982; Ellis 1995), and dominant males are often (though not always) preferred by females (Wong and Candolin 2005).

In house mice, we have shown that females are consistently more attracted to a male after brief investigation of his scent marks than to an equivalent male whose scent is unfamiliar (Cheetham 2006). However, if prevented from mating with a familiar male over an extended period of association (reflecting lack of mating attempts by the familiar male), unmated female house mice show increased interest in unfamiliar males (Patris and Baudoin 1998). Nonetheless, the preference for familiar males is so strong in the closely related mound-builder mouse *Mus spicilegus* that they continue to prefer familiar males even when prevented from mating over an extended period of association, perhaps reflecting monogamous traits in this species (Patris and Baudoin 1998). Female mice (Rich and Hurst 1999), hamsters (Johnston, Sorokin and Ferkin 1997a), and voles (Johnston et al. 1997a) are also more attracted to the owner of scent countermarks or overmarks than to a male whose scent has been countermarked, consistent with the idea that they use scent marks to assess the relative dominance of males. Preference for the odours of familiar males has also been shown in a number of other scent-marking rodents, including prairie voles (Microtus ochrogaster: Newman and Halpin 1988) and lemmings *Dicrostonyx groenlandicus* (Huck and Banks 1979), as well as non-rodents such as the pygmy loris (Nycticebus pygmaeus: Fisher et al. 2003). In the golden hamster (*Mesocricetus auratus*) female sexual receptivity increases and aggression reduces toward familiar males, and females mated with familiar males produce larger litters (Tang-Martinez, Mueller and Taylor 1993) suggesting that females may even adjust their reproductive investment in relation to male quality. Preference for familiarity may not be a trivial effect, as it can produce powerful—if sometimes maladaptive—responses: for example, preexposure to the odours of parasitized male mice reverses females' usual preference for healthy males, causing them to prefer the parasitized, but now familiar, male (Kavaliers, Colwell, Braun and Choleris 2003).

Many examples of females exhibiting a preference for familiar males are linked to the benefits of mating with a dominant animal, but this behaviour would also be expected where the benefits of monogamy outweigh those associated with multiple mating. There is limited evidence in support of the idea that monogamous species prefer to mate with their familiar partner: along with the *Mus* examples discussed above, females of the monogamous common vole (*Microtus arvalis*) prefer their familiar partner to novel males (Ricankova, Sumbera and Sedlacek 2007), as do monogamous *M. ochrogaster* (Shapiro, Austin, Ward and Dewsbury 1986). Polygamous *M. montanus*, in contrast, either show no preference or prefer unfamiliar males (Shapiro et al. 1986).

We can identify a number of other hypothetical benefits arising from a preference for familiar males as mates. Assessing quality, health, and social status might require several encounters with a male or his signals, in which case females will tend to have better information about familiar than unfamiliar males. For example, female house mice are more attracted to the owner of a familiar scent than to a completely unfamiliar one, even when the familiar animal's scent has been countermarked by a competitor male (Cheetham 2006). This suggests that prior familiarity has an important impact on recognition and initial attraction to males. Furthermore, females might only be able to assess some aspects of male quality, such as parental ability, either by reproducing with the male or by observing the male's reproductive success. In subsequent mating bouts, sires offering higher levels of paternal care might be preferred over novel partners with unknown abilities. This could explain why in some birds with biparental care, extrapair copulations tend to be with familiar males (the female's initial social mate) rather than completely novel males (Slagsvold, Johnsen, Lampe and Lifjeld 2001; Walsh, Wilhelm, Cameron-Macmillan and Storey 2006). Prior reproductive success certainly appears to influence subsequent mate choice in birds: for example female canaries (*Serinus canaria*) show a preference for a previous mate when reproductive success was good (at least two chicks hatched), but do not prefer their previous mate when only a single chick resulted from the pairing (Beguin, Leboucher, Bruckert and Kreutzer 2006). Where mate assessment is difficult, females may need to copulate with a male to assess its quality (Hunter, Petrie, Otronen, Birkhead and Moller 1993), or they may use males' responses to copulation solicitation as an indication of male condition (e.g. Lens, VanDongen, VandenBroeck, VanBroeckhoven and Dhondt 1997). This could lead to preference for familiarity since previously-mated males are of known quality while the quality of the unfamiliar male remains unknown until copulation.

26.4 Conclusions

Animals that encounter some individuals more often than others theoretically have the ability to segregate conspecifics into categories of familiar and unfamiliar, although the extent to which they do so is likely to vary according to capacity and necessity. This simple basis of assessment is likely to influence a range of intraspecific interactions, but the effect on mate choice is particularly intriguing because of the apparently opposite effects of familiarity in different circumstances. The effects of familiarity during development on mate choice can be explained by its function as a simple surrogate of relatedness, which may be sufficiently effective in many species to ensure that inbreeding remains rare. Familiarity through mating during adulthood may also modulate the mating system, by allowing promiscuous species to increase fitness through avoidance of partners with which they have already mated, while serving the opposite purpose in monogamous species. Further, familiarity with an individual's sexual signals during adulthood may play an essential role in allowing females to assess the quality and compatibility of the signal owner, resulting in a preference for familiar individuals when the assessment indicates that they are suitable mates. Thus familiarity acquired during different life stages, and in differing social contexts, can have quite disparate but generally predictable effects on subsequent mate choice. Since the ability to distinguish familiarity probably exists in most vertebrates, this potentially widespread factor in mate choice deserves more extensive exploration to understand its specific effects.

References

Andersson, M. (1994) *Sexual selection*. Princeton: Princeton University Press.

Becker, S.D. and Hurst, J.L. (2007) Pregnancy block from a female perspective. In: J. Hurst, R. Beynon, C. Roberts, T. Wyatt (Eds.), *Chemical Signals in Vertebrates* 11. Springer Press, New York, pp. 127–136.

Beguin, N., Leboucher, G., Bruckert, L. and Kreutzer, M. (2006) Mate preferences in female canaries (*Serinus canaria*) within a breeding season. Acta Ethologica 9, 65–70.

Berger, P.J., Negus, N.C. and Day, M. (1997) Recognition of kin and avoidance of inbreeding in the montane vole, *Microtus montanus*. J. Mammal. 78, 1182–1186.

Bruce, H.M. (1959) Exteroceptive block to pregnancy in the mouse. Nature 184, 105–105.

Cheetham, S.A. (2006) Chemical communication in the house mouse: linking biochemistry and behaviour (PhD thesis). Liverpool: University of Liverpool.

Dewsbury, D.A. (1982) Dominance rank, copulatory behavior, and differential reproduction. Quart. Rev. Biol. 57, 135–159.

Eakley, A.L. and Houde, A.E. (2004) Possible role of female discrimination against 'redundant' males in the evolution of colour pattern polymorphism in guppies. Proc. R. Soc. B 271, S299–S301.

Ellis, L. (1995) Dominance and reproductive success among nonhuman animals: a cross-species comparison. Ethol. Sociobiol. 16, 257–333.

Fisher, H.S., Swaisgood, R.R. and Fitch-Snyder, H. (2003) Countermarking by pygmy lorises (*Nycticebus pygmaeus*): do females use odor cues to select mates with high competitive ability? Behav. Ecol. Sociobiol. 53, 123–130.

Gosling, L.M. (1990) Scent-marking by resource holders: alternative mechanisms for advertising the costs of competition. In: D.W. Macdonald, D. Muller-Schwarze, S.E. Natynczuk (Eds.), *Chemical Signals in Vertebrates* 5. Oxford University Press, Oxford, pp. 315–328.

Gregg, B. and Thiessen, D.D. (1981) A simple method of olfactory discrimination of urines for the Mongolian gerbil, *Meriones unguiculatus*. Physiol. Behav. 26, 1133–1136.

Huck, U.W. and Banks, E.M. (1979) Behavioral components of individual recognition in the collared lemming (*Dicrostonyx groenlandicus*). Behav. Ecol. Sociobiol. 6, 85–90.

Hughes, K.A., Du, L., Rodd, F.H. and Reznick, D.N. (1999) Familiarity leads to female mate preference for novel males in the guppy *Poecilia reticulata*. Anim. Behav. 58, 907–916.

Hunter, F.M., Petrie, M., Otronen, M., Birkhead, T. and Moller, A.P. (1993) Why do females copulate repeatedly with one male? Trends Ecol. Evol. 8, 21–26.

Hurst, J.L. (1990) Urine marking in populations of wild house mice *Mus domesticus* Rutty .III. Communication between the sexes. Anim. Behav. 40, 233–243.

Hurst, J.L., Beynon, R.J., Humphries, R.E., Malone, N., Nevison, C.M., Payne, C.E., Robertson, D.H.L. and Veggerby, C. (2001) Information in scent signals of status: the interface between behaviour and chemistry. In: A. Marchlewska-Koj, D. Muller-Schwarze, J. Lepri (Eds.), *Chemical Signals in Vertebrates,* Plenum Press, New York, pp. 43–50.

Jennions, M.D. and Petrie, M. (2000) Why do females mate multiply? A review of the genetic benefits. Biol. Rev. 75, 21–64.

Johnston, R.E., Chiang, G. and Tung, C. (1994) The information in scent over-marks of golden hamsters. Anim. Behav. 48, 323–330.

Johnston, R.E. and Rasmussen, K. (1984) Individual recognition of female hamsters by males–role of chemical cues and of the olfactory and vomeronasal systems. Physiol. Behav. 33, 95–104.

Johnston, R.E., Sorokin, E.S. and Ferkin, M.H. (1997a) Female voles discriminate males' over-marks and prefer top-scent males. Anim. Behav. 54, 679–690.

Johnston, R.E., Sorokin, E.S. and Ferkin, M.H. (1997b) Scent counter-marking by male meadow voles: females prefer the top-scent male. Ethology 103, 443–453.

Jordan, W.C. and Bruford, M.W. (1998) New perspectives on mate choice and the MHC. Heredity 81, 239–245.

Kaba, H., Rosser, A. and Keverne, B. (1989) Neural basis of olfactory memory in the context of pregnancy block. Neuroscience 32, 657–662.

Kavaliers, M., Colwell, D.D., Braun, W.J. and Choleris, E. (2003) Brief exposure to the odour of a parasitized male alters the subsequent mate odour responses of female mice. Anim. Behav. 65, 59–68.

Kelley, J.L., Graves, J.A. and Magurran, A.E. (1999) Familiarity breeds contempt in guppies. Nature 401, 661–662.

Kruczek, M. (2007) Recognition of kin in bank voles (*Clethrionomys glareolus*). Physiol. Behav. 90, 483–489.

Kruczek, M. and Golas, A. (2003) Behavioural development of conspecific odour preferences in bank voles, *Clethrionomys glareolus*. Behav. Process. 64, 31–39.

Lens, L., VanDongen, S., VandenBroeck, M., VanBroeckhoven, C. and Dhondt, A.A. (1997) Why female crested tits copulate repeatedly with the same partner: evidence for the mate assessment hypothesis. Behav. Ecol. 8, 87–91.

Masters, B.S., Hicks, B.G., Johnson, L.S. and Erb, L.A. (2003) Genotype and extra-pair paternity in the house wren: a rare-male effect? Proc. R. Soc. B 270, 1393–1397.

Mateo, J.M. (2004) Recognition systems and biological organization: the perception component of social recognition. Ann. Zool. Fenn. 41, 729–745.

Newman, K.S. and Halpin, Z.T. (1988) Individual odours and mate recognition in the prairie vole, *Microtus ochrogaster*. Anim. Behav. 36, 1779–1787.

Olsson, M. and Shine, R. (1998) Chemosensory mate recognition may facilitate prolonged mate guarding by male snow skinks, *Niveoscincus microlepidotus*. Behav. Ecol. Sociobiol. 43, 359–363.

Parkes, A.S. and Bruce, H.M. (1961) Olfactory stimuli in mammalian reproduction. Science 134, 1049–1054.

Patris, B. and Baudoin, C. (1998) Female sexual preferences differ in *Mus spicilegus* and *Mus musculus domesticus*: the role of familiarization and sexual experience. Anim. Behav. 56, 1465–1470.

Penn, D.J. and Potts, W.K. (1998) MHC-disassortative mating preferences reversed by cross-fostering. Proc. R. Soc. B 265, 1299–1306.

Porter, R.H. (1988) The ontogeny of sibling recognition in rodents: superfamily Muroidea. Behav. Genet. 18, 483–494.

Reby, D., Hewison, M., Izquierdo, M. and Pepin, D. (2001) Red deer (*Cervus elaphus*) hinds discriminate between the roars of their current harem-holder stag and those of neighbouring stags. Ethology 107, 951–959.

Ricankova, V., Sumbera, R. and Sedlacek, F. (2007) Familiarity and partner preferences in female common voles, *Microtus arvalis*. J. Ethol. 25, 95–98.

Rich, T.J. and Hurst, J.L. (1998) Scent marks as reliable signals of the competitive ability of mates. Anim. Behav. 56, 727–735.

Rich, T.J. and Hurst, J.L. (1999) The competing countermarks hypothesis: reliable assessment of competitive ability by potential mates. Anim. Behav. 58, 1027–1037.

Shapiro, L.E., Austin, D., Ward, S.E. and Dewsbury, D.A. (1986) Familiarity and female mate choic ein two species of voles (*Microtus ochrogaster* and *Microtus montanus)*. Anim. Behav. 34, 90–97.

Slagsvold, T., Johnsen, A., Lampe, H.M. and Lifjeld, J.T. (2001) Do female pied flycatchers seek extrapair copulations with familiar males? A test of the incomplete knowledge hypothesis. Behav. Ecol. 12, 412–418.

Smale, L., Pedersen, J.M., Block, M.L. and Zucker, I. (1990) Investigation of conspecific male odours by female prairie voles. Anim. Behav. 39, 768–774.

Solomon, N.G. and Rumbaugh, T. (1997) Odor preferences of weanling and mature male and female pine voles. J. Chem. Ecol. 23, 2133–2143.

Tang-Martinez, Z. (2001) The mechanisms of kin discrimination and the evolution of kin recognition in vertebrates: a critical re-evaluation. Behav. Proc. 53, 21–40.

Tang-Martinez, Z., Mueller, L.L. and Taylor, G.T. (1993) Individual odours and mating success in the golden hamster, *Misocricetus auratus*. Anim. Behav. 45, 1141–1151.

Thom, M.D. and Hurst, J.L. (2004) Individual recognition by scent. Ann. Zool. Fenn. 41, 765–787.

Thom, M.D., Stockley, P., Beynon, R.J. and Hurst, J.L. (2007) Scent, mate choice and genetic heterozygosity. In: J. Hurst, R. Beynon, C. Roberts, T. Wyatt (Eds.), *Chemical Signals in Vertebrates* 11. Springer Press, New York, pp. 269–279.

Thomas, K.J. and Dominic, C.J. (1987) Evaluation of the role of the stud male in preventing male-induced implantation failure (the Bruce effect) in laboratory mice. Anim. Behav. 35, 1257–1259.

Tokarz, R.R. (2006) Importance of prior physical contact with familiar females in the development of a male courtship and mating preference for unfamiliar females in the lizard *Anolis sagrei*. Herpetologica 62, 115–124.

Walsh, C.J., Wilhelm, S.I., Cameron-Macmillan, M.L. and Storey, A.E. (2006) Extra-pair copulations in common murres I: a mate attraction strategy? Behaviour 143, 1241–1262.

Wedell, N., Gage, M.J.G. and Parker, G.A. (2002) Sperm competition, male prudence and sperm-limited females. Trends Ecol. Evol. 17, 313–320.

Wilson, J.R., Kuehn, R.E. and Beach, F.A. (1963) Modification in the sexual behavior of male rats produced by changing the stimulus female. J. Comp. Physiol. Psychol. 56, 636–644.

Wong, B.B.M. and Candolin, U. (2005) How is female mate choice affected by male competition? Biol. Rev. 80, 1–13.

Yamazaki, K., Beauchamp, G.K., Kupniewski, D., Bard, J., Thomas, L. and Boyse, E.A. (1988) Familial imprinting determines H-2 selective mating preferences. Science 240, 1331–1332.

Zajitschek, S.R.K., Evans, J.P. and Brooks, R. (2006) Independent effects of familiarity and mating preferences for ornamental traits on mating decisions in guppies. Behav. Ecol. 17, 911–916.

Zeh, J.A., Newcomer, S.D. and Zeh, D.W. (1998) Polyandrous females discriminate against previous mates. Proc. Natl. Acad. Sci. USA 95, 13732–13736.

Chapter 27
Age of the Subject and Scent Donor Affects the Amount of Time that Voles Self-Groom When They are Exposed to Odors of Opposite-sex Conspecifics

Michael H. Ferkin and Stuart T. Leonard

Abstract Many terrestrial mammals, including voles, self-groom when they encounter odors of opposite-sex conspecifics. Voles also spend different amounts of time self-grooming when they are exposed to odors of reproductively active and reproductively quiescent opposite-sex conspecifics, suggesting that self-grooming may be involved in the behaviors that support reproduction. If self-grooming is affected by the reproductive condition of the donor and the groomer, it is also likely that their ages will influence the amount of time that the groomer will self-groom. The objective of this paper was to test the hypothesis that age of the groomer and the scent donor affects the amount of time that meadow voles, *Microtus pennsylvanicus*, spend self-grooming when exposed to bedding scented by opposite-sex conspecifics. Older males (12–13 mo-old) spent more time self-grooming than younger males did (2–3 mo-old and 8–9 mo-old) when they were exposed to odors of 2–3 mo-old and 8–9 mo-old female voles. Younger males spent similar amounts of time self-grooming in response to odors of 2–3 mo-old, 8–9 mo-old, and 12–13 mo-old female voles. Female voles, independent of their age, spent more time self-grooming in response to odors of 12–13 mo-old males relative to 2–3 mo-old and 8–9 mo-old males. These data demonstrate that voles discriminate between the odors of different age opposite-sex conspecifics and adjust the amount of time they self-groom when exposed to them. The data augment the view that self-grooming is a specialized form of olfactory communication between the sexes.

27.1 Introduction

Many terrestrial mammals self-groom when they encounter odors of opposite-sex conspecifics (Steiner 1973; Brockie 1976; Wiepkema 1979; Harriman and Thiessen 1985; Thiessen and Harriman 1986; Witt, Carter, Carlstead and Read 1988;

Michael H. Ferkin
University of Memphis, Biology Department
mhferkin@memphis.edu

Witt, Carter, Chayer and Adams 1990; Spruijt, Van Hooff and Gispen 1992; Ferkin, Sorokin and Johnston 1996; Wolff, Watson and Thomas 2002; Leonard, Alizadeh-Naderi, Stokes and Ferkin 2005). Collectively, these studies provide a foundation for the hypothesis that self-grooming is involved in olfactory communication between the sexes. This hypothesis would predict that by self-grooming individuals produce odors that are more attractive to opposite-sex conspecifics compared to those pro-duced by opposite-sex conspecifics that did not self-groom (Thiessen 1977). Self-grooming also reduces agonism between the groomer and opposite-sex conspecifics (Harriman and Thiessen 1985; Thiessen and Harriman 1986). In addition, self-grooming increases the odor field that surrounds the groomer, making it more likely that the groomer is detected by nearby opposite-sex conspecifics (Ferkin et al. 1996; Wolff et al. 2002). This hypothesis and its prediction have gained support from numerous studies on meadow voles, *Microtus pennsylvanicus*, and prairie voles, *M. ochrogaster* (Ferkin et al. 1996; Ferkin, Leonard, Heath and Paz-y-Miño 2001; Leonard and Ferkin 2005; Leonard et al. 2005; Ferkin 2006) and in other mammals (Steiner 1973; Brockie 1976; Wiepkema 1979. Spruijt et al. 1992).

The fact that voles and other terrestrial mammals, self-groom when they encounter the odors of particular opposite-sex conspecifics is consistent with the speculation that self-grooming may be a specialized form of olfactory communi-cation (Ferkin 2005). The fact that the reproductive condition of the scent donor affects the amount of time that individuals self-groom strengthens this speculation. Male meadow voles spend more time engaged in self-grooming in response to odors of female voles that were in behavioral estrus and postpartum estrus and less time self-grooming in response to odors of females housed under a short photoperiod or to odors of females that have been ovariectomized (Ferkin et al. 1996; Leonard, Alizadeh-Naderi, Stokes and Ferkin 2005; Ferkin 2006). Other studies indicate that the reproductive condition of the groomer affects the amount of time that it self-grooms when presented with the odors of an opposite-sex conspecific. Groomers that are sexually receptive spend more time self-grooming than those that are not sexually receptive in response to the odors of opposite-sex conspecifics (Leonard et al. 2005; Ferkin 2006). Prairie voles spend more time self-grooming when exposed to their mate than to an unfamiliar male conspecific (Witt et al. 1988). Prairie voles also self-groom more when they encounter the odors of unfamiliar, opposite-sex conspecifics than when they encounter those of like-age, familiar, opposite-sex siblings (Paz-y-Miño, Leonard, Ferkin and Trimble 2002).

Much of the current data show that individuals spend different amounts of time self-grooming when they encounter particular opposite-sex conspecifics or their odors (Ferkin 2005, 2006; Ferkin and Leonard 2005). Thus, it is likely that an individual's age will also influence the amount of self-grooming. For instance, a study on meadow voles indicated that 12–13 mo old males spent more time than 2–3 and 8–9 mo old males investigating the scent marks of female conspecifics (Ferkin 1999). This study also found that and 8–9 mo-old female voles spent more time investigating the scent marks of 12–13 mo-old males than those of 8–9 mo-old and 2–3 mo-old males younger males (Ferkin 1999). These findings suggest that older male voles may be more interested in and attractive to females as compared to younger male voles. This result is interesting in that it suggests that the amount of time that male voles self-groom in response to the odors of females increases as adult males grow older.

In the present study, we investigate whether the age of the subject and the scent donor affects the amount of time an individual self-grooms in response to the odors of opposite-sex conspecifics. In doing so, we have two main objectives. The first objective is to test the hypothesis that age of the groomer affects the amount of time that it spends self-grooming in response to odors of opposite-sex conspecifics (experiment 1). The second objective is to test the hypothesis that age of the scent donor affects the amount of time opposite-sex conspecifics self-groom when exposed to its odors (experiment 2).

27.2 General Methods

27.2.1 Animals

Subjects (n = 36 males and 36 females) and scent donors (n = 16 males and 16 females) were meadow voles maintained from birth under a long photoperiod (14:10h L:D, lights on at 0700h CST), which simulates day length prevalent in the summer breeding season. Meadow voles used in these experiments were third generation captive animals, originally trapped in north, central Kentucky and southern Ohio. At 18 days of age, voles were weaned and then housed with littermates in clear plastic cages (26 × 32 × 31 cm) with wood chip bedding until they were 36 days of age. After this, voles were housed singly in clear plastic cages (13 × 16 × 13 cm) with wood chips. Food (Purina Rodent Diet #5008, PMI Inc., St. Louis, MO, U.S.A.), water, and cotton nesting material were provided ad libitum. We replaced cotton every 7 to 10 days.

The voles used in these experiments were housed singly for three-four weeks before being used as a scent donor or subject. Meadow voles born and reared under a long photoperiod reach sexual maturity by 50 d old (Nadeau 1985), and are sexually receptive and readily mate with sexually receptive opposite-sex conspecifics (delBarco-Trillo and Ferkin 2004; Pierce, Ferkin and Williams 2005). Male and female subjects and scent donors were sexually naïve and not familiar with one another. Female voles do not undergo estrous cycles, rather they are induced into estrous and ovulation (Keller 1985).

27.2.2 Self-grooming Test

Subjects were placed into an empty, covered cage identical in size and appearance to their home cage, except that it contained clean wood chip bedding and 8 g of cotton-nesting material from a particular scent donor. We recorded the amount of time subjects self-groomed during a 5 min exposure to the scented cotton. Typically voles begin self-grooming within 30 s of exposure to scented bedding (Ferkin et al. 1996). We did not provide voles with unscented cotton and therefore did not record the amount of time they spent self-grooming to that stimulus in the present study. Pilot data from this study as well as data from several other studies using the same methods indicate that voles exposed to unscented cotton spend little time, approximately 3.6 ± 1.4 s, self-grooming (Ferkin et al. 1996; Leonard and Ferkin 2005; Leonard et al. 2005; Ferkin 2005, 2006). For meadow voles, the general pattern of self grooming consists of a cephalocaudal progression that begins with rhythmic movements of the paws around the mouth and face, over the ears, descending to the ventrum, flank, anogenital area and tail (Ferkin et al. 1996). Self-grooming was recorded when subjects rubbed, licked or scratched any of these body areas during exposure to the scented cotton bedding. For each exposure, the experimenter was blind to the age of the subject and the sex of the scent donor. After the test was completed, the subject was returned to its home cage. The testing cage was cleaned after each exposure with warm soapy water and alcohol, dried, and the wood chip bedding was replaced prior to each test.

27.2.3 Experimental Procedure

We measured the amount of time that 2–3, 8-9 and 12–13 mo old male and female meadow voles spent self-grooming when exposed to an 8 g piece of cotton nesting material scented by opposite-sex conspecifics that were either 2–3, 8–9 or 12–13 mo old. There were 12 different male and female subjects and 16 different opposite-sex scent donors used for each odor condition. Each subject underwent three self-grooming tests, one test for each of the three age classes of scent donors. The order of the three tests was random. Subjects were not used as scent donors and vice versa. Scent donors and subjects were similar in size to one another (within 7 g) and were not related or familiar to one another.

We selected these age cohorts as they broadly represent the age cohorts that exist in the free-living populations during the breeding season (Negus and Berger 1988; Negus, Berger and Pinter 1992). Voles aged 2–3 mo are sexually mature adults and represent animals born early in the breeding season. Those aged 7–8 mo are sexually mature adults and represent animals that survive the entire breeding season. Voles aged 12–13 mo have not undergone reproductive senescence and represent animals that are born early to mid-breeding season, over-wintered successfully, and continue live as reproductively active adults into the next breeding season (Negus et al. 1992).

27.2.4 Statistics

We analyzed the data by a two-way ANOVA (age of groomer, age of donor) with repeated measures on age of donor to determine if significant differences existed in the time that subjects spent self-grooming in response to the odor of opposite-sex conspecifics. When significant main effects were detected, post-hoc Tukey's HSD tests were used to determine significance differences between the pairwise comparisons. Statistical significance was accepted at a level of a ≤ 0.05 for all tests.

27.3 Results

The amount of time that meadow voles self-groomed in response to the odors of opposite-sex conspecifics differed according to the groomer's age and the scent donor's age. For males, a two-way ANOVA with repeated measures revealed a significant main effect for age of donor ($F_{2,107} = 29.6$, P < 0.001) and for age of groomer ($F_{2,107} = 38.3$, P < 0.001). However, there was also a significant interaction between age of donor and age of groomer ($F_{4,107} = 25.0$, P < 0.001), indicating that the effect of the donor's age varied according to the age of the groomer (Fig. 27.1). To investigate this interaction, data were analyzed via separate one-way repeated measures ANOVAs (age of donor). For male groomers aged 12–13 mo, this analysis revealed a significant main effect of donor age ($F_{2,35} = 71.6$, P < 0.001). Post hoc comparisons indicated that 12–13 mo old males spent significantly more time self-grooming in response to the odors of 2–3 and 8–9 mo old females than to those of 12–13 mo old females (Tukey's HSD, P < 0.001; Fig. 27.1). By contrast, donor

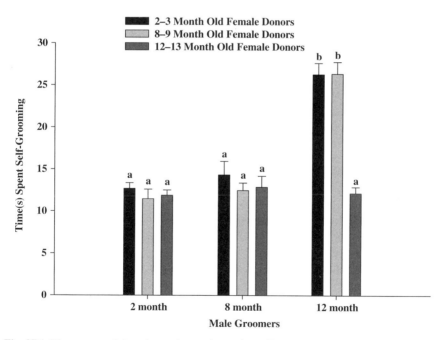

Fig. 27.1 The amount of time that male meadow voles self-groomed in response to the odors of 2–3 mo old, 8–9 mo old, and 12–13 mo old female conspecifics. Histograms capped with different letters are statistically different (P < 0.05)

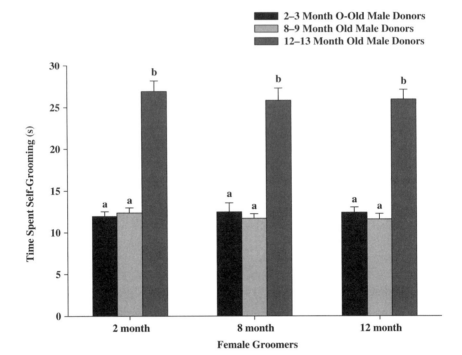

Fig. 27.2 The amount of time that female meadow voles self-groomed in response to the odors of 2–3 mo old, 8–9 mo-old, and 12–13 mo old male conspecifics. Histograms capped with different letters are statistically different (P < 0.05)

age did not affect self-grooming among males aged 2–3 mo ($F_{2,35} = 0.46$, $P = 0.64$) or 8–9 mo ($F_{2,35} = 1.18$, $P = 0.33$).

For females, a two-way ANOVA with repeated measures revealed a significant main effect for age of donor ($F_{2,107} = 247.0$, $P < 0.001$) indicating that the amount of time females self-groomed varied significantly depending upon the scent donor's age (Fig. 27.2). No significant interactions were revealed. Unlike males, the amount of time that female meadow voles self-groomed in response to the odors of opposite-sex conspecifics was only dependant upon the scent donor's age. Post hoc comparisons indicated that 2–3, 8–9 and 12–13 mo old females each spent significantly more time self-grooming in response to the odors of 12–13 mo-old males than to those of 2–3 mo-old and 8–9 mo-old males (Tukey's HSD, P < 0.001; Fig. 27.2).

27.4 Discussion

We found that sex differences existed in the time that 2–3, 8–9, 12–13 mo old voles self-groomed in response to cotton scented by opposite-sex conspecifics. Specifically, 12 mo old males spent significantly more time self-grooming in response to the odors of females aged 2–3 and 8–9 mo than to those aged 12–13 mo. These data support the notion that 12–13 mo old male voles may be more interested in, and attractive to, females than are younger males. This result is interesting in that it suggests that the amount of time that male voles self-groom in response to the odors of females increases as adult male voles grow older.

Why did male meadow voles aged 12–13 mo spend more time self-grooming than younger males when exposed to odors of female conspecifics? One explanation is that older male voles were more attracted to odors of female conspecifics. An earlier study on meadow voles reported that 12–13 mo old males spent more time investigating the scent marks of female conspecifics than did 2–3 or 8–9 mo old males (Ferkin 1999). An alternative interpretation is that older males spend more time self-grooming and investigating the odors of female conspecifics because they have sensory or perceptual deficits that are not present in younger voles. Sensory deficits in older male rodents are not unusual (Sarter and Bruno 1998) and could cause a groomer to take longer to detect, interpret, and respond to olfactory cues. However, we think this explanation is not consistent with the fact that 12–13 mo old males could discriminate between the odors of females in different age groups by spending more time self-grooming in response to odors of females aged 2–3 or 8–9 mo than to those aged 12–13 mo. Perhaps older but not younger males are able to detect some feature found in the odor of older females that makes them less attractive relative to younger females. However, in a previous study male voles spent more time investigating the odors from females aged 8–9 mo relative to those from 2–3 or 12–13 mo old females (Ferkin 1999). Alternatively, younger males may self-groom less than older males because the former may not need to self-groom as much in order to increase the likelihood of being detected and more attractive to the opposite-sex. Thus, 12–13 mo old males may have self-groomed more than younger males in response to female odors because the pelage and sebaceous tissues of the former required more tactile stimulation to release their odiferous substances as compared to the pelage and sebaceous of tissues of younger male voles (Ebling 1977; Johnson 1977).

The data from the current study also indicate that independent of their age, female voles spent more time self-grooming when they were presented with bedding scented by 12–13 mo old males than to bedding scented by younger males. A previous study reported that female voles in the same age categories spent more time investigating the odors of 12–13 mo-old males than those of younger males (Ferkin 1999). An inference drawn from this and previous work suggests older male voles produce odors that are more easily detected by females than do younger males. It is possible that 12–13 mo old male voles produce substances in their odors that are not present in those from 2–3 and 8–9 mo old males. Perhaps, this substance provides a cue to female voles indicating male longevity. Such a cue may make older males more interesting or attractive to females. It is also possible that female voles are attracted to the odors of older male voles and subsequently mate with them to gain indirect benefits of longevity for their sons (Møller and Alatalo 1977).

Overall, the present study supports the hypothesis that age of the groomer and the scent donor affects the amount of time that individuals self-groom in response to odors of opposite-sex conspecifics. If we compare data from this study with those from other studies, we begin to see that the amount of time individuals self-groom when exposed to the odors of opposite-sex conspecifics depends on the age, reproductive state, photoperiodic history, mating history and relatedness of both the groomer and the scent donor (Moore 1986; Witt et al. 1988, 1990; Ferkin et al. 1996; Paz-y-Miño et al. 2002; Ferkin and Leonard 2005; Leonard and Ferkin 2005; Leonard et al. 2005; Ferkin 2005, 2006). An overall finding of these studies is that individuals appear to discriminate between the odors of different opposite-sex conspecifics and adjust the amount of time they self-groom when exposed to them. If self-grooming increases the attractiveness of groomers and makes them easier to detect by opposite-sex conspecifics relative to non-groomers (Ferkin et al. 1996, 2001), self-grooming may be a specialized form of olfactory communication that is involved in the suite of behaviors that surround reproduction (Ferkin and Leonard 2005; Leonard and Ferkin 2005; Ferkin 2006). This would suggest that odors conveyed by self-grooming have different meanings to a groomer's audience and/or contain different messages from the groomer (Ferkin and Leonard 2005).

Acknowledgments We thank Javier delBarco Trillo, Lara LaDage, and Jane Hurst for commenting on earlier drafts of this manuscript. National Science Foundation grants IBN 9421529 and IOB 0444553 and National Institutes of Health grants AG 16594-01 and HD 049525 to M. H. Ferkin supported the research

References

Brockie, R. (1976) Self-anointing by wild hedgehogs, *Erinaceus europaeu*s. Anim. Behav.24, 68–71.

Ebling, F. J. (1977) Hormonal control of mammalian skin glands. In D. Muller-Schwarze, D. and M. M. Mozell (Eds.), *Advances in Chemical Signals in Vertebrates*. Plenum Press, New York: Plenum Press, pp. 17–33.

delBarco-Trillo, J. and Ferkin, M. H. (2004) Male mammals respond to a risk of sperm competition conveyed by odours of conspecific males. Nature 43, 446–449.

Ferkin, M. H. (1999) Attractiveness of opposite-sex odor and responses to it vary with age and sex in meadow voles (*Microtus pennsylvanicus*). J. Chem. Ecol. 4, 757–769.

Ferkin, M. H. (2005) Self-grooming in meadow voles. In R. T. Mason, M. P. LeMaster, and D. Muller-Schwarze (Eds.), *Chemical Signals in Vertebrate*s, Vol. 10, Plenum Press, New York, pp. 64–69.

Ferkin, M. H. (2006) Self-grooming in meadow voles is affected by its reproductive state and that of the top-scent donor. Behav. Processes73, 266–271.

Ferkin, M. H. and Leonard, S. T. (2005) Self-grooming by rodents in social and sexual contexts. Acta Zool. Sinica. 51, 772–779

Ferkin, M. H., Sorokin, E.S., and Johnston, R.E. (1996) Self-grooming as a sexually dimorphic communicative behaviour in meadow voles, *Microtus pennsylvanicus*. Anim. Behav. 51, 801–810.

Ferkin, M.H., Leonard, S.T., Heath, L.A., and Paz-y-Miño C., G. (2001) Self-grooming as a tactic used by prairie voles *Microtus ochrogaster* to enhance sexual communication. Ethology 107, 939–949.

Harriman, A. E and Thiessen, D. D. (1985) Harderian letdown in male Mongolian gerbils (*Meriones unguiculatus*) contributes to proceptive behavior. Horm. Behav. 19, 213–219.

Johnson, E. K. (1977) Seasonal changes in the skin of mammals. Symp. Zool. Soc. London. 39, 373–404.

Keller, B. L. (1985) Reproductive patterns. In R. H. Tamarin (Ed.), *Biology of New World Microtus*. Am Soc. Mammal. Sp. Publ., Vol. 8, Lawrence, Kansas, pp. 725–778.

Leonard, S. T. and Ferkin, M. H. (2005) Seasonal differences in self-grooming in meadow voles, *Microtus pennsylvanicus*. Acta Ethologica. 8, 86–91.

Leonard, S. T., Alizadeh-Naderi, R., Stokes, K., and Ferkin, M. H. (2005) The role of prolactin and testosterone in mediating seasonal differences in the self-grooming behavior of male meadow voles, *Microtus pennsylvanicus*. Physiol. Behav. 85 461–468.

Møller A. P. and Alatalo R V. (1999) Good-genes effects in sexual selection. Proc. R. Soc. Lond. B 266, 85–91.

Moore, C. L. (1986) A hormonal basis for sex differences in the self-grooming of rats. Horm. Behav. 20, 155–165.

Nadeau, J. H. (1985) Ontogeny. In R. H. Tamarin (Ed.), *Biology of New World Microtus*. Am Soc. Mammal. Sp. Publ.,Vol. 8, Lawrence, Kansas, pp. 254–285.

Negus, N. C. and Berger, P. J. (1988) Cohort analysis: environmental cues and diapause in microtine rodents. In M.S. Boyce (Ed.), *Evolution of Life Histories of Mammals*, Theory and Pattern. Yale University Press, New Haven, pp. 65–74.

Negus, N. C., Berger, P. J., and Pinter, A. J. (1992) Phenotypic plasticity of the montane vole (*Microtus montanus*) in unpredictable environments. Can. J. Zool. 70, 2121–2124.

Paz-y-Miño C. G, Leonard, S.T., Ferkin, M.H., and Trimble, J.F. (2002) Self-grooming and sibling recognition in meadow voles (*Microtus pennsylvanicus*) and prairie voles (*M. ochrogaster*). Anim. Behav. 63, 331–338.

Pierce, A. A., Ferkin, M. H., and Williams, T. K. (2005) Food-deprivation-induced changes in sexual behavior of meadow voles, *Microtus pennsylvanicus*. Anim. Behav. 70, 339–348.

Sarter, M. and Bruno, J. P. (1998) Age–related changes in rodent cortical acetylcholine and cognition: main effects of age versus age as an intervening variable. Brain Res. Rev. 27, 143–156.

Spruijt, B.M., Van Hooff, J.A.R.A.M., and Gispen, W. H. (1992) Ethology and neurobiology of grooming behavior. Physiol. Rev. 72, 825–852.

Steiner, A. L. (1973) Self- and allo-grooming behavior in some ground squirrels (Sciuridae), descriptive study. Can. J. Zool. 51, 151–161.

Thiessen D. D. (1977) Thermoenergetics and the evolution of pheromone communication. Prog. Psychobiol. Physiol. Psych. 7, 91–191.

Thiessen, D. D. and Harriman, A. E. (1986) Harderian gland exudates in the male *Meriones unguiculatus* regulate female proceptive behavior, aggression, and investigation. J. Comp. Psychol. 100, 85–87.

Wiepkema, P. R. (1979) The social significance of self-grooming in rats. Netherlands J. Zool. 29, 622–623.

Witt, D. M., Carter, C. S., Carlstead, K., and Read, L. D. (1988) Sexual and social interaction preceding and during male-induced oestrous in prairie voles, *Microtus ochrogaster*. Anim. Behav. 36, 1465–1471.

Witt, D. M. Carter, C. S., Chayer, R., and Adams, K. (1990) Patterns of behavior during postpartum estrous in prairie voles. Anim. Behav. 39, 528–534.

Wolff, J. O., Watson, M.H., and Thomas, S.A. (2002) Is self-grooming by male prairie voles a predictor of mate choice? Ethology 108, 169–179.

Chapter 28
Scent, Mate Choice and Genetic Heterozygosity

Michael D. Thom, Paula Stockley, Robert J. Beynon and Jane L. Hurst

Abstract Females of many species choose to mate with relatively unrelated males in order to ensure outbred, heterozygous offspring. There is some evidence to suggest that the MHC is involved in mate choice decisions, either because MHC heterozygous offspring are more resistant to disease, or because the highly detectable odours associated with this region allow it to act as a marker of general inbreeding. To determine which role the MHC plays it is necessary to disentangle this region from the genetic background, a requirement which has generally proven difficult to achieve. We argue that the emphasis on MHC's role in mate choice has resulted in other potential markers of inbreeding being neglected, and discuss the evidence for MHC disassortative mating, the interaction with genetic background, and a possible role for alternative markers of inbreeding.

28.1 Introduction

Female animals generally favour some males over others as potential mating partners, and those allowed to mate with their preferred male often have greater fitness than when mated with a non-preferred alternative (Drickamer, Gowaty and Holmes 2000; Persaud and Galef 2005). One way in which choosy females may improve their fitness is by selecting males that will produce genetically superior offspring (see Byers and Waits 2006 for a recent example). These males may either provide good genes that improve offspring fitness traits, or they may be relatively more compatible with the choosing female's genotype (Zeh and Zeh 1996; Brown 1997; Tregenza and Wedell 2000; Mays and Hill 2004). While good genes effects are additive and broadly speaking are expressed independently of the maternal genome, compatibility effects are nonadditive and are determined by the interaction between the parental genotypes (Zeh and Zeh 1996). This leads to a key difference between the good genes and compatibility explanations for female choice: while females will usually agree about which males carry good genes, the most compatible male

Michael D. Thom
University of Liverpool, Department of Veterinary Preclinical Sciences
mthom@liv.ac.uk

will vary depending on the genotype of the choosing female. Parental genetic compatibility often improves fitness by increasing the genetic heterozygosity of offspring (but see also Zeh and Zeh 1996), a mechanism which is most familiar from the special case of inbreeding avoidance. In fact, preference for compatible partners and avoidance of inbreeding overlap considerably, since close relatives will often be genetically incompatible (Brown 1997). Here we discuss the detrimental effects of incompatibility, particularly inbreeding, and consider the behavioural mechanisms and signalling systems which may have evolved to avoid these costs.

28.2 Genetic Compatibility and Inbreeding Avoidance

The main consequence of mating between genetically dissimilar partners is increased heterozygosity of offspring, which may itself be the primary advantage of sexual reproduction (Brown 1997). Genetic heterozygosity can influence fitness via two main mechanisms: either because it avoids unmasking any deleterious recessive alleles carried by either parent (the dominance hypothesis), or because heterozygotes are inherently superior to homozygotes (overdominance: Charlesworth and Charlesworth 1987). Whatever the underlying mechanism, homozygosity is frequently detrimental to reproductive success, having been shown to reduce hatching success or embryo survival, litter size, and survival of offspring in captive studies (e.g. Ralls, Ballou and Templeton 1988; Keane, Creel and Waser 1996; e.g. Pusey and Wolf 1996; Bixler and Tang-Martinez 2006). Accumulating data support the idea that genetic similarity among mating partners also has negative effects on offspring number and survival in the wild (Stockley, Searle, Macdonald and Jones 1993; Coltman, Bowen and Wright 1998; Crnokrak and Roff 1999; Keller and Waller 2002; Slate and Pemberton 2002). The detrimental effects of inbreeding may continue into adulthood, with inbred individuals suffering a reduction in survival (Jiménez, Hughes, Alaks, Graham and Lacy 1994; Keller, Arcese, Smith, Hochachka and Stearns 1994; Coltman, Pilkington, Smith and Pemberton 1999), ability to hold territories (Meagher, Penn and Potts 2000), and reproductive success (Keller 1998; Slate, Kruuk, Marshall, Pemberton and Clutton-Brock 2000; Seddon, Amos, Mulder and Tobias 2004).

Given these costs, we would expect the evolution of behavioural mechanisms to reduce the risk of inbreeding. In some species, this is achieved by sex-biased dispersal: members of one sex predominantly leave the natal range and hence minimize the likelihood of encountering relatives as mates (Pusey 1987; Pusey and Wolf 1996). By contrast, species without sex-biased dispersal, and particularly those with aggregated population structures, are likely to risk mating with close relatives. Such species should benefit from kin recognition systems that allow them to recognize and avoid close kin as mates. Active mate preference for non-kin does indeed appear to be widespread, and has been demonstrated in numerous groups including insects (Simmons 1991), marsupials (Parrott, Ward and Temple-Smith 2007), rodents (Krackow and Matuschak 1991; Keane et al. 1996;

Ryan and Lacy 2003), and primates (Smith 1995; Soltis, Mitsunaga, Shimizu, Yanagihara and Nozaki 1999). This preference can be very strongly expressed: females of some species choose not to mate at all when no unrelated males are available (woodpeckers: Koenig, Stanback and Haydock 1999; mole rats: Cooney and Bennett 2000; meerkats: O'Riain, Bennett, Brotherton, McIlrath and Clutton-Brock 2000).

28.3 The Role of MHC in Mate Choice

Behavioural avoidance of inbreeding requires a mechanism for reliably identifying kin, a signal which should be easily detectable, polymorphic and heritable. Much work in vertebrates has focussed on the role of the major histocompatibility complex (MHC), a highly polymorphic region of the vertebrate genome which is associated with specific odours in a range of species. Numerous studies on rodents have demonstrated their ability to discriminate odours associated with the MHC (Yamaguchi, Yamazaki, Beauchamp, Bard, Thomas and Boyse 1981; Yamazaki, Beauchamp, Bard, Thomas and Boyse 1982), including odour changes resulting from only single amino-acid substitutions (Yamazaki, Beauchamp, Egorov, Bard, Thomas and Boyse 1983), although not all differences are detectable (Carroll, Penn and Potts 2002), Sensitivity to MHC-associated odours has also been suggested for other species including sticklebacks (Reusch, Häberli, Aeschlimann and Milinski 2001), salmonids (Olsén, Grahn, Lohm and Langefors 1998), lizards (Olsson, Madsen, Nordby, Wapstra, Ujvari and Wittsell 2003), rats (Schellinck, Slotnick and Brown 1997) and humans (Wedekind, Seebeck, Bettens and Paepke 1995; Thornhill, Gangestad, Miller, Scheyd, McCullough and Franklin 2003). Although many experiments have focussed simply on discriminability of MHC odours, there is also some evidence that these odours affect mate choice in MHC congenic laboratory mouse strains. In some highly inbred pairs of strains that are genetically identical to one another except at MHC, members of both sexes prefer to mate with partners with a different MHC type to self (Yamazaki, Boyse, Miké, Thaler, Mathieson, Abbott, Boyse, Zayas and Thomas 1976; Egid and Brown 1989). The high degree of polymorphism at MHC, together with the observation that this is associated with easily discriminable odours, and evidence for an effect of MHC on mate choice, have led to the suggestion that the MHC may provide a general mechanism of individual (e.g. Brennan and Kendrick 2006) and kin (e.g. Brown and Eklund 1994) recognition across vertebrates used in the context of mate choice. Recent physiological work has highlighted direct detection of MHC peptides as a possible mechanism by demonstrating that receptors capable of detecting MHC peptides exist in the main olfactory epithelium (Spehr, Kelliher, Li, Boehm, Leinders-Zufall and Zufall 2006) and vomeronasal organ (Leinders-Zufall, Brennan, Widmayer, Chandramani, Maul-Pavicic, Jager, Li, Breer, Zufall and Boehm 2004).

It is frequently suggested that animals choose mating partners with complementary MHC type to self because of the resulting increase in offspring genetic heterozygosity specifically at this region (Jordan and Bruford 1998; Tregenza and Wedell 2000; Milinski 2006). As the primary role of the MHC is in the adaptive immune system, relatively MHC heterozygous offspring should experience improved pathogen resistance as a result of their ability to recognize a greater range of foreign peptides (Penn, Damjanovich and Potts 2002). This mechanism of responding to an unpredictable disease environment is hypothesised to be a major force in maintaining the MHC's extraordinary levels of polymorphism (Doherty and Zinkernagel 1975; Hughes and Nei 1988). However the expected relationship between MHC heterozygosity and adaptive immune performance has received mixed support. Some studies have supported the assumption of a link between MHC diversity and fitness, reporting greater resistance (Penn et al. 2002), lower pathogen loads (Wegner, Kalbe, Kurtz, Reusch and Milinski 2003; Kurtz, Kalbe, Aeschlimann, Häberli, Wegner, Reusch and Milinski 2004), and higher survival (Westerdahl, Waldenström, Hansson, Hasselquist, Von Schantz and Bensch 2005) at high levels of heterozygosity. Others have failed to find evidence that MHC heterozygosity influences immunity (Paterson, Wilson and Pemberton 1998; Langefors, Lohm, Grahn, Andersen and von Schantz 2001; Wedekind, Walker and Little 2006), or have found the opposite effect. For example, mice heterozygous for MHC performed less well than the homozygote mean when infected with malaria (Wedekind, Walker and Little 2005). Much of the evidence for a link between MHC heterozygosity and ability to combat disease comes from correlational studies, which are generally hampered by the strong link between MHC and background genetic diversity. Failing to control for this correlation means that observed effects could be due simply to genome-wide heterozygosity rather than to the specific effects of MHC (Penn et al. 2002). Indeed, a recent study on wild fish populations found background genetic heterozygosity to be significantly associated with reduced parasite load, while MHC diversity was not (Rauch, Kalbe and Reusch 2006). This link to background genetic diversity has an influence on other aspects MHC's role, as we discuss further below.

28.4 The Confounding Influence of Genetic Background

In natural populations, the level of MHC sharing between a given pair of individuals is normally correlated with their overall similarity across the whole of the genome. This link could enable animals to use the highly variable and detectable MHC to act as a marker of overall genetic relatedness. Whereas the evidence for MHC homozygosity remains ambiguous, inbreeding can have devastating fitness consequences. Given the costs of mating with close kin, the primary role of MHC in mate choice might be to facilitate avoidance of inbreeding more generally, rather than at the MHC itself (Potts, Manning and Wakeland 1994). High levels of MHC polymorphism ensure that individuals in a normal outbreeding population are likely to share a large proportion of their MHC only with very close relatives; disassortative mating would provide the substantial benefits of inbreeding avoidance, while simultaneously ensuring genetic diversity at the MHC.

Whether the role of MHC in mate choice is related specifically to its involvement in immune defence, or because this region is a reliable signal for overall genetic similarity, the failure of most studies to separate the effects of MHC and genetic background means that any observed effect of MHC could be an artefact of experimental design. Much of the evidence for the role of MHC in mate choice comes from studies demonstrating a deficit of MHC homozygotes in natural (e.g. Landry, Garant, Duchesne and Bernatchez 2001) or artificial (e.g. Potts, Manning and Wakeland 1991) populations, an observation consistent with disassortative mating at MHC. However, because MHC homozygotes are most likely to result from matings between close kin, underproduction of these individuals could be a by-product of general inbreeding avoidance rather than a direct consequence of MHC-based disassortative mating. Assuming that inbreeding avoidance is generally important, the signal used to assess and reject genetically similar mates could be encoded in genes completely unrelated to MHC, with the intrinsic correlation between MHC and the rest of the genome ensuring that MHC homozygotes are rare in normally outbreeding populations. It is only possible to assess whether MHC is important for mate choice by disentangling this from the rest of the genome, something most studies have so far been unable to achieve. Attempts to separate MHC and background generally use laboratory-derived strains, and despite their widespread acceptance in the literature, the results obtained by this method have generally been ambiguous or contradictory. For example, the early studies of Yamazaki et al. (1976; 1978) are usually cited in support of MHC disassortative mating, but these studies also found consistent MHC assortative mating in one strain, and assortative mating has also been reported in another laboratory strain (Andrews and Boyse 1978). A more fundamental limitation in experiments controlling for background genes arises from the tendency to use pairs of highly inbred strains differing only at MHC: it is perhaps unsurprising that, given no other source of genetic variation, females sometimes choose males which differ from them at the only variable loci. These experiments have not adequately tested whether females will respond to *any* variable genetic region against a uniform background. However, evidence from more naturalistic studies involving wild animals is limited by the general failure to control adequately for the correlation between background genetic heterozygosity (or other genetic markers) and MHC. For example, overall genetic relatedness in wild populations is often quantified by counting the number of shared markers at microsatellite loci. This method requires examination of relatively large number of loci to minimize misclassification errors (Blouin, Parsons, Lacaille and Lotz 1996), but many studies actually use very few markers (e.g. 5 in the case of Landry et al. 2001). Other workers have attempted to get around these problems by producing semi-natural populations from captive-bred stock with known genotypes. Potts et al. (1991) tested reproductive success in relation to MHC matching using mice partly derived from wild stock. However in these animals 50% of the genome (including the MHC) originated from highly inbred strains and, because individual parentage could not be assigned, it was not possible to assess whether the deficit in MHC homozygotes found in this study was due to the use of non-MHC signals used to avoid inbreeding.

28.5 Alternative Signals in Inbreeding Avoidance

The almost universal acceptance of MHC as a significant factor in vertebrate mate choice (Jordan and Bruford 1998; Penn 2002; see reviews in: Bernatchez and Landry 2003; Brennan and Kendrick 2006; Milinski 2006), despite inconsistent results and the failure to control adequately for background genes, has resulted in a deficit of research on other markers that might be used to avoid inbreeding. House mice, *Mus musculus domesticus,* have evolved a species-specific polygenic and highly polymorphic gene complex coding for major urinary proteins (MUPs), which are excreted in large quantities in the urine and form a substantial component of this species' scent signals. MUPs are known to play a key role in self-nonself recognition (Hurst, Payne, Nevison, Marie, Humphries, Robertson, Cavaggioni and Beynon 2001; Hurst, Thom, Nevison, Humphries and Beynon 2005) and are used by females to identify individual males from their competitive scent marks (Cheetham, Thom, Jury, Ollier, Beynon and Hurst 2007). MUPs have all the characteristics that make the MHC so intuitively appealing as a marker of kinship, including coding by a complex of tightly linked genes that are inherited together as a haplotype. However MUPs differ from MHC in being produced in large quantities in the liver and excreted at high concentration in urine (Humphries, Robertson, Beynon and Hurst 1999; Beynon and Hurst 2003) before being deliberately deposited extensively in the environment via scent marking. This makes them a widespread and easily detectable signal linked to genome-wide genetic heterozygosity. If mice benefit from avoiding inbreeding, as they clearly appear to (e.g. Meagher et al. 2000), we would expect them to mate disassortatively with respect to MUP in order to avoid mating with close kin, leading inevitably to a parallel reduction in homozygosity at MHC. However if the fitness benefits derived from heterozygosity specifically at the MHC are of greater importance than inbreeding avoidance, MHC odours should be used in preference. MUPs thus provide an ideal control system for identifying experimentally the MHC's role in mate choice, since they provide an equivalent polymorphic, polygenic signal closely linked to background genetic heterozygosity but with no implications for immune function. We are taking advantage of this system and breeding wild-derived populations of mice in which the relative contributions of different genetic components are controlled, allowing us to test for the first time the independent contributions to mate choice of MUP, MHC, and overall relatedness. Allowing these animals to breed in semi-natural enclosures has demonstrated that mice use MUP sharing to avoid mating with close kin. Indeed, the deficit in mating with mice of the same MUP type is sufficient to explain entirely the observed level of inbreeding avoidance, with MHC sharing having no detectable influence on mate choice (Sherborne, Thom, Paterson, Jury, Ollier, Stockley, Beynon and Hurst 2007). Previous studies have been unable to control these three components independently, leaving open the possibility that previously observed deficits in MHC homozygotes in freely-breeding populations of mice could result from the avoidance of mates with the same MUP type. Work on laboratory strains has not investigated the role of MUPs in inbreeding avoidance, probably because variability in MUPs has been largely eliminated in inbred laboratory mice (Robertson, Cox, Gaskell, Evershed and Beynon 1996), and no MUP-congenic strains currently exist.

The evolution of this highly polymorphic and species-specific signal suggests that the use of MHC-associated odours in inbreeding avoidance may not be as widespread as is often assumed, particularly as the main evidence for a direct link between MHC scents and disassortative mating came from mouse studies. This does not rule out the possibility that MHC plays a role in mate choice and this may vary between species, but at present evidence remains ambiguous. A possible exception is the evidence for MHC allele counting in fish (Reusch et al. 2001; Aeschlimann, Häberli, Reusch, Boehm and Milinski 2003). The MHC of fish differs from that of tetrapods as these genes are organised in at least two separate linkage groups rather than occurring in a single tight complex (Sato, Figueroa, Murray, Malaga-Trillo, Zaleska-Rutczynska, Sultmann, Toyosawa, Wedekind, Steck and Klein 2000). Nevertheless, to demonstrate that MHC is of greater importance than the rest of the genome in determining mate choice, these studies still need to control adequately for background genetic relatedness to eliminate the possibility of a correlation between MHC and the marker actually used for inbreeding assessment.

28.6 Conclusions

The negative consequences of inbreeding for survival and reproductive success are well documented, and emphasise the fitness benefits of genetic heterozygosity. While there may be particular advantages to genetic heterozygosity at some highly polymorphic loci, there is currently little evidence to support the idea that animals choose partners specifically to maintain diversity at such regions. Instead, heterozygosity at these polymorphic loci may be maintained by the mechanisms that sustain genome-wide heterozygosity. Nevertheless, such regions may be used to recognize close relatives and avoid inbreeding. The likely evolutionary benefits of MHC heterozygosity, coupled with the highly detectable odours associated with this region, have led to the intuitively obvious conclusion that females should choose mates that will increase offspring MHC heterozygosity. Because this seems such a plausible hypothesis, evidence that the MHC does indeed appear to influence mate choice has rapidly led to the widespread acceptance of MHC disassortative mating. However while it undoubtedly follows that such behaviour should benefit choosing females, the evidence remains decidedly mixed. Experiments using MHC congenic strains of mice have produced conflicting results, and in any case bias the likelihood of detecting an effect by removing all other sources of variation. Attempts to correct for this have generally run into the opposite difficulty of disentangling the tight link between MHC and background. As a result, it remains possible that the disassortative mating behaviours attributed to MHC are in fact a consequence of mate choice on the basis of alternative, largely untested, cues associated with avoiding the substantial costs of inbreeding. MUPs appear to be one such system in house mice, but other signals of inbreeding avoidance may well be widespread. We propose that evidence for MHC effects on mate choice needs to be more critically assessed, and the scope of research into polymorphic signals extended beyond just the MHC.

References

Aeschlimann, P.B., Häberli, M.A., Reusch, T.B.H., Boehm, T. and Milinski, M. (2003) Female sticklebacks *Gasterosteus aculeatus* use self-reference to optimize MHC allele number during mate selection. Behav. Ecol. Sociobiol. 54, 119–126.

Andrews, P.W. and Boyse, E.A. (1978) Mapping of an H-2-linked gene that influences mating preference in mice. Immunogenet. 6, 265–268.

Bernatchez, L. and Landry, C. (2003) MHC studies in nonmodel vertebrates: what have we learned about natural selection in 15 years? J. Evol. Biol. 16, 363–377.

Beynon, R.J. and Hurst, J.L. (2003) Multiple roles of major urinary proteins in the house mouse, *Mus domesticus*. Biochemical Society Transactions 31, 142–146.

Bixler, A. and Tang-Martinez, Z. (2006) Reproductive performance as a function of inbreeding in prairie voles (*Microtus ochrogaster*). J. Mammal. 87, 944–949.

Blouin, M.S., Parsons, M., Lacaille, V. and Lotz, S. (1996) Use of microsatellite loci to classify individuals by relatedness. Mol. Ecol. 5, 393–401.

Brennan, P.A. and Kendrick, K.M. (2006) Mammalian social odours: attraction and individual recognition. Phil. Trans. R. Soc. B 361, 2061–2078.

Brennan, P.A. and Zufall, F. (2006) Pheromonal communication in vertebrates. Nature 444, 308–315.

Brown, J.L. (1997) A theory of mate choice based on heterozygosity. Behav. Ecol. 8, 60–65.

Brown, J.L. and Eklund, A.C. (1994) Kin recognition and the major histocompatibility complex: an integrative review. Am. Nat. 143, 435–461.

Byers, J.A. and Waits, L. (2006) Good genes sexual selection in nature. Proc. Natl. Acad. Sci. USA 103, 16343–16345.

Carroll, L.S., Penn, D.J. and Potts, W.K. (2002) Discrimination of MHC-derived odors by untrained mice is consistent with divergence in peptide-binding region residues. Proc. Natl. Acad. Sci. USA 99, 2187–2192.

Charlesworth, D. and Charlesworth, B. (1987) Inbreeding depression and its evolutionary consequences. Ann. Rev. Ecol. Syst. 18, 237–268.

Cheetham, S.A., Thom, M.D., Jury, F., Ollier, W.E.R., Beynon, R.J. and Hurst, J.L. (2007) MUPs not MHC provide a specific signal for individual recognition in wild house mice. Unpublished manuscript.

Coltman, D.W., Bowen, W.D. and Wright, J.M. (1998) Birth weight and neonatal survival of harbour seal pups are positively correlated with genetic variation measured by microsatellites. Proc. R. Soc. B 265, 803–809.

Coltman, D.W., Pilkington, J.G., Smith, J.A. and Pemberton, J.M. (1999) Parasite-mediated selection against inbred Soay sheep in a free-living, island population. Evolution 53, 1259–1267.

Cooney, R. and Bennett, N.C. (2000) Inbreeding avoidance and reproductive skew in a cooperative mammal. Proc. R. Soc. B 267, 801–806.

Crnokrak, P. and Roff, D.A. (1999) Inbreeding depression in the wild. Heredity 83, 260–270.

Doherty, P.C. and Zinkernagel, R.M. (1975) Enhanced immunological surveillance in mice heterozygous at the H-2 complex. Nature 256, 50–52.

Drickamer, L.C., Gowaty, P.A. and Holmes, C.M. (2000) Free female mate choice in house mice affects reproductive success and offspring viability and performance. Anim. Behav. 59, 371–378.

Egid, K. and Brown, J.L. (1989) The major histocompatibility complex and female mating preferences in mice. Anim. Behav. 38, 548–550.

Hughes, A.L. and Nei, M. (1988) Pattern of nucleotide substitution at major histocompatibility complex class I loci reveals overdominant selection. Nature 335, 167–170.

Humphries, R.E., Robertson, D.H.L., Beynon, R.J. and Hurst, J.L. (1999) Unravelling the chemical basis of competitive scent marking in house mice. Anim. Behav. 58, 1177–1190.

Hurst, J.L., Payne, C.E., Nevison, C.M., Marie, A.D., Humphries, R.E., Robertson, D.H.L., Cavaggioni, A. and Beynon, R.J. (2001) Individual recognition in mice mediated by major urinary proteins. Nature 414, 631–634.

Hurst, J.L., Thom, M.D., Nevison, C.M., Humphries, R.E. and Beynon, R.J. (2005) MHC odours are not required or sufficient for recognition of individual scent owners. Proc. R. Soc. B 272, 715–724.

Jiménez, J.A., Hughes, K.A., Alaks, G., Graham, L. and Lacy, R.C. (1994) An experimental study of inbreeding depression in a natural habitat. Science 266, 271–273.

Jordan, W.C. and Bruford, M.W. (1998) New perspectives on mate choice and the MHC. Heredity 81, 239–245.

Keane, B., Creel, S.R. and Waser, P.M. (1996) No evidence of inbreeding avoidance or inbreeding depression in a social carnivore. Behav. Ecol. 7, 480–489.

Keller, L.F. (1998) Inbreeding and its fitness effects in an insular population of sparrows (*Melospiza melodia*). Evolution 52, 240–250.

Keller, L.F., Arcese, P., Smith, J.N.M., Hochachka, W.M. and Stearns, S.C. (1994) Selection against inbred song sparrows during a natural population bottleneck. Nature 372, 356–357.

Keller, L.F. and Waller, D.M. (2002) Inbreeding effects in wild populations. Trends Ecol. Evol. 17, 230–241.

Koenig, W.D., Stanback, M.T. and Haydock, J. (1999) Demographic consequences of incest avoidance in the cooperatively breeding acorn woodpecker. Anim. Behav. 57, 1287–1293.

Krackow, S. and Matuschak, B. (1991) Mate choice for non-siblings in wild mice - evidence from a choice test and a reproductive test. Ethology 88, 99–108.

Kurtz, J., Kalbe, M., Aeschlimann, P.B., Häberli, M.A., Wegner, K.M., Reusch, T.B.H. and Milinski, M. (2004) Major histocompatibility complex diversity influences parasite resistance and innate immunity in sticklebacks. Proc. R. Soc. B 271, 197–204.

Landry, C., Garant, D., Duchesne, P. and Bernatchez, L. (2001) 'Good genes as heterozygosity': the major histocompatibility complex and mate choice in Atlantic salmon (*Salmo salar*). Proc. R. Soc. B 268, 1279–1285.

Langefors, A., Lohm, J., Grahn, M., Andersen, O. and von Schantz, T. (2001) Association between major histocompatibility complex class IIB alleles and resistance to Aeromonas salmonicida in Atlantic salmon. Proc. R. Soc. B 268, 479–485.

Leinders-Zufall, T., Brennan, P., Widmayer, P., Chandramani, P., Maul-Pavicic, A., Jager, M., Li, X.H., Breer, H., Zufall, F. and Boehm, T. (2004) MHC class I peptides as chemosensory signals in the vomeronasal organ. Science 306, 1033–1037.

Mays, H.L. and Hill, G.E. (2004) Choosing mates: good genes versus genes that are a good fit. Trends Ecol. Evol. 19, 554–559.

Meagher, S., Penn, D.J. and Potts, W.K. (2000) Male-male competition magnifies inbreeding depression in wild house mice. Proc. Natl. Acad. Sci. USA 97, 3324–3329.

Milinski, M. (2006) The major histocompatibility complex, sexual selection, and mate choice. Ann. Rev. Ecol. Evol. Sys. 2006, 159–186.

O'Riain, M.J., Bennett, N.C., Brotherton, P.N.M., McIlrath, G. and Clutton-Brock, T.H. (2000) Reproductive supression and inbreeding avoidance in wild populations of co-operatively breeding meerkats (*Suricata suricatta*). Behav. Ecol. Sociobiol. 48, 471–477.

Olsén, K.H., Grahn, M., Lohm, J. and Langefors, Å. (1998) MHC and kin discrimination in juvenile Arctic charr, *Salvelinus alpinus* (L.). Anim. Behav. 56, 319–327.

Olsson, M., Madsen, T., Nordby, J., Wapstra, E., Ujvari, B. and Wittsell, H. (2003) Major histocompatibility complex and mate choice in sand lizards. Proc. R. Soc. B 270, S254–S256.

Parrott, M.L., Ward, S.J. and Temple-Smith, P.D. (2007) Olfactory cues, genetic relatedness and female mate choice in the agile antechinus (*Antechinus agilis*). Behav. Ecol. Sociobiol., DOI:10.1007/s00265-006-0340-8.

Paterson, S., Wilson, K. and Pemberton, J.M. (1998) Major histocompatibility complex variation associated with juvenile survival and parasite resistance in a large unmanaged ungulate population (*Ovis aires* L.). Proc. Natl. Acad. Sci. USA 95, 3714–3719.

Penn, D.J. (2002) The scent of genetic compatibility: sexual selection and the major histocompatibility complex. Ethology 108, 1–21.

Penn, D.J., Damjanovich, K. and Potts, W.K. (2002) MHC heterozygosity confers a selective advantage against multiple-strain infections. Proc. Natl. Acad. Sci. USA 99, 11260–11264.

Persaud, K.N. and Galef, B.G. (2005) Eggs of a female Japanese quail are more likely to be fertilized by a male that she prefers. J. Comp. Psych. 119, 251–256.

Potts, W.K., Manning, C.J. and Wakeland, E.K. (1991) Mating patterns in seminatural populations of mice influenced by MHC genotype. Nature 352, 619–621.

Potts, W.K., Manning, C.J. and Wakeland, E.K. (1994) The role of infectious disease, inbreeding and mating preferences in maintaining MHC genetic diversity: an experimental test. Phil. Trans. R. Soc. B 346, 369–378.

Pusey, A. and Wolf, M. (1996) Inbreeding avoidance in mammals. Trends Ecol. Evol. 11, 201–206.

Pusey, A.E. (1987) Sex-biased dispersal and inbreeding avoidance in birds and mammals. Trends Ecol. Evol. 2, 295–299.

Ralls, K., Ballou, J.D. and Templeton, A. (1988) Estimates of lethal equivalents and the cost of inbreeding in mammals. Conserv. Biol. 2, 185–193.

Rauch, R., Kalbe, M. and Reusch, T.B.H. (2006) Relative importance of MHC and genetic background for parasite load in a field experiment. Evol. Ecol. Res. 8, 373–386.

Reusch, T.B.H., Häberli, M.A., Aeschlimann, P.B. and Milinski, M. (2001) Female sticklebacks count alleles in a strategy of sexual selection explaining MHC polymorphism. Nature 414, 300–302.

Robertson, D.H.L., Cox, K.A., Gaskell, S.J., Evershed, R.P. and Beynon, R.J. (1996) Molecular heterogeneity in the major urinary proteins of the house mouse *Mus musculus*. Biochem. J. 316, 265–272.

Ryan, K.K. and Lacy, R.C. (2003) Monogamous male mice bias behaviour towards females according to very small differences in kinship. Anim. Behav. 65, 379–384.

Sato, A., Figueroa, F., Murray, B.W., Malaga-Trillo, E., Zaleska-Rutczynska, Z., Sultmann, H., Toyosawa, S., Wedekind, C., Steck, N. and Klein, J. (2000) Nonlinkage of major histocompatibility complex class I and class II loci in bony fishes. Immunogenet. 51, 108–116.

Schellinck, H.M., Slotnick, B.M. and Brown, R.E. (1997) Odors of individuality originating from the major histocompatibility complex are masked by diet cues in the urine of rats. Anim. Learn. Behav. 25, 193–199.

Seddon, N., Amos, W., Mulder, R.A. and Tobias, J.A. (2004) Male heterozygosity predicts territory size, song structure and reproductive success in a cooperatively breeding bird. Proc. R. Soc. B 271, 1823–1829.

Sherborne, A.L., Thom, M.D., Paterson, S., Jury, F., Ollier, W.E.R., Stockley, P., Beynon, R.J. and Hurst, J.L. (2007) The molecular basis of inbreeding avoidance in house mice. Submitted.

Simmons, L.W. (1991) Female choice and relatedness of mates in the field cricket, *Gryllus bimaculatus*. Anim. Behav. 41, 493–501.

Slate, J., Kruuk, L.E.B., Marshall, T.C., Pemberton, J.M. and Clutton-Brock, T.H. (2000) Inbreeding depression influences lifetime breeding success in a wild population of red deer (*Cervus elaphus*). Proc. R. Soc. B 267, 1657–1662.

Slate, J. and Pemberton, J.M. (2002) Comparing molecular measures for detecting inbreeding depression. J. Evol. Biol. 15, 20–31.

Smith, D.G. (1995) Avoidance of close consanguineous inbreeding in captive groups of rhesus macaques. Am. J. Primatol. 35, 31–40.

Soltis, J., Mitsunaga, F., Shimizu, K., Yanagihara, Y. and Nozaki, M. (1999) Female mating strategy in an enclosed group of *Japanes macaques*. Am. J. Primatol. 47, 263–278.

Spehr, M., Kelliher, K.R., Li, X.H., Boehm, T., Leinders-Zufall, T. and Zufall, F. (2006) Essential role of the main olfactory system in social recognition of major histocompatibility complex peptide ligands. J. Neurosci. 26, 1961–1970.

Stockley, P., Searle, J.B., Macdonald, D.W. and Jones, C.S. (1993) Female multiple mating-behaviour in the common shrew as a strategy to reduce inbreeding. Proc. R. Soc. B 254, 173–179.

Thornhill, R., Gangestad, S.W., Miller, R., Scheyd, G., McCullough, J.K. and Franklin, M. (2003) Major histocompatibility genes, symmetry and body scent attractiveness in men and women. Behav. Ecol. 14, 668–678.

Tregenza, T. and Wedell, N. (2000) Genetic compatibility, mate choice and patterns of parentage: invited review. Mol. Ecol. 9, 1013–1027.

Wedekind, C., Seebeck, T., Bettens, F. and Paepke, A. (1995) MHC-dependent mate preferences in humans. Proc. R. Soc. B 260, 245–249.

Wedekind, C., Walker, M. and Little, T.J. (2005) The course of malaria in mice: major histocompatibility complex (MHC) effects, but no general MHC heterozygosity advantage in single-strain infections. Genetics 170, 1427–1430.

Wedekind, C., Walker, M. and Little, T.J. (2006) The separate and combined effects of MHC genotype, parasite clone, and host gender on the course of malaria in mice. BMC Genetics 7, 55.

Wegner, K.M., Kalbe, M., Kurtz, J., Reusch, T.B.H. and Milinski, M. (2003) Parasite selection for immunogenetic optimality. Science 301, 1343.

Westerdahl, H., Waldenström, J., Hansson, B., Hasselquist, D., Von Schantz, T. and Bensch, S. (2005) Associations between malaira and MHC genes in a migratory songbird. Proc. R. Soc. B 272, 1511–1518.

Yamaguchi, M., Yamazaki, K., Beauchamp, G.K., Bard, J., Thomas, L. and Boyse, E.A. (1981) Distinctive urinary odors governed by the major histocompatibility locus of the mouse. Proc. Natl. Acad. Sci. USA 78, 5817–5820.

Yamazaki, K., Beauchamp, G.K., Bard, J., Thomas, L. and Boyse, E.A. (1982) Chemosensory recognition of phenotypes determined by the *Tla* and H-2K regions of chromosome 17 of the mouse. Proc. Natl. Acad. Sci. USA 79, 7828–7831.

Yamazaki, K., Beauchamp, G.K., Egorov, I.K., Bard, J., Thomas, L. and Boyse, E.A. (1983) Sensory distinction between H-2^b and H-2^{bm1} mutant mice. Proc. Natl. Acad. Sci. USA 80, 5685–5688.

Yamazaki, K., Boyse, E.A., Miké, V., Thaler, H.T., Mathieson, B.J., Abbott, J., Boyse, J., Zayas, Z.A. and Thomas, L. (1976) Control of mating preferences in mice by genes in the major histocompatibility complex. J. Exp. Med. 144, 1324–1335.

Yamazaki, K., Yamaguchi, M., Andrews, P.W., Peake, B. and Boyse, E.A. (1978) Mating preferences of F2 segregants of crosses between MHC-congenic mouse strains. Immunogenet. 6, 253–259.

Zeh, J.A. and Zeh, D.W. (1996) The evolution of polyandry I: intragenomic conflict and genetic incompatibility. Proc. R. Soc. B 263, 1711–1717.

Part V
Maternal - Offspring Communication

Chapter 29
Psychobiological functions of the mammary pheromone in newborn rabbits

Gérard Coureaud and Benoist Schaal

Abstract Lactating female rabbits (*Oryctolagus cuniculus*) nurse once daily for 5 min. The pups are thus forced to localize the nipples quickly. The females emit multiple chemosignals to guide them, among which are common odors derived from dietary aromas and one identified pheromone. Thus, the mammary pheromone (MP) released in rabbit milk is highly behaviorally active: it is a potent releaser of suckling-related behaviors in pups. The MP also acts as a strong reinforcer for early odor learning. Both of these functions of the MP are active right after birth and are thus in a position to play a central role in neonatal adaptation. However, the releasing function and the reinforcing functions of the MP follow different time-courses, suggesting that they are controlled by different underlying mechanisms. These data illustrate how a same chemical signal can carry distinct functions in a newborn mammal.

29.1 Rabbit Parenting and Pup Behavioral Adaptations

Rabbit females interact parsimoniously with their litter. They enter the nest once a day to nurse for only 3–5 min (Zarrow, Denenberg and Anderson 1965). During this brief period, the pups have to orient to the female's abdomen, localize the nipples, grasp them and suck efficiently. All these targeted behaviors are expressed in a context of harsh competition under the mother (Drummond, Vázquez, Sanchez-Colón, Martinez-Gómez and Hudson 2000; Bautista, Mendoza-Degante, Coureaud, Martina-Gomez and Hudson 2005). If the pups miss the first 2-3 nursing bouts, their survival is threatened (Coureaud, Schaal, Coudert, Rideaud, Fortun-Lamothe, Hudson and Orgeur 2000). Although the pups cannot rely on vision and hearing in their quest for milk (during the first postnatal week), they are nevertheless able to seize nipples within less than 15 sec. To that aim, they display a typical behavior pattern of probing consisting in horizontal and vertical scanning movements of the head

Gérard Coureaud

Centre des Sciences du Goût, CNRS/Univ. Bourgogne/Inra

coureaud@cesg.cnrs.fr

(searching) ending by the grasping of a nipple by the lips (Schley 1981; Hudson and Distel 1983).

This behavior is under the control of peri-oral somesthesis and olfaction. The critical role of olfaction was demonstrated in a series of studies run over the last 30 years. By altering the olfactory system (Schley 1981; Distel and Hudson 1985; Hudson and Distel 1986), by washing or selectively covering mammary areas (Müller 1978; Hudson and Distel 1983; Coureaud, Schaal, Langlois and Perrier 2001), by contrasting females' physiological states (Hudson and Distel 1984; González-Mariscal, Chirino and Hudson 1994; Coureaud and Schaal 2000), and by manipulating chemosensory experience in the fetus or the newborn (Semke, Distel and Hudson 1995; Schaal, Coureaud, Marlier and Soussignan 2001; Coureaud, Schaal, Hudson, Orgeur and Coudert 2002), it was shown that the nipple search of newborns and the typical responses involved in it were controlled by olfaction.

From these studies it emerged that odor cues released from the female's abdomen to the skin surface or into milk are involved in pup directional behavior (Schley 1981; Hudson and Distel 1983; Keil, von Stralendorff and Hudson 1990; Coureaud et al. 2002). Recently, one of these chemosignals emitted by lactating rabbit females has been identified and characterized for its particular activity in releasing sucking-related behavior in newborn pups.

29.2 A Releasing Factor in Rabbit Milk

Applying gas chromatography coupled with olfactometry to analyze the headspace of fresh rabbit milk, and then assaying individual candidate compounds, an active volatile was identified as 2-methylbut-2-enal (2MB2). This compound efficiently releases the typical searching-grasping behavior in 2-3-day old pups (> 90% of the pups respond to it). 2MB2 appears to be a key compound of rabbit milk, as it reinstates full activity to a sample of milk previously deactivated by evaporation (Coureaud 2001; Schaal, Coureaud, Langlois, Giniès, Sémon and Perrier 2003; Coureaud, Langlois, Perrier and Schaal 2003). 2MB2 was then systematically screened to assess whether its properties qualify as a pheromone.

29.2.1 A Species-Specific Signal

Since the original definition proposed by Karlson and Lüscher (1959) for insects, the concept of pheromone has undergone warm debate (e.g., Bronson 1971; Nordlund and Lewis 1976). To adapt the concept to the cognitive complexity of mammalian behavior and to assess their real existence in mammals, some authors suggested adding criteria to the original definition. They proposed a restricted definition based on a set of five operational criteria (Beauchamp, Doty, Moulton and Mugford 1976; see also Johnston 2000). These criteria were systematically evaluated with regard to 2MB2.

The compound 2MB2 met each of the five criteria (Coureaud 2001; Schaal et al. 2003; Coureaud et al. 2003; Coureaud, Langlois, Sicard and Schaal 2004). First, 2MB2 is a single molecular compound. Second, it triggers well-defined and macroscopically stereotyped responses, resembling those displayed in the natural nursing situation (arousal, orientation to the source, and typical head searching-oral seizing movements). Third, its behavioral activity is selective, both in qualitative (no other volatile from rabbit milk, nor odorants active in other mammals or arbitrarily chosen odorants are as strongly active on pups) and in quantitative terms (2MB2 releasing effect is concentration-dependent). Fourth, 2MB2 activity is species-specific in both reception and emission: *Oryctolagus* newborns respond strongly to it regardless of strain and breed; non-*Oryctolagus* newborns (rat, mouse, cat) and even phylogenetically-close *Lepus europeus* newborns do not respond to it; various milks from heterospecific females (rat, cow, horse, sheep, human) are inactive on rabbit pups. Five, the behavioral activity of 2MB2 appears independent from learning processes occurring pre- or postnatally: pups deprived of contact with the dam or her milk (term or preterm pups withdrawn from the mother at birth) respond to 2MB2 in the same way as pups normally exposed to nursing; further, 2MB2 being undetectable by GC-MS in amniotic fluid and blood of pregnant females, its activity may not derive from prenatal induction.

To sum up, 2MB2 emitted in milk by female rabbits meets the five criteria proposed by Beauchamp et al. (1976) to define a pheromone in the case of mammals. As rabbit milk releases neonatal behavior after full ejection, but not when collected from the mammary tract below the nipples (from the tubes or alveolae), 2MB2 is probably produced *de novo* in the distal part of the mammary apparatus (Moncomble, Coureaud, Quennedey, Langlois, Perrier, Brossut and Schaal 2005). Accordingly, the active compound carried in milk was named "mammary pheromone" (MP).

29.2.2 Developmental and Daily Variations in MP Activity

As the mammary pheromone (MP) is emitted in milk and releases nursing-related behavior in neonate pups, its activity can logically be expected to be restricted in time and constrained by the metabolic needs of the pups.

The behavioral activity of milk and of the MP, and the maternal emission of MP, were investigated along the 30-day lactation period. Pups of fixed age (day 2) were exposed to milk samples collected during early vs. late lactation (postpartum days 2 vs. 23). Early-lactation milk released the typical searching responses at a much higher rate (> 80% of responding pups) than late-lactation milk (< 7%), suggesting that the active factor is altered and/or is less available close to the end of lactation (Coureaud, Langlois, Perrier and Schaal 2006a). A convergent result was obtained for the releasing effect of undefined odor cues emitted from the abdomen of lactating females (among which the MP may play a role), which are less attractive for 2-day old pups in late than in early lactation (Coureaud et al. 2001). It may be hypothesized that active compounds in milk, and perhaps in the effluvium of the female's abdomen, progressively lose their activity when weaning approaches either because their availability drops below pups' sensitivity thresholds, because the pups' perceptual processes are changing relative to their concentration, or because both phenomena occur simultaneously. Data suggest that both causes are influential. First, quantitative GC-MS screening for 2MB2 confirms that its concentration is higher in the headspace of early-lactation milk as compared to late-lactation milk (Coureaud et al. 2006a). Second, typical searching-grasping responses to intensity-constant MP drop from day 15 to completely vanish around weaning (day 28) (Coureaud, Rödel, Kurz, von Holst and Schaal 2006b, and unpublished data).

Pup responsiveness to MP also undergoes circadian fluctuations, suggesting that it is controlled by the interplay of external and internal factors. The well-defined nursing rhythm makes it easy to investigate the role of circadian rhythm of pups (they anticipate the female's nursing visit; Jilge and Hudson 2001) and of factors related to the feeding cycle. A first study consisted thus in following up the oro-cephalic responses of pups exposed to MP from birth to day 10 before and after the daily nursing. When tested at regular times of the 24-hour cycle (i.e., at 6 h, 3 h, 5 min before suckling, and 5 min, 3 h, 6 h and 12 h after suckling) 2-day old pups responded maximally at all measurement times throughout the day. On day 5, their responsiveness to MP remained high in the hours and minutes before milk intake, but decreased significantly immediately after it (< 50% of responding pups). Finally, on day 10 the post-suckling drop in pup responsiveness was deeper and longer (Fig. 29.1). It might then be argued that the response decrease displayed by sated pups on days 5 and 10 is dependent on circadian rhythm factors. However, 5 and 10-day old pups that missed the daily nursing respond strongly to MP in the minutes and hours following the scheduled nursing time (Montigny, Coureaud and Schaal 2006). Thus, the behavioral activity of MP appears to be 'automatic' over the first 2–3 postnatal days, and becomes progressively controlled by feeding-related factors.

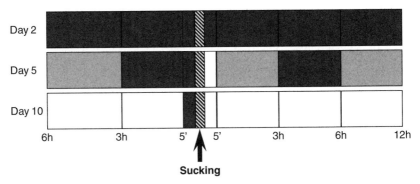

Fig. 29.1 Change in the proportion of pups (> 80%, 60–80%, < 60%: black, grey and white bars, respectively) responding to MP between 6 h before and 12 h after the daily nursing (hatched bars), from day 2 to day 10 (*adapted from* Montigny et al. 2006)

These results highlight that MP is an extremely potent releaser of neonatal behaviors immediately after birth, when it drives directional actions in complete independence of learning. This automatic activity is initially extended over the entire day cycle, but it becomes rapidly restricted to the period of the day preceding nursing. Finally, MP activity decreases in young rabbits and disappears when their exclusive dependence on milk comes to a close. Thus, MP may be considered as a species-specific signal directly involved in the suckling success of altricial pups.

29.3 Mammary Pheromone as a Reinforcing Factor

Pheromones have been classified as a function of their effect. Those triggering immediate behavioral effects were termed releasing pheromones, while those altering slower processes of physiological or developmental nature were termed priming pheromones (Wilson and Bossert 1963). The rabbit MP has a clear releasing function but recent data also show that it has another function that cannot be termed priming in the classical sense.

Immediately after birth, rabbit pups are very efficient learners. Odorants introduced into their prenatal and immediate postnatal environment impact on the development of their sensitivity and preferences (e.g., Bilkó, Altbäcker and Hudson 1994; Coureaud et al. 2002; Kindermann, Gervais and Hudson 1991; Semke et al. 1995). Plastic olfactory learning is thus possible during the period when MP has its highest level of behavioral activity. One may wonder if these plastic and predisposed mechanisms might interfere with each other. Plastic processes might inhibit the predisposed processes, or predisposed processes involving MP could facilitate or fully control plastic processes. To assess the direction of influence between both perceptual processes, newborn rabbits were exposed for 5 min to an odor mixture composed of MP and of a neutral odorant (releasing no orocephalic action). When exposed to the odorant alone 24 h later, they expressed high levels of head searching and oral grasping. Thus, the releasing power of MP may be transferred to a novel odorant to which it is associated, which can itself then act as a reinforcing agent (Coureaud, Moncomble, Montigny, Dewas, Perrier and Schaal 2006c). Importantly, pups do not learn the odorant when they are merely exposed to it for 5 min, indicating that MP is clearly involved in this single-trial conditioning (Fig. 29.2A). Moreover, at certain concentrations of the MP-odor mixture, the initially neutral odorant becomes as efficient in releasing neonatal behavior as the MP itself (Coureaud et al. 2006c; Fig. 29.2B). To sum up, the rabbit MP can be considered as an unconditioned stimulus able to confer to any neutral odorant (conditioned stimulus) the value of a signal. After a consolidation period, the CS becomes then a releaser of a conditioned response similar to the unconditioned one released by MP (Fig. 29.2C).

The reinforcing activity of MP is efficient immediately after birth: when pups are simultaneously exposed to MP and a novel odorant 30 min after delivery, the odorant elicits robust orocephalic responses 24 h later. The proportion of responding pups is then similar to the proportion observed when the conditioning is realized on

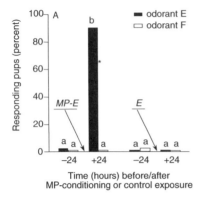

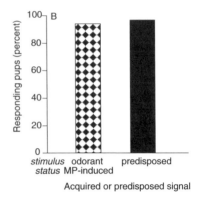

Fig. 29.2 A. Proportion of pups responding by searching movements to neutral odorant E and F, 24h after being exposed to the MP-odorant E mixture (MP-E) or to odorant E alone (E). **B.** Proportion of pups displaying searching to an odorant MP-learned (dotted bar) as compared to the MP (black bar). **C.** Schematic of the reinforcing function of MP (US, CS and UR: unconditioned stimulus, conditioned stimulus and unconditioned response, respectively) (*adapted from* Coureaud et al. 2006c)

postnatal day 2 (Coureaud et al. 2006c). It may be noted that MP does not need to trigger general activation of the pups to be effective in the learning of the odorant. Thus, the ability of MP to recruit the mechanisms supporting odor learning is neither dependent on co-occurring reinforcing factors linked with the mother (i.e., comfort contact, suckling, milk, gastric filling and satiety) nor on general arousal which is reinforcing in itself (Sullivan, Hofer and Brake 1986). Therefore, MP may be considered as a primary reinforcer (Rosenblatt 1983) that operates independently of perinatal experience.

Thus, MP works simultaneously as a releasing and as a reinforcing agent. However, both properties of MP undergo different time-courses during early development. In day-2 pups, MP is highly efficient at releasing oro-cephalic movements regardless of whether pups are hungry or sated; in contrast, MP-induced odor learning is less effective when pups are satiated than when they are fasted (Montigny, Coureaud and Schaal 2005). In addition, the releasing function of MP is high until day 10, but its reinforcing function stops abruptly before the end of the first postnatal week (Coureaud, Montigny, Dewas, Moncomble, Patris and Schaal 2005, and unpublished data). Thus, both releasing and reinforcing functions of the MP are fully active in newborns, but then they follow different lines in early development, suggesting that they may be controlled by different neural/behavioral mechanisms.

29.4 Conclusions

The extensive work conducted on the rabbit during the last three decades has greatly widened our knowledge of the olfactory means coevolved by mammalian mothers and young. On the maternal side, it has demonstrated different mechanisms producing odor cues and signals that act concurrently to boost localization and access of newborns to the milk resource. On the offspring side, it highlights perceptual specializations that promote quasi-immediate adaptive responses immediately after birth. Some of these perceptual processes are automatic, functioning without apparent specification from environmental factors, while others are entirely malleable according to stimuli afforded by the mother or the nest. These automatic and plastic processes interact, in the sense that any odorant simultaneously present with MP in the environment may be learned as a cue for guiding future behavior. Whether such a rapid learning mechanism associating automatic responses to a pheromone, and plastic responses to co-occurring common odorants, is general to mammalian newborns, or whether this is a specificity of the rabbit, remains to be investigated.

Acknowledgments We thank Dominique Langlois, Guy Perrier, Jacques Ponceau, Anne-Sophie Moncomble, Delphine Montigny, Maeva Dewas, Gilles Sicard and André Holley for their various contributions in the work reported above, in either effective realization or active discussion. Our research was supported by CNRS, French Ministry of Research and Technology, Councils of Régions Poitou-Charentes and Bourgogne, Inra, UNAM (Mexico) and Fyssen Foundation.

References

Bautista, A., Mendoza-Degante, M., Coureaud, G., Martina-Gomez, M., and Hudson, R. (2005) Scramble competition in newborn domestic rabbits for an unusually restricted milk supply. Anim. Behav. 70, 997–1002.

Beauchamp, G.K., Doty, R.L., Moulton, D.G. and Mugford, R.A. (1976) The pheromone concept in mammalian chemical communication. In Doty R.L. (Ed.), *Mammalian Olfaction, Reproductive Processes and Behaviour*. Academic Press, New York, pp 143–160.

Bilkó, A., Altbäcker, V. and Hudson, R. (1994) Transmission of food preference in the rabbit: The means of information transfer. Physiol. Behav. 56, 907–912.

Bronson, F.H. (1971) Rodent pheromones. Biol. Reprod. 4, 344–357.

Coureaud, G. (2001) Olfactory regulation of sucking in newborn rabbit: Ethological and chemical characterization of a pheromonal signal. PhD Thesis, University of Paris.

Coureaud, G. and Schaal, B. (2000) Attraction of newborn rabbits to abdominal odors of adult conspecifics differing in sex and physiological state. Dev. Psychobiol. 36, 271–281.

Coureaud, G., Langlois, D., Perrier, G. and Schaal B. (2003) A single key-odorant accounts for the pheromonal effect of rabbit milk: Further test of the mammary pheromone's activity against a wide sample of volatiles from milk. Chemoecology 13, 187–192.

Coureaud, G., Langlois, D., Perrier, G. and Schaal, B. (2006a). Convergent changes in the maternal emission and pup reception of the rabbit mammary pheromone. Chemoecology 16, 169–174.

Coureaud, G., Langlois, D., Sicard, G. and Schaal, B. (2004) Newborn rabbit reactivity to the mammary pheromone: Concentration-response relationship. Chem. Senses 29, 341–350.

Coureaud, G., Montigny, D., Dewas, M., Moncomble, A.S., Patris, P. and Schaal, S. (2005) A sensitive period for pheromone-induced odour learning in newborn rabbits. 29th International Ethological Conference, Budapest, Hungary, 20–27 August.

Coureaud, G., Moncomble, A.S., Montigny, D., Dewas, M., Perrier, G. and Schaal, B. (2006c) A pheromone that rapidly promotes learning in the newborn. Curr. Biol. 16, 1956–1961.

Coureaud, G., Rodel, H.G., Kurz, C., von Holst, D. and Schaal, B. (2006b) Age-related changes in the processing of the rabbit mammary pheromone: comparison in domestic and wild pups. 17th Congress of the European Chemoreception Research Organisation, Granada, Spain, and Chem. Senses 31, 274 (abstract online).

Coureaud, G., Schaal, B., Coudert, P., Rideaud, P., Fortun-Lamothe, L., Hudson, R. and Orgeur, P. (2000) Immediate postnatal sucking in the rabbit: its influence on pup survival and growth. Reprod. Nutr. Dev. 40, 19–32.

Coureaud, G., Schaal, B., Hudson, R., Orgeur, P. and Coudert, P. (2002) Transnatal olfactory continuity in the rabbit: behavioral evidence and short-term consequence of its disruption. Dev. Psychobiol. 40, 372–390.

Coureaud, G., Schaal, B., Langlois, D. and Perrier G (2001) Orientation response of newborn rabbits to odours emitted by lactating females: Relative effectiveness of surface and milk cues. Anim. Behav. 61, 153–162.

Distel, H. and Hudson, R. (1985) The contribution of olfactory and tactile modalities to the performance of nipple-search behaviour in newborn rabbits. J. Comp. Physiol. A 157, 599–605.

Drummond, H. Vázquez, E., Sanchez-Colón, S., Martinez-Gómez, M. and Hudson, R. (2000) Competition for milk in the domestic rabbit: Survivors benefit from littermate deaths. Ethology 106, 511–526.

González-Mariscal, G., Chirino, R. and Hudson, R. (1994) Prolactin stimulates emission of nipple pheromone in ovariectomized New Zealand white rabbits. Biol. Reprod. 50, 373–376.

Hudson, R. and Distel H. (1983) Nipple location by newborn rabbits: behavioural evidence for pheromonal guidance. Behaviour 85, 260–275.

Hudson, R. and Distel, H. (1984) Nipple-search pheromone in rabbits: dependence on season and reproductive state. J. Comp. Physiol. A 155, 13–17.

Hudson, R. and Distel, H. (1986) Pheromonal release of suckling in rabbits does not depend on the vomeronasal organ. Physiol. Behav. 37, 123–129.

Jilge, B. and Hudson, R. (2001) Diversity and development of the circadian rhythms in the European rabbit. Chronobiol. Int. 18, 1–26.

Johnston, R.E. (2000) Chemical communication and pheromones: The types of chemical signals and the role of the vomeronasal system. In Finger T.E., Silver W.L., and Restrepo D. (Eds.), The Neurobiology of Taste and Smell. Wiley-Liss, New York, pp 101–127.

Karlson, P. and Lüscher, M. (1959) "Pheromones": a new term for a class of biologically active substances. Nature 183, 55–56.

Keil, W., von Stralendorff, F. and Hudson, R. (1990) A behavioral bioassay for analysis of rabbit nipple-search pheromone. Physiol. Behav. 47, 525–529.

Kindermann, U., Gervais, R. and Hudson, R. (1991) Rapid odor conditioning in newborn rabbits: Amnesic effect of hypothermia. Physiol. Behav. 50, 457–460.

Moncomble, A.S., Coureaud, G., Quennedey, B., Langlois, D., Perrier, G., Brossut, R. and Schaal, B. (2005) The mammary pheromone of the rabbit: where does it come from? Anim. Behav. 69, 29–38.

Montigny, D., Coureaud, G. and Schaal, B. (2005) Circadian and developmental fluctuations in the releasing and reinforcing potencies of the rabbit mammary pheromone. 16th Congress of the European Chemoreception Research Organisation, Dijon, France, and Chem. Senses 30(2), 180 (abstract online).

Montigny, D., Coureaud, G. and Schaal, B. (2006) Newborn rabbit response to the mammary pheromone: From automatism to prandial control. Physiol. Behav. 89, 742–749.

Müller, K. (1978) Zum Saugverhalten von Kaninchen unter besonderer Berücksichtigung des Geruchsvermögens. PhD Thesis, University of Giessen.

Nordlund, D.A. and Lewis W.J. (1976) Terminology of chemical releasing stimuli in intraspecific and interspecific interactions. J. Chem. Ecol. 2, 211–220.

Rosenblatt, J.S. (1983) Olfaction mediates developmental transition in the altricial newborn of selected species of mammals. Dev. Psychobiol. 16, 347–375.

Schaal, B., Coureaud, G., Langlois, D., Giniès, C., Sémon, E. and Perrier, G. (2003) Chemical and behavioural characterization of the rabbit mammary pheromone. Nature 424, 68–72.

Schaal, B., Coureaud, G., Marlier, L. and Soussignan, R. (2001) Fetal olfactory cognition preadapts neonatal behavior in mammals. In Marchlewska-Koj A., Lepri J. and Müller-Schwarze D. (Eds.), *Chemical Signals in Vertebrates, Volume 9*. Plenum-Klüwer Academic, New York, pp 197–205.

Schley, P. (1981) Olfaction and suckling in young rabbits. Kleintierpraxis 26, 261–263.

Semke E., Distel H. and Hudson, R. (1995) Specific enhancement of olfactory receptor sensitivity associated with foetal learning of food odours in the rabbit. Naturwissenschaften 82, 148–149.

Sullivan, R.M., Hofer, M.A. and Brake, S.C. (1986) Olfactory-guided orientation in neonatal rats is enhanced by a conditioned change in behavioral state. Dev. Psychobiol. 19, 615–623.

Wilson, E.O. and Bossert, W.H. (1963) Chemical communication among animals. Recent Prog. Horm. Res. 19, 673–716.

Zarrow, M. X., Denenberg, V. H. and Anderson, C. O. (1965) Rabbit: frequency of suckling in the pup. Science 150, 1835–1836.

Chapter 30
Rabbit Nipple-Search Pheromone Versus Rabbit Mammary Pheromone Revisited

**Robyn Hudson, Carolina Rojas, Lourdes Arteaga,
Margarita Martínez-Gómez and Hans Distel**

Abstract Among mammals, rabbits (*Oryctolagus cuniculus*) show unusually limited maternal care and only nurse for a few minutes once each day. Successful suckling depends on pheromonal cues on the mother's ventrum, which release a stereotyped and distinctive pattern of nipple-search behaviour in the young, and which have been termed the nipple-search pheromone. The present report summarizes what is currently known about this unusually effective chemical signal and compares this with information in more recent reports of a rabbit mammary pheromone thought to achieve the same function. We draw attention to anomalies in the present state of knowledge regarding the nature and action of these two sets of chemical signals, and thus to the continuing uncertainty as to the chemical nature and source of the cues governing nipple-search behaviour, and thus successful suckling, in the newborn rabbit.

30.1 Introduction

This report is an extension and up-date of work presented at the 1994 Chemical Signals in Vertebrates meeting held in Tübingen, Germany. At that meeting and in an accompanying book chapter (Hudson and Distel 1995) we reviewed the substantial evidence that female rabbits emit odour cues on their ventrum in the form of a rough gradient that release and rapidly guide the altricial young to nipples and enable nipple grasping and attachment. In 1983 we had named these chemically unidentified cues 'nipple-search pheromone' for their vital function in ensuring successful suckling by the young (Hudson and Distel 1983). Since then efforts to identify the substance(s) eliciting the pups' remarkably stereotyped and efficient behaviour have advanced considerably. In 2003 Schaal and colleagues (Schaal, Coureaud, Langlois, Giniès, Sémon and Perrier 2003) reported that a single substance in rabbit milk, 2-methylbut-2-enal (2MB2), which they named 'mammary pheromone', could release search-like head movements and grasping in newborn

Robyn Hudson
Univ. Nac. Autónoma de México, Inst. Invest. Biomédicas
rhudson@biomedicas.unam.mx

rabbits. However, in the subsequent reports on the action of 2MB2 (reviewed below) certain anomalies emerge regarding its equivalence to the nipple-search pheromone and the manner of its action in the natural nursing context, whether alone or together with other behaviourally active substances. It is therefore the purpose of this review to summarize current knowledge about these two pheromones, and how the recent advances contribute to our understanding of this remarkable chemical signal, the exact nature of which remains one of the rabbit's best kept secrets.

30.2 Nursing. The Natural Context

As we reported in the 1995 chapter, rabbits are unusual among mammals in providing their young with remarkably little maternal care. In late pregnancy the mother digs a nursery burrow or a chamber in the colony warren in which she builds a nest of grass, and fur pulled from her own body (reviewed in González-Mariscal and Rosenblatt 1996). After giving birth she leaves the pups immediately (Hudson, Cruz, Lucio, Ninomiya and Martínez-Gómez 1999) and only returns to nurse for three to four minutes approximately every 24 hours (reviewed in Hudson and Distel 1989; Jilge and Hudson 2001; Caldelas, Chimal-Monroy, Martínez-Gómez and Hudson 2005). On entering the nest she simply positions herself over the litter and remains motionless without giving the pups any direct behavioural assistance to suckle (Hudson and Distel 1982; 1983). Moreover, the pups obtain substantial amounts of milk only during the second minute, reducing their effective suckling time to about one minute every 24 hours (Bautista, Mendoza-Degante, Coureaud, Martínez-Gómez and Hudson 2005). Nursing ends abruptly with the mother jumping out of the nest and leaving the pups alone until the following day.

Milk production starts to decline around postpartum day 20 (reviewed in Hudson, Bilkó and Altbacker 1996), and if the mother is pregnant from the usual postpartum mating she will stop nursing the current litter around day 25. Whereas the previous day she shows normal nursing behaviour, 24 hours later she refuses to nurse and under confined laboratory conditions she may even attack young attempting to suckle. With the arrival of the next litter several days later the usual once-daily nursing rhythm is resumed (reviewed in Hudson and Distel 1995; Hudson et al. 1996).

30.3 Suckling. The Natural Context

Rabbit pups anticipate and prepare for the mother's vital, once-daily visit behaviourally and physiologically. An hour or so before the mother's arrival they become increasingly active, increasingly responsive to vibrational and tactile stimuli, and gradually uncover from the nest material (Hudson and Distel 1982). This enables them to reach the mother's ventrum unhindered and to start the rapid search for nipples the moment she stands over them. The anticipatory arousal is accompanied by a rise in pups' (and mothers') body temperature, and in the expression of immediate early and 'clock' genes in hypothalamic and associated brain structures. Anticipation of nursing represents an endogenous circadian rhythm and does not simply depend on emptying of the gut. Thus, litters that miss a feed show again an increase in activity, body temperature and gene expression in anticipation of the next nursing 48 hours later (reviewed in Hudson and Distel 1989; Jilge and Hudson 2001; Caldelas et al. 2005; Caldelas, Tejadilla, González, Montúfar and Hudson 2007).

Immediately the mother enters the nest and adopts the nursing posture the pups rear up and push their muzzles deep into her ventral fur to start the rapid search for nipples. Observing suckling through the floor of a glass-bottomed nest box (Fig. 30.1), we have found that pups take on average only about six seconds to attach to nipples (Hudson and Distel 1983; Bautista et al. 2005). The search behaviour is highly stereotyped and is shown in response to any nursing female at any time of day. Making rapid probing movements with the muzzle, pups scan the doe's ventrum with a sewing machine-like action until a nipple is reached. Surprisingly, they do not remain on one nipple but change them repeatedly, performing the whole search sequence several times within the one nursing session. Although nipple switching reduces the time pups actually spend on nipples to an average of only about 110 seconds per nursing, they are able to drink 25% to 30% of their own weight in this time (Lincoln 1974; own observations).

The transition to solid food begins quite early and even while pups are still dependent on the mother's milk. On about postnatal day 12 they start to nibble faecal pellets deposited by the mother in the nest, and around day 14 they start to eat the nest material (Hudson et al. 1996). As the quantities eaten are small, this probably helps prepare the young for the digestion of solid food, for example by providing appropriate gut flora, rather than being of direct nutritional benefit. Milk production starts to decline around day 20 and several days later the young are able to survive on solid food alone. However, if the mother is not pregnant with a further litter she will continue to nurse for considerably longer (Lincoln 1974; Hudson et al. 1996).

Fig. 30.1 Glass-bottomed observation cage equipped with an angled mirror mounted below used for filming pups' behaviour during nursing as shown in the close-up video image on the right. Criss-crossed lines of silicon on the glass floor were to prevent animals slipping

30.4 The Nipple-Search Pheromone

How, then, are rabbit pups, blind until postnatal day nine and nursed in a dark burrow, able to find nipples and suckle so efficiently? To investigate this we tested pups on does restrained on their back in a U-shaped trough, the ends of which were closed by dividers to form an arena enclosing the six rear nipples (Fig. 30.2; Hudson and Distel 1983; Distel and Hudson 1984; 1985). By testing pups under various conditions we found that odour cues on the doe's ventrum are essential for the release and maintenance of nipple-search behaviour and for nipple attachment (Hudson and Distel 1983; Distel and Hudson 1985). Neither shaving the doe's ventrum nor creating a negative thermal gradient by cooling nipple areas significantly affected pups' behaviour. However, when the shaved ventrum was completely covered with adhesive tape but the nipples left bare, pups did not search and only attached to nipples when they chanced to bring their nose near them.

Furthermore, covering the area around nipples and/or the nipples themselves in various ways (for example as in Fig. 30.2) suggested that the short-ranging odour cues releasing and sustaining the search behaviour increase in strength towards nipples and might help guide pups there; when nipples were covered with adhesive patches pups located the nipples and remained searching vigorously around them although unable to grasp them and attach (Fig. 30.2: Hudson and Distel 1983). The vital nature of these olfactory cues is shown by the fact that newborn rabbits are completely unable to suckle from the mother when made anosmic by removing the olfactory bulbs or irrigating the nasal mucosa with zinc sulphate (Schley 1977; 1981; Distel and Hudson 1985) and even when the accessory olfactory system is left intact (Hudson and Distel 1986). However, olfactory cues alone are not sufficient to ensure suckling. When we denervated pups' muzzles by transecting the subopthalmic branches of the trigeminal nerve, pups searched for nipples but were unable to grasp them and suck (Distel and Hudson 1985).

In conclusion, the reliable, functionally vital and stereotyped nature of the pups' response to the chemical signal governing nipple-search behaviour would seem to qualify this as a true mammalian releasing pheromone (Beauchamp, Doty, Moulton and Mugford 1976) and particularly as it appears to be species-specific; rabbit pups fail to respond to lactating cats, rats, guinea pigs or even hares with nipple-search behaviour or nipple attachment (Müller 1978; Hudson 1985; own observations).

30.4.1 Development of Pups' Responsiveness

It is notable that rabbit pups show the full pattern of nipple-search and suckling behaviour at the very first nursing, as do pups delivered by Caesarean section and thus without postnatal experience (Hudson 1985). However, reports that rabbit pups can learn odours prenatally (Bilkó, Altbäcker and Hudson 1994; Altbäcker, Hudson and Bilkó 1995; Semke, Distel and Hudson 1995) make it difficult to

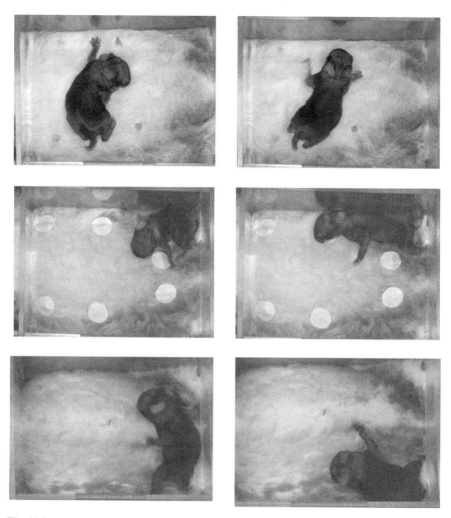

Fig. 30.2 Examples from video recordings of two-day-old pups (1) searching and then sucking on the ventrum of an upturned lactating female (top panel), (2) searching and then locating a nipple area on a lactating female whose nipples were covered with adhesive plastic patches (middle panel), and (3) searching on a female who had just given birth and was therefore lactating but whose nipples had been surgically removed several months previously (bottom panel)

exclude prenatal learning of the pheromonal cues, and particularly as odours present prenatally enhance suckling performance when encountered later on the mother's ventrum (Coureaud, Schaal, Coudert, Rideaud, Fortun-Lamothe, Hudson and Orgeur 2000).

During the first five days of postnatal life pups show a considerable improvement in the speed with which they locate nipples, both during normal nursing and in the test arena (Hudson and Distel 1983; Distel and Hudson 1984). This appears mainly due to an improvement in responding to the pheromone and a consequent reduction in the time taken to initiate searching rather than simply to pups' better motor coordination (Distel and Hudson 1984). The fact that pups hand-raised to day five on an artificial milk formula, and thus without postnatal experience of the pheromone or maternal odours, attach as rapidly as normally nursed pups when held to nipples (Schley 1981), and locate them as quickly when placed on the doe's ventrum (Hudson 1985), suggests that the improvement is due to maturational processes rather than to learning. However, after the first postnatal week, latencies to initiate searching start to increase again after nursing, possibly reflecting motivational changes associated with the developing hunger system (Distel and Hudson 1984).

30.4.2 Hormonal Control of Emission

Somewhat surprisingly, emission of the pheromone is not confined to lactating females and by using the response of pups to regularly test for its presence, all mature females have been found to produce it depending on their reproductive state. Whereas emission is low or absent in winter, it increases in early spring to reach a peak at about the time of the summer solstice (Hudson and Distel 1984). A close relationship between emission and photoperiod can be demonstrated by the experimental reversal of seasonal light conditions. Whereas artificial short days suppress emission, long days stimulate it within one to two weeks (Hudson and Distel 1984; 1990; Hudson, Melo and González-Mariscal 1994). Only sexually mature females produce the pheromone, and at no time of year do pups search for or attach to the nipples of animals younger than three months of age. However, pregnancy and lactation have the strongest influence, overriding the effects of day length. Pregnancy stimulates emission even in winter, so that by parturition and during early lactation pups are able to locate and attach to nipples within a few seconds. Early removal of pups from the mother results in a reduction in emission within a few days (Hudson and Distel 1984).

The correlation between emission and females' reproductive state suggests production of the pheromone to be under hormonal control. This is supported by the results of a series of experiments designed to simulate the changes in emission described above (Hudson and Distel 1984; Hudson and Distel 1990; Hudson et al. 1994; González-Mariscal, Chirino and Hudson 1994), and which may be summarized as follows. Ovariectomized females show virtually no emission even when kept under normally stimulating long day conditions, but show increasingly greater levels when treated with a sequence of estradiol, plus progesterone, plus prolactin. The lack of or weak effects of progesterone or prolactin when administered alone, but strong enhancing effects when injected together with or following other hormone treatments suggest these to act synergistically or as part of a cascade regulating the still unknown physiological processes governing pheromone emission.

30.4.3 Source and Chemical Identity?

Although the nipple-search pheromone is clearly under hormonal control, little is known about where or how it is produced, emitted or comes to be distributed on the mother's ventrum. Nevertheless, the epidermis of the nipples has been suggested to be a likely source (Schley 1976; 1981; Müller 1978; Moncomble, Coureaud, Quennedey, Langlois, Perrier and Schaal 2005). Furthermore, although the active substance (or mixture) does not appear to be contained in the mother's saliva, urine, blood or amniotic fluid (Hudson and Distel 1983; Schaal et al. 2003), it is present in the milk (Schley 1976; Müller 1978; Keil, von Stralendorff and Hudson 1990). When presented with fresh rabbit milk on a cotton bud or glass rod, pups respond with search-like head movements and grasping of the probe. These responses appear specific to rabbit milk and cannot be elicited by milk from rats, cats, cows, sheep, pigs, horses or humans (Schley 1976; Müller 1978; Keil et al. 1990; Schaal et al. 2003).

The specific and robust nature of the pups' response to milk allowed Keil et al. (1990) to develop a behavioural bioassay for use in characterizing the actual pheromonal substance or substances. Using the bioassay it was found that pups' responsiveness declines linearly with the exponent of dilution, and that cues contained in the milk are so potent that milk diluted by as much as 10^{-4} still elicits significantly more responses than cow's milk or other control substances (Keil et al. 1990). However, when left at room temperature milk loses most of its behaviour-releasing quality within about 30 minutes, but retains it for several weeks when stored at $-40°C$ (Müller 1978; Keil et al. 1990).

30.5 The Mammary Pheromone

As mentioned in the Introduction, in 2003 Schaal et al. reported the pheromonally active component of rabbit milk to be 2MB2. More specifically, using gas chromatography-olfaction analysis of the headspace of fresh rabbit milk by holding rabbit pups close to a sniff-port, they found that pups showed strong search-like head movements and grasping of the port in response to a peak identified as 2MB2. They further established in an extensive series of experiments, largely similar to those performed on the nipple-search pheromone, (1) that pups' responsiveness is dependent on the concentration of 2MB2; (2) that the content of 2MB2 in milk decreases when milk is left at room temperature for extended periods of time; (3) that 2MB2 is effective in pups of all rabbit breeds tested but not in newborn hares, rats, mice or cats; (4) that 2MB2 is not present in the blood of pregnant or lactating female rabbits, nor in cows' milk; and finally (5) that pups delivered by Caesarean section and thus without postnatal experience respond to 2MB2 with searching movements and grasping (Schaal et al. 2003).

However, with time a number of anomalies have emerged among the reports on the mammary pheromone itself as well as between these reports and those on the nipple-search pheromone. First, the natural concentration of 2MB2 in rabbit milk, first reported by Coureaud, Langlois, Perrier and Schaal in 2006, is apparently below 10^{-9} g/ml (Coureaud, Langlois, Perrier and Schaal 2006), that is, three log units below the concentration used for comparing the behavioural effectiveness of 2MB2 and rabbit milk (Schaal et al. 2003; Moncomble et al. 2005), and one to two log units below the experimentally determined threshold concentrations of perception of 2MB2 by rabbit pups (Coureaud, Langlois, Sicard and Schaal 2004). This would suggest that other behaviourally active substances in the milk also contribute to the pups' response, a possibility acknowledged by Coureaud et al. (2006), and plausible given the number of unidentified and therefore untested compounds in the head space of fresh rabbit milk to which pups were reported to respond almost as strongly as to 2MB2 in the original 2003 report.

An additional doubt concerns the naming of 2MB2 as mammary pheromone and its functional equivalence to the nipple-search pheromone. This name was presumably based on the presence of 2MB2 in milk and the assumption that the source will be ultimately found in the mammary apparatus. However, Moncomble et al. (2005) were only able to report that rabbit pups respond to tissue taken from nipples and to the first drops of milk pressed from nipples, but not to samples of tissue taken from beneath the nipples, to mammary tissue itself, or to alveolar milk. Nevertheless, these findings are compatible with the notion that the nipple-search pheromone is present on the mother's ventrum as a gradient with the highest concentration at nipples (Hudson and Distel 1983; Distel and Hudson 1984; Coureaud, Schaal, Langlois and Perrier 2001). However, as is clear from the experiments concerning the endocrine basis of pheromone emission reported in 4.2, milk is not the only or even a necessary source of the nipple-search pheromone; pups search vigorously and locate nipples on non-lactating oestrous, on pregnant, and on hormonally-treated ovariectomized females. Nor do nipples appear to be the only source of pheromonal nipple-search cues, although the cues certainly appear to be more concentrated there. Thus, experiments currently in progress show that in three-minute tests in the standard nipple-search arena, pups ($n = 16$, 4 litters) searched for a mean 102.3 sec $\pm$ 53.5 s SD on lactating does whose nipples were covered with 1.5 cm diameter plastic patches, and searched ($n = 20$, 5 litters) for a mean 79.3 s $\pm$ 65.1 s SD on lactating does whose nipples had been surgically removed several months prior to mating (Fig. 30.2).

30.6 Conclusion

In short, the present situation regarding these two, differently named pheromones can be briefly summarized as follows. With the nipple-search pheromone we have a chemical signal with a clear survival function in a biologically relevant context but still chemically unidentified. With the mammary pheromone we have a single substance which undoubtedly releases specific responses within a certain range of concentrations but the function of which, in a biologically relevant context, still has to be demonstrated. Such is the complexity and challenge of trying to understand the nature of chemical signals, and even those as striking in their effect as the still mysterious cues governing nipple-search behaviour and suckling in the newborn rabbit.

Acknowledgments We thank Amando Bautista for help with the figures.

References

Altbäcker, V., Hudson, R. and Bilkó, Á. (1995) Rabbit-mothers' diet influences pups' later food choice. Ethology 99, 107–116.

Bautista, A., Mendoza-Degante, M., Coureaud, G., Martínez-Gómez, M. and Hudson, R. (2005) Scramble competition in newborn domestic rabbits for an unusually restricted milk supply. Anim. Behav. 70, 1011–1021.

Beauchamp, G.K., Doty, D.G., Moulton, D.G. and Mugford R.A. (1976). The pheromone concept in mammalian chemical communication: A critique. In: R.L. Doty (Ed.), *Mammalian Olfaction*. Academic Press, New York, pp. 143–160.

Bilkó. Á., Altbäcker, V. and Hudson, R. (1994) Transmission of food preference in the rabbit: The means of information transfer. Physiol. Behav. 56, 907–912.

Caldelas, I., Chimal-Monroy, J., Martinez-Gomez, M. and Hudson, R. (2005) Non-photic circadian entrainment in mammals: A brief review and proposal for study during development. Biol. Rhythm Res. 36, 23–37.

Caldelas, I., Tejadilla D., González B., Montúfar R. and Hudson R. (2007) Diurnal pattern of clock gene expression in the hypothalamus of the newborn rabbit. Neurosci. 144, 395–401.

Coureaud, G., Schaal, B., Coudert, P., Rideaud, P., Fortun-Lamothe, L., Hudson, R. and Orgeur, P. (2000) Immediate postnatal suckling in the rabbit: Its influence on pup survival and growth. Reprod. Nutr. Develop. 40, 19–32.

Coureaud, G., Schaal, B., Langlois, D. and Perrier, G. (2001) Orientation response of newborn rabbits to odours of lactating females: Relative effectiveness of surface and milk cues. Anim. Behav. 61, 153–162.

Coureaud, Langlois, Sicard, and Schaal (2004) Newborn rabbit responsiveness to the mammary pheromone is concentration dependent. Chem. Senses 294, 341–350.

Coureaud, G., Langlois, D., Perrier, G. and Schaal, B. (2006) Convergent changes in the maternal emission and pup reception of the rabbit mammary pheromone. Chemoecol. 16: 169–174.

Distel, H. and Hudson, R. (1984) Nipple-search performance by rabbit pups: Changes with age and time of day. Anim. Behav. 32, 501–507.

Distel, H. and Hudson, R. (1985) The contribution of the olfactory and tactile modalities to the nipple-search behaviour of newborn rabbits. J. Comp. Physiol. A 157, 599–605.

González-Mariscal, G., Chirino, R. and Hudson, R. (1994) Prolactin stimulates emission of nipple pheromone in ovariectomized New Zealand white rabbits. Biol. Reprod. 50, 373–376.

González-Mariscal, G. and Rosenblatt, J.S. (1996) Maternal behavior in rabbits. A historical and multidisciplinary perspective. In: J.S. Rosenblatt and C.T. Snowdon (Eds.), *Advances in the Study of Behavior, Vol. 25, Parental Care: Evolution, Mechanisms and Adaptive Significance*. Academic Press, New York, pp. 333–360.

Hudson, R. (1985) Do newborn rabbits learn the odor stimuli releasing nipple-search behavior? Dev. Psychobiol. 18, 575–585.

Hudson, R. and Distel, H. (1982) The pattern of behaviour of rabbit pups in the nest. Behaviour 79, 255–272.

Hudson, R. and Distel, H. (1983) Nipple location by newborn rabbits: Behavioural evidence for pheromonal guidance. Behaviour 85, 260–275.

Hudson, R. and Distel, H. (1984) Nipple-search pheromone in rabbits: Dependence on season and reproductive state. J. Comp. Physiol. A 155, 13–17.

Hudson, R. and Distel, H. (1986) Pheromonal release of suckling in rabbits does not depend on the vomeronasal organ. Physiol. Behav. 37, 123–129.

Hudson, R. and Distel, H. (1989) Temporal pattern of suckling in rabbit pups: A model of circadian synchrony between mother and young. In: S.M. Reppert (Ed.), *Development of Circadian Rhythmicity and Photoperiodism in Mammals*. Perinatology Press, Boston, pp. 83–102.

Hudson, R. and Distel, H. (1990) Sensitivity of female rabbits to changes in photoperiod as measured by pheromone emission. J. Comp. Physiol. A 167, 225–230.

Hudson, R., González-Mariscal, G. and Beyer, C. (1990) Chin marking behavior, sexual receptivity, and pheromone emission in steroid-treated, ovariectomized rabbits. Hormones Behav. 24, 1–13.

Hudson, R., Melo, A.I. and González-Mariscal G. (1994) Effect of photoperiod and exogenous melatonin on correlates of estrus on the domestic rabbit. J. Comp. Physiol. A 175, 573–593.

Hudson, R. and Distel, H. (1995) On the nature and action of the rabbit nipple-search pheromone: A review. In: R. Apfelbach, D. Müller-Schwarze, K. Reuter and E. Weiler (Eds.), *Chemical Signals in Vertebrates VII*. Elsevier Science, London, pp. 223–232.

Hudson, R., Müller, A. and Kennedy, G.A. (1995) Parturition in the rabbit is compromised by daytime nursing: The role of oxytocin. Biol. Reprod. 53, 519–524.

Hudson, R., Bilkó, Á. and Altbäcker, V. (1996) Nursing, weaning and the development of independent feeding in the rabbit (*Oryctolagus cuniculus*). Z. Säugetierkunde 61, 39–48.

Hudson, R., Cruz, Y., Lucio, R.L., Ninomiya, J. and Martínez-Gómez, M. (1999) Temporal and behavioral patterning of parturition in rabbits and rats. Physiol. Behav. 66, 599–604.

Jilge, B. and Hudson, R. (2001) Diversity and development of circadian rhythms in the European rabbit. Chronobiol. Internat. 18, 1–26.

Keil, W., von Stralendorff, F. and Hudson, R. (1990) A behavioral bioassay for analysis of rabbit nipple-search pheromone. Physiol. Behav. 47, 525–529.

Lincoln, D.W. (1974) Suckling: A time constant in the nursing behaviour of the rabbit. Physiol. Behav. 13, 711–714.

Moncomble, R-S., Coureaud, G., Quennedey, B., Langlois, D., Perrier, G. and Schaal, B. (2005) The mammary pheromone of the rabbit: From where does it come? Anim. Behav. 69, 29–38.

Müller (1978) Zum Saugverhalten von Kaninchen unter besonderer Berücksichtigung des Geruchsvermögens. Doctoral Thesis, Universität Giessen.

Schaal, B., Coureaud, G., Langlois, D., Giniès, C., Sémon, E. and Perrier, G. (2003) Chemical and behavioural characterisation of the rabbit mammary pheromone. Nature 424, 68–72.

Schley, P. (1976) Untersuchung zur künstlichen Aufzucht von Hauskaninchen. Habilitation Thesis, Universität Giessen.

Schley, P. (1977) Die Ausschaltung des Geruchsvermögens und sein Einfluß auf das Saugverhalten von Jungkaninchen. Berl. Münch. Tieräztl. Wschr. 90, 382–385.

Schley, P. (1981) Geruchssinn und Saugverhalten bei Jungkaninchen. Kleintierpraxis 26, 197–264.

Semke, E., Distel, H. and Hudson, R. (1995) Specific enhancement of olfactory receptor sensitivity associated with foetal learning of food odors in the rabbit. Naturwissensch. 82, 148–149.

Chapter 31
The Human Breast as a Scent Organ: Exocrine structures, Secretions, Volatile Components, and Possible Functions in Breastfeeding Interactions

Benoist Schaal, Sébastien Doucet, Robert Soussignan, Matthias Rietdorf, Gunnar Weibchen and Wittko Francke

Abstract Milk and the nipples of mammalian females have long been known to release attraction in conspecific newborns. This applies also to humans, in whom breast odour cues control infant state and directional responses. Such cues from the whole breast as well as from the isolated areola delay crying onset and stimulate positive orientation and oral actions in infants. Native secretions from areolar glands are especially salient to newborns in which they intensify oral-facial actions and respiration. Thus, odorous compounds from areolar glands may be in a position to play a role, among many other determinants, in establishing the processes pertaining to milk production, transfer and intake by infants.

31.1 Introduction

For naïve, newly born mammals, finding a nipple for the first time is not to be taken for granted. In general (but with notable exceptions), newborns initially exhibit lengthy and strenuous searching in the mother's ventral fur, before localising the mammary area, seizing a nipple and finally sucking to withdraw colostrum. Even human infants who are fully assisted in finding the breast often have difficulty in feeding optimally over the first postnatal days (Dewey, Nommsen, Heinig and Cohen 2003). As rapid intake of colostrum is critical to engage neonatal metabolism, immunity and learning, strong selective pressure may have driven the evolution in mammalian females of nipples that are conspicuous to their newborns. Among the species studied so far (e.g., rat, mouse, rabbit, cat, dog, sheep, pig, human), it appears that the most conserved strategy to increase the saliency of nipples relies on somesthesis and olfaction. Regardless of the newborn's position in the altricial-precocial continuum and of the species-specific pattern of nursing behaviour, mam-

Benoist Schaal
CNRS, Centre Européen des Sciences du Goût, Dijon
schaal@cesg.cnrs.fr

J.L. Hurst et al., *Chemical Signals in Vertebrates 11.*
© Springer 2008

malian females emit some sort of chemical signal in, on or around their nipples, and newborns clearly react to them.

Anatomists named the mammalian female's suckable appendages *papilla* or *mamilla*, surrounded (at least in humans) by a patch of glabrous, pigmented skin named *areola*. According to species and morphology, the papilla is more commonly labeled *nipple* (in the case of small-sized erectile structures) or *teat* (larger pendulous structures). In all cases, the papillae represent the minimal areas of the mammalian females' body that obligatorily and recurrently enter in tightest contact with the newborns' mouth and nose. The structural and functional properties of these suckable appendages are devised to optimise success in canalizing milk to newborns and to govern the timely and commensurate production of milk in females. Nipples are thus certainly shaped to capture by all possible sensory means the "receiver's psychology" of conspecific newborns.

Much as in other mammals, human females concentrate in their mammary regions features potentially related with infant-directed communication. The potential semiochemical significance of the human breast will be examined here in terms of: 1) morphology and secretory activity of areolar skin glands; 2) infant responses when exposed to the effluvium of their mother's breast and areolae, and to the odour of related secretions; 3) volatile compounds present in pure areolar secretions as compared with those of milk; 4) relations of maternal areolar gland endowment to adaptive outcomes in the infant and the mother.

31.2 Human Breast Structures of Semiochemical Interest

In human females, the nipple/areolae concentrate numerous features whose function is supposedly confined to pregnancy and lactation. These regions are densely supplied with the full range of skin gland types. The hairless nipple abounds in apocrine and sebaceous glands with ducts opening on its tip to give off secretions during lactation (Perkins and Miller 1926; Montagna and MacPherson 1974). On the areolae themselves, eccrine sweat glands and enlarged sebaceous glands can be found (Montagna and MacPherson 1974). Additionally, the areolar surface is scattered with small prominences, called 'follicles' or 'tubercles' by Morgagni (1719) or 'protuberances' by Roederer (1753). Montgomery (1837) described their macrostructure in more detail and attested to their gestational/lactational turgidity (they were accordingly named Montgomery's tubercles or glands). The secretory activity of these skin tubercles was nevertheless recognized since Morgagni (1719), who reported "I have seen lactiferous tubes going to each of these tubercles and expanding within them, so that in fact their formation was in a great degree caused by the dilatation of these ducts and their prominence beyond the surface of the areola" (quoted in Montgomery, 1837). The dual nature of Montgomery's glands was recognized early, although the apocrine or sebaceous nature of the unit paired with the lactiferous unit was long discussed. The areolar tubercles were finally confirmed to coalesce sebaceous glands with miniature mammary acini (Montagna and Yun 1972;

Smith, Peters and Donegan 1982). Hereafter, structures contrasting morphologically from the surrounding areolar skin (in elevation and/or pigmentation) will be named areolar glands (AG), without further histological consideration.

After Montgomery's (1837) rough estimate (ie, 12 to 20 AG/areola), a first quantitative assessment of AG prevalence, distribution and activity was run in a sample of 64 women during the period of breastfeeding establishment (Schaal, Doucet, Sagot, Hertling and Soussignan 2006). The majority of subjects (98.4%) bore more than 1 AG, and 83% bore 1 to 15 AG per areola. The oozing of areolar fluid during nursing seen by all attentive observers was confirmed in about 1 out of 5 women, in which at least one AG gave off a whitish, latescent fluid between nursing bouts. This proportion may be underestimated, however, as this first AG census was made amid two nursing sessions without distal (crying) or proximal (suckling) infant stimuli known to trigger milk release. In any case, it is clear that the glandular phenotype of human areolae is highly polymorphic in the Caucasian sample studied.

Any potential communicative role of AG may be secondary to other basic functions. First, the increased productivity of the AG during late pregnancy and early lactation (in both the lacteal and sebaceous units; Burton, Shuster, Cartlidge, Libman and Martell 1973) is a means of epidermal and ductal protection from pathogens. Second, these greasy secretions act to preserve the skin from corrosive nurslings' saliva and sucking-related friction. Finally, AG secretions combine with infant saliva to realise the hermetic seal between the infant's lips and areolar skin without which sucking would be ineffective. In sum, the AG endowment of the human nipple-areolae fulfils multiple mechanical, protective, and communicative functions.

The polymorphism in AG number and activity may be balanced in part by their non-random distribution on the areolae. The protective, mechanical and communicative functions of AG may relate differentially to their spatial arrangement. While protective functions would imply uniform distribution over the areolae, communicative functions would favour a distributional bias on zones to which the infants' noses are mostly directed. The data indicate that AG appear, on average, unevenly distributed (Schaal et al. 2006). Some scatter the whole areolar disk, but more (particularly, the secretory AG; Fig. 31.1) dot the upper and lateral quadrants. If a communicative function is supported by the areolae, one may further expect an increase in the secretory output of the AG after delivery and just before each nursing bout. Only suggestive data were obtained so far on these points, as more women tended to evince secretory AG on days 1-3 than on days 15 or 30. But more definitive conclusions on these points await longitudinal investigations in a large sample of women.

Colostrum and milk released at the nipples from the main lactiferous ducts add their intrinsic olfactory qualities to the areolae, viz. odorants reflecting the mother's diet, metabolism or immunogenetic constitution (Schaal 2005). Together these varied sources of substrates create a complex odour cocktail. The lipid fraction of this mixture, sebum originating from free sebaceous glands and from AG, as well as fatty acids from milk, may act as odour fixatives that improve the stability of the olfactory complex formed on the areolae. The intricate arrangement of sebaceous and lacteal sources within AG (Smith et al. 1982) may indeed favour the mingling of sebum and

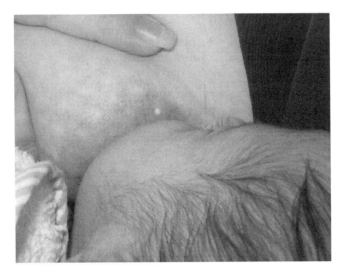

Fig. 31.1 An areolar gland giving off a drop of milk-like secretion during breastfeeding (Photograph: S Doucet)

areolar milk during sucking episodes. In addition, salivary enzymes deposited by the suckling infant may affect the local release of odour-active compounds (Buettner 2002). Finally, Haller's subareolar vascular plexus (Mitz and Lalardie 1977) confers to this region a higher surface temperature as compared to the nipple and to the remainder of the breast, a property that may regulate the evaporation rate of local odorants. Interestingly, this thermal feature is anticipatorily triggered by the crying infant (Vuorenkoski, Wasz-Hockert, Koivisto and Lind 1969), resulting in optimal conditions for odour release when the infant is offered the breast.

In sum, if a localized structure such as the areola, combining multiple sources of odorants, of lipidic fixatives, of enzymes, and a heat-based diffusion device, would have been described in any other mammalian species, it would have been ascribed a chemocommunicative function without much hesitation. This was already suggested by early anatomists P. Schiefferdecker (1922) and J. Schaffer (1937) for whom the breast is to be considered as a *Duftorgan* (a scent organ) involved in the nursling's guidance to the breast.

31.3 Breast-Areolar Secretions are Perceived by Neonates

Over the past three decades, several studies have assessed neonatal responses to odors naturally emitted from the breasts by lactating women (e.g., Macfarlane 1975; Russell 1976; Schaal, Montagner, Hertling, Bolzoni, Moyse and Quichon 1980; Schaal 1986; Makin and Porter 1989). The odour of the whole breast, collected on a cotton pad applied on the areola, reduces arousal in active newborns (Schaal et al. 1980; Schaal 1986; Sullivan and Toubas 1998) and increases it in somnolent ones (Russell 1976; Soussignan, Schaal, Marlier and Jiang 1997; Sullivan and Toubas 1998). It also elicits positive head (nose) orientation (Macfarlane 1975; Schaal et al. 1980; Makin and Porter 1989) and stimulates oral activity (Russell 1976; Soussignan et al. 1997). It may further stimulate neonatal approach behaviour in the form of directional crawling (Varendi and Porter 2001).

In the same way as volatiles from the whole breast, odorants carried in human colostrum/milk are arousing and attractive to newborns (Mizuno, Mizuno, Shinohara and Noda 2004; Marlier and Schaal 2005). Interestingly, neonatal responsiveness to these milk cues does not seem to depend on breastfeeding experience as term-born infants exclusively fed formula (Marlier and Schaal 2005), and premature infants (Bingham et al. 2003), react to them in the same way as regularly breast-fed infants.

Another major mammary source of potential odour cues is the areola. A recent study (Doucet, Soussignan, Sagot and Schaal 2007a) addressed whether morphologically differentiable areas of the breast surface could bear distinct functional activity for newborns. These areas were 'fractionated' in applying an odour-free air-proof film directly onto the breast, various openings in it allowing subtraction of the areolar contribution to the whole breast odour and separation of the areolar odour from that of the nipple or milk. To test the relative efficacy of the isolated areola to elicit responses, waking newborns were then positioned close to such a masked, or selectively unmasked, breast in an unconstrained nursing posture before a feed. We found that natural odour cues from the breast of lactating women modulate arousal states and promote appetitive oral activity in 3-day-olds. Thus, the scene of the breast is different to newborns if corresponding odours are present or if they are masked. In the former case, infants began crying later (by an average of 13 sec in the present conditions), and displayed feeding-preparatory oral responses earlier (by about 12 sec) and for longer (by 12 sec). Breast odour cues could thus make a transitory, but momentous, difference in controlling an infant's frustration and in boosting oral function in a real feed. However, when comparing whole breast vs. areola vs. nipple vs. milk odours, the infants responded similarly, suggesting that the odours of the different breast areas have equivalent attractive values, all equalling the reinforcing value of human milk. The behavioural uniformity of these different breast stimuli may be explained either by the presence of similar compounds due to shared exocrine sources or to cross-contamination, or finally by perceptually-distinct compounds having acquired similar attractiveness due to similar conditions of reward (Delaunay-El Allam, Marlier and Schaal 2006). In this later case, finer-grained behavioural and psycho-physiological analyses may disclose differentiable properties of these stimuli.

Such a finer-grained study was run by Doucet, Soussignan, Sagot and Schaal (2007b) in alternately exposing 3-day old newborns to the odour of native secretion from AG and to the odour of its components, milk and sebum (i.e., main milk and forehead sebum, respectively). Neonatal reactivity to these breast-related stimuli were tested against several control stimuli (e.g., water, vanilla, cow milk, formula). It came out that the pure AG secretion elicits higher levels of oral-facial responses than all other stimuli. In the same way, during stimulus presentation, the

AG secretion increased inspiratory amplitude in the infants more than all stimuli, but human milk. Thus, AG cues appear to carry the same general behavioural impact as milk, indicated by the intensification of respiration, but they have distinctively higher appetitive impact than milk and sebum, as attested by oro-facial responses. Interestingly, the mode of feeding, and hence the rate of previous exposure to breast-related stimuli, did not affect the infant responses to the odour of AG secretion.

Further investigation is needed to confirm these first results, and to substantiate whether milky AG secretions and mainstream milk convey similar or distinct information to infants. Indeed, studies in other species revealed that milk can carry odorant cues of varying origins (e.g., dietary aromas, pheromones) bearing differentiable meanings to newborns (Schaal, Coureaud, Langlois, Giniès, Sémon and Perrier 2003; Schaal 2005).

31.4 Volatiles of Colostrum and of Areolar Secretions

Olfactorily active compounds carried in human milk have rarely been subjected to chemical analysis (Stafford, Horning and Zlatkis 1976; Pellizari, Hartwell, Harris III, Waddell, Whitaker and Erickson 1982; Shimoda, Ishikawa, Hayakawa and Osajima 2000; Bingham, Sreven-Tuttle, Lavin and Acree 2003), and so far no attempt was undertaken to characterize the volatile composition of human AG secretions. Thus, the degree of their chemical overlap or specificity cannot be evaluated. Here we present preliminary data from such comparative analyses of colostrum and AG secretions from the same women.

Colostrum or AG-secretions of three different women were separately mixed with Florisil (Merck; 1:3 w/w). This material (6 samples) was transferred to a small column that already contained about the same amount of Florisil and enough methylene chloride to cover the two Florisil-batches. Subsequently, organic components adsorbed on the Florisil were eluted with solvents of increasing polarities (hexane to methanol). The fractions were concentrated and analysed by coupled gas chromatography/ mass spectrometry (Francke 1988).

As might be expected, the series of "classical" fatty acids, C13, C14, C16, C18, oleic acid, linoleic acid, and linolenic acid were present in both colostrum and AG secretions. In addition, low boiling aldehydes such as octanal, nonanal, and decanal, as well as the fatty alcohols hexadecanol and octadecanol, could be identified in both. The three aldehydes had already been reported earlier as components of mother's milk (Pellizari et al. 1982; Shimoda et al. 2000). We also found oxygenated terpenes like phellandral in colostrum as well as in AG-secretions, and the same was true for 1-methyl-4-(1-methylethenyl)-1,2-cyclohexanediol, which may be the product of a bishydroxylation of limonene, which the mother may have sequestered with the food. The presence of terpenes in mother's milk has been reported earlier (Pellizari et al. 1982). In colostrum, we found several oxygen-containing heterocycles (Fig. 31.2). Most of these substances that contain 5 or 6 carbon atoms may be derived from carbohydrates and formed through enzymatic types of the

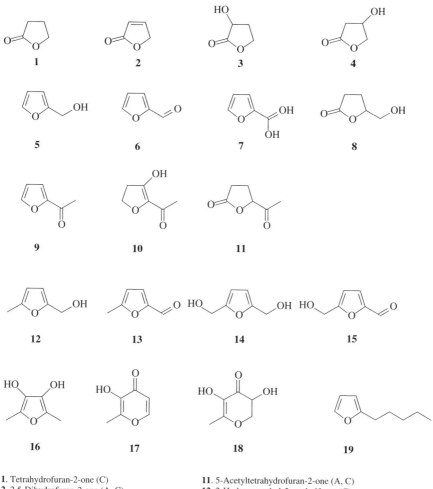

Fig. 31.2 Oxygen-containing heterocyclic compounds found in AG-secretions (A) and/or in colostrum (C)

1. Tetrahydrofuran-2-one (C)
2. 2,5-Dihydrofuran-2-one (A, C)
3. 3-Hydroxytetrahydrofuran-2-one (A, C)
4. 4-Hydroxytetrahydrofuran-2-one (A, C)
5. 2-Hydroxymethylfuran = Furfurylalcohol (A, C)
6. 2-Formylfuran = Furfural (A, C)
7. Furan-2-carboxylic acid (C)
8. 5-Hydroxymethyltetrahydrofuran-2-one (A, C)
9. 2-Acetylfuran (C)
10. 5-Acetyl-2,3-dihydro-4-hydroxyfuran (C)
11. 5-Acetyltetrahydrofuran-2-one (A, C)
12. 2-Hydroxymethyl-5-methylfuran (C)
13. 2-Formyl-5-methylfuran (C)
14. 2,5-Dihydroxymethylfuran (C)
15. 2-Formyl-5-hydroxymethylfuran (C)
16. 3,4-Dihydroxy-2,5-dimethylfuran (C)
17. 3-Hydroxy-2-methylpyran-4-one (C)
18. 2,3-Dihydro-3,5-dihydroxy-6-methylpyran-4-one (C)
19. 2-Pentylfuran (C)

Maillard reaction. We are sure, however, they do not represent artefacts, formed upon work up or during the analytical procedure. Several of the heterocycles have been described earlier as volatile constituents of mother's milk (Pellizari et al. 1982; cf. Figure 31.2: compounds 5, 6, 9, 13, 17, 19 among other unknown oxygen containing compounds with unsaturated and/or cyclic structures). Interestingly, we also detected the structurally closely related lactones 2, 3, and 4, the furan derivatives fur-furylalcohol (5) and furfural (6), as well as the lactones 8 and 11 in both colostrum and AG-secretions. Some of the heterocycles (1–19) are characterised by a smell of roasted bread or caramel/syrup, which is particularly true for 16, 17, and 18.

Analyses of traces of volatiles from colostrum and in particular from AG-secretions are hampered by the large excess of fat and by the small amounts of available material. Whether more of the above mentioned compounds are present in colostrum as well as in AG-secretions, whether additional compounds will be identified, and whether "cross-contamination" between the two sources of volatiles can be excluded, await further investigation.

31.5 Potential Roles of Areolar Glands in Mother-Infant Co-adaptation

If the AG endowment of the human nipple-areolae fulfils the multiple functions mentioned above, it is not unreasonable to expect that areolae which bear more profuse arrangements of AG should facilitate initial breastfeeding interactions. This was first tested by relating areolar morphological features of post-parturient mothers to indicators of adaptation in the mother-infant dyad (Schaal et al. 2006; Doucet, Soussignan, Sagot and Schaal 2007c). On the infant's side, adaptation was meant in terms of weight gain, considered as the most integrative outcome of the constraints of initial breastfeeds (ability to grasp a nipple, to coordinate sucking-swallowing-respiration, to achieve digestion, to save energy). On the mother's side, adaptation was meant as the time elapsing from delivery to the onset of lactation.

Results reveal that the mother's AG number correlated with two neonatal outcomes. First, the mothers' perception of their infant's suckling responses was associated with their endowment in AG, in that infants from mothers having more profuse AG were reported to latch on to the nipple more rapidly, and to be more active suckers after latching. Secondly, the total AG number was positively associated with the infants' weight gain between birth and day 3, lending some support to the hypothesis that a higher AG number is associated with early thriving. However, this maternal AG-infant weight gain association was modulated by parity. Only the infants from primiparous women bearing higher AG counts displayed higher weight gain during the first 3 days (Schaal et al. 2006).

Likewise, the inception of lactation was associated with the maternal endowment in AG, but also only in first-time mothers, suggesting that infants of new mothers reacted more, as a group, to a higher AG number. First-born infants being exposed to less favourable conditions of lactational physiology (e.g., Ingram, Woolridge, Greenwood and McGrath 1999) and to less expert maternal guidance to the breast (e.g., Fleming 1990), it may be hypothesised that they would need to rely more on their own sensory and motor keenness to achieve successful nipple localisation and latching. In this context, primiparous mothers endowed with more AG (compared with primiparae endowed with less AG) may offer their infant more helpful (chemo)sensory guidance to the nipple.

The causal pathways linking maternal AG endowment to infant weight variation during the first postpartum days remain to be determined. AG may affect directly the organisation of neonatal behaviour at the breast, and indirectly maternal behaviour and lactational physiology. On the one hand, infants of mothers bearing higher AG numbers were reported to latch-on more easily and to suck more intensely, all variables logically resulting in more efficient colostrum intake. Such facilitation may in turn be beneficial to mothers (especially primiparae) as self-reassurance in their breastfeeding competence. On the other hand, facilitated sucking of areolae bearing higher AG numbers provokes effective nipple stimulation which is linked with earlier onset of lactation. Although this nexus of relationships remains to be substantiated, the above studies offer some clues about the possible involvement of skin glands of the lactating human breast in adaptive transactions at the onset of the postnatal mother-infant relationship.

The final processes of mammalian reproduction (parental investment in lactation and bonding, and filial attachment) are based on a complex web of morphological, biochemical, physiological, sensory, cognitive, behavioural, and social adaptations that are evolutionarily co-opted in maternal and neonatal organisms. Within this intricate multifactorial web, mammary odour cues may play subtle, perhaps ephemeral, but vital, roles at the very first postnatal encounters of females and young. These cues can be predicted to confer success in the offspring's approach and exploration of the maternal body surface, effective initial feeds, and rapid learning of maternal identity. But the strength and duration of their impact are certainly variable in newborns of species exposed to contrasting life history patterns during early development. The impact of areolar chemo-stimuli in short- and long-term neonatal fitness remains to be positively demonstrated in any mammal (cf. Blass and Teicher 1980), and hence also in our own species. If such cues are involved in the initial interactions of human infants with the mother's breasts, their effect should be measurable in terms of infant feeding success, growth, and viability.

Despite the fact that the feeding-related motor competence of human newborns is often considered to be automatic and reflex-like, their sucking behaviour is easily perturbed by adverse maternal, infantile or environmental conditions (Winberg 1995). The newborns' success in localising and adequately seizing the nipple is thus not to be taken for granted. Delays in the establishment of optimal colostrum/milk transfer in human mother-infant pairs are indeed far from uncommon. For example, a Californian study revealed that 49% and 22% of normal term-born newborns were non optimal in breastfeeding on the day of birth and on day 3, respectively; these unfavourable outcomes were strongly associated with primiparity (Dewey et al. 2003; for similar results in a French sample, see Michel, Gremmo-Feger, Oger and Sizun 2006). In conditions presumed to bear resemblance with those that prevailed during human evolutionary history, a one-day postponement in the initiation of breastfeeding explained 16% of neonatal deaths during home births in rural Ghana; moreover, a delay of only one hour post-birth to engage breastfeeding explained 22% of neonatal mortality (Edmond, Zandoh, Quigley, Amenga-Etego, Owusu-Agyei and Kirkwood 2006). Thus, any sensory means that can speed up the localisation and acceptance of the breast by the newly born infant should have been beneficial to infant viability and adaptive life onset over evolutionary time. Milky areolar secretions and mainstream milk may carry such cues capable of boosting adequate responsiveness in infants' offered the breast for the first times.

Acknowledgments We would like to thank Paul Sagot and Elisabeth Hertling for their help in the infant experiments reported above. We are also grateful to Kathryn G. Dewey for discussion and to the Editors for correcting the English in a previous draft. These studies were supported by CNRS and French Ministry of Research and Technology.

References

Ackerman, B., and Penneys, N.S. (1971) Montgomery's tubercles. Obstet. Gynecol. 38, 924–927.

Bingham, P.M., Sreven-Tuttle, D. Lavin, E., and Acree, T. (2003). Odorants in breast milk. Arch. Pediatr. Adolesc. Med. 157, 1031.

Blass, E.M., and Teicher, M.H. (1980). Suckling. Science 210, 15–22.

Buettner, A. (2002). Influence of human saliva on odorant concentrations: 2. J. Agric Food Chem. 50, 7105–7110

Burton, J.L., Shuster, S., Cartlidge, M., Libman, L.J., and Martell, U. (1973). Lactation, sebum excretion and melanocyte-stimulating hormone. Nature 243, 349–350.

Delaunay-El Allam, M., Marlier, L., and Schaal, B. (2006). Learning at the breast: preference formation for an artificial scent and its attraction against the odor of maternal milk. Infant Behav. Dev. 29, 308–321.

Dewey, K.G., Nommsen, L.A., Heinig, M.J., and Cohen, R.J. (2003). Risk factors for suboptimal infant breastdeefing behavior, delayed onset of lactation, and excess neonatal weight loss. Pediatrics 112, 607–619.

Doucet, S., Soussignan, R., Sagot, P., and Schaal, B. (2007a). The 'smellscape' of mother's breast: effects of odour masking and selective unmasking on neonatal arousal, oral and visual responses. Dev. Psychobiol., 49, 129–138.

Doucet, S., Soussignan, R., Sagot, P., and Schaal, B. (2007b). Human areolar glands emit odorants affecting behaviour and autonomous responses in newborns. Unpublished manuscript.

Doucet, S., Soussignan, R., Sagot, P., and Schaal, B. (2007c). The areolar glands in postparturient women and their links with breastfeeding, lactation onset and early infant growth. Unpublished manuscript.

Edmond, K.M., Zandoh, C., Quigley, M.A., Amenga-Etego, S., Owusu-Agyei, S. and Kirkwood, B.R. (2006). Delayed breastfeeding initiation increases risk of neonatal mortality. Pediatrics 117, e380–386.

Fleming, A. S. (1990). Hormonal and experiential correlates of maternal responsiveness in human mothers. In: N.A. Krasnegor, and R.S. Bridges (Eds.), Mammalian Parenting: Biochemical, Neurobiological and Behavioral determinants. Oxford University Press, New York, pp 184–208.

Francke, W. (1988). Techniken der Strukturzuordnung von Spurensubstanzen. Fresenius Z. Anal. Chem. 330, 320–321.

Ingram, J.C., Woolridge, M.W., Greenwood, R.J., and McGrath, L. (1999). Maternal predictors of early breast milk output. Acta Paediatr. 88, 493–499.

Macfarlane, A.J. (1975). Olfaction in the development of social preferences in the human neonate. Ciba Found. Symp., 33, 103–117.

Makin, J.W., and Porter, R.H. (1989). Attractiveness of lactating female's breast odors to neonates. Child Dev. 60, 803–810.

Marlier, L., and Schaal, B. (2005). Human newborns prefer human milk: conspecific milk odor is attractive without postnatal exposure. Child Dev. 76, 155–168.

Michel, M.P., Gremmo-Feger G., Oger, E., and Sizun, J. (2006). Etude des difficultés de mise en place de l'allaitement maternel en maternité chez des nouveau-nés à terme. Proceedings of 3rd Journée Nationale de l'Allaitement, CoFAM-Co-naître, Brest, pp. 26–30.

Mitz, V., and Lalardie, J.P. (1977). A propos de la vascularisation et de l'innervation sensitive du sein. Senologia 2, 33–39.

Mizuno, K., Mizuno, N., Shinohara, T., and Noda, M. (2004). Mother-infant skin-to-skin contact after delivery results in early recognition of own mother's milk odour. Acta Paediatr. 93, 1640–1645.

Montagna, W., and MacPherson, E.E. (1974). Some neglected aspects of the anatomy of human breasts. J. Invest. Dermatol. 63, 10–16.

Montagna, W., and Yun, J.S. (1972). The glands of Montgomery. Br. J. Dermatol. 86, 126–133.

Montgomery, W.F. (1937). An Exposition of the Signs and Symptoms of Pregnancy, the Period of Human Gestation, and Signs of Delivery. Sherwood, Gilber, and Piper, London.

Morgagni, G.B. (1719). Adversaria Anatomica Omnia. Padua

Pellizari, E.D., Hartwell, T.D., Harris III, B.S.H., Waddell, R.D., Whitaker, D.A., and Erickson, M.D. (1982). Purgeable organic compounds in mother's milk. Bull. Environm. Contam. Toxicol. 28, 322–328.

Perkins, O.M., and Miller, A.M. (1926). Sebaceous glands in the human nipple. Am. J. Obstet. 11, 789–794.

Roederer, X. (1753). Elementa Artis Obstetriciae in Usum Praelectionum Academicarum. Göttingen.

Russell, M.J. (1976). Human olfactory communication. Nature 260, 520–522.

Schaal, B. (1986). Presumed olfactory exchanges between mother and neonate in humans. In J. Le Camus and J. Cosnier (Eds.), Ethology and psychology. Privat, I.E.C, Toulouse, pp. 101–110.

Schaal, B. (2005). From amnion to colostrum to milk: odor bridging in early developmental transitions. In B. Hopkins and S. P. Johnson (Eds.), Prenatal development of postnatal functions. Praeger, London, pp. 51–102.

Schaal, B., Coureaud, G., Langlois, D., Giniès, C., Sémon, E., and Perrier, G. (2003). Chemical and behavioural characterization of the rabbit mammary pheromone. Nature 424, 68–72.

Schaal, B., Doucet, S., Sagot, P., Hertling, E., and Soussignan, R. (2006). Human breast areolae as scent organs: Morphological data and possible involvement in maternal-neonatal coadaptation. Dev. Psychobiol. 48, 100–110.

Schaal B., Montagner H., Hertling E., Bolzoni D., Moyse R., and Quichon R. (1980). [Olfactory stimulations in mother-infant relations]. Reprod. Nutr. Dev. 20, 843–858.

Schaffer, J. (1937). Die Duftorgane des Menschen. Wiener Klin. Wochenschr. 20, 790–796.

Schiefferdecker, P. (1922). Die Hautdrüsen des Menschen und der Säugetiere. Zoologica 27, 1–154.

Shimoda, Y.T., Ishikawa H., Hayakawa I., and Osajima Y. (2000). Volatile compounds of human milk. J. Fac. Agr. Kyushu. Univ. 45, 199–206.

Smith, D.M., Peters, T.G., and Donegan, W.L. (1982). Montgomery's areolar tubercle. A light microscopic study. Arch. Pathol. Lab. Med. 106, 60–63.

Soussignan, R., Schaal, B., Marlier, L., and Jiang, T. (1997). Facial and autonomic responses to biological and artificial olfactory stimuli in human neonates: re-examining early hedonic discrimination of odors. Physiol. Behav. 62, 745–758.

Stafford, M., Horning, M.G., and Zlatkis, A. (1976). Profiles of volatile metabolites in body fluids. J. Chromatogr. B 126, 495–502.

Sullivan, R.M., and Toubas, P. (1998). Clinical usefulness of maternal odor in newborns: soothing and feeding preparatory responses. Biol. Neonate 74, 402–408.

Varendi, H., and Porter, R.H. (2001). Breast odour as the only maternal stimulus elicits crawling towards the odour source. Acta Paediatr. 90, 372–375.

Vuorenkoski, V., Wasz-Hockert, O., Koivisto, E., and Lind, J. (1969). The effect of cry stimulus on the temperature of the lactating breast of primipara. Experientia 25, 1286–1287.

Winberg, J. (1995). Examining breast feeding performance: forgotten influencing factors. Acta Paediatr. 84, 465–467.

Chapter 32
Responses of Pre-term Infants to the Odour of Mother's Milk

Richard H. Porter, Chantal Raimbault, Anne Henrot and Elie Saliba

Abstract Fourteen premature infants (born at 30–33 weeks gestational age) were tested for their responses to the odours of different categories of human breast milk that are commonly available in intensive care nurseries: viz. their own mother's fresh and frozen milk, and frozen / pasteurized donor milk. Freezing reduced, but did not eliminate the stimulating effect of breast milk odour. The pattern of results obtained at 36 weeks post-conceptual age suggests that preterm infants may recognize the odour of their own mother's milk, and/or pasteurisation reduces the salience of milk odour. Breast milk odour elicits heightened sucking activity, which could have a positive effect on the development of breastfeeding behaviour.

32.1 Introduction

Maternal olfactory cues play an important role in the breastfeeding behaviour of newborn full-term infants. During tests conducted within the first hour after birth, and on postnatal days 3–4, neonates preferentially oriented towards and sucked from their mother's unwashed breast, rather than the alternative breast that had been washed and rinsed to remove its natural odours (Varendi, Porter and Winberg 1994, 1997). Similar attraction to maternal breast odours is seen when newborns are tested with pairs of odourised stimulus pads (Macfarlane 1975; Schaal, Montagner, Hertling, Bolzoni, Moyse and Quichon 1980; Porter and Winberg 1999). At 2 weeks of age, even infants that had been exclusively bottle-fed since birth responded discriminatively to the odour of breast pads worn by (unfamiliar) nursing mothers (Makin and Porter 1989; Porter, Makin, Davis and Christensen 1991). Heightened rates of sucking/mouthing movements are also observed when babies are exposed to the odour of their mother's breasts or milk (Russell 1976; Mizuno and Ueda 2004). These sucking activities are an important component of the stereotyped sequence of prefeeding behaviour that is displayed by healthy newborn infants (Widstrom, Ransjo-Arvidson, Christensson, Matthiesen, Winberg and Uvnas-Moberg 1987).

Richard H. Porter
UMR 6175 CNRS-INRA, 37380 Nouzilly, France
porter.rh@gmail.com

To date, the possible involvement of maternal odours in the early feeding behaviour of premature infants has received little research attention. Sucking skills tend to be poorly developed in pre-term newborns, but improve with increasing postconceptual age and sucking experience (Nyqvist, Sjoden and Ewald 1999; Hafstrom and Kjellmer 2000). Interventions that provide the opportunity for non-nutritive sucking have a positive effect on subsequent feeding performance and the transition from tube to bottle feeding (Pinelli and Symington 2005). Preliminary studies indicate that the frequency of non-nutritive sucking increases when premature infants are stimulated with the odour of maternal breast milk (Meza, Powell and Covington 1998; Bingham, Abassi and Sivieri 2003).

Human breast milk is the best nutrient for preterm infants and is often delivered via gavage feeding methods before the baby's sucking skills mature. Because sufficient quantities of the mother's fresh breast milk may not be available, milk may be frozen for later consumption or be provided by donor mothers. To protect against the risk of disease transmission, donor milk is usually pasteurized. In the present study, we assessed preterm infants' responses to the odours of different categories of milk that are routinely available in neonatal intensive care nurseries.

32.2 Methods

32.2.1 Subjects

Fourteen pre-term neonates (5 boys / 9 girls) who met the following criteria were included in the study: stable condition; spontaneous ventilation; no evidence of congenital malformations, asphyxia or grade III/IV intraventricular hemorrhage; mother intended to breastfeed her baby. The infants' mean gestational age at birth was 31.4 weeks (range = 30–33.4 weeks) and mean birth weight was 1980 g (range: 1405–2470 g). For the first 8–10 postnatal days, the participating infants were fed pasteurized donor milk via orogastric tube; they received their own mother's milk after this initial period. Preliminary breastfeeding attempts began at 35 weeks postconceptual age (PCA), but all infants continued to be tube fed throughout the study.

32.2.2 Testing Procedures

The infants were tested for their responses to the following odour stimuli: (1) their own mother's fresh breast milk; (2) a thawed sample of their mother's breast milk that was stored at $-20\,^{\circ}$C; (3) breast milk contributed by several donor women that was pooled, pasteurized (heated to $63\,^{\circ}$C for 30 min.) and frozen at $-20\,^{\circ}$C; (4) water (odourless control).

Each infant participated in two series of tests; at 34 and 36 weeks PCA. Tests were conducted in the neonatal intensive care unit, with the baby laying face up with the head held in place by rolled cotton towels. During each of these test series, infants were exposed to all 4 odour stimuli in a random order, with an interval of 5 min. between successive trials. All stimulus liquids were warmed to 37 °C for the tests. For a given test trial, a cotton-tipped applicator moistened with the odour stimulus was held ~1 cm from the infant's nostrils for 120 s. Infants were in a quiet, calm state at the beginning of each test trial. When necessary, the 5 min inter-trial interval was lengthened until the infant subject returned to a quiet state. The entire test series was filmed with a tripod-mounted VCR focused on the infant's face, and the test stimuli were identified by a code number.

An individual who was "blind" in regard to the odour stimulus presented at each trial repeatedly viewed each test session and scored the following behavioural measures: number of lateral head movements, number of hand-to-mouth movements, and frequency of sucking movements with the lips / mouth.

The infant's cardiac rate was also recorded via a neonatal monitor at 60 s intervals beginning 60 s before the stimulus odour was presented.

32.3 Results

Separate analyses were performed for the test series conducted at 34 and 36 weeks PCA. For each behavioural measure the results were compared across the 4 odour stimuli with Friedman's tests. A similar series of within-subject comparisons was performed for the changes in baseline cardiac rates following odour presentations.

No significant overall effects of stimulus odour were found at either testing age for the number of lateral head movements, the number of hand-to-mouth movements, or changes in cardiac rate. At 34 weeks PCA, there was no significant difference in the frequencies of sucking movements across the 4 stimulus conditions (Fig. 32.1), however, significant differences in sucking rates between the odour stimuli were found at 36 weeks (Friedman's test, $\chi^2 = 18.6$, df = 3, p < 0.0005; Fig. 32.2). Post-hoc paired comparisons (Wilcoxon's signed-ranks test) revealed that at 36 weeks PCA the odour of mother's fresh milk elicited significantly more sucking movements than did either pooled donor milk (p < 0.005) or water (p < 0.005). The sucking frequency was also greater in response to mother's fresh milk compared to mother's frozen milk, but the difference was not statistically significant (p = 0.09). Higher sucking rates were observed during the presentations of frozen mother's milk and pooled donor milk compared to the water trials (p < 0.05), however sucking behaviour did not differ between the mother's frozen milk and pooled donor milk conditions.

32.4 Discussion

As seen in Fig. 32.1 and Fig. 32.2, the frequency of sucking movements differed across the 4 odour stimuli for the tests conducted at 36 weeks PCA, but not at 34 weeks PCA. At the later test age, each of the 3 milk odours elicited greater sucking

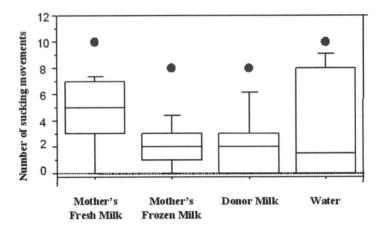

Test Stimulus

Fig. 32.1 Responses of pre-term infants to the odours of different milk stimuli at 34 weeks PCA. Horizontal lines represent the 10th, 25th, 50th (median, inside the box), 75th and 90th percentiles. Scores above the 90th percentile (outliers) are plotted as individual points

activity than did water. Therefore, freezing and pasteurization did not completely eliminate the stimulating effect of the odour of breast milk. Nevertheless, the overall pattern of results suggests that freezing may have reduced the salience of the odour of mother's milk. Whereas the sucking rate was significantly greater during exposure to mother's fresh milk compared to pooled milk, there was no reliable difference in sucking in response to mother's frozen milk vs. pooled milk. Thus, discriminative responses to own mother's milk vs. pooled milk were not observed when the mother's milk had been frozen.

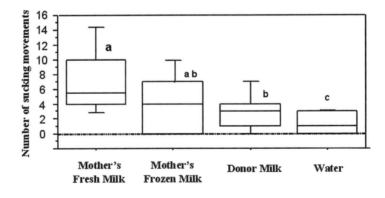

Test Stimulus

Fig. 32.2 Responses of pre-term infants to the odours of different milk stimuli at 36 weeks PCA. Boxplots that do not share the same super-script letter differed significantly

Although the infants also displayed greater sucking rates in response to their mother's fresh milk than her frozen milk, this difference did not reach statistical significance. However, there was a clear difference in the sucking activity elicited by mother's fresh milk vs. pooled milk. This combination of results suggests that the pasteurization process may have had a negative effect on the donor milk odour. Pasteurization is known to alter the nutritional and immunological properties of breast milk (Wardell, Wright, Bardsley and D'Souza 1984; Wight 2001), which could result in detectable olfactory changes. It is also possible that the infants recognized their own mother's milk odour at 36 weeks PCA. Previous studies determined that full-term infants respond preferentially to their mother's axillary, breast and milk odours over comparable olfactory stimuli from other mothers (Russell 1976; Cernoch and Porter 1985; Mizuno, Mizuno, Shonohara and Noda 2004; Marlier and Schaal 1997). Perhaps pre-term infants are similarly capable of discriminating odours produced by their own mother. It will be recalled that infants in our study were tube-fed their own mother's milk after the first 8–10 days and also began their initial breastfeeding attempts at 35 weeks PCA. Thus, they were all exposed repeatedly to the odours of their mother's breasts and milk prior to the tests at 36 weeks, and therefore had the opportunity to become familiar with those scents. The lack of discriminative responses to mother's milk at 34 weeks PCA could reflect limited pre-test contact and familiarisation with those cues, or deficient sucking behavior at that earlier age. Further research is needed to assess these hypotheses.

The differential sucking rates observed at 36 weeks PCA might be explained by qualitative differences between the odour stimuli and/or differences in their perceived intensity. In any case, our data indicate that human breast milk, especially fresh milk produced by the infant's own mother, is an effective stimulus for eliciting sucking responses in pre-term infants. These results are clinically relevant since (as discussed above) enhanced sucking experience fosters the development of competent feeding behavior in premature newborns (Pinelli and Symington 2005). Exposure to the odour of mother's milk is a non-invasive technique for stimulating sucking activity that is cost-free and has no apparent risks. This simple olfactory intervention is consistent with the natural history of our species; according to current beliefs, the evolved pattern of mother-infant interactions included frequent or continual access to the breasts (Blurton-Jones 1972; Konner and Worthman 1980) and therefore concomitant exposure of neonates to milk and other odorous secretions emanating from the nipple / areola region (Porter, Varendi and Winberg 2001).

References

Bingham, P.M. Abassi, S. and Sivieri, E. (2003) A pilot study of milk odor effect on nonnutritive sucking by premature newborns. Arch. Pediatr. Adolesc. Med. 157, 72–75.

Blurton-Jones, N. (1972) Comparative aspects of mother-child contact. In: N. Blurton-Jones (Ed.), *Ethological Studies of Child Behaviour*. Cambridge University Press, Cambridge, pp. 305–328.

Cernoch, J.M. and Porter, R.H. (1985) Recognition of maternal axillary odors by infants. Child Develop. 56, 1593–1598.

Hafstrom, M. and Kjellmer, I. (2000) Non-nutritive sucking in the healthy pre-term infant. Early Human Dev. 60, 13–24.

Konner, M. and Worthman, C. (1980) Nursing frequency, gonadal function, and birth spacing among !Kung hunter-gatherers. Science 207, 788–791.

Macfarlane, A. (1975) Olfaction in the development of social preferences in the human neonate. In: R. Porter and M. O'Connor (Eds.), *Parent-Infant Interactions*, Ciba Found. Symp. 33. Elsevier, New York, pp. 103–113.

Makin, J.W. and Porter, R.H. (1989) Attractiveness of lactating females' breast odors to neonates. Child Develop. 60, 803–810.

Marlier, L. and Schaal, B. (1997) Familiarité et discrimination olfactive chez le nouveau-né: influence differentielle du mode d'alimentation? Enfance 1, 47–61.

Meza, C.V., Powell, N.J. and Covington, C. (1998) The influence of olfactory intervention on non-nutritive sucking skills in a premature infant. Occup. Ther. J. Res. 18, 71–83.

Mizuno, K., Mizuno, N., Shonohara, T. and Noda, M. (2004) Mother-infant skin-to-skin contact after delivery reults in early recognition of own mother's milk odor. Acta Paediatr. 93, 1640–1645.

Mizuno, K. and Ueda, A. (2004) Antenatal olfactory learning influences infant feeding. Early Human Dev. 76, 83–90.

Nyqvist, K.H., Sjoden, P.-O. and Ewald, U. (1999) The development of preterm infants' breast-feeding behavior. Early Human Dev. 55, 247–264.

Pinelli, J. and Symington, A. (2005) Non-nutritive sucking for promoting physiologic stability and nutrition in preterm infants. Cochrane Database Syst. Rev. 19, CD001071.

Porter, R.H., Makin, J.W., Davis, L.B. and Christensen, K.M. (1991) An assessment of the salient olfactory environment of formula-fed infants. Physiol. Behav. 50, 907–911.

Porter, R.H., Varendi, H. and Winberg, J. (2001) The role of olfaction in the feeding behavior of human neonates. In: A. Marchlewska-Koj, J.J. Lepri and D. Muller-Schwarze (Eds.), *Chemical Signals in Vertebrates 9*. Kluwer Academic/Plenum, New York, pp. 417–422.

Porter, R.H. and Winberg, J. (1999) Unique salience of maternal breast odors for newborn infants. Neurosci. Biobehav. Rev. 23, 439–449.

Russell, M.J. (1976) Human olfactory communication. Nature 260, 520–522.

Schaal, B., Montagner, H., Hertling, E., Bolzoni, D., Moyse, A. and Quichon, R. (1980) Les stimulations olfactives dans les relations entre l'enfant et la mère. Reprod. Nutr. Dev. 20, 843–858.

Varendi, H., Porter, R.H. and Winberg, J. (1994) Does the newborn baby find the nipple by smell? The Lancet 344, 989–990.

Varendi, H., Porter, R.H. and Winberg, J. (1997) Natural odour preferences of newborn infants change over time; Acta Paediatr. 86, 985–990.

Wardell, J.M., Wright, A.J., Bardsley, W.G. and D'Souza, S.W. (1984) Bile salt-stimulated lipase and esterase activity in human milk after collection, storage, and heating: nutritional implications. Pediatr. Res. 18, 382–386.

Widstrom, A.-M., Ransjo-Arvidson, A.B., Christensson, K., Matthiesen, A.-S., Winberg, J. and Uvnas-Moberg, K. (1987) Gastric suction in healthy newborn infants. Acta Paediatr. Scand. 76, 566–572.

Wight, N.E. (2001) Donor milk for preterm infants. J. Perinatol. 21, 249–254.

Part VI
Communication between Species, Predators and Prey

Chapter 33
Patterns of Tongue-Flicking by Garter Snakes (*Thamnophis sirtalis*) during Presentation of Chemicals under Varying Conditions

Takisha G. Schulterbrandt, John Kubie, Hans von Gizycki, Ido Zuri and Mimi Halpern

Abstract Tongue-flicking is a sensory-gathering behavior used by snakes to deliver odorants to the vomeronasal organ. In the present study we provide a detailed description of environmental control and motor patterns of tongue-flicking in garter snakes, *Thamnophis sirtalis*. Tongue-flicks were monitored during prey extract trailing, foraging, delivery of air-borne odors and in an open field. Tongue-flick rates increased during airborne odor delivery and as a function of prey extract concentration during trailing, as previously reported. Motivation and prey consumption appeared to modify tongue-flick patterns since 1. tongue-flick rates were higher under foraging conditions than in an open field where no prior prey consumption had occurred and no prey odors were present; and 2. tongue-flick rates were elevated after prey consumption. The number of oscillations and the duration of tongue extensions were significantly reduced following tongue-substrate touches, suggesting that tongue contact with the substrate is the immediate stimulus for tongue retraction.

33.1 Introduction

It is well known that the chemical senses play a critical role in the behavior of snakes (Halpern, 1987, 1992; Mason 1992 and Schwenk 1995). Tongue-flicking, a chemosensory behavior pattern unique to snakes and lizards (Gove 1979; Schwenk 1993), serves as the primary vehicle for transfer of chemical substances to the vomeronasal organ (Burghardt and Pruitt 1975; Graves and Halpern 1989; Halpern and Kubie 1980; Kahmann 1932; Wilde 1938). Snakes have well-developed vomeronasal systems and flick their tongues in response to odorants perceived in their environment.

Takisha G. Schulterbrandt
SUNY Downstate Medical Center, Department of Anatomy and Cell Biology
schultert@mail.nih.gov

J.L. Hurst et al., *Chemical Signals in Vertebrates 11.*
© Springer 2008

INTERNAL FACTORS   **EXTERNAL FACTORS**

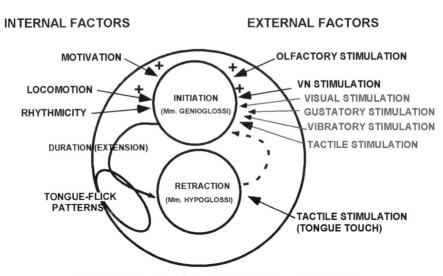

TONGUE-FLICK CONTROL CENTER

Fig. 33.1 Model of factors controlling tongue-flicking in snakes. Several external factors influence initiation of tongue-flicking, however, only olfactory and vomeronasal factors were investigated in the present study. Tactile stimulation influences retraction of the tongue when the tongue contacts the substrate. Internal factors influencing initiation of tongue-flicking are motivation, locomotion and rhythmicity. The duration of a tongue-flick event varies as a function of the tongue-flick pattern that is evoked by the environment and activity of the snake. (+ indicates positive effects on tongue-flick initiation as demonstrated in this manuscript or previously described)

Tongue-flicking is a sensory-gathering behavior, the pattern of which varies under differing environmental conditions (see Halpern 1992 for review). Here a model for control of tongue-flick behavior is proposed (Fig. 33.1) consisting of a tongue-flick control center composed of the brainstem neural elements controlling tongue protrusion (the genioglossi muscles), tongue retraction (the hypoglossi muscles) (Meredith and Burghardt 1978) and pattern generators (at present unidentified). A number of external and internal factors are known to influence the initiation portion of this tongue-flick control center, including olfactory (Halpern et al. 1997; Halpern and Kubie 1983) and vomeronasal (Burghardt and Pruitt 1975; Halpern and Frumin 1979; Kubie and Halpern 1979) stimuli. Internal factors such as locomotion (Chiszar and Carter 1975; Kubie and Halpern 1975), motivation (e.g., hunger) (Burghardt 1970), and expectations of food reward based on a prior history may also influence tongue-flick initiation. Since the tongue, in addition to being a vehicle for odorant delivery to the VNO, is a tactile organ, it is capable of providing immediate feedback to the snake that it has contacted an environmental substrate. As the tongue is especially effective in picking up odorants of low volatility it is reasonable to hypothesize that when the tongue contacts a substrate the snake will respond by retracting the tongue into the mouth for sampling the contents of the contacted substrate (Gove 1979; Ulinski 1972).

Tongue-flicking in snakes has been extensively explored. However, few studies (e.g., Gove and Burghardt 1983) have compared tongue-flick behavior patterns of the same animal across multiple conditions such as foraging, airborne odorant delivery, prey trailing, open field exploration, social and defensive conditions. An analysis of tongue-flick patterns and the environmental conditions under which each pattern is exhibited is critical to understanding the contexts in which different tongue-flick types are expressed. This information can be used to eventually identify the neural circuits that generate different tongue-flick patterns.

In the present study, video recordings were made of snakes during performance of four tasks: 1. trailing earthworm extracts of varying concentrations; 2. exploring an unfamiliar odor-free open field; 3. exploring an odor-free maze to obtain food rewards (i.e., foraging task); and 4. tongue-flicking to odorants delivered in an airstream. In addition to video taping snakes from overhead, as was done previously (Halpern and Kubie 1983; Kubie and Halpern 1979), we video taped the snakes from the side to determine the relationship between tongue-substrate contact and tongue retraction into the mouth.

33.2 Methods

33.2.1 Subjects

Six juvenile garter snakes (*Thamnophis sirtalis* and *Thamnophis radix*), three of each sex, were used in this study. The snakes were housed individually in ventilated plastic containers. During earthworm extract trailing and the foraging task snakes were fed earthworms only in the maze. During airborne delivery of odors and open field exploration snakes were fed earthworms once a week, just before a two day weekend during which they were not tested. Water was available *ad libitum.*

33.2.2 Prey Trailing

A four-choice star maze, similar to that used by Kubie and Halpern (1978, 1979), was used as a two-choice Y maze for training the snakes. Snakes were trained to follow earthworm extract trails ($1X = 6$ gm earthworm per 20 ml dH$_2$O) in the apparatus until their performance exceeded chance behavior. (See Kubie and Halpern 1978 for details). At the termination of testing, snakes were videotaped during trailing at each earthworm concentration (1X, 1/9X, 1/81X, Dry). Trials were repeated until a minimum of one minute of good, analyzable film had been obtained at each concentration. This usually involved one to three trials. For filming of animals from the side, a clear plastic maze of similar dimensions was used.

33.2.3 Foraging

Snakes were trained to look for earthworm bits in a plastic 5-arm radial maze (similar to that used by Kubie and Halpern 1978, 1979). At the beginning of every trial, the snake was placed in an open-ended cylinder placed in the center of the foraging apparatus. After 1 minute, the cylinder was lifted and the snake was allowed to explore the apparatus. During training, the goal box at the end of each arm of the maze was baited with earthworm bits to encourage the animal to enter each arm in search of food. Tongue-flick data was recorded before and after prey consumption.

33.2.4 Airborne Odorant Delivery

Snakes were tested in a 1-L glass cylinder (the same used by Halpern et al. 1997). At the start of every test session, the snake was placed in the testing apparatus for 5 minutes to acclimate. The snake was videotaped for 1 minute before, 1 minute during, and 1 minute after odorant was delivered, with three-minute intervals between subjects and thirty minutes between odorants. The odors tested were: distilled water, amyl acetate, lemon extract, earthworm extract, fish water, and live earthworms. The order of odorants delivered was not randomized.

33.2.5 Open Field Exploration

A square field (80×80 cm), with a milk-glass floor, marked off in 5 cm squares, opaque walls and a clear Plexiglas top was used. Snakes were placed in an open-ended cylinder and placed in the middle of the apparatus and the snake was allowed 1 minute to adapt. The cylinder was lifted and the snake was allowed to explore the apparatus for one minute during which tongue-flicks were videotaped. No prey or prey odor was ever placed in this apparatus.

33.2.6 Filming Protocol and Analysis

For each task, the snakes were filmed (using a Sony DCR VX1000 Digital videocamera; 30 frames per second) for 1 minute from overhead and, in the trailing task, for 1 minute from the side. During overhead videotaping, the following measures (see Kubie and Halpern 1978 for details) were analyzed for each session: a) tongue-flick rate, b) duration of each tongue-flick, c) between tongue-flick interval, d) tongue-flick initiation interval, e) maximum extension of the tongue (in cm), and f) the the distance between the two tongue tips at the tongue's greatest extension (in cm). Similar measurements were recorded from the side view (trailing trials only). In side view, the number of oscillations, i.e., the number of up-and-down movements of the tongue, was measured and the time (in number of frames) before and after tongue contact with the substrate and retraction of the tongue into the mouth was determined. Films were viewed and analyzed using the Apple Mac version of Motion Plus, Adobe Premiere and Apple Video Player.

33.2.7 Statistical Analysis

We used a factorial analysis of variance with between and within subjects factors. In all statistical tests, a p value of $\longleftarrow$ 0.05 was considered significant.

33.2.8 Autocorrelations

Autocorrelations were calculated on the time-series of tongue-flick duration scores during earthworm extract trailing. This was done for a series of tongue-flick-number shifts. So, for example, for a shift of 1 the autocorrelation paired each tongue-flick duration with the subsequent tongue flick duration. For a shift of two, the correlation paired each tongue flick duration with the second to follow.

33.3 Results and Discussion

The present study has examined tongue usage in four experimental situations: prey extract trailing, foraging, open-field exploration and during delivery of airborne odors. The same animals were tested in each experimental situation to facilitate comparisons of tongue-flick parameters in the different tasks.

A portion of the present study consisted of a partial replication of prior studies conducted in our laboratory investigating tongue-flick patterns during prey extract trailing (Halpern and Kubie 1983; Kubie and Halpern 1978, 1979) and during delivery of airborne odors (Halpern et al. 1997; Zuri and Halpern 2003), therefore they are not presented in detail here. The fundamental observations of those studies have been confirmed, i.e., that tongue-flick rate increases as a function of prey extract concentration during trailing and during airborne delivery of prey and non-prey odors (see Table 33.1 in which the significant results are summarized).

33.3.1 Trailing

As previously reported (Halpern and Kubie 1983; Kubie and Halpern 1978) tongue-flick rates varied as a function of extract concentration ($F_{1,2,6.2} = 12.85$, p = 0.009) with tongue-flick rates greatest when the snakes were following high concentration trails (1X) and lowest at the weakest concentration (1/81X) and dry trails (Table 33.1). As expected from the above, tongue-flick initiation intervals, the inverse of tongue-flick rates, also varied as a function of earthworm extract con-

Table 33.1 Summary of Significant Results. Abbreviations: AA = amyl acetate; BTFI = between tongue-flick interval; dH2O = distilled water; EE = earthworm extract; FW = fish water; LEW = live earthworm; MTFE = maximum tongue-flick extension; > significantly greater than; < significantly less than

A. Trailing of Earthworm Extract		
TF Rate		$1X > 1/9X > 1/81X = $ dry
TF Duration		$1X < 1/9X < 1/81X = $ dry
TF Initial Interval		$1X < 1/9X < 1/81X = $ dry
Max. TF Ext.		$1X < 1/9X = 1/81X = $ dry
B. Foraging: Before v After Prey Consumption		
TF Rate		Before < After
TF Duration		Before > After
TF Initial Interval		Before > After
Max. TF Ext.		Before > After
C. Foraging v Trailing (before prey consumption)		
TF Rate		Foraging < 1X
TF Duration		Foraging > 1X
TF Initial Interval		Foraging > 1X
D. Foraging v Open Field		
TF Rate		Foraging > Open Field
TF Duration		Foraging < Open Field
TF Initial Interval		Foraging < Open Field
Max. TF Ext.		Foraging < Open Field
Between TF Interval		Foraging < Open Field
E. Tongue Substrate Contact Before v After Tongue Touch		
Number of Oscillations		Before > After
TF Duration		Before > After
F. Airborne Odorants: Before (B), During (D) and After (A) Delivery		
TF Rate	EE	During > Before, After
	FW	During > Before, After
	LEW	During > Before, After
	dH$_2$O	During > Before
	AA	During > Before
TF Duration	EE	During > Before, After
	FW	During > Before, After
	LEW	During > Before, After
Max TF Extension	FW	During > Before
	AA	During > Before, After
Width of Tongue Tip	LEW	During > After
G. TF Rate as a function of time of exposure: 1-10s > 20–60s		

centration ($F_{3,15} = 16.33$, p < 0.001). The main component contributing to tongue-flick rate is duration of tongue-flick (the amount of time the tongue is outside the mouth). Overall, tongue-flick durations varied as a function of earthworm extract concentration ($F_{3,15} = 37.9$, p < 0.001), with the shortest tongue-flick durations occurring when the snakes were following high concentration trails (1X) and longest at the weakest concentration (1/81X) and dry trails (Table 33.1).

This description encourages the formulation of a systems approach to the analysis of tongue flicking, where potential signals and outputs can be separately analyzed (Fig. 33.1). Thus, trail odor concentration, detected primarily by the vomeronasal system (Kubie and Halpern 1979), appears to affect the neural mechanisms that determine how long the tongue will remain out of the mouth and how far the tongue will be extended. Since the information on trail concentration must be obtained from a prior tongue-flick, we can posit that the sensory information obtained on a given tongue-flick (or tongue-flicks) sets the pattern generator to be appropriate for the type of trail being followed.

The degree to which a tongue flick's duration predicts the duration of the subsequent tongue flick is shown by the autocorrelation with a shift of one. There was considerable individual variation in the pattern of the autocorrelations. At 1X earthworm extract: 6 of 7 snakes have significant correlation with one displacement. They also all show a decline in correlation as a function of shift number from the predictor tongue flick, although it is not monotonic in all cases. S6 is a particularly good example (Fig. 33.2). Only one snake (S8) displayed a "random" pattern of tongue-flicks when trailing 1X extract, indicating that for this animal tongue-flick duration was a poor predictor of tongue-flick duration on subsequent tongue-flicks. One possible determinant of tongue-flick rate is that there is an oscillator in the brain that sets tongue-flick rate. For six of seven snakes the autocorrelations were high and appear to support the central oscillator hypothesis. Furthermore, since autocorrelations remain high for four or five shifts, it appears that the central oscillator does not rapidly change its frequency. A substantial frequency shift appears to take several seconds.

33.3.2 Foraging

Traversing a maze in which no proximate odor cues are present, but in which snakes have previously received rewards was here defined as a foraging task. Such a task permits one to analyze tongue-flick patterns before and after prey consumption.

We compared six tongue-flick parameters before and after prey consumption in the foraging task (See Table 33.1). The overall MANOVA was significant ($F_{4,2} = 557$, $p = 0.002$). Mean tongue-flick rate was significantly lower before prey consumption than after prey consumption ($F_{1,5} = 83.4$, $p < 0.001$). The mean flick initiation interval was significantly longer before than after prey consumption ($F_{1,5} = 96.1$, $p < 0.001$) and tongue-flick duration was significantly longer before than after prey consumption ($F_{1,5} = 908$, $p < 0.001$). During foraging the snakes are motivated, based on hunger and prior experience, to seek prey. This motivation should have a positive effect on tongue-flick initiation. Furthermore, they are moving through the maze, which should also have a positive effect on tongue-flick initiation and tongue-flick rate. Our finding that tongue-flick rate increases after prey consumption can be understood in terms of stimulus intensity and motivation

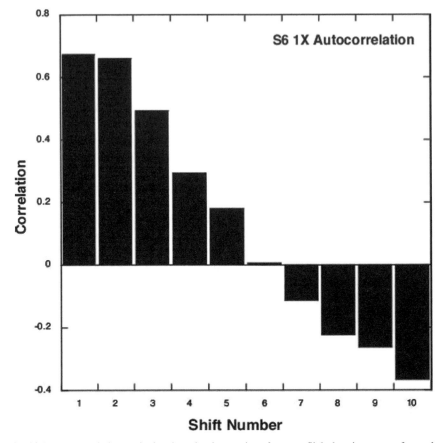

Fig. 33.2 Autocorrelations calculated on the time-series of tongue-flick duration scores for snake S6 during trailing of 1X earthworm extract

(See Fig. 33.1). Prior to prey ingestion, there is no prey stimulus in the maze, whereas during and after prey attack and ingestion, a very intense sensory stimulus is present. In addition, once prey is found in the apparatus, a snake is probably highly motivated to find more prey. Thus, both stimulus intensity and motivation may converge to increase tongue-flick rates following prey ingestion.

33.3.3 Foraging Compared to Trailing and Open Field Testing

The foraging task permits dissociation of the pattern of tongue-flicking observed during movement in the presence or absence of chemical cues derived from a prey trail under conditions where a prey reward has previously been delivered (foraging versus trailing) and in the absence of a history of prey reward (foraging versuss open field testing). Tongue-flick rates were highest during extract trailing, intermediate during foraging (before prey consumption) and lowest in the open field. These findings are a reflection of the dual importance of vomeronasal stimulation and motivation (see Fig. 33.1). During prey extract trailing, both motivation and vomeronasal stimuli could act on the tongue-flick initiation center and alter the tongue-flick pattern generator to produce frequent and rapid tongue-flicks. During foraging (prior to prey consumption), motivational factors (expectations based on prior history of obtaining prey in the apparatus) may be acting on the tongue-flick initiation center and pattern generator to produce intermediate frequency tongue-flicks. In contrast, in the open field, no prey stimuli are present and there is no history of prey delivery, so only exploratory drive is motivating locomotion and the accompanying low level of tongue-flicking.

Tongue-flick behavior in the foraging task is more similar to tongue-flick behavior on the trailing task than to tongue-flick behavior in the open field. Tongue-flick rate, tongue-flick initiation interval, tongue-flick duration and maximum tongue-flick extension differed significantly between foraging and open-field testing (Table 33.1). When tongue-flick variables during foraging trials (prior to prey consumption) are compared with those during trailing 1X earthworm extract, tongue-flick rates, tongue-flick duration and tongue-flick initiation interval all differ significantly. Thus foraging tongue-flick variables before prey consumption are similar to those exhibited during low concentration prey extract trailing, whereas these same variables in the foraging task after prey consumption are more similar to those exhibited when snakes follow 1X extract trails. We have noted previously (Halpern 1988; Kubie and Halpern 1979) that stimulation of the vomeronasal organ of snakes with prey extracts is reinforcing. The decrease in the interval between tongue-flicks during prey trailing compared to foraging may reflect the reinforcing effect of stimulus delivery during the trailing task and its effect on the initiation of the next tongue-flick.

33.3.4 Tongue substrate contact

Filming tongue-flicks from the side permitted us to analyze the number of oscillations and the duration of individual tongue-flicks before and after the tongue touched the substrate during trailing. For five snakes we analyzed the tongue-flick data recorded from the side to determine if the tongue touching the substrate would expedite the return of the tongue to the mouth, thus ending a tongue-flick bout.

Significantly more oscillations were observed during tongue-flick events before the tongue touched the substrate than after ($F_{1,4} = 156$, $p < 0.001$) overall and at each earthworm extract concentration/condition tested (1X, $p = 0.006$; 1/9, $p = 0.002$; 1/81X, $p = 0.003$; dry, $p = 0.004$). The duration of tongue-flicks before tongue-touching the substrate were significantly longer than after tongue-touching the substrate ($F_{1,4} = 345$, $p < 0.001$), overall and at each earthworm extract concentration/condition tested (1X, $p = 0.005$; 1/9, $p < 0.001$; 1/81X, $p = 0.002$; dry, $p = 0.004$). These findings suggest that contact of the tongue with the substrate

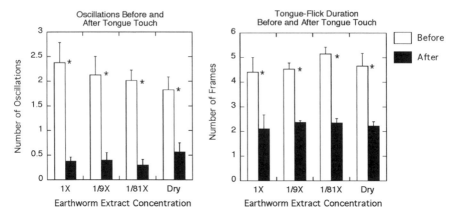

Fig. 33.3 Mean number (+SEM) of oscillations and mean duration (number of frames) (+SEM) of tongue-extension before and after tongue-substrate touches during trailing of 1X, 1/9X, 1/81X and Dry earthworm extract trails. Asterisks denote significant differences before and after tongue touch

is a stimulus for retracting the tongue into the mouth/termination of a tongue-flick (Fig. 33.1).

33.3.5 *Airborne odorants*

Six snakes were tested with airborne odors from earthworm wash, fish water, live earthworm, dH_2O, amyl acetate and lemon extract (Table 33.1). Tongue-flick rates increase during odorant delivery compared to before or after odorant delivery (Halpern et al. 1997; Halpern and Kubie 1983; Zuri and Halpern 2003).

To determine when, during odor exposure, the increase in tongue-flicks occurred, we divided the one min of odor exposure into 6 ten sec intervals and tallied the number of tongue-flicks occurring in each interval for each airborne odor delivered (Table 33.1). There were significant differences in the number of tongue-flicks emitted during different odor exposures ($F_{5,20} = 2.74$, p $= 0.049$) and during the different time intervals ($F_{5,20} = 6.66$, p $= 0.001$).

The observation that most tongue-flicks occur during the first ten seconds of odor delivery suggests that airborne odors initiate tongue-flicking, but that they are not sufficient to sustain tongue-flicking throughout the stimulus delivery period. The increased width of the tips of the tongue during this period may reflect an attempt to locate the source of the stimulus.

33.3.6 *General Discussion*

We have previously suggested (Halpern and Kubie 1983) that tongue-flicking in snakes can be classified as a "sense-seeking behavior," similar to sniffing, ear and eye movements. A major function of tongue-flicking is certainly to collect information about the outside world, particularly its chemical environment. This idea is supported by a number of observations (see Halpern 1992 for review). Tongue-flicking delivers chemical substances to the vomeronasal organ; increases in the presence of volatile and nonvolatile chemicals; is elevated after withdrawal of prey stimuli; is correlated with prey-seeking behavior; and is a concomitant to locomotion during exploration by snakes. Furthermore, it is clear that different tongue-flick patterns are elicited under these different conditions and therefore somewhat different neural circuits are probably involved in generating these different response patterns.

References

Burghardt, G.M. (1970). Intraspecific geographical variation in chemical food cue preferences of newborn garter snakes (*Thamnophis sirtalis*). Behav. 36, 246–257.

Burghardt, G. M. and Pruitt, C. H. (1975) Role of the tongue and senses in feeding of naive and experienced garter snakes. Physiol. Behav.14, 85–194.

Chiszar, D. and Carter, T. (1975) Reliability of individual differences between garter snakes (*Thamnophis radix*) during repeatd exposures to an open field. Bull. Psychon. Soc. 5, 507–509.

Gove, D. (1979) A comparative study of snake and lizard tongue-flicking, with an evolutionary hypothesis. Zeit. Tierpsychol. 51, 58–76.

Gove, D. and Burghardt, G.M. (1983) Context-correlated parameters of snake and lizard tongue-flicking. Anim. Behav. 31, 718–723.

Graves, B. M. and M. Halpern, M. (1989) Chemical access to the vomeronasal organs of the lizard, Chalcides ocellatus. J. Exper. Zool. 249,150–157.

Halpern, M. (1987) The organization and function of the vomeronasal system. Ann. Rev. Neurosci. 10, 325–362.

Halpern, M. (1992) Nasal chemical senses in reptiles: Structure and function. In C. Gans and D. Crews (Eds.), *Hormones, Brain and Behavior. Biology of the Reptilia.Vol. 18, Physiology E.* (pp. 423-523) Chicago :University of Chicago Press.

Halpern, M. (1988) Vomeronasal system functions: Role in mediating the reinforcing properties of chemical stimuli. In W. K. Schwerdtfeger and W. J. A. J. Smeets (Eds), *The Forebrain of Reptiles.*Karger, Basel, pp. 142–150.

Halpern, M. and Frumin, N. (1979) Roles of the vomeronasal and olfactory systems in prey attack and feeding in adult garter snakes. Physiol. Behav. 22, 1183–1189.

Halpern, M., Halpern, J., Erichsen, E. and Borghjid, S. (1997) The role of nasal chemical senses in garter snake response to airborne odor cues from prey. J. Comp. Physiol. Psychol. 111, 251–260.

Halpern, M. and Kubie, J.L. (1980) Chemical access to the vomeronasal organs of garter snakes. Physiol. Behav. 24, 367–371.

Halpern, M. and Kubie, J.L. (1983) Snake tongue flicking behavior: Clues to vomeronasal system functions. In R.M. Silverstein and D. Müller Schwarze, (Eds.), *Chemical Signals III*. Plenum Publishing Corp, New York, pp. 45–72.

Kahmann, H. (1932).Sennesphysiologische studien an Reptilien. I. Experimentelle Untersuchungen über das Jakobonische Organ der Eidechesen und Schlangen. Zool. Jahrbuch., Abt. für Allgemeine Zool. Physiol. der Tiere 51, 173–238.

Kubie, J. and Halpern, M. (1975) Laboratory observations of trailing behavior in garter snakes. J. Comp. Physiol. Psychol. 89, 667–674.

Kubie, J.L. and Halpern, M. (1978) Garter snake trailing behavior: effects of varying prey extract concentration and mode of prey extract presentation. J. Comp. Physiol. Psychol. 92, 362–373.

Kubie, J.L. and Halpern, M. (1979) The chemical senses involved in garter snake prey trailing. J. Comp. Physiol. Psychol. 93, 648–667.

Mason, R.T. (1992) Reptilian Pheromones. In C. Gans and D. Crews (Eds.), *Hormones, Brain and Behavior. Biology of the Reptilia. Vol. 18, Physiology E*. University of Chicago Press, Chicago, pp. 114–228.

Meredith, M. and Burghardt, G.M. (1978) Electrophysiological studies of the tongue and accessory olfactory bulb in garter snakes. Physiol. Behav. 21, 1001–1008.

Schwenk, K. (1993) The evolution of chemoreception in squamate reptiles: a phylogenetic approach. Brain Behav. Evol. 41, 124–137.

Schwenk, K. (1995) Of tongues and noses: chemoreception in lizards and snakes. Trends Ecol. Evol. 10, 7–12.

Ulinski, P.S. (1972) Tongue movements in the common boa (*Constrictor constrictor*). Anim. Behav. 20, 373–382.

Wilde, W.S. (1938) the role of Jacobson's organ in the feeding reaction of the common garter snake, *Thamnophis sirtalis sirtalis*. J.Exper. Zool. 77, 445–465.

Zuri, I. and Halpern, M. (2003) Differential effects of lesions of the vomeronasal and olfactory nerves on garter snake (*Thamnophis sirtalis*) response to airborne chemical stimuli. Behav. Neurosci. 117, 169–183.

Chapter 34
Multi-Contextual use of Chemosignals by *Liolaemus* Lizards

Antonieta Labra

Abstract Squamata reptiles are often divided in two major groups based on the main sensory modality that they use: the chemical/visual and the visual taxa. Although *Liolaemus* lizards belong to the visual taxon Iguania, I show that they may depend heavily on chemosignals in many different aspects of their life. The combined information from *Liolaemus* and other "visual genera" that use chemosignals, urges us to reconsider the classical dichotomous segregation of Squamata in terms of sensory modality. In addition, further work is also necessary to understand the role of chemical signals in "visual" lizards.

34.1 Introduction

Squamata reptiles (lizards and snakes) depend primarily on chemical and visual sensory modalities to communicate and to explore the environment (e.g. Mason 1992; Ord and Martins 2006). These sensory modalities, however, are not equally relevant for the different species. The two major Squamata clades, Scleroglossa and Iguania, are recognized as chemical/visual and visual, respectively (Vitt and Pianka 2005). This inter-clade difference is correlated with a variety of other characters, and is particularly evident in foraging modes. For example, Autarchoglossa lizards, one of the two major clades of Scleroglossa (e.g. *Varanus*), are active foragers that depend heavily on chemical signals[1] to detect prey trails through the vomeronasal organ, located in the roof of the mouth (Vitt and Pianka 2005). This group has well developed characters that improve chemoreception, such as a long and bifurcate tongue, which allows animals to collect and transport more chemosignals to the vomeronasal organ (Schwenk 1995). In addition, their vomeronasal organ has a large number of sensory neurons (for a review see Cooper 1997, 2003). In contrast, Iguania lizards (e.g. *Anolis*) are mainly ambush predators, which depend largely on vision to detect

---

Department of Biology, University of Oslo
antonieta.labra@bio.uio.no

[1] For the purpose of this chapter, chemical signals or chemosignals, refer to characteristics that convey information, and "intentionality" is not a requirement.

prey. These lizards may not have the ability to detect chemical signals, as their chemoreceptive machinery is poorly developed (Cooper 1997, 2003).

The view of Iguania as "visual lizards" is, however, not supported when we considered the use of chemosignals by *Liolaemus* lizards. This highly specious (> 160 spp) Iguania genus from South America has been predicted to have reduced or no abilities to use chemosignals, at least to detect prey signals (Cooper 2003). However, a number of observations contradict this view of just "visual lizards". First, the structure of *Liolaemus* visual displays is remarkably simple compared with displays produced by well known "visual Iguania", such as *Anolis* and *Scelo-porus* (Jenssen 1978; Martins, Labra, Halloy and Thompson 2004). Although these displays are important during conspecific interactions (Trigosso-Venario, Labra and Niemeyer 2002), their simplicity suggests that *Liolaemus* may not be "extremely visual oriented" lizards. Second, under a variety of circumstances, different *Liolae-mus* species make tongue flicks (i.e. protrusions and rapid retractions of the tongue), a behavior considered to be an indicator of chemical exploration (Mason 1992). These observations have motivated studies to determine the chemoreception abilities of this genus. Here, I summarize the available knowledge on chemoreception in *Liolaemus*, and present some new information.

34.2 Trophic Interactions

34.2.1 Prey Recognition

The exact phylogenetic position of *Liolaemus* is controversial, but there is no doubt that it belongs to Iguania (Etheridge 1995; Frost, Etheridge, Janies and Titus 2001); hence these lizards should not use chemosignals to detect prey (Cooper 1995). In fact, De Perno and Cooper (1993) found that *L. zapallarensis* did not detect prey chemosignals. However, these results contrast with recent data obtained for *L. lem-niscatus*; this species is able to discriminate substrate labeled with prey chemosig-nals, and would use this information to determine the permanence in a prey patch (Labra 2007).

34.2.2 Predator Recognition

Predation constitutes a strong selective pressure for prey to evolve mechanisms to detect signals of predator presence. Thus, if *Liolaemus* use chemosignals to orient their behavior, they should be able to detect predators such as snakes, which release chemosignals (Mason 1992). If so, this ability may depend on the predation pressure to which the animals are normally exposed. This was tested by comparing the behavior of three *Liolaemus* species submitted to different levels of predation pressure by a saurophagous snake (*Philodryas chamissonis*). The species tested were sympatric (*L. lemniscatus*), parapatric (*L. nigroviridis*), and allopatric (*L. fitzgeraldi*) to the snake. Additionally, two populations of *L. lemniscatus* that experience different snake predation pressure were compared. Lizards were exposed to chemosignals from the snake, from conspecifics, and to a control, without chemosignals (Labra and Niemeyer 2004).

Liolaemus lizards chemo-assessed snake chemosignals, but their responses were dependent on the predation pressure they experienced in their natural habitats. When exposed to snake chemosignals, the sympatric *L. lemniscatus* showed less chemical exploratory behavior, and a high frequency of antipredator behaviors that may help to reduce its detectability by a predator. In contrast, the parapatric *L. nigroviridis* showed similar levels of chemical exploration under all conditions, but exhibited antipredator behaviors when confronted with snake chemosignals. This is a clear indication of predator signal recognition. The allopatric *L. fitzgeraldi* showed no sign of predator recognition. The two populations of *L. lemniscatus* showed similar abilities to detect predation risk, but the population subject to a lower predation level showed less antipredator behaviors when confronted with snake signals (Labra and Niemeyer 2004).

34.3 Social Interactions

34.3.1 Recognition and Modulation

Conspecific recognition has a central role in social behavior and is closely related to self-recognition (i.e. discrimination of self-signals from those of conspecifics). Most Squamata use chemosignals for social recognition (Mason 1992) and *Liolaemus* is no exception. In fact, chemical self-recognition may be widespread in this genus. Adults of different species, ascribed to different phylogenetic clades, and from different geographic areas, show self-recognition (Labra, Cortéz and Niemeyer 2003; Labra, Escobar, Aguilar and Niemeyer 2002; Labra, Beltrán and Niemeyer 2001). Data from new species (Fig. 34.1) further strengthen this hypothesis.

Social recognition is modulated by different factors. Ontogeny for example, seems to modulate self-recognition in *L. bellii*. During the post-hibernation season, three age classes of *L. bellii* behave differently when confronted with their own secretions than when confronted with a control. As in other species, adults showed the pattern of less chemical exploration toward their own secretions (e.g. Fig. 34.1). By contrast, juveniles apparently make no distinction between the two conditions, while neonates may have self-recognition as they show different exploration in experimental conditions, but they explored their own secretions more (Labra et al. 2003). These differences across age classes suggest that *L. bellii* may require a "learning period" to internalize their own signals. As adults they can then have "easy self-recognition", i.e. a low number of tongue flicks toward self-secretions.

Through conspecific chemical recognition, *Liolaemus* are able to recognize the sex of the sender, and both conspecific and self-recognition are modulated by sea-

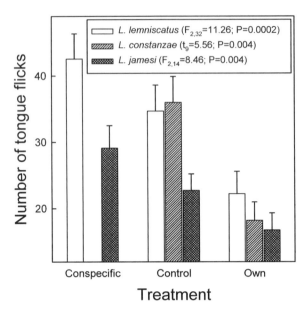

Fig. 34.1 Self-recognition in three *Liolaemus* species. Data of *L. lemniscatus* in conspecific and control treatments come from Labra and Niemeyer (2004)

sonality. Conspecific recognition usually is exhibited during breeding seasons, and self-recognition may or may not be present during the whole activity season (Labra and Niemeyer 1999; Labra, Beltrán and Niemeyer 2001; Labra et al. 2003). The only species that apparently does not have chemically based social recognition is *L. fitzgeraldi*, as it did not show conspecific recognition during post-hibernation (Labra and Niemeyer 2004). However, it is unclear if this is a seasonal response, or if this species lacks conspecific recognition, unique among the *Liolaemus* studied.

34.3.2 An Example of Fine-tuned Conspecific Recognition

Clearly, *Liolaemus* can recognize conspecifics, but are they able to obtain detailed information about conspecifics from their chemosignals? This was investigated in the context of fighting risk. During the breeding season, male *Liolaemus* sometimes engage in fights, and the ability to accurately assess the fighting characteristics of a potential opponent may be very relevant.

Chemosignals offer the advantage that they can be detected in the absence of the signaler. In the context of fighting, a receiver confronted with the signals of a potential opponent can avoid antagonistic interactions or, if these are unavoidable, can prepare itself for the conflict. If receivers can extract detailed information from chemosignals, their responses will depend on signaler characteristics. This was tested in males of *L. monticola*, a highly territorial species (Fox and Shipman 2003). Males were exposed to secretions of potential opponents of various body sizes (an indirect measure of fighting abilities); "intruders" (receivers) were placed in the terrarium of unfamiliar "residents" (signalers) in the absence of the latter, and their behaviors recorded. Simple regressions between the different behavioral variables measured and the body sizes of intruder and resident, or their relative difference in body size, were examined. Although resident characteristics were a good predictor of receiver behavior, the relative body-size difference was the best predictor. This was negatively correlated with behaviors associated with activity (motion time), chemo-exploration (tongue flicks) and social interactions (head bobs). This indicates that *L. monticola* may be able to extract detailed information from senders' chemosignals, allowing decisions during a "pre-confrontation" stage to be based on a balance between the relative fighting abilities (i.e. relative body size) of both opponents (Labra 2006).

34.3.3 Source of Chemosignals

Where are the chemosignals involved in *Liolaemus* intraspecific communication produced? The first source explored was feces. During the pre-hibernation season, individuals of both sexes of *L. tenuis* achieve self-recognition through the information contained in feces (Labra et al. 2002). In parallel, the secretions from precloacal pores were studied. These pores are located at the anterior border of the cloaca shield, and are mainly present in males (Donoso-Barros 1966). Results indicate that *L. tenuis* males cannot achieve self-recognition using precloacal secretions. However, these secretions produce a different behavioral pattern of exploration than did the control condition, which suggests that secretions carry some information that might be relevant in different contexts (Labra et al. 2002).

Later, we explored the possibility that the different secretions play a more significant role for recognition by the opposite sex. For this, *L. tenuis* was collected at Codegua and Las Viscachas localities, during the pre-hibernation season of 2004. Secretions were from skin, feces, and precloacal pores. Skin secretions were obtained by holding lizards by the head, in a vertical position, and dropping 5 ml of dichloromethane on their backs. The solvent plus secretions of each lizard were individually collected and spread over tiles. Feces and precloacal secretions were dissolved in 5 ml of dichloromethane and spread over tiles (see Labra et al. 2002). Tiles were used 14 h after preparation. Lizards were placed in an enclosure with sand and a tile with secretion; their behavior was recorded for 15 min. Males and females did not discriminate the different secretions (they had similar motion time, latency to the first tongue flick, number of tongue flicks to the tile, and time spent over the tile). The only behavioral difference across treatments was in the total chemical exploration (number of tongue flicks) made by females; they explored more tiles covered with precloacal secretions than with skin extracts from males (Fig. 34.2A). Ten, eight, and nine females out of 13 responded by showing other behaviors when they were confronted with secretions of skin, feces, or precloacal pores, respectively. The different sources of chemosignals induced different behaviors in females. Skin

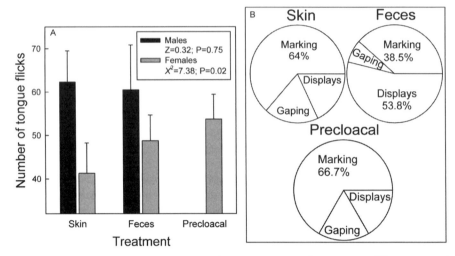

Fig. 34.2 Response of males (n = 10) and females (n = 13) of *L. tenuis* to different types of secretion (skin, feces and precloacal) from individuals of the opposite sex presented on a tile. **A.** Mean number of tongue flicks ± SE. **B.** The behaviors exhibited by females when exposed to male secretions; marking behaviors (defecation, cloaca and face rub), displays (head bobs, arm and tail waving), and mouth gaping, a behavior that may facilitate the uptake of chemosignals

and precloacal secretions mainly triggered marking behaviors, while feces mainly triggered displays (Fig. 34.2B). This suggests that male secretions may have different functions in conveying information to females. However, the absence of differences in other behaviors across treatments, suggests that females may require the whole set of secretions to assess male characteristics.

It is unknown which compounds are involved in *Liolaemus* communication. However, information available from the chemical composition of the lipidic fraction of the precloacal secretions suggests that individuals, populations and species can be discriminated by the chemical profile of these secretions (Escobar, Escobar, Labra and Niemeyer 2003; Escobar, Labra and Niemeyer 2001). This, in addition to probable chemical differences in the other sources of chemosignals, may allow individual, population and species recognition (Labra et al. 2001). Based on the individual variation observed in the composition of the secretions of *L. bellii* (Escobar et al. 2001), it is likely that lizards can achieve self-recognition, and also that they can extract detailed information about sender characteristics, as we have seen for *L. monticola* (Labra 2006).

34.3.4 Cellular basis of conspecific recognition

In an attempt to establish the neural bases of social recognition, isolated neurons from the vomeronasal organ of female and male *L. bellii*, were studied during the potential breeding season. The physiological properties of these cells were determined, as well as their ability to respond to different types of chemosignals from individuals of both sexes. Neurons were stimulated with precloacal secretions, feces and skin extracts. Neurons from both sexes responded to different secretions from both sexes (Labra, Brann and Fadool 2005), indicating that chemoreception of these secretions is mediated by the vomeronasal organ.

34.4 Congeneric Recognition and Speciation

The mechanisms that generate and maintain the high diversity of *Liolaemus* are unknown. Based on data from *L. monticola*, it has been proposed that this high diversity may be a consequence of a high rate of Robertsonian mutations (chromosomal mutations), maintained by geographic barriers. *Liolaemus monticola* has at least six chromosomal races, isolated by rivers (Lamborot, Eaton and Carrasco 2003). So far, however, chromosomal races have only been described in this and no other species with a similar geographic distribution (e.g. Vidal 2002). Furthermore, there is no explanation for the maintenance of many species in sympatry. Up to six or seven *Liolaemus* species can live in sympatry, some of which are highly syntopic with similar morphology (Jaksic, Núñez and Ojeda 1980), ecology (Donoso-Barros 1966), and karyotype (Lamborot and Alvarez-Sarret 1989; Lamborot, Espinoza and Alvarez 1979).

Analyses of the barriers to gene flow between closely related species suggest that behavioral changes are frequently key events in initiating speciation (Butlin and Ritchie 1994). Thus, considering that *Liolaemus* use chemosignals in a variety of contexts, and that the chemical profiles of precloacal secretions are species-specific (Escobar et al. 2001), chemosignals and their recognition may play some role in the diversification of this genus as well as in maintaining this diversity.

The role of chemosignals in the maintenance of species isolation is supported by observations that the species *L. jamesi*, *L. bellii*, and *L. lemniscatus*, are able to discriminate between conspecifics and congeneric individuals (Labra et al. 2001; unpublished data). Moreover, these species do not discriminate between sympatric and allopatric congeneric species. In sympatry, the recognition of an individual as a non-conspecific reduces the possibility of hybridization among congeneric species.

Responses at the population level may clarify the potential role of chemosignals in the diversification of *Liolaemus*. The chemical profiles of the precloacal secretions of two populations of *L. fabiani* indicate that they show some level of differentiation (Escobar et al. 2003). However, populations of *L. lemniscatus* and *L. tenuis* did not show clear patterns of recognition of their own population vs. other populations; apparently, chemosignals may not play a key role in the speciation process of *Liolaemus* (unpublished data).

34.5 Concluding Comments

We can conclude that *Liolaemus* lizards use chemosignals extensively, and that they can extract substantial information about the sender from its chemosignals.

Furthermore, the response to a particular set of chemosignals can be modulated by various factors, such as ontogeny and seasonality in the case of social recognition, and predation pressure in the case of predator recognition. This leads to the question of whether *Liolaemus* is unique among Iguania in their use of chemosignals. Data indicate that they are not. Studies show that other well known "visual lizards" use chemosignals, such as *Sceloporus* (e.g. Punzo and Parker 2006), *Dipsosaurus* (e.g. Bealor and Krekorian 2002; Pedersen 1992), and *Iguana* (e.g. Alberts and Werner 1993). Therefore, in addition to a revision of the categorization of Squamata into chemical/visual and visual taxa, studies should examine the diversity of use of chemical signals in this "visual" taxon.

Acknowledgments Funds come from the IFS 2933-1/2 (Sweden), FONDECYT 3990021 (Chile), and Aurora (IS EJ/hsm Aur 06-14; Norway). Thanks to all who collaborated in the field and laboratory work (P. Aguilar, E. Aguilera, D. Benítez, S. Cortéz, P. Espejo, R. Irrizari, J. Labra, A. Lillo, B. López, D. Macari, E. Mikeles, L. Ovalle, M. Soto-Gamboa, and R. Trigosso-Venario); to M. Penna who generously provided space to develop part of this work, and to T.F. Hansen for comments on the early manuscript. Studies were carried out under authorization of Servicio Agrícola y Ganadero, the Chilean agency that regulates the capture and maintenance of native fauna.

References

Alberts, A. C. and Werner, D. I. (1993) Chemical recognition of unfamiliar conspecifics by green iguanas: functional significance of different signals components. Anim. Behav. 46, 197–199.

Bealor, M. T. and Krekorian, C. O. (2002) Chemosensory identification of lizard-eating snakes in the desert iguana, *Dipsosaurus dorsalis* (Squamata: Iguanidae). J. Herpetol. 36, 9–15.

Butlin, R. K. and Ritchie, M. G. (1994) Behaviour and speciation. In: P. J. B. Slater and T. R. Halliday (Eds.), *Behaviour and Evolution*. Cambridge University Press, Cambridge, pp. 43–79.

Cooper, W. E. (1995) Foraging mode, prey chemical discrimination, and phylogeny in lizards. Anim. Behav. 50, 973–985.

Cooper, W. E. (1997) Correlated evolution of prey chemical discrimination with foraging, lingual morphology and vomeronasal chemoreceptor abundance in lizards. Behav. Ecol. Sociobiol. 41, 257–265.

Cooper, W. E. (2003) Correlated evolution of herbivory and food chemical discrimination in iguanian and ambush foraging lizards. Behav. Ecol. 14, 409–416.

De Perno, C. S. and Cooper, W. E. (1993) Prey chemical discrimination and strike-induced chemosensory searching in the lizard *Liolaemus zapallarensis*. Chemoecology 4, 86–92.

Donoso-Barros, R. (1966) *Reptiles de Chile*. Universidad de Chile, Santiago, Chile.

Escobar, C., Escobar, C. A., Labra, A. and Niemeyer, H. M. (2003) Chemical composition of precloacal secretions of two *Liolaemus fabiani* populations: are they different? J. Chem. Ecol. 29, 629–638.

Escobar, C. A., Labra, A. and Niemeyer, H. M. (2001) Chemical composition of precloacal secretions of *Liolaemus* lizards. J. Chem. Ecol. 27, 1677–1690.

Etheridge, R. (1995) Redescription of *Ctenoblepharys adspersa* Tschudi, 1845, and the taxonomy of Liolaeminae (Reptilia: Squamata: Tropiduridae). Amer. Mus. Nat. Hist. 3142, 1–34.

Fox, S. F. and Shipman, P. A. (2003) Social behavior at high and low elevations: environmental release and phylogenetic effects in *Liolaemus*. In: S. F. Fox, J. K. McCoy and T. A. Baird (Eds.), *Lizard Social Behavior*. John Hopkins University Press, New York, pp. 310–355.

Frost, D. R., Etheridge, R., Janies, D. and Titus, T. A. (2001) Total evidence, sequence alignment, evolution of polychrotid lizards, and a reclassification of the Iguania (Squamata: Iguania).

Amer. Mus. Novitates, 3343, 1–38

Jaksic, F. M., Núñez, H. and Ojeda, F. P. (1980) Body proportions, microhabitat selection, and adaptive radiation of *Liolaemus* lizards in central Chile. Oecologia 45, 178–181.

Jenssen, T. A. (1978) Display diversity in anoline lizards and problems in interpretation. In: N. Greenberg and D. Maclean (Eds.), *Behavior and Neuroethology of Lizards*. National Institute of Mental Health, Washington, D.C. pp. 269–285.

Labra, A. (2006) Chemoreception and the assessment of fighting abilities in the lizard *Liolaemus monticola*. Ethology 112, 993–999.

Labra, A. (2007) The peculiar case of an insectivorous iguanid lizard that detects chemical cues from prey. Chemoecology 17, *in press*.

Labra, A. and Niemeyer, H. M. (1999) Intraspecific chemical recognition in the lizard *Liolaemus tenuis*. J. Chem. Ecol. 25, 1799–1811.

Labra, A. and Niemeyer, H. M. (2004) Variability in the assessment of snake predation risk by *Liolaemus* lizards. Ethology 110, 649–662.

Labra, A., Brann, J. H. and Fadool, D. A. (2005) Heterogeneity of voltage- and chemosignal-activated response profiles in vomeronasal sensory neurons. J. Neurophysiol. 94, 2535–2548.

Labra, A., Beltrán, S. and Niemeyer, H. M. (2001) Chemical exploratory behavior in the lizard *Liolaemus bellii*. J. Herpetol. 35, 51–55.

Labra, A., Cortéz, S. and Niemeyer, H. M. (2003) Age and season affect chemical discrimination of *Liolaemus bellii* own space. J. Chem. Ecol. 29, 2615–2620.

Labra, A., Escobar, C. A., Aguilar, P. M. and Niemeyer, H. M. (2002) Sources of pheromones in the lizard *Liolaemus tenuis*. Rev. Chil. Hist. Nat. 75, 141–147.

Labra, A., Escobar, C. A. and Niemeyer, H. M. (2001) Chemical discrimination in *Liolaemus* lizards: comparison of behavioral and chemical data. In: A. Marchelewska-Koj, J. J. Lepri and D. Müller-Schwarze (Eds.), *Chemical Signals in Vertebrates IX*. Kluwer Academic/Plenum Publishers, New York, pp. 439–444.

Lamborot, M. and Alvarez-Sarret, E. (1989) Karyotypic characterization of some *Liolaemus* lizards in Chile (Iguanidae). Genome 32, 393–403.

Lamborot, M., Eaton, L. and Carrasco, B. A. (2003) The Aconcagua river as another barrier to *Liolaemus monticola* (Sauria: Iguanidae) chromosomal races of central Chile. Rev. Chil. Hist. Nat. 76, 23–24.

Lamborot, M., Espinoza, A. and Alvarez, E. (1979) Karyotypic variation in Chilean lizards of the genus *Liolaemus* (Iguanidae). Experientia 35, 593–595.

Martins, M. P., Labra, A., Halloy, M. and Thompson, J. T. (2004) Repeated large scale patterns of signal evolution: an interspecific study of *Liolaemus* lizards headbob displays. Anim. Behav. 68, 453–463.

Mason, R. T. (1992) Reptilian pheromones. In: C. Gans and D. Crews (Eds.), *Hormones, Brain and Behavior. Biology of Reptilia*. The University Chicago Press, Chicago, Illinois, pp. 114–228.

Ord, T. J. and Martins, E. P. (2006) Tracing the origins of signal diversity in anole lizards: phylogenetic approaches to inferring the evolution of complex behaviour. Anim. Behav. 71, 1411–1429.

Pedersen, J. M. (1992) Field observations on the role of tongue extrusion in the social-behavior of the desert iguana (*Dipsosaurus dorsalis*). J. Comp. Psych. 106, 287–294.

Punzo, F. and Parker, L. G. (2006) Food-deprivation affects tongue extrusions as well as attractivity and proceptivity components of sexual behavior in the lizard, *Sceloporus jarrovii*. Amphibia-Reptilia 27, 377–383.

Schwenk, K. (1995) Of tongue and noses: chemoreception in lizards and snakes. TREE 10, 7–12.

Trigosso-Venario, R., Labra, A. and Niemeyer, H. N. (2002) Interactions between males of the lizard *Liolaemus tenuis*: roles of familiarity and memory. Ethology 108, 1057–1064.

Vidal, M. A. (2002) Variación morfológica, cromosómica e isoenzimática en *Liolaemus tenuis*. Master Thesis in Zoology, Universidad de Concepción, Chile.

Vitt, L. J. and Pianka, E. R. (2005) Deep history impacts present-day ecology and biodiversity. PNAS 102, 7877–7881.

Chapter 35
Selective Response of Medial Amygdala Subregions to Reproductive and Defensive Chemosignals from Conspecific and Heterospecific Species

Michael Meredith, Chad Samuelsen, Camille Blake and Jenne Westberry

Abstract In hamsters and inbred mice, pheromone-containing chemosensory signals originating from the animal's own species (conspecific) and other species (heterospecific) produce differential patterns of immediate early gene (IEG = Fos/ FRAs) expression in the medial amygdala. In males of both species, conspecific stimuli, regardless of gender or putative function, activated neurons in both anterior and posterior medial amygdala (MeA, MeP). With heterospecific stimuli, MeA was activated but MeP appeared to be suppressed. MeP neurons expressing GABA-receptor were selectively suppressed by heterospecific stimuli at the same time as the GABAergic caudal intercalated nucleus (ICNc) of the amygdala was activated, suggesting suppression of MeP by ICN. We propose that information on conspecific chemosignals with preprogrammed meaning (pheromones) is analyzed by MeP neurons, probably influenced by gonadal steroid status. Information about heterospecific stimuli that activate anterior medial amygdala via the vomeronasal organ appears to have restricted access to MeP. Signals from conspecific males that are potentially threatening elicit different patterns of activation in MeP than other conspecific signals. In hamsters, male flank gland secretion activates predominantly GABA-immunoreactive neurons and mainly in ventral MeP (MePv). Male mouse urine also activates predominantly MePv in mice. This region responds to predator odors in rats and is reported to do so in mice. These findings, with other data, support a division of labor in medial amygdala according to the reproductive or defense-related potential of the stimuli. There is some evidence for a convergence of information on conspecific and heterospecific threatening stimuli but, so far, the details are not entirely consistent. In our experiments with hamsters and mice, stimuli from potential predators (cat urine, cat collar) like other heterospecific stimuli, activated MeA and not MeP. Others studying mice found activation in ventral MeA (MeAv) during male-male interactions and in MePv by cat collar

Florida State University, Prog. Neuroscience and Dept. Biol. Sci.
mmered@neuro.fsu.edu

stimuli. *Since the submission of this paper we have also found activation in mouse MePv by stronger cat collar stimuli (see note at end of text).*

35.1 Introduction

For chemosensory signals used for social and reproductive communication within a species, often called pheromones, the meaning of the signal generally does not have to be learned. Appropriate physiological and behavioral responses elicited from recipients of same species (conspecific) are unlearned, unconditioned, and pre-programmed. In mammals, the vomeronasal organ is a primary detector for these signals. Its high sensitivity and selectivity are well adapted for recognition of particular molecules or combinations with pre-programmed meanings. The main olfactory system is generally of primary importance for learning complex odor signatures of individuals, places, foods, etc., so they may be recognized later, without the individual components or even particular combinations having any ***intrinsic*** meaning. However, in some species, the main olfactory system may also be a primary detector for some pheromones (Keller, Douhard, Baum, Bakker 2006) but it is not yet clear whether it is essential for this information to be routed to the amygdala, as it is for vomeronasal input. There are also special cases where the vomeronasal system may be a primary detector for learned chemosensory signals, as in the Bruce effect (Brennan, Kaba and Keverne 1990). Many species also rely on chemosensory systems for information about potential predators, prey and competitors, taking advantage of their ability to detect signals used for intraspecific communications of the other (heterospecific) species. The main olfactory system could discriminate any of these signals, once learned, but a surprising number of the signals used for intraspecific communication also activate the accessory olfactory bulbs and the vomeronasal systems of other species, without prior exposure or learning (Meredith and Westberry 2004). Thus, both olfactory and vomeronasal systems can potentially carry information important for (unlearned) recognition of conspecific mating partners or competitors and heterospecific competitors, predators or prey. These signals are distinguished and identified by central chemosensory circuits, especially the medial amygdala, the principal target of vomeronasal input. The medial amygdala is also the site of convergence of vomeronasal and olfactory input and the source of onward projections to basal forebrain areas important for reproductive, defensive and other critical behaviors (Meredith 1998). Thus, it forms the nexus between sensory input and central executive functions, where unconditioned sensory signals can be interpreted and decisions about their relevance for different behaviors may be made. The amygdala is also important for assigning affective/ motivational value to sensory signals of all types. Thus, it is an interface where unconditioned stimuli with pre-programmed meaning and value, and conditioned signals with variable meaning and value, converge. Here we report on ongoing experiments to define the medial amygdala contribution to analysis of unconditioned stimuli.

35.2 Methods

General methods were similar to previous reports (see Meredith and Westberry 2004). Briefly, we exposed male hamsters and mice to cotton-tipped swabs containing chemosensory stimuli from males and females of the same or different species, for 15 min in a clean cage with clean bedding, renewing the swab at 3 min intervals. After a further 45 min, the animal was anesthetized and perfused for immuno-cyto-chemical analysis of brain sections for immediate-early gene (IEG) protein and/or cellular markers of cell phenotype. An increase in the protein products of the IEG, *c-fos* (FOS) and of *Fos-related* genes (Fos related antigens or FRAs) occurs when neurons are activated. FRAs expression is activated in some neurons that do not show significant increases in Fos expression, as in the GABA-ergic intercalated nucleus cells described here. Stimuli used here included the following (see also Table 35.1): (1) Conspecific for hamsters: female Hamster Vaginal Fluid (HVF; diluted 1:10), female and male Flank Gland Secretion (fFGS, mFGS; wiped from the gland). (2) Conspecific for mice: female and male urine (fMU, mMU; 1:10). (3) Heterospecific for both: male Cat Urine (mCU; 1:10), pieces of cat collar (CC). (4) Clean Swab control (CS). These stimuli included potential reproductive signals for males (female stimuli), indicators of conspecific competitive threat (male stimuli) and indicators of heterospecific predatory threat (cat stimuli). Mouse stimuli are heterospecific for hamsters and vice versa, and both may indicate potential competitors for resources. In other experiments referred to here, we explored the source of sensory information about these stimuli by removing vomeronasal organs from sexually naïve or experienced male hamsters, by ablating olfactory sensory neurons with intranasal infusion of $ZnSO_4$ solution, or by artificially stimulating sensory pathways, using methods described previously (see below).

35.3 Vomeronasal Input Analyzed Centrally by MeA/MeP Circuit

In male hamsters, we find that investigation of conspecific and heterospecific chemosensory stimuli elicits categorically different responses in medial amygdala of sexually-naïve intact animals. The primary sensory input to the anterior and posterior medial amygdala (MeA and MeP) is from the vomeronasal organ (VNO) via the accessory olfactory bulb (AOB) (Meredith 1998). MeA and MeP are strongly and reciprocally connected, but there is a denser afferent sensory projection to MeA from the AOB, suggesting a predominant flow of sensory information from AOB to MeA to MeP (Fig. 35.1). Removal of vomeronasal organs (VNX) in sexually inex-perienced (naïve) male hamsters severely impairs mating behavior as well as activa-tion of medial amygdala by reproductive chemosignals (Fewell and Meredith 2002). Activation of medial amygdala by other conspecific and by heterospecific chemosig-nals is also lost after VNX, suggesting that behaviorally important signals do not have the ability to activate medial amygdala via the main olfactory system in naïve

Table 35.1 IEG Activation in Amygdala Subareas

HAMSTER (Meredith, Westberry, Blake)			MeAd	MeAv	MePd	MePv	ICNc
Hamster Vaginal Fluid	HVF	Repro./ Con-specific			+++		reduced
Female Fl-Gland Secr.	fFGS	Repro./ Con-specific			++		0
Male Fl-Gland Secr.	mFGS	Defense/ Con-sp.			+	+++	++
Female Mouse Urine	fMU	Neutral/ Hetero-sp.			0		++
Male Mouse Urine	mMU	Neutral/ Hetero- sp.			0		++
Male Cat Urine	mCU	Defense/ Hetero- sp.			0		NA
Cat Collar (weak)	CC	Defense/ Hetero- sp.			0		++
VNO Electrical stimulation		Artificial		++			NA
AOB Drug stimulation		Artificial	Artificial	++			

MOUSE (Meredith, Samuelsen)			MeAd	MeAv	MePd	MePv	ICNc
Hamster Vaginal Fluid	HVF	Neutral/ Hetero-sp.	++	++	0	0	+
Female Fl-Gland Secr.	fFGS	Neutral/ Hetero-sp.		+			NA
Male Fl-Gland Secr.	mFGS	Neutral/ Hetero-sp.		+			+
Female Mouse Urine	fMU	Repro./ Con-specific	++	++	++	++	NA
Male Mouse Urine	mMU	Defense/ Con-sp.	++	++	0	+	0
Cat Collar (weak)	CC	Defense/ Hetero-sp.	0	++	0	0	NA

MOUSE (Choi et al. 2005)			MeAd	MeAv	MePd	MePv	ICNc
Female Mouse Urine	fMU	Repro./ Con-specific	NA	NA	*Lhx6++	+	NA
Male Mouse Urine	mMU	Defense/ Con-sp.	NA	**Lhx5 ++	0	NA	NA
Cat Collar (strong)	CC	Defense/ Hetero-sp.	NA	NA	0	** ++	NA

*projects to VMHvl   **projects to VMHdm

male hamsters (Westberry and Meredith unpublished data). Elimination of main olfactory input by intranasal infusion of $ZnSO_4$ does not affect mating or activation of medial amygdala in hamsters (Meredith and Westberry 2004). Thus, vomeronasal input is necessary for normal mating in naïve males and olfactory input is not sufficient, suggesting that high sensitivity and selectivity of vomeronasal sensory neurons (Leinders-Zufall, Lane, Puche, Ma, Novotny, Shipley and Zufall 2000) may provide a pre-programmed signal identifying appropriate mating partners. Some regions of amygdala are activated during reproductive and agonistic behavior (Kollack-Walker and Newman 1995) and could be a route for pre-programmed recognition of conspecific competitors as well as mates.

35.4 Conspecific Signals are Analyzed in MeP in Hamsters and Mice

In male hamsters, we presented conspecific chemosignals from female or male hamsters on cotton swabs (Meredith and Westberry 2004) and found they activate immediate early gene expression (Fos or FRAs) in both anterior medial amygdala (MeA) and posterior medial amygdala (MeP). Chemosignals used for the same kinds of communication but by other (heterospecific) species, e.g male or female mouse urine or cat urine, all activated MeA and suppressed MeP in male hamsters (see Fig. 35.1 in Meredith and Westberry 2004; Table 35.1). With hetero-specific stimulation, the caudal part of the medial intercalated nucleus (ICNc), a group of GABA-immunoreactive (ir) cells (Meredith and Westberry 2004) adjacent to MeP

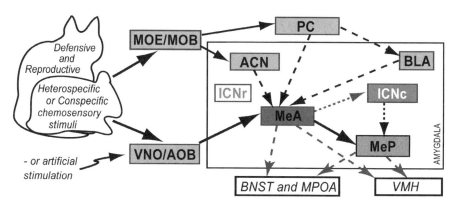

Fig. 35.1 Simplified diagram of chemosensory circuit in amygdala. Vomeronasal input via accessory olfactory bulb (VNO/ AOB) is analyzed in anterior and posterior medial amygdala (MeA, MeP). MeP appears to be inhibited by intercalated nucleus (ICNc) for heterospecific and artificial stimuli. MOE/ MOB: Main olfactory epithelium/Main olfactory bulb. ACN: Anterior Cortical Nucleus. PC: Piriform Cortex. BLA: Basolateral amygdala. ICNr: rostral part of medial intercalated nucleus. ICNc: caudal part of ICN. MPOA: Medial Preoptic Area. VMH: Ventro-medial hypothalamus

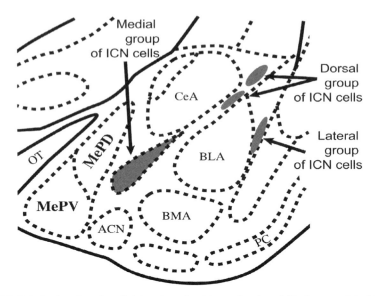

Fig. 35.2 Relative location of medial, basolateral and intercalated nuclei in the amygdale. Modified from drawings in the Morin and Wood (2001) atlas

(Fig. 35.2), was also activated and could be responsible for suppression of MeP. Artificial stimulation of the VNO sensory pathway by electrical stimulation in VNO or by drug-stimulation in AOB (Nolte and Meredith 2005), also activates MeA but not MeP. VNO stimulation, also activates ICNc (Meredith and Westberry 2004). Either type of artificial stimulation would produce a general, synchronous activation of input neurons (to AOB or to MeA/MeP) without regard to their normal response to any chemosensory stimulus (non-sense signals). We concluded (Meredith and Westberry 2004) that the amygdala circuit categorizes natural communication stimuli as socially-relevant (conspecific) regardless of their particular social function, or as not socially relevant (heterospecific and artificial). This categorization is not seen in the responses at the AOB level in hamsters (Meredith and Westberry 2004). Responses in the AOB of mice (Luo, Fee and Katz 2003) clearly discriminate between multiple criteria within the range of strain and gender stimuli present in mouse chemosensory signatures. It is not clear whether this is a higher order analysis (categorization) or simply a response to different combinations of chemicals in the different stimulus sources.

All socially relevant information appears to be routed to MeP but different conspecific social signals clearly activate different groups of neurons in MeP, which are distinguished to some extent by location within MeP and by cell phenotype. Thus, MeP could be important for deciding appropriate social response. For example, within MeP two conspecific reproduction-related stimuli (HVF, fFGS) strongly activate cells predominantly in the dorsal portion (MePd) and predominantly cells that are GABA-Receptor (GABA-R)-ir (Table 35.1). HVF activates few GABA-ir cells but fFGS activates more (Westberry and Meredith unpublished data). Responses to these signals in MeP contrast with responses to male flank-gland secretion (mFGS), which activates predominantly GABA-ir cells in MeP, predominantly in MePv. Male hamsters flank mark competitively to signal their presence to females (Cohen, Johnston and Kwon 2001), so mFGS from other males signals a competitive threat and belongs in a conspecific threat-related or defense-related sub-category. All three of these conspecific stimuli activate MeP (in addition to cells in MeAd and MeAv, Table 35.1), but the defense-related and reproductive stimuli activate different cells and in different but overlapping subareas of MeP. All stimuli we used were unfamiliar to the test animals so these responses relate to categories of animals not to individuals.

We reasoned that hamster and mouse chemosignals should activate medial amygdala in those two species in reciprocal ways. In our preliminary experiments with mice, we find that they do. In mice, conspecific (mouse) stimuli activate MeA and MeP but heterospecific (hamster) stimuli activate only MeA. Female mouse urine (fMU), a conspecific-female reproductive stimulus, activates MeA (both dorsal and ventral) and strongly activates posterior dorsal Me (MePd) in male mice. The equivalent hamster reproductive stimulus, HVF, activates MeA but not MeP in mice; as fMU does in hamster. The conspecific-male defense- or threat-related stimulus for mice, equivalent to mFGS for hamsters, is male mouse urine (mMU). We find significant activation of MePv by this defense-related conspecifc-male stimulus, as well as significant activation in MeAd and MeAv (see Table 35.1). We do not yet know whether these cells have the same transmitter phenotype as those activated by mFGS in male hamsters. We do have preliminary evidence that heterospecific stimuli activate ICN in mice as in hamsters. Figure 35.3 shows activation of IEGs (FRAs) in medial amygdala and ICN in serial sections through the anterior-posterior extent

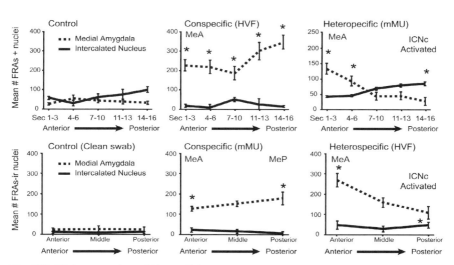

Fig. 35.3 FRAs expression within medial amygdala and ICN averaged over sets of three serial sections from MeA to MeP - for hamster (top) and mouse (below). The same two stimuli were used, each conspecific for one species, where it activates MeA and MeP, and heterospecific for the other, where it activates MeA and ICN. Control values for hamster were taken from animals perfused 15 min after exposure to HVF, before significant FRAs expression begins.
*p < 0.05 compared to control at that site (ANOVA)

of medial amygdala (MeA and MeP) in hamsters and mice. The results in mice are preliminary, involving few animals, but are clearly similar to those in hamster. The same two stimuli (HVF, mMU) were used, each conspecific for one species and heterospecific for the other. In both cases, anterior sections showed activation with both stimuli in MeA and posterior sections showed activation in MeP only for the stimulus that was conspecific (HVF for hamsters, mMU for mice). In the same sections, posterior sections showed activation of caudal ICN (ICNc) only for the heterospecific stimulus (mMU for hamsters, HVF for mice).

35.5 MeA/MeP is Essential for Evaluation of Conspecific Signals

The accessory olfactory bulb (AOB), supplies neural input to medial amygdala circuits from two distinct sets of vomeronasal sensory neurons (VSNs) but shows no evidence of the categorization of responses seen in MeA/MeP (Meredith and Westberry 2004). The amygdala response appears to be a second stage analysis of the chemosensory information and a place for integration with olfactory input. We suggest MeA, MeP and ICN constitute a preprogrammed neural circuit that selects signals for further (tertiary) analysis by more central circuits, and directs reproductive signals for analysis by the steroid-receptor rich circuits of MeP and Bed Nucleus of Stria Terminalis (BNST) (Wood, Brabec, Swann and Newman 1992).

Lesions of MeA that spare the medially located (superficial) AOB input to MeP eliminate mating behavior in male hamsters, whereas lesions of MeP alone disrupt, but do not eliminate, mating in laboratory tests (Lehman and Winans 1982). However, the greater deficit after MeA lesions may result in part from the disconnection of MeA input to MeP. So, both areas may actually contribute more equally to mating behavior. Microinjections of steroid hormones in MeP are sufficient to restore mating behavior in castrated male hamsters, but only if the area supplied with hormone also has intact chemosensory input (Wood and Coolen 1997). Lesions of medial amygdala also affect defensive behavior (Luiten, Koolhaas, de Boer and Koopmans 1985), which is also modulated by steroid levels. Thus, control of chemosensory input to MeP, which appears to be one function of the categorical circuit, may ensure that species-typical social behavior in naïve animals is limited to appropriate situations signalled by conspecific stimuli.

35.6 Reproductive and Defensive Functions of Medial Amygdala

The differential activation of subareas of MeP by reproductive and conspecific threat/defensive stimuli in hamster is generally consistent with the proposals of Canteras and colleagues (Canteras 2002, Petrovich, Canteras and Swanson 2001). From tracing experiments in rats, they suggest that medial amygdala is divided into subareas concerned with reproductive and defensive stimuli, each connected to subareas of the hypothalamus with the same function. Following these proposals, Choi,

Dong, Murphy, Valenzuela, Yancopoulos, Swanson and Anderson (2005) investigated responses in medial amygdala of male mice, in cells expressing different LIM-Homeo domain proteins. These proteins are critically involved in axon guidance in other areas of the brain. In a technically wide-ranging report, Choi et al. (2005) found strong IEG (*c-fos* or Fos) activity in Lhx-6 positive cells in response to a conspecific reproductive stimulus (fMU). In agreement with our studies and those of others in mice, hamsters and rats (Baum and Everitt 1992), these activated cells are located in dorsal posterior medial amygdala (MePd). The majority of activated cells in this area were Lhx-6-positive. Some of these projected to the ventrolateral (reproductive) part of the Ventromedial Hypothalamus (VMHvl), as predicted by the Canteras hypothesis, but the proportion of activated cells that fell into this category is unclear. Unlike the cells activated in hamster by reproductive stimuli, these Lhx-6 positive cells are glutamic acid decarboxylase (GAD)-positive and probably GABAergic. A discrete group of cells in ventral MeA express another LIM marker, Lhx5-ir and project to the dorsomedial part of the Ventromedial Hypothalamus (VMHdm), a component of the putative defensive circuit. These cells were apparently not examined after chemosensory stimulation alone, but were strongly activated during male-male agonistic encounters in mice. They express vGlut2, a glutamate transmitter transporter, and do not express GAD. So, they are not the displaced equivalents of the cells activated in MePv in hamsters. We have recently acquired the antibody to Lhx5 (courtesy of Dr. Tom Jessel) and also find a patch of cells positive for Lhx5 in hamster MeAv. We do not yet know whether these Lhx5 cells in hamster respond selectively to mFGS, the conspecific-male defence-related stimulus for hamsters. They are located within the extensive region of MeA activated by mFGS, but this region also responds to most other natural stimuli tested. Lhx5+ cells are widespread in the hamster brain, so it is not a unique marker for cells responsive to conspecific-male stimuli or VMHdm projecting cells.

Table 35.1 includes data from Choi et al. (2005) for comparison with our studies on activation of medial amygdala and subareas in hamsters and mice. They concentrated on whether cells with an identified phenotype respond to particular stimuli and project to the hypothalamic subnuclei predicted by the Canteras hypothesis, but did not systematically report on other cells or other areas.

In Choi et al.'s (2005) experiments, a group of cells located in MePv and projecting to (defensive) VMHdm, but not labeled with a LIM phenotype, was activated by predator stimuli (pieces of collar worn by a cat - CC). Activation of cells in this area by cat odor is also well established in rats (McGregor, Hargreaves, Apfelbach and Hunt 2004). We did not see activation in this area of hamsters by male cat urine (mCU) (Meredith and Westberry 2004) but this may not be the most potent predator related stimulus from cats (McGregor et al. 2004). We tested pieces of cat collar as stimuli for both hamsters and mice but found only the standard heterospecific pattern of amygdala response, with activation of MeA but not MeP. A separate statistical analysis of dorsal and ventral subdivisions of MeA and MeP did not find any activation even in MePv (Fig. 35.4). However, our cat collars were worn by the cat for only 12 hrs compared with 2 weeks for collars used by Choi et al. and in other previous experiments. We believe the shorter time should be sufficient to

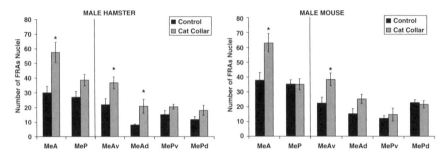

Fig. 35.4 2.5-cm pieces from a cat-collar worn by a male cat for 12 hrs activate MeAv and MeAd in hamster (left) and MeAv in mouse (right) but did not activate MeP (* Hamster, MeA p < 0.001; MeAv p < 0.02; MeAd p < 0.04; Mouse, MeA, MeAv p < 0.01 – all vs Control; ANOVA)

acquire a distinguishable odor but there is evidence for an intensity-related factor in the behavioral response to cat odor stimuli by rats (Takahashi, Nakashima, Hong and Watanabe 2005) and our stimulus may have been too weak.

Summary

There is agreement across studies in designating response in dorsal posterior medial amygdala (MePd) as likely to be reproduction-related, but there appears to be a mismatch between the transmitter phenotypes of the cells predominantly activated in hamsters (non-GABA-ir) and reported in mice (GAD-positive). Ventral medial amygdala (MeAv, MePv) may be an area where both conspecific and heterospecific defence-related stimuli are represented. These stimuli are not equivalent, but some of the behavior they elicit is similar. MePv is activated by conspecific threat-related stimuli, mFGS in hamsters and mMU in mice, and by heterospecific threat stimuli (cat collar, CC) in mice. The cells in MePv that predominate in responses to conspecific threat in hamsters (GABA-ir) and in responses to heterospecific threat in mice (non-GAD-ir) could be two co-existing populations. Resolution of this question will require further work. Choi et al. (2005) did not report on activation in MePv by conspecific male stimuli and our failure to see activation there with cat stimuli may be related to intensity of stimulation. The strong activation of Lhx5 cells in MeAv during male-male agonistic interactions reported by Choi et al. (2005) in mice is not clearly matched in our experiments by a selective activation of this area by conspecific-male defence-related stimuli (mFGS) in hamsters or by mMU in mice. In future experiments we will use a wider range of stimuli and more extensive tests to evaluate cell phenotypes; and we will complement IEG methods with electrophysiology in order to resolve these conflicts. This will provide a more complete account of medial amygdala function in discriminating between stimuli and, perhaps, in deciding on an appropriate response.

Note added in proof: In work completed after the submission of the manuscript we find that stimuli from cat collars worn for 2 weeks do activate ventral posterior medial amygdala (MePv) in mice, as reported by Choi et al. 2005. We confirm that collars worn for 12 hrs do not activate MePv (Samuelsen, C. and Meredith, M., unpublished data).

Acknowledgments Supported by NIDCD grant DC 5813 from the U.S. National Institutes of Health.

References

Baum, M. J. and Everitt, B. J. (1992) Increased expression of c-fos in the medial preoptic area after mating in male rats: role of afferent inputs from the medial amygdala and midbrain central tegmental field. Neuroscience 50, 627–646.

Brennan, P., Kaba, H. and Keverne, E.B. (1990) Olfactory recognition, a simple memory system. Science 250, 1223–6

Canteras, N. S. (2002) The medial hypothalamic defensive system: hodological organization and functional implications. Pharmacol. Biochem. Behav. 71, 481–491.

Choi, G. B., Dong, H.W., Murphy, A.J. Valenzuela, D.M., Yancopoulos, G.D., Swanson, L.W. and Anderson, D.J. (2005) Lhx6 delineates a pathway mediating innate reproductive behaviors from the amygdala to the hypothalamus. Neuron 46, 647–60.

Cohen, A.B., Johnston, R. and Kwon, A. (2001) How golden hamsters (Mesocricetus auratus) discriminate top from bottom flank scents in over marks. J. Comp. Psychol. 115, 241–247.

Fewell, G.D. and Meredith, M. (2002) Experience facilitates vomeronasal and olfactory influence on Fos expression in medial preoptic area during pheromone exposure or mating in male hamsters, Brain Res. 941, 91–106.

Keller, M., Douhard, O., Baum, M..J, Bakker, J. (2006) Sexual experience does not compensate for the disruptive effects of zinc sulfate-lesioning of the main olfactory epithelium on sexual behavior in male mice. Chem. Senses 31,753–762

Kollack-Walker, S. and Newman, S. (1995) Mating and agonistic behavior produce different patterns of Fos immunolabeling in male Syrian hamster brain. Neuroscience 66, 721–36.

Lehman, M.N, and Winans, S.S. (1982) Vomeronasal and olfactory pathways to the amygdala controlling male hamster sexual behavior: autoradiographic and behavioral analyses, Brain Res. 240, 27–41.

Leinders-Zufall, T., Lane, A.P., Puche, A.C., Ma, W., Novotny, M.V., Shipley, M.T. and Zufall, F. (2000) Ultrasensitive pheromone detection by mammalian vomeronasal neurons, Nature 405, 792–6.

Luiten, P., Koolhaas, J., de Boer, S. and Koopmans, S. (1985) The cortico-medial amygdala in the central nervous system organization of agonistic behavior. Brain Res. 332, 283–297.

Luo, M., Fee, M.S. and Katz, L.C. (2003) Encoding pheromonal signals in the accessory olfactory bulb of behaving mice, Science 299, 1196–201.

McGregor, I. S., Hargreaves, G. A., Apfelbach, R. and Hunt, G.E. (2004) Neural correlates of cat odor-induced anxiety in rats: region-specific effects of the benzodiazepine midazolam. J. Neurosci. 24, 4134–44.

Meredith, M. (1998) Vomeronasal, olfactory, hormonal convergence in the brain. Cooperation or coincidence? Ann N Y Acad. Sci. 855, 349–61.

Meredith, M. and Westberry, J.M. (2004) Distinctive responses in the medial amygdala to same-species and different-species pheromones, J Neurosci. 24, 5719–25.

Morin, L.P. and Wood, R.I. (2001) *A stereotaxic atlas of the golden hamster brain*. San Diego: Academic Press 2001.

Nolte, C, and Meredith, M. (2005). mGluR2 activation of medial amygdala input impairs vomeronasally-mediated behavior. Physiol. Behav. 86, 314–323.

Petrovich, G.D., Canteras, N.S. and Swanson, L.W. (2001) Combinatorial amygdalar inputs to hippocampal domains and hypothalamic behavior systems. Brain Res. Rev. 38, 247–289

Takahashi, L. K., Nakashima, B. R., Hong, H. and Watanabe, K. (2005) The smell of danger: a behavioral and neural analysis of predator odor-induced fear. Neurosci. Biobehav. Rev. 29, 1157–1167.

Wood, R..I, Brabec, R.K., Swann, J.M. and Newman, S.W. (1992) Androgen and estrogen concentrating neurons in chemosensory pathways of the male Syrian hamster brain. Brain Res. 596, 89–98.

Wood, R,I, and Coolen, L.M. (1997) Integration of chemosensory and hormonal cues is essential for sexual behaviour in the male Syrian hamster: role of the medial amygdaloid nucleus, Neuroscience 78, 1027–35.

Chapter 37
Seasonal Responses to Predator Faecal Odours in Australian Native Rodents Vary Between Species

R. Andrew Hayes

Abstract Small mammals are subject to predation from mammalian, avian and reptilian predators. There is an obvious advantage for prey species to detect the presence of predators in their environment, enabling them to make decisions about movement and foraging behavior based on perceived risk of predation. One technique commonly exploited to assess this risk is to use the odours of the faeces and/or urine of their predators to determine presence/absence and the length of time since a predator passed through an area. I examined the effect of faecal odours from marsupial and eutherian predators, and a native reptilian predator, on the behavior of three endemic Australian rodent species (the Fawn-footed Melomys *Melomys cervinipes*, the Bush Rat *Rattus fuscipes* and the Giant White-tailed Rat *Uromys caudimaculatus*) in rainforest remnants on the Atherton Tableland, North Queensland, Australia. Infrared camera traps were used to assess visit rates of rodents to odour stations containing faecal and control odours. Rodents avoided odour stations containing predator faeces, but did not avoid herbivore or control odours. The responses of the three prey species differed: in the late wet season *U. caudimaculatus* avoided predator odours, while *R. fuscipes* and *M. cervinipes* did not. In contrast, in the late dry season all three species avoided odour stations containing predator odours. I speculate that these differential responses may result from variations in life history traits between the species.

37.1 Introduction

Small mammals are frequently at risk of being eaten by native and introduced predators, mammalian, avian and reptilian. The prey species use a variety of mechanisms to avoid being caught - morphological, life-historical and behavioural (Apfelbach, Blanchard, Blanchard, Hayes and McGregor 2005). Preventing predation is, however, potentially costly in terms of missed opportunities for mating, foraging etc. A

R. Andrew Hayes
Queensland University of Technology, School of Natural Resource Sciences, University of Queensland, Institute for Molecular Bioscience
r.hayes@imb.uq.edu.au

species can maximise its benefits, while reducing its costs, if accurate assessments are made about the risk of a predation event. Predation risk can be assessed and reduced by small mammals through their detection and avoidance of locations containing the faeces of their predators. This is referred to as the scat-avoidance hypothesis (Banks, Hughes and Rose 2003). The response to predator odours, including faecal odours, has been tested in a variety of species, particularly rodents (reviewed in Apfelbach et al. 2005; Kats and Dill 1998). The response of these prey species tends to be an innate and generalised avoidance of predator odours seen even in predator-naive herbivores, rather than a conditioned response to a specific predator. This generalised response has been suggested to be due to similarities in the chemical composition of these odours, such as sulfurous compounds resulting from protein breakdown found in urine and faeces (Dickman and Doncaster 1984; Nolte, Mason, Epple, Aronov and Campbell 1994).

A prey animal's response to a predator cue may be influenced by a variety of factors. Familiarity with a predator may be important if the two species have a long evolutionary history of contact and it might be supposed that a cue would be more recognizable than if the species have only recently come into contact, or are allopatric (Catarell and Chanel 1979; Dickman 1992; Müller-Schwarze 1972; Sullivan, Nordstrom and Sullivan 1985). The age of an animal may also influence its response, or responses may alter seasonally (Borowski 1998, 2002).

This study describes the response of three native Australian rodent species in North Queensland rainforests to predator faecal odours. We examined the questions:

1. do native rodents respond to predator faecal odours, and, if so,
2. do they respond differently to marsupial and eutherian predators (Dickman 1993);
3. does familiarity (sympatry) affect the response to a predator odour (Catarell and Chanel 1979; Müller-Schwarze 1972; Sullivan et al. 1985);
4. is there a difference in response to reptilian and mammalian predators; and
5. do responses vary seasonally (Borowski 1998, 2002)?

37.2 Methods

37.2.1 Study Species and Locations

This study was conducted in two remnant rainforest patches on private land on the Atherton Tableland, North Queensland, Australia. Location 1 was the property of S. and M. Ogun (145°39'31"E, 17°15'47"S) and location 2 was Fur'N'Feathers Rainforest Tree Houses (145° 36'27"E, 17°24'36"S). Vegetation at both locations comprises complex mesophyll vine forest (Type 1b) (Tracey 1982). All Australian rodents belong to the family Muridae; one group, the 'old endemics' invaded the continent 15 million years ago, another group, the 'new endemics', invaded about one million years ago and a third group arrived with European settlement about

220 years ago (Strahan 1995). At both study locations three species dominate the rodent fauna: the fawn-footed melomys, *Melomys cervinipes*, the bush rat, *Rattus fuscipes*, and the giant white-tailed rat, *Uromys caudimaculatus*. *Melomys cervinipes* and *U. caudimaculatus* are 'old endemics', while *R. fuscipes* is a member of the 'new endemics'.

Life history characteristics of these species differ: both *M. cervinipes* and *R. fuscipes* are small (~ 100–$150\,$g) and relatively short-lived, rarely surviving into a second year, *while U. caudimaculatus* is larger (~ 1000–$1500\,$g) and regularly lives for more than four years. Both of the old endemics are univoltine, while *R. fuscipes* is multivoltine (Covacevich and Easton 1974; Lunney 1995; Moore 1995; Redhead 1995; Watts and Aslin 1981). The principal mammalian predators at the study locations are the dingo, *Canis familiarus dingo*, and the spotted-tailed quoll, *Dasyurus maculatus*. Scat analyses show that the three study species comprise around 30% of dingo and 25% of spotted-tailed quoll diets, respectively, in North Queensland (Burnett 2001; Vernes, Dennis and Winter 2001). Primary reptilian predators are the carpet python, *Morelia spilota variegata*, and the red-bellied black snake, *Pseudechis porphyriacus*.

I studied the response of these rodents to seven odour types, including herbivore faeces, a strong novel odour ($\sim 2\%$ acetic acid) and a variety of predator faeces (Table 37.1). The novel odour, acetic acid, detected by *Rattus norvegicus* (Laing, Panhuber and Slotnick 1989; Seelke and Blumberg 2004), would not normally be encountered in the study locations and was used to confirm that behaviours observed were in response to the test odours and not a response to any strong, novel odour (reviewed in Bramley and Waas 2001; Mappes, Koskela and Ylonen 1998; Wolff and Davis-Born 1997). A strip of cardboard ($1 \times 2\,$cm), was soaked in the acetic acid solution for at least 3 hr to produce the odour source, and was still detectable to the human nose after more than 12 hr when the strip was removed from the field, so was assumed to be detectable to rodents for the duration of the exposure time. Predator faecal samples were collected from captive animals (Table 37.1) fed on a diet that included murid rodents. Predators were chosen that varied in respect to familiarity and also whether they were marsupial or eutherians. Faecal samples were collected within 12 h of production and stored at $-20\,°$C in airtight vials with Teflon-lined lids (120 ml, Alltech Associates Australia, Sydney, Australia) until used in the field. Faeces were presented so as to reproduce field encounter conditions by prey, with each sample consisting of the entire faecal deposit produced in a single bowel motion by the predator. The mean mass of faeces used for each species is given in Table 37.1.

37.2.2 Design of Odour Station

I used infrared-triggered digital still cameras with infrared flash (Olympus Camedia, C-830L) at odour stations containing predator faeces, herbivore faeces, novel odour and blanks. Camera traps have the advantage over box traps that they are not

Table 37.1 Familiarity and evolutionary history with target prey species, faecal mass (mean ±
SE), and collection details of odour sources used in camera traps. [a](Strahan 1995), [b](Cogger 1994),
[c](Jones, Rose and Burnett 2001), [d](Kruuk and Jarman 1995), [e](Corbett 1995), [f](Butler 1969; Calaby and Lewis 1977; Calaby and White 1967; Dawson 1982; Horton 1977), [g]P. Latch, personal
communication, [h]BFP = Brisbane Forest Park, The Gap, Queensland; LPKS = Lone Pine Koala
Sanctuary, Fig Tree Pocket, Queensland; CSIRO = CSIRO Sustainable Ecosystems, Crace, ACT

Odour source	Prey familiarity	Odour type	Evolutionary history	Mass of deposit (g)	Collection details[h]
Whiptail wallaby, *Macropus parryi*	Familiar	Mammalian herbivore	Since murid arrival[a]	7.7 ± 0.36	LPKS
Acetic acid (~ 2%)	Unfamiliar	Novel odour	N/A	N/A	Sigma
Carpet python, *Morelia spilota variegata*	Familiar	Reptilian predator	Since murid arrival[b]	6.2 ± 0.75	BFP
Spotted-tailed quoll, *Dasyurus maculatus*	Familiar	Marsupial predator	Since murid arrival[c]	25.2 ± 2.20	BFP
Dingo, *Canis familiaris dingo*	Familiar	Eutherian predator	~6000 yrs[e]	42.6 ± 2.37	LPKS
Tasmanian devil, *Sarcophilus harrisii*	Unfamiliar	Marsupial predator	Until ~3000 yrs ago[f]	19.9 ± 1.05	LPKS
Red fox, *Vulpes vulpes*	Unfamiliar	Eutherian predator	No sympatry[g]	22.9 ± 2.27	CSIRO

removed from data collection when set off, and can therefore detect visits by
multiple individuals (or multiple visits by the same individuals) in a night. The
stations, consisting of a camera, sensors and caged odour source and attractant,
were constructed and distributed as described previously (Hayes, Nahrung and Wilson 2006). I could easily identify species and record number of visits by the three
rodent species of interest from the resultant photographs (Fig. 37.1).

Although camera positions were static, blank and odour treatments were randomly assigned between stations each night. Three to five blank stations (attractant
only) were placed in the study location every night. I conducted 15 replicates of
each odour station (linseed oil + novel odour, herbivore or predator faeces) in the
late dry season (October) 2003 (155 trap-nights), 25 replicates in the late wet season
(May) 2004 (218 trap-nights) and 15 replicates in the late dry season (November)
2004 (120 trap-nights).

37.2.3 Statistical Analysis

To compare between days and locations I standardised visitation rates for each
species by calculating visits to each odour type as a proportion of mean visits to
the blank stations by that species, on that night, and at that location. By making the
data proportional I removed effects of differing population size and density, as well
as variation between nights, for example in moon phase. These data were checked
for normality and symmetry, and subsequently analysed with a two-tailed Wilcoxon
signed ranks test (Zar 1999) to determine whether visitation rate differed from a
value of one (indicating no difference between visitation rate to an odour source

Fig. 37.1 An odour station being visited by a *U. caudimaculatus,* showing the odour source in cage at front of picture and IR sensors on each side

and a blank, on that night at that location). No Bonferroni, or similar, correction was applied to the data as the inflation of risk to Type II error rate was considered to be too great and the procedure has been argued to lack biological justification (Moran 2003).

37.3 Results

The standardised visitation rates of each of the three rodent species to each odour type is given for each of the three sampling seasons (Table 37.2). Rodents did not avoid novel or herbivore odours at any of the three sampling times, except *R. fuscipes* in May and November 2004. No rodent species avoided the faeces of the reptilian predator (carpet python) in any season. In general, in October 2003 and November 2004, visitation rate by *R. fuscipes* and *M. cervinipes* was not affected by odour type. The only exception was that quoll faeces (familiar marsupial predator) were avoided by *M. cervinipes* in October 2003 and by *R. fuscipes* in November 2004. In May, however, both *M. cervinipes* and *R. fuscipes* avoided odour stations containing mammalian predator faeces. In contrast, *U. caudimaculatus* reduced its visitation rate to odour stations containing the scents of known and unknown mammalian predators (except the Tasmanian devil) in October 2003. No data are available for *U. caudimaculatus* from November 2004 as the capture rate was too small to

Table 37.2 Standardised visit rates of rodents to odor stations (mean ± SEM), a value of one indicates no difference between the visits to an odor station and to a blank. Shaded cells indicate a value significantly different to one ($\alpha = 0.05$).

	M. cervinipes			R. fuscipes			U. caudimaculatus	
	Oct	May	Nov	Oct	May	Nov	Oct	May
Proportion immature	Low	High	Low	Low	High	Low	N/A	N/A
Novel	1.07 ± 0.16	1.08 ± 0.17	0.86 ± 0.10	0.95 ± 0.14	0.98 ± 0.15	0.75 ± 0.10	1.27 ± 0.16	1.12 ± 0.08
Herbivore	0.96 ± 0.15	0.99 ± 0.19	0.91 ± 0.08	0.88 ± 0.13	1.26 ± 0.20	1.31 ± 0.37	1.06 ± 0.14	1.12 ± 0.10
Reptile	1.27 ± 0.22	1.08 ± 0.20	0.67 ± 0.11	1.12 ± 0.27	0.88 ± 0.15	0.96 ± 0.18	0.79 ± 0.08	0.89 ± 0.16
Dingo (Known eutherian)	1.07 ± 0.24	0.77 ± 0.08	0.81 ± 0.11	0.88 ± 0.31	0.77 ± 0.08	1.03 ± 0.14	0.62 ± 0.07	0.79 ± 0.07
Fox (Unknown eutherian)	1.35 ± 0.21	0.68 ± 0.07	0.73 ± 0.12	1.05 ± 0.21	0.78 ± 0.08	0.79 ± 0.15	0.76 ± 0.08	0.86 ± 0.09
Quoll (Known marsupial)	0.86 ± 0.06	0.76 ± 0.05	1.03 ± 0.08	1.09 ± 0.21	1.01 ± 0.08	0.72 ± 0.09	0.76 ± 0.09	0.73 ± 0.05
Devil (Unknown marsupial)	1.24 ± 0.21	0.74 ± 0.07	0.97 ± 0.14	0.96 ± 0.10	0.78 ± 0.07	0.72 ± 0.14	0.89 ± 0.11	0.87 ± 0.07

allow statistical analysis. The data for May show similar trends: *U. caudimaculatus* avoided several mammalian predators (Table 37.2).

37.4 Discussion

Rodent visitation rates to odour stations decreased in the presence of predator faeces, compared to blank and novel odours. The response to different predators varied seasonally, and encompassed avoidance of familiar and unfamiliar faecal odours from marsupial and eutherian predators. This study supports the scat avoidance hypothesis (*sensu* Banks et al. 2003), which predicts that faeces are a useful indicator of predator presence, and I thus argue that, at least in the present context, faecal odours could be a good cue of predator presence for rainforest species.

Not all faeces were avoided: no species significantly avoided the herbivore faeces tested (whiptail wallaby) or the reptilian predator. Their avoidance of mammalian predator faeces, however, suggests that there may be an intrinsic component to such faeces that elicits a response. The novel odour (acetic acid) was never avoided by *M. cervinipes* and *U. caudimaculatus*; however, *R. fuscipes* avoided this odour twice, its most consistent avoidance response, which I am at a loss to explain. Quoll faeces (familiar marsupial predator) were avoided on 75% of sampling occasions, the familiarity and long co-evolutionary history between rodents and quolls suggests that these are the predator faeces most likely to be avoided.

The seasonal response to the faeces varied between the prey species. In the two smaller species, the late wet season was when they showed most avoidance. At this time, even fox and Tasmanian devil faeces (unfamiliar predators) were avoided, supporting the idea of common chemical 'triggers' in carnivore faeces (Dickman and Doncaster 1984). In *U. caudimaculatus*, response to familiar predators appeared stronger than to unfamiliar predators: quoll and dingo faeces were always avoided.

Life-history differences seem to be more important in determining the response of these rodents than is phylogeny. *M. cervinipes* and *U. caudimaculatus* are more closely related. However, it is *M. cervinipes* and *R. fuscipes* that display the same pattern. The two short-lived species, *R. fuscipes* and *M. cervinipes*, avoided predator faeces only in the late wet season (May). There are significantly more immature individuals in the populations of these species in May, and significantly fewer in October–November (D. Elmouttie, K. Horskins & C. Straetfield, unpublished data). Young animals that have their reproductive time ahead of them may distinguish and avoid predator odours and thus survive. By the end of the year, the population is mostly composed of adults, few of which will survive to breed the following year. There is thus an advantage for those animals that now "throw caution to the wind", and maximise their breeding opportunities by not avoiding the predator odours. The response of root voles, *Microtus oeconomus*, to least weasel, *Mustela nivalis*, odour also varies between breeding and nonbreeding seasons, with a reduction in avoidance by older individuals (Borowski 2002).

The other study species, *U. caudimaculatus*, has a longer life span than the smaller species. Animals can live for more than four years, and there is not the disparity in the proportion of the population that is mature at any given time of year, as in the other two species. This also means that the majority of the adult population is not pressured to find a final mating opportunity towards the end of the year; they will get another chance later. The animals can then maximise their lifetime breeding success by surviving to breed over several years, and will likely continue their caution and avoidance of predator odours later in the year.

My study shows the value of sampling throughout the year and across different seasons. This seasonal variation in response would obviously not have been detected had the study been conducted at only one time of the year, even if the time of year was replicated. Although Borowski (1998, 2002) showed that responses of voles to predators are variable throughout the year, many subsequent studies have been restricted to a single season. Valuable information about species' behaviour is likely to be missed by such an approach and I believe that a single season approach is fundamentally flawed.

Acknowledgments The study was approved by the Queensland University of Technology University Animal Ethics Committee and Queensland Environmental Protection Agency. I thank S. Ogun, M. Ogun and S. Walker for allowing fieldwork to be conducted on their land; P. Latch of QPWS for loan of surveillance equipment; J. Brumm of Lone Pine Koala Sanctuary, M. Fingland, E. Thyer and R. Richards of Brisbane Forest Park and Dr A. Braid of CSIRO Sustainable

Ecosystems for supply of predator faecal samples; K. Horskins and D. Elmouttie of QUT for help in species identification from photographs; H. F. Nahrung for help with statistical analyses and the late Dr. J. C. Wilson for encouragement, advice and help with experimental design.

References

Apfelbach, R., Blanchard, D. C., Blanchard, R. J., Hayes, R. A. and McGregor, I. S. (2005) The effects of predator odors in mammalian prey species: A review of field and laboratory studies. Neurosci. Biobehav. Rev. 29, 1123–1144

Banks, P. B., Hughes, N. K. and Rose, T. A. (2003) Do native Australian small mammals avoid faeces of domestic dogs? Responses of *Rattus fuscipes* and *Antechinus stuartii*. Aust. Zool. 32, 406–409.

Borowski, Z. (1998) Influence of weasel (*Mustela nivalis* Linneaus, 1776) odour on spatial behaviour of root voles (*Microtus oeconomus* Pallas, 1776). Can. J. Zool. 76, 1799–1804.

Borowski, Z. (2002) Individual and seasonal differences in antipredatory behaviour of root voles - a field experiment. Can. J. Zool. 80, 1520–1525.

Bramley, G. N. and Waas, J. R. (2001) Laboratory and field evaluation of predator odors as repellents for kiore (*Rattus exulans*) and ship rats (*Rattus rattus*). J. Chem. Ecol. 27, 1029–1047.

Burnett, S. (2001) Ecology and conservation status of the northern spot-tailed quoll, *dasyurus maculatus*, with reference to the future of Australia's marsupial carnivores. PhD thesis, James Cook University, Townsville, Australia.

Butler, W. H. (1969) Remains of *Sarcophilus* the "Tasmanian" devil (Marsupialia, Dasyuridae) from coastal dunes south of Scott River, Western Australia. West. Aust. Nat. 11, 87–88.

Calaby, J. H. and Lewis, D. J. (1977) The Tasmanian devil in Arnhem Land rock art. Mankind 11, 150–151.

Calaby, J. H. and White, C. (1967) The Tasmanian devil (*Sarcophilus harrisi*) in Northern Australia in recent times. Aust. J. Sci. 29, 473–475.

Catarell, M. and Chanel, R. (1979) Influence of some biologically meaningful odorants in the vigilance state of the rat. Physiol. Behav. 23, 831–838.

Cogger, H. G. (1994) *Reptiles and Amphibians of Australia*. Reed Books, Sydney.

Corbett, L. K. (1995) *The Dingo in Australia and Asia*. UNSW Press Kensington NSW.

Covacevich J. and Easton A. (1974) *Rats and mice in QueenslandEds.*. Queensland Museum, Brisbane.

Dawson, L. (1982) Taxonomic status of fossil devils (Sarcophilus, Dasyuridae, Marsupialia) from late Quaternary eastern Australian localities. In:Apfelbach, M. Archer (Eds.), *Carnivorous Marsupials*. Royal Zoological Society of New South Wales, Sydney, pp. 517–525.

Dickman, C. R. (1992) Predation and habitat shift in the house mouse, *Mus domesticus*. Ecol. 73, 313–322.

Dickman, C. R. (1993) Raiders of the last ark: cats in inland Australia. Aust. Nat. Hist. 24, 44–52.

Dickman, C. R. and Doncaster, C. P. (1984) Responses of small mammals to red fox (*Vulpes vulpes*) odour. J. Zool. Lond. 204, 521–531.

Hayes, R. A., Nahrung, H. F. and Wilson, J. C. (2006) The response of native Australian rodents to predator odours varies seasonally: a by-product of life-history variations? Anim. Behav. 71, 1307–1314.

Horton, D. R. (1977) A 10,000-year-old Sarcophilus from Cape York. Search 10, 374–375.

Jones, M. E., Rose, R. K. and Burnett, S. (2001) *Dasyurus maculatus*. Mammal. Spec. 676, 1–9.

Kats, L. B. and Dill, L. M. (1998) The scent of death: chemosensory assessment of predation risk by prey animals. Ecosci. 5, 361–394.

Kruuk, H. and Jarman, P. J. (1995) Latrine use by the spotted-tailed quoll (*Dasyurus maculatus*: Dasyuridae, Marsupialis) in its natural habitat. J. Zool. Lond. 236, 345–349.

Laing, D. G., Panhuber, H. and Slotnick, B. M. (1989) Odor masking in the rat. Physiol. Behav. 45, 689–694.

Lunney, D. (1995) Bush rat. Rattus fuscipes. In: R. Strahan (Eds.), *The Mammals of Australia*. Reed Books, Sydney, pp. 651–653.

Mappes, T., Koskela, E. and Ylonen, H. (1998) Breeding suppression in the bank vole under predation risk of small mustelids: laboratory or methodological artifact? Oikos 82, 365–369.

Moore, L. A. (1995) Giant white-tailed rat. *Uromys caudimaculatus*. In: R. Strahan (Eds.), *The Mammals of Australia*. Reed Books, Sydney, pp. 638–640.

Moran, M. D. (2003) Arguments for rejecting the sequential Bonferroni in ecological studies. Oikos 100, 403–405.

Müller-Schwarze, D. (1972) The responses of young black-tailed deer to predator odors. J. Mammal. 53, 393–394.

Nolte, D. L., Mason, J. R., Epple, G., Aronov, E. and Campbell, D. L. (1994) Why are predator urines aversive to prey? J. Chem. Ecol. 20, 1505–1516.

Redhead, T. D. (1995) Fawn-footed melomys. *Melomys cervinipes*. In: R. Strahan (Eds.), *The Mammals of Australia*. Reed Books, Sydney, pp. 636–637.

Seelke, A. M. H. and Blumberg, M. S. (2004) Sniffing in infant rats during sleep and wakefulness. Behav. Neurosci. 118, 267–273.

Strahan, R. (1995) *The Mammals of Australia*. Reed Books, Sydney.

Sullivan, T. P., Nordstrom, L. O. and Sullivan, D. S. (1985) Use of predator odors as repellents to reduce feeding damage by herbivores. I. Snowshoe hares (*Lepus americanus*). J. Chem. Ecol. 11, 903–919.

Tracey, J. G. (1982) *The Vegetation of the Humid Tropical Region of North Queensland*. CSIRO, Melbourne.

Vernes, K., Dennis, A. and Winter, J. (2001) Mammalian diet and broad hunting strategy of the dingo (*Canis familiarus dingo*) in the Wet Tropical Rain Forests of northeastern Australia. Biotrop. 33, 339–345.

Watts, C. H. S. and Aslin, H. J. (1981) *The rodents of Australia*. Angus and Robertson, Sydney.

Wolff, J. O. and Davis-Born, R. (1997) Response of gray-tailed voles to odours of a mustelid predator: a field test. Oikos 79, 543–548.

Zar, J. H. (1999) *Biostatistical Analysis*. Prentice Hall, Upper Saddle River, NJ.

Part VII
Applications

Chapter 37
A Critical Review of Zoo-based Olfactory Enrichment

Fay Clark and Andrew J. King

Abstract Olfactory stimuli are frequently integrated into zoo enrichment programs. This 'olfactory enrichment' can stimulate reproduction or naturalistic behaviour, enhance enclosure exploration, or reduce inactivity. However, not all scents achieve their desired goals, and can in fact bring about undesirable behaviour such as increased levels of stereotypy. Few attempts have been made to quantify the impact of introducing olfactory stimuli to zoo enclosures, and there are inherent difficulties when designing, implementing and evaluating olfactory enrichment. Firstly, it is difficult without appropriate chemical analyses to anticipate what information a scent conveys, and therefore whether it will be received as an excitatory or aversive stimulant. Second, more practical difficulties are encountered. Consideration needs to be given to (i) the choice of scent used, its relevance and motivation, (ii) how to present the scent in time and space, (iii) individual variation in response rates and neophobia (fear of novelty) to scents, and finally (iv) the health implications linked to the use of olfactory stimuli. This paper reviews the olfactory stimuli used in zoos as enrichments and their reported effects. Practical suggestions are made to encourage and stimulate more empirical quantification of olfactory stimulation in zoo animals.

37.1 An Introduction to Environmental Enrichment in Zoos

Environmental enrichment can be described as an improvement in the biological functioning of captive animals resulting from modifications to their environment (Newberry 1995). In recent years there has been a shift from controlled and theoretical studies of laboratory animal enrichment, toward more practical solutions tackling captivity-induced problems in zoo species (Renner and Lussier 2002). The goals of enrichment remain broad and debatable, but it is generally agreed that these include (1) increasing an animal's behavioural diversity or time spent performing species-specific behaviours, (2) promoting an animal's appropriate interaction with

Fay Clark
Institute of Zoology, Zoological Society of London
prolixiat@aol.com

its environment, and (3) increasing an animal's ability to cope with environmental change (Shepherdson 1989; Chamove and Moodie 1990). Enrichment is now considered a fundamental part of zoo animal husbandry, becoming more complex in its form and mode of presentation (Young 2003). Goals of enrichment should be objective, assessing whether environmental modifications do in fact enrich the subjects.

37.2 Zoo-based Olfactory Enrichment

Olfactory enrichment refers to the addition of scents or scented material to an animal's enclosure (Swaisgood and Shepherdson 2005). Typical enrichment items include food scent, essential oils, herbs and spices, feces, urine and other scents derived from animals, commercial lures, and artificial scents. Olfactory enrichment is not to be confused with feeding enrichment, where the provision of food items allows consumption, nor studies of scent discrimination trials, where olfactory awareness is the primary aim.

Despite a relatively large number of published enrichment studies appearing in peer-reviewed zoo science and welfare journals, there are comparably few olfactory enrichment studies. A review of the current literature shows a distinct bias towards providing olfactory enrichment to charismatic zoo species such as large felids (46% of studies to date), 16% of studies have been undertaken on primates, most commonly prosimians, while reptiles and canids are each represented in only 3% of studies.

This paper reviews the practical difficulties inherent in implementing olfactory enrichment outside of a controlled laboratory setting. Considerations discussed include: (i) choice of scent, relevance and motivation; (ii) how to present scent in time and space; (iii) individual variation in response rates and neophobia; and (iv) health implications.

37.2.1 Choices of Scent, Relevance and Motivation

Many scents are chosen on the basis of assumed relevance to the test subject. Interest in the scent of a predator, prey, mate or natural environment will be affected by the animal's familiarity (previous experience) with the scent, and the motivation to investigate it.

An innate ability to recognize natural predators is frequently tested in zoo animals, often offering contradictory results (e.g. Buchanan-Smith, Anderson and Ryan 1993; Boon 2003). Some authors only advocate the use of scents that are relevant to a zoo animal in the wild: for example, natural prey and adversary species faeces, as used for olfactory enrichment in African lions (Baker, Campbell and Gilbert 1997; Schuett and Frase 2001). Most scents used occur in some form in the natural environment, although not necessarily the natural environment of the test subjects. Faecal and urine scents are the most common natural scents used (53%), followed by herbs, spices, essential oils and vegetation (37%). Given the array of scents used as enrichment, it is surprising that justification for using a particular scent is often absent in the literature. It appears that many scents are used with no previous empirical testing, using anecdotal evidence from other zoos or success in other species.

37.2.2 How to Present Scent in Time and Space

The zoo environment is semi-controlled; limited by necessary husbandry procedures and regulations that the zoo may impose. Reviewed literature reveals that quantities or concentrations of scents are often subjective or entirely absent (e.g. Leach, Young and Waran 1998; Trager and Germanton 1997), with large diversity in methods of presentation. However, even where specific quantification isn't possible, presentations of scent can be classified more generally as (i) concentrated, (ii) semi-concentrated and (iii) dispersed, typically ranging from providing a discrete scented object, to scented air.

Concentrated scent provision involves providing the scent in or on a receptacle, such as logs (e.g. Williams, Chapman and Plowman 1999), cups (e.g. Hawkins 1997), or cloth (e.g. Wells and Egli 2004). The benefit of scent provision in this manner is that most scent remains on the receptacle, and this has been successfully used in an attempt to increase investigation of infrequently used enclosure areas by large felids (Williams et al. 1999; Knight 2004).

Semi-concentrated scent provision is often presented as loose faecal or other enclosure material contained inside a sack or other receptacle. The sack is often quickly ripped apart and subsequently scent dissipated throughout the enclosure by physical action (Noonan 1999; Clark, Melfi and Mitchell 2005).

Dispersed scent presentation refers to filling the enclosure with scent. This may take the form of applying scent to numerous enclosure items, or scenting the air within the enclosure using a vaporizer or spray. Spielman (2000) used the synthetic analogue of feline facial pheromone, Feliway, as olfactory enrichment for lions and tigers. The entire scented enclosure induced a significant reduction in head rubbing and spray marking in tigers, but had no significant effect in lions. Similarly, Struthers and Campbell (1996) provided peppermint scent via a vaporizer to chimpanzees, the effect of which was an increase in activity levels. Dispersed methods do not provide test subjects opportunity to retreat away from the scent if found aversive, and this should be taken into account when presenting potentially aversive scents.

There is also no explicit way to tell how far apart in space and time two or more scents should be provided in order to prevent their interaction. The order effect of scents refers to the notion that the presence of scent A may affect the response of a subject to scent B, in terms of habituation, sensitization, arousal, fatigue, or mixing and accumulation of scents. To overcome this effect, the order of scent provision should be randomized thoroughly, and a sufficient gap between scents should be left to allow the previous scent to dissipate. The interval length between enrichment uses, rather than repetition number, has been stated as the factor determining habituation of enrichment, and the suggested interval between the uses of any single item of enrichment is three weeks (Melfi, Plowman, Knowles and Roynon 2007).

A last consideration when presenting scents is accumulation of scent. Providing scents on existing enclosure furniture, or use of a single receptacle throughout a study removes the confounding factor of a novel scent receptacle, but augments scent accumulation. Providing each scent on a new receptacle reduces this problem, and receptacles can be removed after observations (Ostrower and Brent 1997; Wells and Egli 2004).

37.2.3 Individual Variations in Response Rates

Enrichment items should be considered as putative and can only be termed enriching with hindsight, once their desired effects (e.g. a reduction in stereotypical behaviour) have been confirmed. This requires careful and considerate experimental design, implementation and evaluation (reviewed by Clark et al. 2005). Unexpected results of enrichment have included an increase in stereotypical pacing in captive jaguars in response to primate faecal scent (Clark 2004), and contrasting species-specific responses in zoo-housed felids to catnip (Hill, Pavlik, Smith, Burghardt and Coulson 1976). There can also be individual differences within species to olfactory enrichment due to gender or age differences. For example, in felids reproductive-age adults are more sensitive than aged or immature animals in their responses to catnip (Hill et al. 1976).

Other effects of enrichment are perhaps biologically relevant but nevertheless undesired in captivity. These include displacement and anxiety related behaviours in lemurs exposed to conspecific faeces (McCusker and Smith 2002), and 'stress' and 'jumpiness' in tapirs exposed to jaguar urine (Calderisi 2005). Small amounts of stress are considered beneficial for animals (Breazile 1987; Moodie and Chamove 1990), but whilst these results are interesting and highlight a degree of apparent olfactory awareness, it is questionable whether the stress induced can be perceived as enriching.

Neophobia, or the fear of novelty, is sometimes observed in captive animals which have received little or no previous novel sensory stimulation (Mason 1991). Neophobia appears to have been exhibited in Goeldi's monkeys exposed to peppermint oil (Boon 2003), and a young tiger exposed to catnip (Todd 1963). Neophobia is a biologically relevant reaction, protecting animals from perceived threats both in the wild and in captivity. Therefore neophobia may be considered a 'correct' response to provisioned scents, if the aim of enrichment is to increase the expression of species-specific behaviours. It is questionable whether neophobic reactions are desirable in captivity, when the aims of enrichment may be to decrease levels of stress and increase perceived wellbeing. Using a visual, non-olfactory stimulus as a control substance (e.g. Schuett and Frase 2001) can be a useful tool for testing whether enrichment is likely to be stressful for a subject used to low stimulus diversity.

37.2.4 Health Implications

Olfactory enrichment should provide stimulation and choice whilst minimizing health risks. However, relatively few studies have reported their health and safety considerations. The most commonly used forms of olfactory stimuli are faeces and urine, and yet associated risks, for example when items are consumed during provision (Schaap 2002), are hardly mentioned in the literature. Pathogen and parasite testing will decrease risk of disease transmission, and pathogens and scents may be removed by exposing items to extremes of temperature; Burr (1997) microwaved felid hair before providing it to reptiles.

Over a quarter of olfactory enrichment in the reviewed literature involves essential oils. Although no health complications have been reported in zoos due to the introduction of essential oils as stimuli, lethal and non-lethal toxicity to essential oils have been reported in domestic felids, with side effects including skin sensitivity and irritation, eye problems and vomiting (Richardson 1999; Foss 2002). Additionally, anecdotal evidence suggests several perfumes to have had 'euphoric' effects on small felids, and individual reactions can vary greatly to a given scent; extreme responses of felids to catnip have been reported, and prosimian deaths due to use of nightshade plants (Engel 2002).

Zoos strive to maintain high levels of sanitation to prevent disease transmission and meet public expectation for visually clean enclosures (Kleiman, Allen, Thompson and Lumpkin 1998). This results in a direct trade-off between cleanliness and naturalism. A considerable amount of information is contained within the scent marks animals produce (e.g. prosimian scent marks: Drea, this volume) and by cleaning zoo enclosures, an animal's own scent marks are removed (Kitchener 1991; Mellen 1993; Robinson 1997). Over-cleanliness not only removes scents but can have health risks: there have been cases of attraction (self anointing) and even deaths due to the use of disinfectants, e.g. coatis (Trager and Germanton 1997). A solution to the problem may be to clean half the enclosure at a time as recommended for laboratory mice (Saibaba, Sales, Stodulski and Hau 1996; Van Loo, Kruitwagen, Van Zutphen, Koolhaas and Baumans 2000), allowing at least part accumulation of natural scent marks.

37.3 Conclusions

Scent provision as a form of environmental enrichment is a complex issue in a zoo setting. Unfortunately, few attempts have been made to quantify the impact of introducing olfactory stimuli into zoo enclosures; certainly there is a distinct lack of research in this area published in zoo and animal welfare journals. This may be due to the inherent difficulties of its study in a zoo environment; namely difficulties in presenting scents in time and space.

An animal's needs for stimulus diversity are difficult both to define and quantify (Carlstead 1996). Enrichment is often anthropomorphically rather than ecologically relevant (Chamove 1989), and due to human sensory biases we may fail to realize the importance of olfaction to less charismatic species, or those species with lower perceived levels of olfactory awareness (Hancox 1990; Somerville and Broom 1998).

Animals appear to need a motivation, stimulus and reward in order to exhibit their most natural behaviour patterns. It seems appropriate then, that a scent stimulus should be paired with a reward for investigation, otherwise motivation may be lost, for example pairing scent with other stimuli should allow large felids to exhibit their natural stalk, rush and kill repertoire (Leyhausen 1979). An animal investigating a novel environment may also be comforted when surrounded by its own odour (Halpin 1986). Therefore, allowing animals to effectively enrich themselves with natural scent marks, through less frequent enclosure cleaning, or using an animal's own scents when transporting it to another enclosure may reduce levels of stress and promote welfare.

Using olfactory stimuli to tackle key issues like these should now be an integral part of modern zoo research and management. Taking a more empirical approach, whilst maintaining an awareness of the inherent problems outlined here, future researchers can move toward a more thorough understanding of olfactory stimuli as a tool for enhancing captive species behaviour and welfare.

Acknowledgments We would like to thank the Science Department at Paignton Zoo Environmental Park and the Department of Conservation and Wildlife Management at Marwell Zoological Park.

References

Baker, W.K. Jr., Campbell, R. and Gilbert, J. (1997) Enriching the pride: scents that make sense. *The Shape of Enrichment* 6, 1–3.

Boon, M. (2003) Goeldi's monkeys (*Callimico goeldii*): olfactory enrichment to stimulate natural behaviour and greater activity. In: T. Gilbert (Ed.), *Proceedings of the 5th Annual Symposium on Zoo Research, Marwell Zoo*, pp. 212–224.

Breazile, J.E. (1987) Physiological basis and consequences of distress in animals. *J. Am. Vet. Med. Assoc.* 191, 1212–1215.

Buchanan-Smith, H.M., Anderson, D.A. and Ryan, C.W. (1993) Responses of cotton-top tamarins (*Saguinus oedipus*) to faecal scents of predators and non-predators. *Anim. Welf.* 2, 17–32.

Burr, L.E. (1997) Reptile enrichment: scenting for a response. *Animal Keepers Forum.* 24, 122–23.

Burrell, K., Wehnelt, S. and Rowland, H. (2004) The effect of whole carcass feeding and novel scent on jaguars (*Panthera onca*) at Chester Zoo. *British and Irish Association of Zoos and Aquariums Zoo Federation Research Newsletter* 5, 4.

Calderisi, D. (1997) Different scents for different responses in predator-prey relationships as a form of enrichment in captive animals. In: V. Hare and K. Worley (Eds.), *Proceedings of the Third International Conference on Environmental Enrichment,* Sea World, Florida, pp. 155–161.

Carlstead, K. (1996) Effects of captivity on the behavior of wild mammals. In: D. G. Kleiman, M. E. Allen, K. V. Thomson and S. Lumpkin (Eds.), *Wild Mammals in Captivity: Principles and Techniques,* University of Chicago Press: Chicago, pp. 317–333.

Chamove, A.S. (1989) Environmental enrichment: a review. *Anim. Tech.* 40, 155–79.

Clark, F. (2004) Olfactory enrichment for the captive jaguar (*Panthera onca*). Unpublished BSc. thesis, University of Southampton.

Clark, F., Melfi, V. and Mitchell, H. (2005) Wake up and smell the enrichment: a critical review of an olfactory enrichment study. In: N. Clum, S. Silver and P. Thomas (Eds.), *Proceedings of the Seventh International Conference on Environmental Enrichment*, Wildlife Conservation Society, New York City, pp. 178–185.

Eisenberg, J.F. and Kleiman, D.G. (1972) Olfactory communication in mammals. *Ann. Rev. Ecol. Syst.* 3, 1–32.

Engel, C. (2002) *Wild health: how animals keep themselves well and what we can learn from them.* Houghton Mifflin, Boston.

Foss, T.S. (2002) Liquid potpourri and cats - Essence of Trouble. *Vet. Tech.* 23, 686–689.

Hancox, M. (1990) Smell as a factor is mammalian behaviour. *Int. Zoo News.* 224, 19–20.

Halpin, Z.T. (1986) Individual odors among mammals: origins and functions. *Adv. Study Behav.* 16, 39–70.

Hawkins, M. (1997) Effects of Olfactory Enrichment on Australian Marsupial Species. In: V. Hare and K.Worley (Eds.*), Proceedings of the Third International Conference on Environmental Enrichment,* Sea World, Florida, pp. 135–149.

Hayes, M.P., Jennings, M.R. and Mellen, J.D. (1998) Beyond mammals: environmental enrichment for amphibians & reptiles. In: Shepherdson, D.J., Mellen, J.D. and Hutchins, M. (Eds) *Second Nature: Environmental Enrichment for Captive Animals.* Smithsonian Institute Press, Washington, pp. 205–235.

Hill, J.O., Pavlik, E.J., Smith, G.L., Burghardt, G.M and Coulson, P.B. (1976) Species-characteristic responses to catnip by undomesticated felids. *J. Chem. Ecol.* 2, 239–253.

Kitchener, A. (1991) *The Natural History of the Wild Cats.* Comstock, New York.

Kleiman, D.G., Allen, M.E., Thompson, K.V. and Lumpkin, S. (1998) *Wild Animals in Captivity: Principles and Techniques.* University of Chicago Press.

Knight K. (2004) Does the presence of a novel olfactory stimulus affect the behaviour of captive Jaguars (*Panthera onca*): implications for enrichment. Unpublished BSc. thesis, Liverpool John Moores University.

Leach, M., Young, R. and Waran, N. (1998) Olfactory enrichment for Asian elephants: is it as effective as it smells? *Int. Zoo News* 54 (5): 285–290.

Leyhausen, P. (1979) *Cat Behaviour: The Predatory and Social Behaviour of Domestic and Wild Cats.* Garland Press, New York.

Mason, G.J. (1991) Stereotypies and suffering. *Behav. Process.* 25, 103–115.

McCusker, C. and Smith, T.E. (2002) The potential of biologically relevant odour cues to function as a novel form of enrichment in captive ring-tailed lemurs, *Lemur catta. British and Irish Association of Zoos and Aquariums Federation Research Newsletter* 3, 3.

Melfi, V., Plowman, A., Knowles, L. and Roynon, J. (2007) Constructing an environmental enrichment programme with minimal habituation: a case study using Sumatran tigers. Unpublished manuscript.

Mellen, J. (1993) A comparative analysis of scent-marking, social and reproductive behaviour in twenty species of small cats *(Felis). Am. Zool.* 33, 151–166.

Moodie, E.M., and Chamove, A.S. (1990) Brief threatening events beneficial for captive tamarins? *Zoo. Bio.* 9, 275–286.

Newberry, R.C. (1995) Environmental enrichment: increasing the biological relevance of captive environments. *Appl. Anim. Behav. Sci.* 44, 229–244.

Noonan, B. (1999) Enrichment for African lions. *The Shape of Enrichment* 8, 6–7.

Ostrower, S. and Brent, L. (1997) Olfactory enrichment for captive chimpanzees: responses to different odours. *Laboratory primate newsletter* 36, 8–10.

Powell, D.M. (1995) Preliminary investigation of environmental enrichment techniques for African lions *(Panthera leo)*. *Anim. Welf.* 4, 361–70.

Renner, M.J. and Lussier, J.P. (2002) Environmental Enrichment for the captive spectacled bear *(Tremarctos ornatus)*. *Pharmacol. Biochem. Behav.* 73, 279–283.

Richardson, J.A. (1999) Potpourri hazards in cats. *Vet. Med.* 94, 1010–1012.

Robinson, M.H. (1998) Enriching the lives of zoo animals, and their welfare: where research can be fundamental. *Anim. Welf.* 7, 151–175.

Saibaba, P., Sales, G.D., Stodulski, G. and Hau, J. (1996) Behaviour of rats in their home cages: daytime variations and effects of routine husbandry procedures analysed by time sampling techniques. *Lab. Anim.* 30, 13–21.

Schaap, D. (2002) Enriching the Devil: The Tasmanian Devil. *The Shape of Enrichment.* 11, 1–4.

Schuett, E.B. and Frase, B.A. (2001) Making scents: using the olfactory senses for lion enrichment. *The Shape of Enrichment.* 10, 1–3.

Shepherdson, D. (1989) Stereotypic behaviour: what is it and how can it be eliminated or prevented? *Journal of the Association of British Wild Animal Keepers.* 16, 100–5.

Sommerville, B.A. and Broom, D.M. (1998) Olfactory awareness. *Appl. Anim. Behav. Sci.* 57, 269–286.

Spielman, J.S (2000) Olfactory enrichment for captive tigers *(Panthera tigris)* and lions *(Panthera leo)*, using a synthetic analogue of feline facial pheromone. Unpublished MSc. thesis, University of Edinburgh.

Struthers, E. J. and Campbell, J. (1996) Scent-specific behavioral response to olfactory enrichment in captive chimpanzees *(Pan troglodytes). Presented at the XVIth Congress of the International Primatological Society and the XIXth Conference of the American Society of Primatology, Wisconsin.*

Swaisgood, R.R. and Shepherdson, D.J. (2005) Scientific approaches to enrichment and stereotypies in zoo animals: what's been done and where should we go next? *Zoo. Biol.* 24, 499–518.

Testa, D. (1997) Paws to play: enrichment ideas for lynxes. *The Shape of Enrichment.* 6, 1–2.

Todd, N.B. (1963) The catnip response. Unpublished PhD. thesis, Harvard University.

Trager, G.C. and Germanton, H. (1997) Coatimundis enrich their own lives in nature by putting on the perfume, so why not let them do it in captivity? In: V. Hare and K.Worley (Eds.), *Proceedings of the Third International Conference on Environmental Enrichment,* Sea World, Florida, pp. 150–154.

Van Loo, P.L.P., Kruitwagen, C.L.J.J., Van Zutphen, L.F.M., Koolhaas, V. and Baumans, V. (2000) Modulation of aggression in male mice: influence of cage cleaning regime and scent marks. *Anim. Welf.* 9, 281–295.

Wells, D.L. and Egli, J.M. (2004) The influence of olfactory enrichment on the behaviour of black-footed cats, *Felis nigripes. Appl. Anim. Behav. Sci.* 85, 107–119.

Williams, N., Chapman, J., and Plowman, A. (1999) Olfactory enrichment for big cats, *Panthera leo persica* and *Panthera tigris sumatrae*. In: V. Hare and K.Worley (Eds.), *Proceedings of the Fourth International Conference on Environmental Enrichment,* Edinburgh Zoo, Edinburgh, pp. 300–303.

Young, R.J. (2003) *Environmental enrichment for captive animals.* Blackwell Science Publishing, Oxford.

Chapter 38
Pig Semiochemicals and Their Potential for Feral Pig Control in NE Australia

Sigrid R. Heise-Pavlov, James G.Logan and John A. Pickett

Abstract Preliminary investigations have been conducted to identify the chemical composition of carpal gland secretions from feral pigs in the lowland tropical rainforest of NE Australia. Carpal glands are located along the inner part of the front legs and their secretion may be distributed to the surrounding vegetation and bedding sites. Secretions were collected from the external surface of the glands by swabbing with filter paper discs. Compounds were then extracted in distilled ether and analysed by capillary gas chromatography. Secretions from boars and reproductive females contained more compounds than those from non-reproductive females. Only seventeen compounds were found in the secretions collected from boars and reproductive females, while one compound was only present in secretions from reproductive females. In boars compounds were at higher concentrations than in reproductive females. Lowest concentrations were found in non-reproductive females. Boars and reproductive females have more compounds with a higher molecular weight than non-reproductive females. The similarity in the chemical composition of carpal gland secretions from boars and reproductive females may be responsible for the same response that these animals trigger in conspecifics, i.e. avoidance behaviour. Further investigations will focus on the potential these secretions may have as repellents in feral pig control.

38.1 Introduction

Pigs (Artiodactyla, Suidae) possess a well-developed olfactory system and show the typical flehmen behaviour which is connected with the involvement of the vomeronasal organ in processing the odour cues (Güntherschulze 1979; Gosling 1985; Martyrs 1977; Mueller-Schwarze 1979). Pigs also possess a range of odour producing organs including glands which suggest that odour cues play an important role in the social life and habitat use. The functions and chemistry of the majority of these organs are presently unknown (Albone 1984). However,

Sigrid R. Heise-Pavlov
PavEcol Wildlife Management Consultancy, Australia
ryparosa@bigpond.com.au

some have been studied and the results are commercially utilised in piggeries for improved rearing of piglets (McGlone and Anderson 2002) and artificial insemination (Melrose, Reed and Patterson 1971).

The preputial gland, which is an invagination of the skin rather than a typical gland, is known to produce a mixture of seminal fluid, urine, ammonia, p-cresol and trace amounts of 16-androstene (Albone 1984). Such excretions are also involved in advertising dominance in boars as they are known to scent-mark more frequently when exposed to sows than subordinate boars (Mayer and Brisbin 1986). Other glands, including the *glandulae mentalis* (Mental organ), the antorbital and proctodeal (anal) glands have not yet been investigated although they are known to be used when individuals mark objects and conspecifics (Byers 1978). Several species of the Suidae family also possess a range of body glands, often called dorsal or lateral glands. These sweat glands have been found to contain a Δ 16-androgen-steroid 3 α-hydroxy-5α-androst-16-en, described as a musk-smelling steroid (Stinson and Patterson 1972). Studies on the chemistry of the dorsal glands of Tayassu peccari, which use these glands intensively to mark conspecifics, reveal that 3α androstenol and 5α androstenone are major components, but that unsaturated and saturated C7 to C10 carboxyl acids and esters (mainly farnesene esters) are also present (Waterhouse, Hudson, Pickett and Weldon 2001).

Recently, studies have been conducted on the function of carpal glands in feral pigs of Australia (Heise-Pavlov, Heise-Pavlov and Bradley 2005). These glands are situated along the inner part of the front legs and are characterized by 4 or 5 openings leading to ducts which are connected to secretory cells. The glands are larger in boars than in females, larger in reproductive than non-reproductive females and are slightly smaller during the summer wet season than the winter dry season. This suggests that their secretion could be involved in advertising the reproductive status of females and in distributing a group-specific odour (Heise-Pavlov et al. 2005).

Carpal glands' secretion may be distributed to the surrounding vegetation while the animal is moving, or to a bedding site while the animal is lying with the legs outstretched, thus advertising the presence of the animal (Gosling 1985). Histochemical investigations showed that secretions of the carpal glands seem to be a mucous substance containing glycoproteins and glycolipids (Gargiulo, Pedini and Ceccarelli 1989; Hraste and Stojkovic 1995). Lipid-rich substances have been described as carrier-components for substances of a higher evaporation rate (Alberts 1992).

Only few attempts have been made to investigate whether these secretions could be used to control wild or/and feral populations of pigs. Pig semiochemicals have great potential to be used in control technologies. For example, McIlroy and Gifford (2005) demonstrated in a pilot study that more feral pigs can be caught in traps with oestrous sows rather than anoestrous sows as lures.

Our study follows on from the study by Heise-Pavlov et al. (2005) and focuses on the chemical composition of the secretion of these glands. As differences in the size of carpal glands are known between reproductive and non-reproductive females, and between boars and females, we expect differences in the chemical composition of secretions collected from different sexes and females of different reproductive status (Hypothesis 1). The larger gland sizes known to be present in boars and reproductive females suggest that these glands could be used by boars and reproductive females to advertise their presence.

Based on the results of previous studies on pheromones in wild boars we expect that secretions of carpal glands from feral pigs contain a range of low molecular weight compounds which allow a fast diffusion from the skin surface, but also compounds of higher molecular weight which may function as carrier compounds facilitating the slow release of highly volatile compounds (Hypothesis 2).

Here we present the results of a preliminary study to investigate the chemical composition of secretions from carpal glands of feral pigs living in densely vegetated, humid tropical lowland rainforest habitats with a view to exploiting semiochemicals to improve trapping efficiency and to identify repellents.

38.2 Method

38.2.1 Study area

The study area is situated in the tropical lowland rainforest between the Daintree River (145°26.36'E 16°17'S) and Cape Tribulation (145°27'E and 16°04.7'S) along the coast of north eastern Australia.

38.2.2 Sampling

From June 2005 to January 2006 feral pigs were caught regularly in cage-traps, baited with split coconuts or other available fruit (for details see Heise-Pavlov et al. 2005). Caught animals were killed by a head-shot according to the guidelines "Model Code of Practice for the Welfare of Animals" (SCA Technical Report SeriesNo. 34 1996).

The animals were sexed and the reproductive status of females (pregnant / lactating) was recorded during a routine dissection. Secretions of the carpal glands were often visible by a darker surface around the openings of the ducts of the glands compared to the rest of the skin along the inner side of the front legs. The secretion was collected by rubbing a circular piece of Whatman Filterpaper (GF/B, 21 mm diameter) on the surface using a pair of autoclaved forceps. The filter paper was then placed into a 20 ml autoclaved vial with screw cap and stored in a freezer at $-20\,°C$. Vials were shipped to Rothamsted Research, UK on solid CO_2 and stored at $-20\,°$.

38.2.3 Analytical methods

The samples of three reproductive, three non-reproductive females and five boars were extracted with 4 ml redistilled ether twice for 15 minute periods. The two extracts were pooled, dried with $MgSO_4$ and concentrated to 200µl under a flow of N_2. We injected 4 µl of each sample into a Hewlett Packard 6890 Gas-Chromatograph (GC). The GC was fitted with a non-polar ultra methyl silicone (HP1) cross-linked capillary column (50 m×32 mm ID, film thickness: 0.52 µ m) with hydrogen as the carrier gas. A cool on-column injector and a flame ioni-sation detector (FID) were used. The GC oven temperature was held at 30 °C for 1 min and programmed to increase at 5 °C min^{-1} to 150 °C then 10 °C min^{-1} to 240 °C. The integration of the peak areas was performed using Chemstation software.

38.2.4 Quantitative and qualitative analysis of all chemicals within extracts

The identification of all chemicals within the extracts were unknown, therefore accurate quantifications were not performed for all chemicals. As a preliminary investigation, approximate amounts were calculated using the following method. A 1 µ L injection of 100 ng µL^{-1} solution of n-alkanes (C7-25) in hexane was analysed by GC on an HP1 column. The approximate amounts of chemicals within the extract were calculated using the following equation:

(Amount (ng) of alkanes* area of compound of interest)/Mean area of alkanes

$$(38.1)$$

A retention index (RI) was used to allow comparison of compound retention data between columns of different equipment, e.g. GC and GC-MS. Straight chain hydro-carbons (alkanes) were assigned an index of 100 times the number of carbon atoms in the molecule. The RI's of all compounds within all the extracts were calculated by using the difference between the retention indices of the alkane eluting before and after that compound:

$$RI = 100 \cdot \left(\log Rt_X - \frac{\log Rt_n}{\log Rt_{n+1}} - \log Rt_n \right) + 100n \qquad (38.2)$$

where RI = retention index; Rt = retention time; X = compound of interest; n = alkane before the compound of interest; n + 1 = alkane after the compound of interest. The RI's were used to compare between the samples from reproductive females, non-reproductive females and boars. The concentrations of compounds with the same RI were averaged over the samples. Only compounds which were present in more than 50% of the samples from reproductive, non-reproductive females and boars respectively were compared. This reduced the likelihood of contaminations being included in the analysis. Compounds specific only for one type of animal were selected for further identification by GC-MS. A quantitative analysis of compounds present in reproductive, non-reproductive females and boars was performed.

38.3 Results

Extracts collected from reproductive females and boars contained on average more compounds than those from non-reproductive females (106, 125 and 61 respectively). There were a range of volatile compounds although there were many at a high molecular weight. A quantitative analysis of the compounds present in more than 50% of the animals per type revealed a high similarity between the compounds found in reproductive females and boars. Two major compounds of all extracts have a retention time of 8.45–8.51 (RI = 800) and 14.93–14.95 (RI = 992).

The amount of compounds present in extracts collected from all animals was compared between reproductive and non-reproductive females and between females and boars. Of the 38 compounds common for all females, 30 showed a higher concentration in reproductive than in non-reproductive females. Of the 35 compounds found in females and boars, 28 showed a higher concentration in boars compared to females. Common compounds were found at a higher amount in boars than in females and at lower concentration in non-reproductive than reproductive females.

Only one reproductive female-specific compound with the retention index of 2348 was detected. Non-reproductive female-specific compounds and boar-specific compounds were not detected in the extracts.

Seventeen compounds were only found in boars and reproductive females. However, some also occurred in one of the non-reproductive females and of those in 83% of all cases in the same non-reproductive female.

38.4 Discussion

Pigs have a range of glands or gland like structures which are used in communication (Albone 1984) as well as well developed main and accessory olfactory bulbs (Guentherschulze 1979). It is likely that feral pigs, living in the tropical rainforest habitats of north-eastern Australia, use specific odour cues to communicate. Regularly visited rubbing sites in the forest, such as termite mounds or trees, indicate that chemical communication plays a role in these pig populations. Observations suggest that feral pigs in rainforests live in matriarchal groups while males are mainly solitarily and join groups when oestrous sows are present (Spencer, Lapidge, Hampton and Pluske 2005). Carpal glands, located along the inner side of the front legs, have been shown to differ in size between sexes. Boars have larger carpal glands than females and reproductive females have larger glands than non-reproductive females (Heise-Pavlov et al. 2005) suggesting that their secretion is used to advertise the presence of boars and reproductive (pregnant/lactating) females. Carpal gland size also differs between seasons with slightly smaller glands in the summer wet season (Heise-Pavlov et al. 2005) indicating that their activity is linked with the distribution of a group-specific odour and varies depending on the availability of food and the area a group is using for feeding.

The differences in sizes of carpal glands between sexes and between females of different reproductive status are reflected in our results on the chemical composition of their secretions. The extracts from boars and reproductive females contained more compounds than those from non-reproductive females. Furthermore, the composition of compounds is more similar between reproductive females and boars than between the two types of females.

Seventeen compounds were present only in boars and reproductive females giving their complex odour-profile a specific character. However, some of these compounds have been found in one of the non-reproductive females. This female also shows a much higher number of compounds compared to the other two non-reproductive females which suggests that this animal may have been at an early (undetected) stage of pregnancy.

The composition of carpal gland secretions from boars and reproductive females is distinct from that of non-reproductive females with respect to the number of compounds as well as the presence of specific components. Both animals may produce compounds that initiate a similar response in other animals, such as avoidance behaviour. Boars and reproductive females are known to be aggressive and tend to be rather isolated from other groups or other animals of the group (in reproductive females in particular shortly before farrowing, Meynhardt 1982). Secretions inducing avoidance behaviour may have the potential to be used as repellents in feral pig control in fruit orchards or other cropping areas where pigs cause regularly damage to agriculture. However, our study did not involve behavioural observations of animals. Some studies have utilised the odours of oestrous sows as lures to increase trapping of feral pigs in Western Australia (McIlroy and Gifford 2005). Other investigations have used delta-decanolactone and the pheromone dodecen-acetate as lures in trapping feral goat populations in New Zealand (Veltman, Cook, Drake and Devine 2001). However, these studies show a low success rate in attracting enough animals to traps to achieve efficient control. On the other hand, odour cues from predators have been used to repel deer and rodents from farms (Wyatt 2003). Species-specific odours used as repellents may have a higher success rate. To our knowledge, our study is the first attempt to investigate this option with secretions of the carpal glands of feral pigs in rainforest habitats of NE Australia where the efficiency of pig control by trapping is reduced due to the relatively high food availability all year round. Here more efficient aids to control these feral pig populations are required (Heise-Pavlov and Heise-Pavlov 2003)

Boars and reproductive females differ in their secretions with respect to the amount of compounds they have in common rather than the presence of type-specific compounds. Eighty percent of compounds common in boars and reproductive females occur at a higher concentration in boars than in reproductive females. Seventy eight percent of all compounds common to reproductive and non-reproductive females are at a lower concentration in non-reproductive females than in reproductive females. Reproductive females and boars may compensate the high diffusion rate of their volatile compounds by excreting them at a higher concentration compared to non-reproductive females. This may increase their persistence and advertising the presence of a reproductive female or boar for a longer time inducing avoidance behaviour in nearby animals. Only one specific compound could be selected for reproductive females which is not present in the secretions of boars and non-reproductive females (RI = 2348 with the retention time of 49.98).

Our results reveal that volatile and non-volatile compounds are found in secretions from carpal glands of feral pigs. Non-volatile compounds can increase the longevity of volatile, low-molecular compounds by acting as carrier-components (Wyatt 2003). Carrier components are known to be lipid rich substances or proteins (often called odorant-binding proteins). They have been found in salivary secretions of the boar and are named pheromaxin (Booth and White 1988) although other odour binding proteins of the lipocalin family may also play a role in the pheromone transduction of these glands (Marchese, Pes, Scaloni, Carbone and Pelosi 1998). Carrier compounds such as urinary proteins or axillary odorant binding proteins have been found in several species and humans (Robertson, Cox, Gaskell, Evershed and Beynon 1996; Zeng, Spelman, Vowels, Leyden, Biemann and Preti 1996). These carrier substances may be of importance in animals living in highly humid habitats such as the tropical lowland rainforests. High temperature and humidity increase evaporation rates and therefore reduces the persistence of chemical signals (Wyatt 2003). In order to increase the longevity of chemical signals in these environments we expected a range of non-volatile compounds in our analysis. It was suggested that highly volatile pheromones should have a size of not more than 20 carbon atoms and a molecular weight of between 80 and 300 (Wyatt 2003). Reproductive females and boars had a higher number of compounds with more than 20 carbon atoms (retention index of about 1970) (n = 8 and 9 respectively) than non-reproductive females (n = 3). The only specific compound that was found in the secretion of carpal glands from reproductive females had a retention time of 49.98 and is likely to have approximately 23 carbon atoms. Further work using GC-MS is underway to identify the compounds of interest. Future studies will clarify the potential role of non-volatile compounds as carrier components for volatile compounds in the secretions of carpal glands from pigs. Because feral pigs inhabit very different environments in Australia such as tropical foodplains, semiarid habitats, rainforest and cool temperate environments (Choquenot, McIlroy and Korn 1996) the composition of their semiochemicals can be expected to vary with the habitat.

Acknowledgments We wish to thank Peter Heise-Pavlov for his assistance in sampling gland swabs, Ron and Sue Stannard at Julatten, Australia, for the storage of the samples and Lynda Ireland at Rothamsted Research, UK, for the technical assistance during the analyses of the samples.

References

Alberts, AC (1992) Constraints on the design of chemical communication systems in terrestrial vertebrates. Amer. Nat., Supplement 1992, 62–89.

Albone, E.S. (1984) *Mammalian Semiochemistry. Chichester*. John Wiley & Sons Limited, New York, Brisbane, Toronto, Singapore.

Booth, W.D. and White, C.A. (1988) The isolation, purification and some properties of phero-maxin, the pheromonal steroid-binding protein, in porcine submaxillary glands and saliva. J. Endocrin. 118, 47–57.

Byers, J.A. (1978) Probable involvement of the preorbital glands in two social behavioural patterns of the collared peccary, *Dicotyles tajacu*. J. Mammal. 59, 855–856.

Choquenot, D., McIlroy, J. and Korn, T. (1996) *Managing Vertebrate Pests: Feral Pigs*. Bureau of Resource Sciences/ Australian Government Publishing Service, Canberra.

Gargiulo, A.M.; Pedini, V. and Ceccarelli, P. (1989) Histology, ultrastructure and carbohydrate histochemistry of pig carpal glands. Anat. Histol. Embryol. 18, 289–296.

Gosling, L.M. (1985) The even-toed ungulates: order Artiodactyla. In: R.E. Brown and D.W. Mac-donald (Eds.), *Social odours in mammals. Vol 2*, Clarendon Press, Oxford, pp. 550–618.

Güntherschulze, J. (1979) Studies about Regio olfactoria of wild boar (*Sus scrofa*) and the domestic pig (Sus scrofa domestica). Zool. Anz. Jena 202, 256–279.

Heise-Pavlov, P.M. and Heise-Pavlov, S.R. (2003) Feral pigs in tropical lowland rainforest in north eastern Australia: ecology, zoonoses and management. Wildl. Biol. 9, 29–36.

Heise-Pavlov, S., Heise-Pavlov, P. and Bradley, A. (2005) Carpal Glands in feral pigs (*Sus domesticus*) in tropical lowland rainforest in NE Queensland, Australia. J. Zool. 266, 73–80.

Hraste, A. and Stojkovic, R. (1995) Histomorphologic and histochemical chararcteristics of carpal glands (glandulae carpeae) in domestic swine (*Sus scrofa domesticus*) and wild swine (*Sus scrofa ferus*). Anatomia, Histologia, Embryologia 24 (3), 209.

Marchese, S. Pes, D.; Scaloni, A. Carbone, V. and Pelosi, P. (1998): Lipocalins of boar salivary glands binding odours and pheromones. Eur. J. Biochem. 252, 563–568.

Martys, M. (1977) Das flehmen der schweine, Suidae - The flehmen behaviour of pigs. Zool. Anz. Jena 199: 433–440.

Mayer, J.J. and Brisbin, I.L. (1986) A note on the scent-marking behaviour of two captive-reared feral boars. Appl. Anim. Behav. Sci. 16, 85–90.

McGlone, J.J. and Anderson, D.L. (2002) Synthetic maternal pheromone stimulates feeding behaviour and weight gain in weaned pigs. J. Anim. Sci 80, 3179–83.

McIlroy, J.C. and Gifford, E.J. (2005) Are oestrous feral pigs, *Sus scrofa*, useful as trapping lures?. Wildl. Res. 32 (7), 605–608.

Melrose, D.R. Reed, H.C.B. and Patterson, R.L.S. (1971) Androgen steroids associated with boar odour as an aid to the detection of oestrus in pig artificial insemination. Brit. Vet. J. 127, 497–501.

Meynhardt, H. (1982) *Schwarzwild-Report.*, Verlag J. Neumann-Neudamm Melsungen, Berlin, Basel, Wien.

Mueller-Schwarze, D. (1979) Flehmen in the context of mammalian urine communication.- In: F.J. Ritter (Ed.), *Chemical ecology: Odour communication in animals*, Elsevier/North Holland Biomedical Press, Amsterdam, pp. 85–96.

Pedini, V., Scocco, P., Dall'Aglio, C. and Gargiulo, A.M. (1999) Detection of glycosidic residues in carpal glands of wild and domestic pig revealed by basic and lectin histochemistry. Anat. Anz. 181 (3), 269–274.

Robertson, D.H., Cox, K.A., Gaskell, S.J., Evershed, R.P. and Beynon, R.J. (1996) Molecular heterogeneity in the Major Urinary Proteins of the house mouse *Mus musculus*. Biochem. J. 316, 265–272.

SCA Technical Report Series, No. 34. (1996) *Feral Livestock Animals – Destruction or Capture, Handling and Marketing; Standing Committee on Agriculture*, Animal Health Committee—Model Code of Practice for the Welfare of Animals. Canberra.

Spencer, P.B.S., Lapidge, S.J., Hampton, J.O. and Pluske, J.R. (2005) The sociogenetic structure of a controlled feral pig population. Wildl. Res. 32, 297–304.

Stinson, C.G. and Patterson, R.L.S. (1972) $C_{19} - \Delta^{16}$ steroids in boar sweat glands. Brit. Vet. J. 128, 245–267.

Waterhouse, J.S., Hudson, M., Pickett, J.A. and Weldon, P.J. (2001) Volatile components in dorsal gland secretions of the white-lipped peccary, *Tayassu pecari*, from Bolivia. J. Chem. Ecol. 27, 2459–2469.

Veltman, C.J., Cook, C.J., Drake, K.A. and Devine, C.D. (2001) Potential of delta-decanolactone and (Z)-7-dedecen-1-yl acetate to attract feral goats (*Capra hirsus*). Wildl. Res. 28, 589–597.

Wyatt, T.D. (2003) Pheromones and Animal Behaviour: Communication by Smell and Taste. Cambridge University Press.

Zeng, C., Spelman, A.I., Vowels, B.R., Leyden, J.J., Biemann, K. and Preti, G. (1996) A human axillary odorant is carried by apolipoprotein D. Proc. Natl. Acad. Sci. USA 93, 6626–6630.

Chapter 39
Use of Chemical Ecology for Control of the Cane Toad?

R. Andrew Hayes, Alexis Barrett, Paul F. Alewood, Gordon C. Grigg and Robert J. Capon

Abstract In 1935, 101 cane toads, *B. marinus,* were introduced into north Queensland, Australia in an attempt to control the greyback cane beetle, *Dermolepida albohirtum*, a pest of sugar cane fields. The cane toad was, however, completely unable to control the beetles and itself became a successful pest. Since their arrival, cane toads have been implicated in the population declines of many native frog species and mammalian and reptilian predators. These effects are through predation, competition and the toxic secretions produced by the toad, poisoning potential predators. While the toxic nature of their secretions has been long known, only a part of the chemical complexity of the secretion has been identified to a molecular level. Our study aims to look at how diverse the chemical composition of cane toad skin secretions is, as well as its variability across life-history stages, between individuals and also whether different populations of toads may show differences in their chemistry. Beyond this, the chemical ecology of the toad, which probably includes pheromonal communication, may offer opportunities for control of this pest.

39.1 Introduction

The cane toad, *Bufo marinus*, is a very successful species that has invaded the Australian environment, since its deliberate introduction in 1935. The species was introduced to northern Queensland to control the greyback cane beetle, *Dermolepida albohirtum*, a pest of sugar cane fields. The toad did not, however, have any effect on the beetles as the species inhabit completely different parts of the crop. The cane toad has since spread south and west throughout Queensland and across state borders into New South Wales, the Northern Territory and will soon spread into Western Australia. The species produces toxic skin secretions, and has been implicated in population declines of other amphibians (through competition and predation) and also reptilian and mammalian predators of frogs. Recent data (Grigg *et al.* 2006)

R. Andrew Hayes
University of Queensland, Institute for Molecular Bioscience
r.hayes@imb.uq.edu.au

has suggested the impacts on other frogs may not be as severe as initially thought, but the threats to toad predators seem to be significant. The cane toad is widely regarded as a pest species, and a range of emotions, from understandable concern to alarm laced with hysteria, greets its imminent arrival in Western Australia. A method for control or eradication of the species is being sought.

As anurans have a pronounced calling behaviour, it was long assumed that chemical communication played little or no role in their behaviour and ecology (e.g. (Houck and Sever 1993). Recent results have changed this view (Brizzi, Delfino and Pellegrini 2002; Lee and Waldman 2002; Pearl, Cervantes, Chan, Ho, Shoji and Thomas 2000; Wabnitz, Bowie, Tyler, Wallace and Smith 1999; Waldman and Bishop 2004), and it is now clear that many, if not all, anurans use chemical signals to modulate their behaviour. This presents an ideal opportunity to explore the development of a method of control for a pest species based on undermining the species' own signaling system, using the animal's own chemical ecology against it.

39.2 Cane Toad Chemical Ecology?

The first step in a strategy for developing effective control of a pest by chemical ecology must be to determine if the species uses a chemical signaling system. If intra-specific signaling is shown to occur, the next step is to identify and synthesise the chemical agents utilized, with a view to successful application under field conditions. The most likely source of communication chemicals in amphibians is the secretions of the skin glands.

Hypothesis 1: Chemical signaling between individuals is important to cane toad behavior.
HPLC, LC-MS, LC-NMR and GC-MS will be used to examine the skin secretions of the cane toad. Making no presupposition about the composition of the secretion, we will develop analytical methods to elucidate: volatiles and non-volatiles, organic and water-soluble, stable and non-stable, and major and minor components.

Chemicals that cycle in abundance are obvious candidates for consideration as pheromones. Armed with effective analytical techniques we can address the issue of cane toad chemical plasticity with respect to different variables (Table 39.1). We will also investigate the impact of crude and fractionated secretions on the behavior of cane toads, and compare these responses between the same groupings of animals, i.e. sex, life-stage, season etc. We have identified three stages in the cane toad life cycle that are likely controlled by chemical signals, and are an ideal place to start attempting to disrupt their chemical ecology.

Hypothesis 2: Cane toads produce a chemosignal that controls / facilitates reproduc-tion, either through increasing aggregation or attraction.
Auditory communication, such as croaking, has limitations over distance, both geo-graphically and temporally, and a chemical signal that can travel further and

Table 39.1 Variables which may contribute to cane toad (B. marinus) chemical plasticity

Source	Are there differences between collection locations on the body?
Sex	Are males different from females?
Life-stage	Does the chemistry vary between stages?
Season	Do some signals increase in mating season?
Environment	What is the effect of specific cues such as, changes in temperature, humidity, male calling etc?
Geography	Do animals from different locations (e.g. QLD or NT) vary in their chemical composition?

persist through time may be used in conjunction with the overt vocal signal to assist in mate attraction. Both aggregation signals and mating signals are a good potential basis to use in local control of a pest species, as they are likely to be species-specific. Traps scented with either of these should be more effective at trapping animals than an unscented trap. Sex pheromones have been identified in anurans (Brizzi et al. 2002; Pearl et al. 2000; Wabnitz et al. 1999), if a similar substance exists for cane toads this would be an extremely productive tool for local control.

Hypothesis 3: Females avoid laying eggs into a water body already containing cane toad eggs / tadpoles, they detect this through a chemical cue in the water.
Breeding site selection vitally influences reproductive success and thus evolutionary fitness. Data from other anurans, including other *Bufo* species, establishes that order of laying in a pond influences survival and competition between tadpoles (Alford and Wilbur 1985). Cannibalism by individuals from older cohorts is suggested to be the highest risk to juvenile *B. marinus* (Crump 1983; Hearnden 1991). It has been suggested (Shine, R. personal communication) that *B. marinus* is less likely to lay in a pond already containing eggs. The basis of this avoidance may be a chemical stimulus released by the eggs, or by the mating female or male. If such a signal were isolated it could be used to disrupt reproduction in specific, sensitive water bodies, or even be used in a more widespread manner throughout the range of the species.

Hypothesis 4: An alarm signal exists which reduces time to metamorphosis and mass at metamorphosis of cane toad tadpoles.
Anecdotal evidence suggests that cane toad tadpoles living in a pond into which the odour of a crushed conspecific is added will move away from the odour source, metamorphose at a smaller mass, and do so more quickly than those living in pools without this cue (Alford 1994; Hearnden 1991). This response is presumably a predator avoidance behaviour. This type of response has been reported for other anurans, but the underlying chemistry of the signal is completely unknown (Chivers and Smith 1998; Kraft, Wilson and Franklin 2005; Summey and Mathis 1998; Wilson, Kraft and van Damme 2005). This cue would have applications for speeding up metamorphosis and increasing the proportion of the most predator-sensitive section of the population (Cohen and Alford 1993; Hearnden 1991), thus impacting survival of the pest.

39.3 Skin Secretions of *Bufo marinus*

The skin of amphibians is highly glandular and from these glands many bioactive compounds are secreted (Erspamer 1994; Toledo and Jared 1995). These glands may be grouped together into agglomerations of glands, such as the parotoid gland found in the shoulder region of many anurans. (Hutchinson and Savitzky 2004; Toledo and Jared 1995).

Three of the classes of compounds found from the skin of amphibians have been identified from the skin of bufonids, including *Bufo marinus*. These are: steroids (bufadienolides), biogenic amines (catecholamines, indolylalkylamines and alkaloids) and bioactive peptides and proteins.

39.3.1 Steroids

Bufadienolides are a class of cardioactive steroids that increase the contractile force of the heart by inhibiting Na^+/K^+-ATPase enzyme activity. These compounds are C24 steroids, featuring a C-17 β pyrone side chain, *cis*-fusion of the A/B and C/D rings and a C-3 β hydroxyl group that may or may not be further substituted. Other variation is primarily in the number and position of hydroxyl groups (Dewick 1997). Although bufadienolides are widely reported from many animal and plant sources, and were originally described from toads, of the eighty three bufadienolides identified from *Bufo* species (Steyn and van Heerden 1998), only five of these have been identified from cane toad skin (Chen and Osuch 1969; Matsukawa, Akizawa, Ohigashi, Morris, Butler Jr and Yoshioka 1997) (Fig. 39.1a), It is likely, however, that many more are present *e.g.* five more have been identified from cane toad eggs (Akizawa, Mukai, Matsukawa, Yoshioka, Morris and Butler 1994) (Fig. 39.1b).

Toad bufadienolides occur not only by themselves but also in a conjugated form, sulfates, dicarboxylic esters and amino acid-dicarboxylic acid esters have all been reported (Steyn and van Heerden 1998). Because of the activity of the bufadienolides in inhibiting active monovalent cation transporters, it is suggested that these compounds have a role in maintaining sodium homeostasis in toads that migrate between fresh and salt water environments (Flier, Edwards, Daly and Myers 1980).

A novel substance with a bufadienolide-related chemical structure has been isolated from *B. marinus* skin. This substance is 3β-hydroxy-11, 12-seco-5β, 14β-bufa-20, 22-dienolide-11, 14-olides-12oic acid (called marinoic acid) and shows the Na^+/K^+-ATPase inhibitory activity characteristic of the bufadienolides (Matsukawa, Akizawa, Morris, Butler Jr and Yoshioka 1996).

39.3.2 Biogenic Amines

Cane toad biogenic amines fall into three classes, the catecholamines, the indolylalkylamines and an alkaloid. The catecholamine adrenaline is a significant portion (6–11%) of the skin secretions (Erspamer 1994; Gregerman 1952). Parotoid and

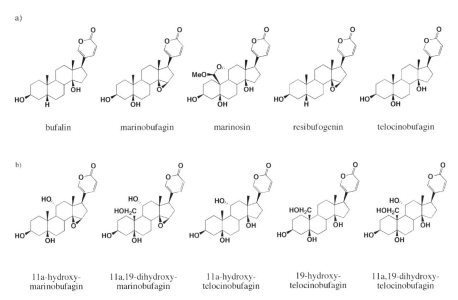

Fig. 39.1 Bufadienolides identified from the a) skin and b) eggs of *B. marinus*

skin secretions of *B. marinus* also contain hydroxytyramine, and noradrenaline (Clarke 1997) (Fig. 39.2a).

More than twenty indole derivatives have been identified from bufonid skin extracts. The indolylalkylamines bufotenidine, bufotenine, de-hydrobufotenine, bufotionine and serotonin (5-hydroxytryptamine) (Fig. 39.2b) have been identified in skin secretions of *Bufo marinus*, while the latter four have been detected in parotoid gland secretions (Erspamer 1994; Maciel, Schwartz, Pires Jr, Sebben, Castro, Sousa, Fontes and Schwartz 2003). The concentration of serotonin in the dried secretion of *B. marinus* was found to equate to approximately 0.1% of the total composition and primarily acts as a vasoconstrictor (Gregerman 1952; Toledo and Jared 1995).

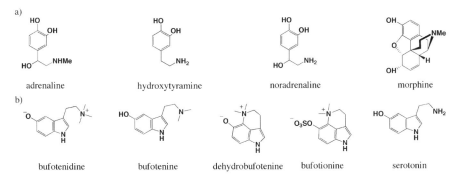

Fig. 39.2 Biogenic amines from *B. marinus* skin, a) catecholamines and an alkaloid, b) indolylalkylamines

Alkaloids (basic, heterocyclic nitrogen-containing compounds) have been less frequently reported in anuran skin secretions, but several studies confirm findings of morphine in the skin of *B. marinus* (Erspamer 1994; Oka, Kantrowitz and Spector 1985; Toledo and Jared 1995) (Fig. 39.2a). This substance was not isolated or characterized in these studies, and its assignment is based on pharmacological rather than molecular studies. It has been suggested that its vasodilatory affect on the skin may be associated with the thermoregulation of the toad (Oka et al. 1985).

39.3.3 Bioactive Peptides and Proteins

Although peptides and proteins are common in anuran skin, it was originally believed that they were either absent or present in small amounts in *Bufo* species (Clarke 1997). However, more recently, skin secretions of three species of North American and European toads (*Bufo boreas, B. viridis* and *B. bufo*) have been found to contain an array of peptides and polypeptides and even some small proteins (Clarke 1997). In addition, two proteins have been reported from the skin secretions of the toad *Bufo andrewsi*, an irreversible serine protease inhibitor, termed baseprin (Zhao, Jin, Lee and Zhang 2005), and a novel 22-kDa protein, trypsin inhibitor called BATI (Zhao, Jin, Lee and Zhang 2005). Currently the only identified anuran pheromones are peptidic, such as the sex pheromone identified from the magnificent tree-frog, *Litoria splendida* (Wabnitz et al. 1999). Findings like this demonstrate the likelihood that peptides and proteins are more readily present in the skin secretions *Bufo* species than originally thought.

39.4 Preliminary Results

Preliminary HPLC studies performed on cane toad parotoid secretion have revealed a far more complex molecular picture than recognised previously. We have found at least fifteen compounds with a UV absorbance at 297 nm, which is characteristic for the bufadienolides (Fig. 39.3). Even focusing only on this absorbance, we have already detected many more than the five compounds so far reported. These compounds occur over a range of polarities, from a very polar, water-soluble fraction, a group with intermediate polarity and some very non-polar components. It is likely that this range of polarities is produced by conjugation of the basic bufadienolide structure with sugars, amino acids or carboxylic acids. Additionally, we have identified that there are many proteinaceous compounds in this secretion; the first time that peptides or proteins have been described for *B. marinus*. These findings provide hope that a useful signal will be found that can be used for control of the species.

39.5 Conclusion

The use of a species chemical ecology to control pest animals allows an environmentally sensitive control method. Chemical signals within a species are usually

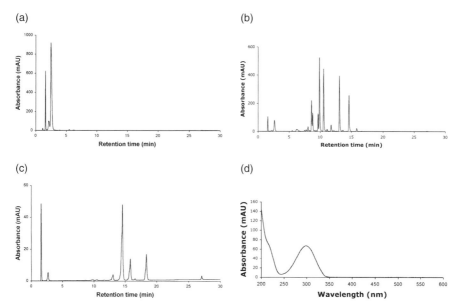

Fig. 39.3 LC PDA analysis, UV (297 nm) trace (using 1 mL/min gradient elution from 90% H₂O/MeCN (0.01% TFA) to MeCN (0.01% TFA) over 30 minutes followed by a 5 minute flush with MeCN on a Phenomenex Onyx C₁₈ 100 × 4.6 mm column) showing a) non-polar, b) intermediate, c) polar fractions and d) characteristic absorbance profile of parotoid secretion from *B. marinus.*

highly species-specific and have no effect on even closely related species. These techniques can be used to subvert messages passed within a species, and thus inhibit their natural behaviour and ecology. Our research team brings together behavioural ecologists, natural products chemists and protein chemists with expertise that will allow these questions to be answered. In addition, we have developed links with other researchers throughout Australia that will expand on the skills base within our team. While eradication is unlikely in the short to medium term, controlling the spread and impact of cane toads is of vital concern. The addition of chemical ecological studies will broaden the arsenal of control techniques and increase the possibility of having an impact on this important pest species

References

Akizawa, T., Mukai, T., Matsukawa, M., Yoshioka, M., Morris, J. F. and Butler Jr, V. P. (1994) Structures of novel bufadienolides in the eggs of a toad, *Bufo marinus.* Chem. Pharm. Bull. 42, 754–756.

Alford, R. A. (1994) Interference and exploitation competition in larval *Bufo marinus.* . In: P. C. Mishra, N. Behara, B. K. Sevapati and B. C. Guru (Eds.), *Advances in Ecology and Environmental Sciences.* Ashish Press, New Dehli, pp. 297–306.

Alford, R. A. and Wilbur, H. M. (1985) Priority effects in experimental pond communities: competition between *Bufo* and *Rana.* Ecol. 66, 1097–1105.

Brizzi, R., Delfino, G. and Pellegrini, R. (2002) Specialized mucous glands and their possible adaptive role in the males of some species of *Rana* (Amphibia, Anura). J. Morph. 254, 328–341.

Chen, C. and Osuch, M. V. (1969) Biosynthesis of bufadienolides – 3ßhydroxycholonates as precursors in *Bufo marinus* bufadienolides synthesis. Biochem. Pharmacol. 18, 1797–1802.

Chivers, D. P. and Smith, R. J. F. (1998) Chemical alarm signalling in aquatic predator-prey systems: a review and prospectus. Ecosci. 5, 338–352.

Clarke, B. T. (1997) The natural history of amphibian skin secretions, their normal functioning and potential medical applications. Biol. Rev. 72, 365–379.

Cohen, M. P. and Alford, R. A. (1993) Growth, survival and activity patterns of recently metamorphosed *Bufo marinus*. Wildl. Res. 20, 1–13.

Crump, M. L. (1983) Opportunistic cannibalism by amphibian larvae in temporary aquatic environments. Am. Nat. 121, 281–289.

Dewick, P. M. (1997) *Medicinal Natural Products. A biosynthetic approach*. John Wiley & Sons, Chichester.

Erspamer, V. (1994) Bioactive secretions of the amphibian integument. In: H. Heatwole and G. T. Barthalmus (Eds.), *Amphibian Biology*. Surrey Beatty & Sons, Sydney, pp. 179–350.

Flier, J., Edwards, M. W., Daly, J. W. and Myers, C. W. (1980) Widespread occurrence in frogs and toads of skin compounds interacting with the ouabain site of Na^+/K^+-ATPase. Science 208, 503–505.

Gregerman, R. I. (1952) Adrenalin and hydroxytyramine in the parotoid gland venom of the toad *Bufo marinus*. J. Gen. Physiol. 35, 483–487.

Grigg, G. C., Taylor, A., McCallum, H. and Fletcher, L. (2006) Monitoring the impact of cane toads (*Bufo marinus*) on Northern Territory frogs - a progress report. In: K. L. Molloy and W. R. Henderson (Eds.), *Science of Cane Toad Invasion and Control, Proceedings of the Invasive Animals CRC/CSIRO/QLD NRM&W Cane Toad Workshop, June 2006, Brisbane*. Invasive Animals Cooperative Research Centre, Canberra, pp. 47–54

Hearnden, M. N. (1991) *The reproductive and larval ecology of* Bufo marinus *(Anura: Bufonidae)*. PhD thesis. James Cook University, Townsville, Australia. p. 245.

Houck, L. D. and Sever, D. M. (1993) Role of the skin in reproduction and behavior. In: H. Heatwole and G. T. Barthalmus (Eds.), *Amphibian Biology*. Surrey Beatty & Sons, Sydney, pp. 351–381.

Hutchinson, D. A. and Savitzky, A. H. (2004) Vasculature of the parotoid glands of four species of toads (Bufonidae: *Bufo*). J. Morph. 260, 247–254.

Kraft, P. G., Wilson, R. S. and Franklin, C. E. (2005) Predator-mediated phenotypic plasticity in tadpoles of the striped marsh frog, *Limnodynastes peronii*. Aust. Ecol. 30, 558–563.

Lee, J. S. F. and Waldman, B. (2002) Communication by fecal chemosignals in an archaic frog, *Leiopelma hamiltoni*. Copeia 2002, 679–686.

Maciel, N. M., Schwartz, C. A., Pires Jr, O. R., Sebben, A., Castro, M. S., Sousa, M. V., Fontes, W. and Schwartz, E. N. F. (2003) Composition of indolalkylamines of *Bufo rubescens* cutaneous secretions compared to six other Barzilian Bufonids with phylogenetic implications. Comp. Biochem. Physiol. B. 134, 641–649.

Matsukawa, M., Akizawa, T., Morris, J. F., Butler Jr, V. P. and Yoshioka, M. (1996) Marinoic acid, a novel bufadienolide-based substance in the skin of the giant toad, *Bufo marinus*. Chem. Pharm. Bull. 44, 255–257.

Matsukawa, M., Akizawa, T., Ohigashi, M., Morris, J. F., Butler Jr, V. P. and Yoshioka, M. (1997) A novel bufadienolide, marinosin, in the skin of the giant toad, *Bufo marinus*. Chem. Pharm. Bull. 45, 249–254.

Oka, K., Kantrowitz, J. D. and Spector, S. (1985) Isolation of morphine from toad skin. PNAS USA 82, 1852–1854.

Pearl, C. A., Cervantes, M., Chan, M., Ho, U., Shoji, R. and Thomas, E. O. (2000) Evidence for a mate-attracting chemosignal in the dwarf African clawed frog *Hymenochirus*. Horm. Behav. 38, 67–74.

Steyn, P. S. and van Heerden, F. R. (1998) Bufadienolides of plant and animal origin. Nat. Prod. Rep. 15, 397–413.

Summey, M. R. and Mathis, A. (1998) Alarm responses to chemical stimuli from damaged conspecific by larval Anurans: tests of three Neotropical species. Herpetol. 54, 402–408.

Toledo, R. C. and Jared, C. (1995) Cutaneous granular glands and amphibian venoms. Comp. Biochem. Physiol. 111A, 1–29.

Wabnitz, P. A., Bowie, J. H., Tyler, M. J., Wallace, J. C. and Smith, B. P. (1999) Aquatic sex pheromone from a male tree frog. Nature 401, 444–445.

Waldman, B. and Bishop, P. J. (2004) Chemical communication in an archaic anuram amphibian. Behav. Ecol. 15, 88–93.

Wilson, R. S., Kraft, P. G. and van Damme, R. (2005) Predator-specific changes in the morphology and swimming performance of larval *Rana lessonae*. Func. Ecol. 19, 238–244.

Zhao, Y., Jin, Y., Lee, W.-H. and Zhang, Y. (2005) Isolation and preliminary characterization of a 22-kDa protein with trypsin inhibitory activity from toad *Bufo andrewsi* skin. Toxicon. 46, 277–281.

Zhao, Y., Jin, Y., Wei, S.-S., Lee, W.-H. and Zhang, Y. (2005) Purification and characterization of an irreversible serine protease inhibitor from skin secretions of *Bufo andrewsi*. Toxicon. 46, 635–640.

Index

Printed in the United States of America